RapidMiner

Data Mining Use Cases and Business Analytics Applications

Chapman & Hall/CRC
Data Mining and Knowledge Discovery Series

AIMS AND SCOPE

This series aims to capture new developments and applications in data mining and knowledge discovery, while summarizing the computational tools and techniques useful in data analysis. This series encourages the integration of mathematical, statistical, and computational methods and techniques through the publication of a broad range of textbooks, reference works, and handbooks. The inclusion of concrete examples and applications is highly encouraged. The scope of the series includes, but is not limited to, titles in the areas of data mining and knowledge discovery methods and applications, modeling, algorithms, theory and foundations, data and knowledge visualization, data mining systems and tools, and privacy and security issues.

PUBLISHED TITLES

ADVANCES IN MACHINE LEARNING AND DATA MINING FOR ASTRONOMY
Michael J. Way, Jeffrey D. Scargle, Kamal M. Ali, and Ashok N. Srivastava

BIOLOGICAL DATA MINING
Jake Y. Chen and Stefano Lonardi

COMPUTATIONAL INTELLIGENT DATA ANALYSIS FOR SUSTAINABLE DEVELOPMENT
Ting Yu, Nitesh V. Chawla, and Simeon Simoff

COMPUTATIONAL METHODS OF FEATURE SELECTION
Huan Liu and Hiroshi Motoda

CONSTRAINED CLUSTERING: ADVANCES IN ALGORITHMS, THEORY, AND APPLICATIONS
Sugato Basu, Ian Davidson, and Kiri L. Wagstaff

CONTRAST DATA MINING: CONCEPTS, ALGORITHMS, AND APPLICATIONS
Guozhu Dong and James Bailey

DATA CLUSTERING: ALGORITHMS AND APPLICATIONS
Charu C. Aggarawal and Chandan K. Reddy

DATA CLUSTERING IN C++: AN OBJECT-ORIENTED APPROACH
Guojun Gan

DATA MINING FOR DESIGN AND MARKETING
Yukio Ohsawa and Katsutoshi Yada

DATA MINING WITH R: LEARNING WITH CASE STUDIES
Luís Torgo

FOUNDATIONS OF PREDICTIVE ANALYTICS
James Wu and Stephen Coggeshall

GEOGRAPHIC DATA MINING AND KNOWLEDGE DISCOVERY, SECOND EDITION
Harvey J. Miller and Jiawei Han

HANDBOOK OF EDUCATIONAL DATA MINING
Cristóbal Romero, Sebastian Ventura, Mykola Pechenizkiy, and Ryan S.J.d. Baker

INFORMATION DISCOVERY ON ELECTRONIC HEALTH RECORDS
Vagelis Hristidis

INTELLIGENT TECHNOLOGIES FOR WEB APPLICATIONS
Priti Srinivas Sajja and Rajendra Akerkar

INTRODUCTION TO PRIVACY-PRESERVING DATA PUBLISHING: CONCEPTS AND TECHNIQUES
Benjamin C. M. Fung, Ke Wang, Ada Wai-Chee Fu, and Philip S. Yu

KNOWLEDGE DISCOVERY FOR COUNTERTERRORISM AND LAW ENFORCEMENT
David Skillicorn

KNOWLEDGE DISCOVERY FROM DATA STREAMS
João Gama

MACHINE LEARNING AND KNOWLEDGE DISCOVERY FOR ENGINEERING SYSTEMS HEALTH MANAGEMENT
Ashok N. Srivastava and Jiawei Han

MINING SOFTWARE SPECIFICATIONS: METHODOLOGIES AND APPLICATIONS
David Lo, Siau-Cheng Khoo, Jiawei Han, and Chao Liu

MULTIMEDIA DATA MINING: A SYSTEMATIC INTRODUCTION TO CONCEPTS AND THEORY
Zhongfei Zhang and Ruofei Zhang

MUSIC DATA MINING
Tao Li, Mitsunori Ogihara, and George Tzanetakis

NEXT GENERATION OF DATA MINING
Hillol Kargupta, Jiawei Han, Philip S. Yu, Rajeev Motwani, and Vipin Kumar

RAPIDMINER: DATA MINING USE CASES AND BUSINESS ANALYTICS APPLICATIONS
Markus Hofmann and Ralf Klinkenberg

RELATIONAL DATA CLUSTERING: MODELS, ALGORITHMS, AND APPLICATIONS
Bo Long, Zhongfei Zhang, and Philip S. Yu

SERVICE-ORIENTED DISTRIBUTED KNOWLEDGE DISCOVERY
Domenico Talia and Paolo Trunfio

SPECTRAL FEATURE SELECTION FOR DATA MINING
Zheng Alan Zhao and Huan Liu

STATISTICAL DATA MINING USING SAS APPLICATIONS, SECOND EDITION
George Fernandez

SUPPORT VECTOR MACHINES: OPTIMIZATION BASED THEORY, ALGORITHMS, AND EXTENSIONS
Naiyang Deng, Yingjie Tian, and Chunhua Zhang

TEMPORAL DATA MINING
Theophano Mitsa

TEXT MINING: CLASSIFICATION, CLUSTERING, AND APPLICATIONS
Ashok N. Srivastava and Mehran Sahami

THE TOP TEN ALGORITHMS IN DATA MINING
Xindong Wu and Vipin Kumar

UNDERSTANDING COMPLEX DATASETS: DATA MINING WITH MATRIX DECOMPOSITIONS
David Skillicorn

RapidMiner

Data Mining Use Cases and Business Analytics Applications

Edited by

Markus Hofmann
Institute of Technology
Blanchardstown, Dublin, Ireland

Ralf Klinkenberg
Rapid-I / RapidMiner
Dortmund, Germany

CRC Press is an imprint of the
Taylor & Francis Group, an **informa** business
A CHAPMAN & HALL BOOK

CRC Press
Taylor & Francis Group
6000 Broken Sound Parkway NW, Suite 300
Boca Raton, FL 33487-2742

Printed on acid-free paper
Version Date: 20130919

International Standard Book Number-13: 978-1-4822-0549-7 (Hardback)

Visit the Taylor & Francis Web site at
http://www.taylorandfrancis.com

and the CRC Press Web site at
http://www.crcpress.com

Dedication

To my beautiful wife, Glenda and my three boys, Killian, Darragh, and Daniel, for their love and support.

M.H.

To my parents Brigitte and Herbert, for providing me with an outstanding education.

R.K.

Contents

Foreword

Case Studies Are for Communication and Collaboration

Data mining or data analysis in general has become more and more important, since large amounts of data are available and open up new opportunities for enhanced empirical sciences, planning and control, and targeted marketing and information services. Fortunately, theoretically well-based methods of data analysis and their algorithmic implementations are available. Experienced analysts put these programs to good use in a broad variety of applications. However, the successful application of algorithms is an art! There is no mapping from application tasks to algorithms, which could determine the appropriate chain of operators leading from the given data to the desired result of the analysis—but there are examples of such processes. Case studies are an easy way of communicating smart application design. This book is about such case studies in the field of data analysis.

Analysts are interested in the work of others and curiously inspect new solutions in order to stay up to date. Case studies are an excellent means to communicate best-practice compositions of methods to form a successful process. Cases are also well-suited to storing results for further use. A case is then used as a blueprint for further applications.[1] This eases the development of an application to quite some extent.

Another good use of case studies is to ease the communication between application domain experts and data mining specialists. The case shows what could already be achieved and inspire future cases[2]. This allows to frame new applications and to illustrate possible results. For those experts who want to set up a data mining project, it is a valuable justification.

Finally, teaching means communication. Teaching data mining is not complete without reference to case studies, either. Rapid-I offers at their website, `http://rapid-i.com`, video tutorials (webinars), white papers, manuals—a large variety of documentation with many illustrations. Offering case studies is now a further step into communicating not only the facilities of the system, but its use in real-world applications. The details of complex data analysis processes help those who want to become data analysts.

In summary, case studies support the collaboration of data analysts among themselves, the communication of data analysts and application experts, the interaction between experienced and beginners. Now, how can complex data mining processes be communicated, exchanged, and illustrated? An easy-to-understand view of the process is to abstract away the programming details. As is explained in the following, RapidMiner offers this.

[1] T. Euler. Publishing operational models of data mining case studies. In B. Kitts, G. Melli, and K. Rexer, editors, *Procs. of the 1st Workshop on Data Mining Case Studies*, held at IEEE ICDM, pages 99-106, 2005.

[2] G. Melli, X Wu, P. Beinat, F. Bonchi, L. Cao, Rong Dan, C. Faloutsos, B. Kitts, B. Goethals, G. McLachlan, J. Pei, A. Srivastava, and O. Zaiane. Top10 data mining case studies. *Int. J. Information Technology and Decision Making*, 11(2):389-400, 2012.

RapidMiner

RapidMiner is a system which supports the design and documentation of an overall data mining process. It offers not only an almost comprehensive set of operators, but also structures that express the control flow of the process.

- Nesting operator chains were characteristic of RapidMiner (Yale) from the very beginning. This allows us to have a small number of operators at one level, each being expanded at the level below by simply clicking on the lower right area of the operator box.
- An example set can be multiplied for different processes that are executed in parallel and then be unified again. Sets of rows of different example sets of the same set of attributes can be appended. Hence, the example set that is used by some learning method can flexibly be modified.
- The cross validation is one of the most popular nested operators. The training set is split into n parts and, in a loop, $n-1$ parts are put to training and the remaining part to testing, so that the performance on the test set can be averaged over a range of different example sets from the same domain. The operator X-VALIDATION is used in most of the case studies in order to achieve sensible performance evaluations.
- Several loop operators can be specified for an application. The LOOP PARAMETERS operator repeatedly executes some other operators. The parameters of the inner operators as well as conditions controlling the loop itself tailor the operator to the desired control flow.
- Wrapper approaches wrap a performance-based selection around a learning operator. For instance, those feature sets are selected for which a learner's performance is best. The wrapper must implement some search strategy for iterating over sets and for each set call a learning algorithm and its evaluation in terms of some performance measure.

These structures are similar to notions of programming languages, but no programming is necessary – it is just drag, drop, and click! Visually, the structures are shown by boxes which are linked or nested. This presentation is easy to understand.

Only a small (though decisive) part of an overall data mining process is about model building. Evaluating and visualizing the results is the concluding part. The largest part is the pre-processing.

- It starts with reading in the data and declaring the meta data. RapidMiner supports many data formats and offers operators for assigning not only value domains of attributes (attribute type), but also their role in the learning process.
- The inspection of the data through diverse plots is an important step in developing the case at hand. In many case studies, this step is not recorded, since after the exploration it is no longer necessary. The understanding of the analysis task and the data leads to the successful process.
- Operators that change the given representation are important to bridge the gap between the given data and the input format that a learner needs. Most analysts have a favorite learning method and tweak the given data until they suit this algorithm well. If frequent set mining is the favorite, the analyst will transform m nominal values of

one attribute into m binomial attributes so that frequent set mining can be applied. If the attribute type requirements of a learner are not yet fulfilled, RapidMiner proposes fixes.

- The discretization of real-valued attributes into nominal- or binomial-valued attributes is more complex and, hence, RapidMiner offers a variety of operators for this task.

- Beyond type requirements, features extraction and construction allow learners to find interesting information in data which otherwise would be hidden. A very large collection of operators offers the transformation of representations. The text processing plug-in, the value series plug-in, and the image processing plug-in are specifically made for the pre-processing of texts, time series or value series in general, and images.

- The feature selection plug-in automatically applies user-specified criteria to design the best feature set for a learning task. Moreover, it evaluates the selected features with respect to stability. For real-world applications, it is important that good performance is achieved at any sample of data. It is not sufficient that the selected features allow a good performance on average in the cross-validation runs, but it must be guaranteed that the features allow a sufficiently good performance on every data sample.

Given the long operator chains and nested processes in data mining, the aspect of documentation becomes indispensable. The chosen parameters of, e.g., discretization, the particular feature transformations, and the criteria of feature selection are stored with the RapidMiner process. The metadata characterize the data at any state in the process. Hence, its result is explainable and reproducible.

In this book, case studies communicate how to analyze databases, text collections, and image data. The favorite learning tasks are classification and regression with the favorite learning method being support vector machines followed by decision trees. How the given data are transformed to meet the requirements of the method is illustrated by pictures of RapidMiner. The RapidMiner processes and datasets described in the case studies are published on the companion web page of this book. The inspiring applications may be used as a blueprint and a justification of future applications.

Prof. Dr. Katharina Morik *(TU Dortmund, Germany)*

Preface

Data and the ability to make the best use of it are becoming more and more crucial for today's and tomorrow's companies, organizations, governments, scientists, and societies to tackle everyday challenges as well as complex problems and to stay competitive. *Data mining, predictive analytics*, and *business analytics* leverage these data, provide unprecedented insights, enable better-informed decisions, deliver forecasts, and help to solve increasingly complex problems. Companies and organizations collect growing amounts of data from all kinds of internal and external sources and become more and more data-driven. Powerful tools for mastering data analytics and the know-how to use them are essential to not fall behind, but to gain competitive advantages, and to increase insights, effectiveness, efficiency, growth, and profitability.

This book provides an *introduction to data mining and business analytics*, to the most powerful and flexible open source software solutions for data mining and business analytics, namely *RapidMiner and RapidAnalytics*, and to many application *use cases* in scientific research, medicine, industry, commerce, and diverse other sectors. RapidMiner and RapidAnalytics and their extensions used in this book are all freely available as open source software community editions and can be downloaded from
`http://www.RapidMiner.com`

Each chapter of this book describes an application *use case*, how to approach it with *data mining methods*, and how to implement it with RapidMiner and RapidAnalytics. These application-oriented chapters do not only provide you with the necessary analytics know-how to solve these problems and tasks, but also with easy-to-follow reproducible step-by-step descriptions for accomplishing this with RapidMiner and RapidAnalytics. The datasets and RapidMiner processes used in this book are available from the companion web page of this book:
`http://www.RapidMinerBook.com`

This application-oriented analytics use case collection will quickly enable you to solve similar problems effectively yourself. The case studies can serve as blueprints for your own data mining applications.

What Is Data Mining? What Is It Good for, What Are Its Applications, and What Does It Enable Me to Do?

While technology enables us to capture and store ever larger quantities of *data*, finding relevant *information* like underlying patterns, trends, anomalies, and outliers in the data and summarizing them with simple understandable and robust quantitative and qualitative models is a grand challenge. Data mining helps to discover underlying structures in the data, to turn *data* into *information*, and information into *knowledge*. Emerged from mathematics, statistics, logic, computer science, and information theory, data mining and machine learning and statistical learning theory now provide a solid theoretical foundation and powerful methods to master this challenge. *Data mining* is the extraction of implicit, previously unknown, and potentially useful information from data. The automatically extracted models provide insight into customer behavior and into processes generating data, but can also be applied to, for example, automatically classify objects or documents or images into given categories, to estimate numerical target variables, to predict future values of observed time series data, to recommend products, to prevent customer churn, to optimize direct marketing campaigns, to forecast and reduce credit risk, to predict and prevent machine failures before they occur, to automatically route e-mail messages based on their content and to automatically detect e-mail spam, and to many other tasks where data helps to make better decisions or even to automate decisions and processes. Data mining can be applied not only to structured data from files and databases, but text mining extends the applicability of these techniques to unstructured data like texts from documents, news, customer feedback, e-mails, web pages, Internet discussion groups, and social media. Image mining, audio mining, and video mining apply these techniques to even further types of data.

Why Should I Read This Book? Why Case Studies? What Will I Learn? What Will I Be Able to Achieve?

This book introduces the most important machine learning algorithms and data mining techniques and enables you to use them in real-world applications. The open source software solutions RapidMiner and RapidAnalytics provide implementations for all of these algorithms and a powerful and flexible framework for their application to all kinds analytics tasks. The book and these software tools cover all relevant steps of the data mining process from data loading, transformation, integration, aggregation, and visualization via modeling, model validation, performance evaluation, model application and deployment, to automated feature selection, automated parameter and process optimization, and integration with other tools like, for example, the open source statistics package R or into your IT infrastructure via web services. The book and the tools also extensively cover the analysis of unstructured data including text mining and image mining.

The application-oriented focus of this book and the included use cases provide you with the know-how and blueprints to quickly learn how to efficiently design data mining processes and how to successfully apply them to real-world tasks. The book not only introduces you to important machine learning methods for tasks like clustering, classification, regression, association and recommendation rule generation, outlier and anomaly detection, but also to the data preprocessing and transformation techniques, which often are at least as crucial for success in real-world applications with customer data, product data, sales data, transactional data, medical data, chemical molecule structure data, textual data, web page data, image data, etc. The use cases in this book cover domains like retail, banking, marketing, communication, education, security, medicine, physics, and chemistry and tasks like direct marketing campaign optimization, product affinity scoring, customer churn prediction and prevention, automated product recommendation, increasing sales volume and profits by cross-selling, automated video lecture recommendation, intrusion detection, fraud detection, credit approval, automated text classification, e-mail and mobile phone text message spam detection, automated language identification, customer feedback and hotel review analysis, image classification, image feature extraction, automated feature selection, clustering students in higher education and automated study program recommendation, ranking school applicants, teaching assistant evaluation, pharmaceutical molecule activity prediction, medical research, biochemical research, neutrino physics research, and data mining research.

What Are the Advantages of the Open Source Solutions RapidMiner and RapidAnalytics Used in This Book?

RapidMiner and RapidAnalytics provide an integrated environment for all steps of the data mining process, an easy-to-use graphical user interface (GUI) for the interactive data mining process design, data and results visualization, validation and optimization of these processes, and for their automated deployment and possible integration into more complex systems. RapidMiner enables one to design data mining processes by simple drag and drop of boxes representing functional modules called operators into the process, to define data flows by simply connecting these boxes, to define even complex and nested control flows, and all without programming. While you can seamlessly integrate, for example, R scripts or Groovy scripts into these processes, you do not need to write any scripts, if you do not want to. RapidMiner stores the data mining processes in a machine-readable XML format, which is directly executable in RapidMiner with the click of a button, and which along with the graphical visualization of the data mining process and the data flow serves as an automatically generated documentation of the data mining process, makes it easy to execute, to validate, to automatically optimize, to reproduce, and to automate.

Their broad functionality for all steps of the data mining process and their flexibility make RapidMiner and RapidAnalytics the tools of choice. They optimally support all steps of the overall data mining process and the flexible deployment of the processes and results within their framework and also integrated into other solutions via web services, Java API, or command-line calls. The process view of data mining eases the application to complex real-world tasks and the structuring and automation of even complex highly nested data mining processes. The processes also serve as documentation and for the reproducibility and reusability of scientific results as well as business applications. The open source nature of RapidMiner and RapidAnalytics, their numerous import and export and web service interfaces, and their openness, flexibility, and extendibility by custom extensions, operators, and scripts make them the ideal solutions for scientific, industrial, and business applications. Being able to reproduce earlier results, to reuse previous processes, to modify and adapt them or to extend them with customized or self-developed extensions means a high value for research, educational training, and industrial and business applications. RapidMiner allows you to quickly build working prototypes and also quickly deploy them on real data of all types including files, databases, time series data, text data, web pages, social media, image data, audio data, web services, and many other data sources.

What Is the Structure of This Book and Which Chapters Should I Read?

The *first chapter* of this book introduces the *basic concepts of data mining and machine learning*, common terms used in the field and throughout this book, and the decision tree modeling technique as a machine learning technique for classification tasks. The *second chapter* gives you an introductory tour through the *RapidMiner graphical user interface (GUI)* and how to use it to define data mining processes. In case you are already familiar with data mining and RapidMiner, you can skip these two chapters. However, if you are a novice in the field or regarding the software, these first two chapters are highly recommended and will give you a quick start in both data mining and RapidMiner. All following chapters provide a use case each and introduce additional data mining concepts and RapidMiner operators needed to solve the task at hand.

The *Chapters 3 to 6* describe *classification use cases* and introduce the *k-nearest neighbors (k-NN)* and *Naïve Bayes learning algorithms. Chapter 3* applies k-NN for the *evaluation of teaching assistants.* In *Chapter 4* k-NN is used to *classify different glass types based on chemical components* and the RapidMiner process is extended by Principal Component Analysis (PCA) to better preprocess the data and to improve the classification accuracy. *Chapter 5* explains Naïve Bayes as an algorithm for generating *classification* models and uses this modeling technique to generate a *credit approval* model to decide whether a credit loan for which a potential or existing customer applies should be approved or not, i.e. whether it is likely that the customer will pay back the credit loan as desired or not. *Chapter 6* uses Naïve Bayes to *rank applications* for nursery schools, introduces the RapidMiner operator for importing Excel sheets, and provides further explanations of Naïve Bayes.

Chapter 7 addresses the task of *product affinity-based marketing* and *optimizing a direct marketing campaign.* A bank has introduced a new financial product, a new type of current (checking) account, and some of its customers have already opened accounts of the new type, but many others have not done so yet. The bank's marketing department wants to push sales of the new account by sending direct mail to customers who have not yet opted for it. However, in order not to waste efforts on customers who are unlikely to buy, they would like to address only those customers with the highest affinity for the new product. Binary classification is used to predict for each customer, whether they will buy the product, along with a confidence value indicating how likely each of them is to buy the new product. Customers are then ranked by this confidence value and the 20% with the highest expected probability to buy the product are chosen for the campaign.

Following the *CRoss-Industry Standard Process for Data Mining (CRISP-DM)* covering all steps from business understanding and data understanding via data preprocessing and modeling to performance evaluation and deployment, this chapter first describes the task, the available data, how to extract characteristic customer properties from the customer data, their products and accounts data and their transactions, which data preprocessing to apply to balance classes and aggregate information from a customer's accounts and transactions into attributes for comparing customers, modeling with binary classification, evaluating the predictive accuracy of the model, visualizing the performance of the model using Lift charts and ROC charts, and finally ranking customers by the predicted confidence for a purchase to select the best candidates for the campaign. The predictive accuracy of several learning algorithms including Decision Trees, Linear Regression, and Logistic Regression is compared and visualized comparing their ROC charts. Automated attribute weight and parameter optimizations are deployed to maximize the prediction accuracy and thereby the customer response, sales volume, and profitability of the campaign. Similar processes can be used for *customer churn prediction* and addressing the customers predicted to churn in a campaign with special offers trying *to prevent them from churning.*

Chapters 8 to 10 describe three different approaches to *building recommender systems. Product recommendations in online shops* like Amazon increase the sales volume per customer by *cross-selling*, i.e., by selling more products per customer by recommending products that the customer may also like and buy.

The recommendations can be based on product combinations frequently observed in market baskets in the past. Products that co-occurred in many purchases in the past are assumed to be also bought together frequently in the future. *Chapter 8* describes how to generate such *association rules* for product recommendations from shopping cart data using the FP-Growth algorithm. Along the way, this chapter also explains how to import product sales data from CSV files and from retailers' databases and how to handle data quality issues and missing values.

Chapter 9 introduces the *RapidMiner Extension for Recommender Systems.* This extension allows building more sophisticated recommendation systems than described in the previous chapter. The application task in this chapter is to *recommend appropriate video lectures* to potential viewers. The recommendations can be based on the content of the lectures or on the viewing behavior or on both. The corresponding approaches are called content-based, collaborative, and hybrid recommendation, respectively. Content-based recommendations can be based on attributes or similarity and collaborative recommendation systems deploy neighborhoods or factorization. This chapter explains, evaluates, and compares these approaches. It also demonstrates how to make RapidMiner processes available as *RapidAnalytics web services*, i.e., how to build a *recommendation engine* and make it available for *real-time recommendations* and easy *integration* into web sites, online shops, and other systems via web services.

A third way of building recommender systems in RapidMiner is shown in *Chapter 10*, where *classification algorithms* are used to *recommend the best-fitting study program for higher-education students* based on their predicted success for different study programs at a particular department of a particular university. The idea is an early analysis of students' success on each study program and the recommendation of a study program where a student will likely succeed. At this university department, the first year of study is common for all students. In the second year, the students select their preferred study program among several available programs. The attributes captured for each graduate student describe their success in the first-year exams, their number of points in the entrance examination, their sex, and their region of origin. The target variable is the average grade of the student at graduation, which is discretized into several categories. The prediction accuracy of several classification learning algorithms, including Naïve Bayes, Decision Trees, Linear Model Tree (LMT), and CART (Classification and Regression Trees), is compared for the prediction of the student's success as measured by the discretized average grade. For each student, the expected success classes for each study program is predicted and the study program with the highest predicted success class is recommended to the student. An optimization loop is used to determine the best learning algorithm and automated feature selection is used to find the best set of attributes for the most accurate prediction. The RapidMiner processes seamlessly integrate and compare learning techniques implemented in RapidMiner with learning techniques implemented in the open source data mining library Weka, thanks to the *Weka extension for RapidMiner* that seamlessly integrates all Weka learners into RapidMiner.

Chapter 11 provides an introduction to *clustering*, to the *k-Means* clustering algorithm, to several *cluster validity measures*, and to their *visualizations*. Clustering algorithms group cases into groups of similar cases. While for classification, a training set with examples with predefined categories is necessary for training a classifier to automatically classify new cases into one of the predefined categories, clustering algorithms need no labeled training examples with predefined categories, but automatically group unlabeled examples into clusters of similar cases. While the predictive accuracy of classification algorithms can be easily measured by comparing known category labels of known examples to the categories predicted by the algorithm, there are no labels known in advance in the case of clustering. Hence it is more difficult to achieve an objective evaluation of a clustering result. Visualizing cluster validity measures can help humans to evaluate the quality of a set of clusters. This chapter uses k-Means clustering on a *medical dataset to find groups of similar E-Coli bacteria with regards to where protein localization occurs in them* and explains how to judge the quality of the clusters found using visualized cluster validity metrics. Cluster validity measures implemented in the open source statistics package R are seamlessly integrated and used within RapidMiner processes, thanks to the *R extension for RapidMiner.*

Chapter 12 applies *clustering* to automatically *group higher education students.* The dataset corresponds to the one already described in Chapter 10, but now the task is to find groups of similarly performing students, which is achieved with automated clustering techniques. The attributes describing the students may have missing values and different scales. Hence data preprocessing steps are used to replace missing values and to normalize the attribute values to identical value ranges. A parameter loop automatically selects and evaluates the performance of several clustering techniques including k-Means, k-Medoids, Support Vector Clustering (SVC), and DBSCAN.

Chapters 13 to 15 are about *text mining* applications. *Chapter 13* gives an *introduction to text mining,* i.e., the application of data mining techniques like classification to text documents like e-mail messages, mobile phone text messages (SMS = Short Message Service) or web pages collected from the World-Wide Web. In order to detect text message spam, preprocessing steps using the *RapidMiner text processing extension* transform the unstructured texts into document vectors of equal length, which make the data applicable to standard classification techniques like Naïve Bayes, which is then trained to *automatically separate legitimate mobile phone text messages from spam messages.*

The second text mining use case uses *classification to automatically identify the language of a text* based on its characters, character sequences, and/or words. *Chapter 14* discusses character encodings of different European, Arabic, and Asian languages. The chapter describes different text representations by characters, by tokens like words, and by character sequences of a certain length also called n-grams. The transformation of document texts into document vectors also involves the weighting of the attributes by term frequency and document frequency-based metrics like TF/IDF, which is also described here. The classification techniques Naïve Bayes and Support Vector Machines (SVM) are then trained and evaluated on four different multi-lingual text corpora including for example dictionary texts from Wikipedia and book texts from the Gutenberg project. Finally, the chapter shows how to make the RapidMiner *language detection* available as *web service for the automated language identification* of web pages via *RapidAnalytics web services.*

Chapter 15 analyses *hotel review texts* and ratings by customers collected from the TripAdvisor web page. Frequently co-occurring words in the review texts are found using FP-Growth and association rule generation and visualized in a *word-association graph.* In a second analysis, the review texts are *clustered* with k-Means, which reveals *groups of similar texts.* Both approaches provide insights about the hotels and their customers, i.e., about *topics of interest and of complaints,* quality and service issues, likes, dislikes, and preferences, and could similarly be applied to all kinds of *textual reviews and customer feedback.*

Chapter 16 describes a data mining use case in *astroparticle physics,* the application of automated *classification* and automated *feature selection* in *neutrino astronomy* to separate a small number of neutrinos from a large number of background noise particles or signals (muons). One of the main scientific goals of neutrino telescopes is the detection of neutrinos originating from astrophysical sources as well as a precise measurement of the energy spectrum of neutrinos produced in cosmic ray air showers in the Earth's atmosphere. These so-called atmospheric neutrinos, however, are hidden in a noisy background of atmospheric muons produced in air showers as well. The first task in rejecting this background is the selection of upward-going tracks since the Earth is opaque to muons but can be traversed by neutrinos up to very high energies. This procedure reduces the background by roughly three orders of magnitude. For a detailed analysis of atmospheric neutrinos, however, a very clean sample with purity larger than 95% is required. The main source of remaining background at this stage are muon tracks, falsely reconstructed as upward going. These falsely reconstructed muon tracks still dominate the signal by three orders of magnitude and have to be rejected by the use of straight cuts or multivariate methods. Due to the ratio of noise (muons) and signal (neutrinos), about 10,000 particles need to be recorded in order to catch about 10 neutrinos. Hence, the amount of data delivered by these experiments is enormous and it must be processed and analyzed within a proper amount of time. Moreover, data in these experiments are delivered in a format that contains more than 2000 attributes originating from various reconstruction algorithms. Most of these attributes have been reconstructed from only a few physical quantities. The direction of a neutrino event penetrating the detector at a certain angle can, for example, be reconstructed from a pattern of light that is initiated by particles produced by an interaction of the neutrino close to or even in the detector. Due to the fact that all of the 2000 reconstructed attributes are not equally well suited for classification, the first task in applying data mining techniques in neutrino astronomy lies in finding a good and reliable representation of the dataset in fewer dimensions. This is a task which very often determines the quality of the overall data analysis. The second task is the training and evaluation of a stable learning algorithm with a very high performance in order to separate signal and background events. Here, the challenge lies in the biased distribution of many more background noise (negative) examples than there are signals (positive) examples. Handling such skewed distributions is necessary in many real-world problems. The application of RapidMiner in neutrino astronomy models the separation of neutrinos from background as a two-step process, accordingly. In this chapter, the *feature or attribute selection* is explained in the first part and the *training of selecting relevant events* from the masses of incoming data is explained in the second part. For the feature selection, the *Feature Selection Extension for RapidMiner* is used and a wrapper cross-validation to evaluate the performance of the feature selection methods. For the selection of the relevant events, *Random Forests* are used as classification learner.

Chapter 17 provides an introduction to *medical data mining*, an overview of methods often used for classification, regression, clustering, and association rules generation in this domain, and two application use cases with data about patients suffering from *carpal tunnel syndrome* and *diabetes*, respectively.

In the study of the *carpal tunnel syndrome (CTS)*, thermographic images of hands were collected for constructing a predictive classification model for CTS, which could be helpful when looking for a non-invasive diagnostic method. The temperatures of different areas of a patient's hand were extracted from the image and saved in the dataset. Using a RapidMiner preprocessing operator for aggregation, the temperatures were averaged for all segments of the thermal images. Different machine learning algorithms including *Artificial Neural Network* and *Support Vector Machines (SVM)* were evaluated for generating a classification model capable of diagnosing CTS on the basis of very discrete temperature differences that are invisible to the human eye in a thermographic image.

In the study of *diabetes*, various research questions were posed to evaluate the level of knowledge and overall perceptions of diabetes mellitus type 2 (DM) within the older population in North-East Slovenia. As a chronic disease, diabetes represents a substantial burden for the patient. In order to accomplish good self-care, patients need to be qualified and able to accept decisions about managing the disease on a daily basis. Therefore, a high level of knowledge about the disease is necessary for the patient to act as a partner in managing the disease. Various research questions were posed to determine what the general knowledge about diabetes is among diabetic patients 65 years and older, and what the difference in knowledge about diabetes is with regard to the education and place of living on (1) diet, (2) HbA1c, (3) hypoglycemia management, (4) activity, (5) effect of illness and infection on blood sugar levels, and (6) foot care. A hypothesis about the level of general knowledge of diabetes in older populations living in urban and rural areas was predicted and verified through the study. A cross-sectional study of older (age >65 years), non-insulin dependent patients with diabetes mellitus type 2 who visited a family physician, DM outpatient clinic, a private specialist practice, or were living in a nursing home was implemented. The Slovenian version of the Michigan Diabetes Knowledge test was then used for data collection. In the data preprocessing, missing values in the data were replaced, before *k-means clustering* was used to find groups of similar patients, for which then a *decision tree* learner was used to find attributes discriminating the clusters and generate a classification model for the clusters. A grouped *ANOVA (ANalysis Of VAriances) statistical test* verified the hypothesis that there are differences in the level of knowledge about diabetes in rural populations and city populations in the age group of 65 years and older.

Chapter 18 covers a use case relevant in *chemistry* and the *pharmaceutical industry*. The RapidMiner Extension *PaDEL (Pharmaceutical Data Exploration Laboratory)* developed at the University of Singapore is deployed to *calculate a variety of molecular properties from the 2-D or 3-D molecular structures of chemical compounds*. Based on these molecular property vectors, RapidMiner can then generate predictive models for *predicting chemical, biochemical, or biological properties based on molecular properties*, which is a frequently encountered task in theoretical chemistry and the pharmaceutical industry. The combination of RapidMiner and PaDEL provides an open source solution to generate prediction systems for a broad range of biological properties and effects.

One application example in *drug design* is the *prediction of effects and side effects* of a new drug candidate before even producing it, which can help to avoid testing many drug candidates that probably are not helpful or possibly even harmful and thereby help to focus research resources on more promising drug candidates. With PaDEL and RapidMiner, properties can be calculated for any molecular structure, even if the compound is not physically accessible. Since both tools are open source and can compute the properties of a molecular structure quickly, this allows significant reduction in cost and an increase in speed of the development of new chemical compounds and drugs with desired properties, because more candidate molecules can be considered automatically and fewer of them need to be actually generated and physically, chemically, or biologically tested.

The combination of data mining (RapidMiner) and a tool to handle molecules (PaDEL) provides a convenient and user-friendly way to generate accurate relationships between chemical structures and any property that is supposed to be predicted, mostly biological activities. Relationships can be formulated as qualitative structure-property relationships (SPRs), qualitative structure-activity relationships (SARs) or quantitative structure-activity relationships (QSARs). SPR models aim to highlight associations between molecular structures and a target property, such as lipophilicity. SAR models correlate an activity with structural properties and QSAR models quantitatively predict an activity. Models are typically developed to predict properties that are difficult to obtain, impossible to measure, require time-consuming experiments, or are based on a variety of other complex properties. They may also be useful to predict complicated properties using several simple properties. The PaDEL extension enables RapidMiner to directly read and handle molecular structures, calculate their molecular properties, and to then correlate them to and generate predictive models for chemical, biochemical, or biological properties of these molecular structures. In this chapter *linear regression* is used as a *QSAR modeling* technique to *predict chemical properties* with RapidMiner *based on molecular properties* computed by PaDEL.

Chapter 19 describes a second *Quantitative Structure-Activity Relationship (QSAR)* use case relevant in *chemistry* and the *pharmaceutical industry, the identification of novel functional inhibitors of acid sphingomyelinase (ASM).* The use case in this chapter is based on the previous chapter and hence you should first read Chapter 18 before reading this chapter. In the data preprocessing step, the *PaDEL (Pharmaceutical Data Exploration Laboratory)* extension for RapidMiner described in the previous chapter is again used to compute molecular properties from given molecular 2-D or 3-D structures. These properties are then used to *predict ASM inhibition.* Automated *feature selection* with *backward elimination* is used to reduce the number of properties to a relevant set for the prediction task, for which a *classification* learner, namely *Random Forests*, generates the predictive model that captures the *structure- and property-activity relationships.*

The process of drug design from the biological target to the drug candidate and, subsequently, the approved drug has become increasingly expensive. Therefore, strategies and tools that reduce costs have been investigated to improve the effectiveness of drug design. Among them, the most time-consuming and cost-intensive steps are the selection, synthesis, and experimental testing of the drug candidates. Therefore, numerous attempts have been made to reduce the number of potential drug candidates for experimental testing. Several methods that rank compounds with respect to their likelihood to act as an active drug have been developed and applied with variable success. In silico methods that support the drug design process by reducing the number of promising drug candidates are collectively known as virtual screening methods. Their common goal is to reduce the number of drug candidates subjected to biological testing and to thereby increase the efficacy of the drug design process.

This chapter demonstrates an in silico method to predict biological activity based on RapidMiner data mining work flows. This chapter is based on the type of chemoinformatic predictions described in the previous chapter based on chemoinformatic descriptors computed by PaDEL. *Random Forests* are used as a predictive model for predicting the molecular activity of a molecule of a given structure, for which PaDEL is used to compute molecular structural properties, which are first reduced to a smaller set by automated *attribute weighting* and *selecting the attributes* with the highest weights according to several weighting criteria and which are reduced to an even smaller set of attributes by automated *attribute selection* using a *Backward Elimination* wrapper. Starting with a large number of properties for the example set, a *feature selection* vastly reduces the number of attributes before the systematic backward elimination search finds the most predictive model for the feature generation. Finally, a *validation* is performed to avoid over-fitting and the benefits of *Y-randomization* are shown.

Chapter 20 introduces the *RapidMiner IMage Mining (IMMI) Extension* and presents some introductory *image processing* and *image mining* use cases. *Chapter 21* provides more advanced image mining applications.

Given a set of images in a file folder, the image processing task in the first use case in *Chapter 20* is to adjust the contrast in all images in the given folder and to store the transformed images in another folder. The IMMI extension provides RapidMiner operators for *reading and writing images*, which can be used within a RapidMiner loop iterating over all files in the given directory, *adjusting the contrast* of each of these images, for example, using a histogram equalization method. Then the chapter describes image *conversions between color and gray-scale images* and different *feature extraction* methods, which convert image data in unstructured form into a tabular form. Feature extraction algorithms for images can be divided into three basic categories: local-level, segment-level, and global-level feature extraction.

The term *local-level* denotes that information is mined from given points (locations) in the image. *Local-level feature extraction* is suitable for *segmentation*, *object detection* or *area detection*. From each point in the image, it is possible to extract information like pixel gray value, minimal or maximal gray value in a specified radius, value after applying kernel function (blurring, edge enhancements). Examples of utilization of such data are the *trainable segmentation of an image*, *point of interest detection*, and *object detection*.

The term *segment-level* denotes feature extraction from segments. Many different segmentation algorithms exist, such as k-means, watershed, or statistical region merging. Segment level feature extraction algorithms extract information from the whole segments. Examples of such features are mean, median, lowest and highest gray value, circularity, and eccentricity. In contrast to local-level features, it does not take into consideration only a single point and its neighborhood, however, it considers the whole segment and its properties like shape, size, and roundness. With the use of knowledge about the size or shape of target objects, it is for example possible to *select or remove objects according to their size or shape*.

The *global-level* denotes *feature extraction from the whole image*, for example, mean color, dominant color, maximal gray value, minimal gray value, variance of pixels, number of edges etc. Unlike the local or segment level, the global level segmentation is not suitable for points or areas identification or segmentation. Rather, it is suitable *for classification of images* and determining *properties of the image as a whole*.

Chapter 20 provides examples demonstrating the use of local-level, segment-level, and global-level feature extraction. *Local-level feature extraction* is used for *trainable image segmentation* with radial-basis function (RBF) Support Vector Machines (SVM). *Segment-level feature extraction* and trainable segment selection reveal interesting segment properties like size and shape for image analysis. With the help of *global-level feature extraction*, images are classified into pre-defined classes. In the presented use case, two *classes of images* are *distinguished automatically*: images containing birds and images containing sunsets. To achieve this, global features like dominant color, minimal intensity, maximal intensity, percent of edges, etc. are extracted and based on those, an image classifier is trained.

Chapter 21 presents *advanced image mining applications* using the *RapidMiner IMage Mining (IMMI) Extension* introduced in the previous chapter. This chapter demonstrates several examples of the use of the IMMI extension for image processing, image segmentation, feature extraction, pattern detection, and image classification. The first application *extracts global features* from multiple images to enable *automated image classification*. The second application demonstrates the *Viola-Jones algorithm for pattern detection*. And the third process illustrates the *image segmentation* and *mask processing*.

The *classification of an image* is used to identify which group of images a particular image belongs to. An automated image classifier could, for example, be used to distinguish different scene types like nature versus urban environment, exterior versus interior, images with and without people, etc. *Global features* are usually used for this purpose. These features are calculated from the whole image. The key to a correct classification is to find the features that differentiate one class from other classes. Such a feature can be, for example, the dominant color in the image. These features can be calculated from the original image or from an image after pre-processing like Gaussian blur or edge detection.

Pattern detection searches known patterns in images in the images, where approximate fits of the patterns may be sufficient. A good algorithm for detection should not be sensitive to the size of the pattern in the image or its position or rotation. One possible approach is to use a histogram. This approach compares the histogram of the pattern with the histogram of a selected area in the image. In this way, the algorithm passes step by step through the whole image, and if the match of histograms is larger than a certain threshold, the area is declared to be the sought pattern. Another algorithm, which is described in this chapter, is the Viola-Jones algorithm. The classifier is trained with positive and negative image examples. Appropriate features are selected using the AdaBoost algorithm. An image is iterated during pattern detection using a window with increasing size. Positive detections are then marked with a square area of the same size as the window. The provided example application uses this process to detect the cross-sectional artery in an ultrasound image. After detection, the images can be used to measure the patient's pulse if taken from a video or stream of time-stamped images.

The third example application demonstrates *image segmentation* and *feature extraction*: Image segmentation is often used for the *detection of different objects* in the image. Its task is to split the image into parts so that the individual segments correspond to objects in the image. In this example, the identified segments are combined with *masks* to remove the background and focus on the object found.

Chapter 22 introduces the *RapidMiner Extension for Instance Selection and Prototype-based Rule (ISPR) induction*. It describes the *instance selection* and *prototype construction* methods implemented in this extension and applies them *to accelerate* 1-NN classification on large datasets and to perform *outlier elimination* and *noise reduction*. The datasets analyzed in this chapter include several *medical datasets* for classifying patients with respect to certain medical conditions, i.e., *diabetes, heart diseases, and breast cancer*, as well as an *e-mail spam detection* dataset. The chapter describes a variety of *prototype selection algorithms* including k- Nearest-Neighbors (k-NN), Monte-Carlo (MC) algorithm, Random Mutation Hill Climbing (RMHC) algorithm, Condensed Nearest-Neighbor (CNN), Edited Nearest-Neighbor (ENN), Repeated ENN (RENN), Gabriel Editing proximity graph-based algorithm (GE selection), Relative Neighbor Graph algorithm (RNG selection), Instance-Based Learning (IBL) algorithm (IB3 selection), Encoding Length Heuristic (ELH selection), and combinations of them and compares their performance on the datasets mentioned above. *Prototype construction methods* include all algorithms that produce a set of instances at the output. The family contains all prototype-based clustering methods like k-Means, Fuzzy C-Means (FCM), and Vector Quantization (VQ) as well as the Learning Vector Quantization (LVQ) set of algorithms. The price for the speed-up of 1-NN by instance selection is visualized by the drop in predictive accuracy with decreasing sample size.

Chapter 23 gives an overview of a large range of *anomaly detection* methods and introduces the *RapidMiner Anomaly Detection Extension*. Anomaly detection is the process of finding patterns in a given dataset which deviate from the characteristics of the majority. These outstanding patterns are also known as anomalies, outliers, intrusions, exceptions, misuses, or fraud. Anomaly detection identifies single records in datasets which significantly deviate from the normal data. Application domains among others include *network security, intrusion detection, computer virus detection, fraud detection, misuse detection, complex system supervision*, and *finding suspicious records in medical data*. Anomaly detection for fraud detection is used to *detect fraudulent credit card transactions* caused by stolen credit cards, *fraud in Internet payments*, and *suspicious transactions in financial accounting data*. In the medical domain, anomaly detection is also used, for example, for *detecting tumors in medical images* or *monitoring patient data* (electrocardiogram) to get early warnings in case of life-threatening situations. Furthermore, a variety of other specific applications exists such as *anomaly detection in surveillance camera data, fault detection in complex systems* or *detecting forgeries in the document forensics*. Despite the differences of the various application domains, the basic principle remains the same. Multivariate normal data needs to be modeled and the few deviations need to be detected, preferably with a *score* indicating their "outlierness", i.e., a score indicating their extent of being an outlier. In case of a univariate data, such an *outlier factor* could for example be the number of standard deviations by which an outlier differs from the mean of this variable.

The overview of anomaly detection method provided in this chapter distinguishes three different types of anomalies, namely (1) *point anomalies*, which are single data records deviating from others, (2) *contextual anomalies*, which occur with respect to their context only, for example, with respect to time, and (3) *collective anomalies*, where a bunch of data points causes the anomaly. Most anomaly detection algorithms detect point anomalies only, which leads to the requirement of transforming contextual and collective anomalies to point anomaly problems using an appropriate pre-processing and thus generating processable data views. Furthermore, anomaly detection algorithms can be categorized with respect to their operation mode, namely (1) *supervised algorithms* with training and test data as used in traditional machine learning, (2) *semi-supervised algorithms* with the need of anomaly-free training data for one-class learning, and (3) *unsupervised approaches* without the requirement of any labeled data. Anomaly detection is, in most cases, associated with an unsupervised setup, which is also the focus of this chapter. In this context, all available unsupervised algorithms from the RapidMiner anomaly detection extension are described and the most well-known algorithm, the *Local Outlier Factor (LOF)* is explained in detail in order to get a deeper understanding of the approaches themselves. The *unsupervised anomaly detection algorithms* covered in this chapter include Grubbs' outlier test and noise removal procedure, k-NN Global Anomaly Score, Local Outlier Factor (LOF), Connectivity-Based Outlier Factor (COF), Influenced Outlierness (INFLO), Local Outlier Probability (LoOP), Local Correlation Integral (LOCI) and aLOCI, Cluster-Based Local Outlier Factor (CBLOF), and Local Density Cluster-Based Outlier Factor (LDCOF). The *semi-supervised anomaly detection algorithms* covered in this chapter include a one-class Support Vector Machine (SVM) and a two-step approach with clustering and distance computations for detecting anomalies.

Besides a simple example consisting of a two-dimensional mixture of Gaussians, which is ideal for first experiments, two real-world datasets are analyzed. For the unsupervised anomaly detection the player statistics of the NBA, i.e., a dataset with the *NBA regular-season basketball player statistics from 1946 to 2009*, are analyzed for outstanding players, including all necessary pre-processing. The *UCI NASA shuttle dataset* is used for illustrating how semi-supervised anomaly detection can be performed in RapidMiner to find suspicious states during a NASA shuttle mission. In this context, a *Groovy script* is implemented for a simple *semi-supervised cluster-distance-based anomaly detection* approach, showing how to easily extend RapidMiner by your own operators or scripts.

Chapter 24 features a complex *data mining research use case*, the *performance evaluation and comparison of several classification learning algorithms* including Naïve Bayes, k-NN, Decision Trees, Random Forests, and Support Vector Machines (SVM*) across many different datasets.* Nested process control structures for loops over datasets, loops over different learning algorithms, and cross validation allow an automated validation and the selection of the best model for each application dataset. Statistical tests like t-test and ANOVA test (ANalysis Of VAriance) determine whether performance differences between different learning techniques are statistically significant or whether they may be simply due to chance. Using a custom-built Groovy script within RapidMiner, meta-attributes about the datasets are extracted, which can then be used for *meta-learning*, i.e., for learning to predict the performance of each learner from a given set of learners for a given new dataset, which then allows the selection of the learner with the best expected accuracy for the given dataset. The performance of fast learners called landmarkers on a given new dataset and the meta-data extracted from the dataset can be used for meta-learning to predict the performance of another learner on this dataset. The RapidMiner Extension for Pattern Recognition Engineering (PaREn) and its Automatic System Construction Wizard perform this kind of meta-learning for automated learner selection and a parameter optimization for a given dataset.

The *index at the end of the book* helps you to find explanations of data mining concepts and terms you would like to learn more about, use case applications you may be interested in, or reference use cases for certain modeling techniques or RapidMiner operators you are looking for. The *companion web page for this book* provides the RapidMiner processes and datasets deployed in the use cases:
`http://www.RapidMinerBook.com`

About the Editors

Markus Hofmann

Dr. Markus Hofmann is currently a lecturer at the Institute of Technology Blanchardstown in Ireland where he focuses on the areas of data mining, text mining, data exploration and visualisation as well as business intelligence. He holds a PhD degree from Trinity College Dublin, an MSc in Computing (Information Technology for Strategic Management) from the Dublin Institute of Technology and a BA in Information Management Systems. He has taught extensively at undergraduate and postgraduate level in the fields of Data Mining, Information Retrieval, Text/Web Mining, Data Mining Applications, Data Pre-processing and Exploration and Databases. Dr. Hofmann published widely at national as well as international level and specialised in recent years in the areas of Data Mining, learning object creation, and virtual learning environments. Further he has strong connections to the Business Intelligence and Data Mining sector both on an academic as well as industry level. Dr. Hofmann has worked as technology expert together with 20 different organisations in recent years including companies such as Intel. Most of his involvement was on the innovation side of technology services and products where his contributions had significant impact on the success of such projects. He is a member of the Register of Expert Panelists of the Irish Higher Education and Training Awards council, external examiner to two other third level institutes and a specialist in undergraduate and post graduate course development. He has been internal as well as external examiner of postgraduate thesis submissions. He was also local and technical chair of national and international conferences.

Ralf Klinkenberg

Ralf Klinkenberg holds Master of Science degrees in computer science with focus on machine learning, data mining, text mining, and predictive analytics from the Technical University of Dortmund in Germany and Missouri University of Science and Technology in the USA. He performed several years of research in these fields at both universities before initiating the RapidMiner open source data mining project in 2001, whose first version was called Yet Another Learning Environment (YALE). Ralf Klinkenberg founded this software project together with Dr. Ingo Mierswa and Dr. Simon Fischer. In 2006 he founded the company Rapid-I together with Ingo Mierswa. Rapid-I now is the company behind the open source software solution RapidMiner and its server version RapidAnalytics, providing these and further data analysis solutions, consulting, training, projects, implementations, support, and all kinds of related services. Ralf Klinkenberg has more than 15 years of experience in consulting and training large and small corporations and organizations in many different sectors how to best leverage data mining and RapidMiner based solutions for their needs. He performed data mining, text mining, web mining, and business analytics projects for companies like telecoms, banks, insurances, manufacturers, retailers, pharmaceutical companies, healthcare, IT, aviation, automotive, and market research companies, utility and energy providers, as well as government organizations in many European and North American countries. He provided solutions for tasks like automated direct marketing campaign optimization, churn prediction and prevention, sales volume forecasting, automated online media monitoring and sentiment analysis to generate customer insights, market insights, and competitive intelligence, customer feedback analysis for product and service optimization, automated e-mail routing, fraud detection, preventive maintenance, machine failure prediction and prevention, manufacturing process optimization, quality and cost optimization, profit maximization, time series analysis and forecasting, critical event detection and prediction, and many other data mining and predictive analytics applications.

List of Contributors

Editors

- **Markus Hofmann**, Institute of Technology Blanchardstown, Ireland
- **Ralf Klinkenberg**, Rapid-I, Germany

Chapter Authors

- **Ingo Mierswa**, Rapid-I, Germany
- **M. Fareed Akhtar**, Fastonish, Australia
- **Timm Euler**, viadee IT-Consultancy, Münster/Köln (Cologne), Germany
- **Matthew A. North**, The College of Idaho, Caldwell, Idaho, USA
- **Matej Mihelčić**, Electrical Engineering, Mathematics and Computer Science, University of Twente, Netherlands; Rudjer Boskovic Institute, Zagreb, Croatia
- **Matko Bošnjak**, University of Porto, Porto, Portugal; Rudjer Boskovic Institute, Zagreb, Croatia
- **Nino Antulov-Fantulin**, Rudjer Boskovic Institute, Zagreb, Croatia
- **Tomislav Šmuc**, Rudjer Boskovic Institute, Zagreb, Croatia
- **Milan Vukićević**, Faculty of Organizational Sciences, University of Belgrade, Belgrade, Serbia
- **Miloš Jovanović**, Faculty of Organizational Sciences, University of Belgrade, Belgrade, Serbia
- **Boris Delibašić**, Faculty of Organizational Sciences, University of Belgrade, Belgrade, Serbia
- **Milija Suknović**, Faculty of Organizational Sciences, University of Belgrade, Belgrade, Serbia
- **Andrew Chisholm**, Institute of Technology, Blanchardstown, Dublin, Ireland
- **Neil McGuigan**, University of British Columbia, Sauder School of Business, Canada
- **Eduarda Mendes Rodrigues**, University of Porto, Porto, Portugal
- **Luis Sarmento**, Sapo.pt - Portugal Telecom, Lisbon, Portugal
- **Gurdal Ertek**, Sabancı University, Istanbul, Turkey
- **Dilek Tapucu**, Sabancı University, Istanbul, Turkey

- **Inanc Arin**, Sabancı University, Istanbul, Turkey
- **Tim Ruhe**, TU Dortmund, Dortmund, Germany
- **Katharina Morik**, TU Dortmund, Dortmund, Germany
- **Wolfgang Rhode**, TU Dortmund, Dortmund, Germany
- **Mertik Matej**, Faculty of information study Novo mesto, Slovenia
- **Palfy Miroslav**, University Medical Centre Maribor, Slovenia
- **Markus Muehlbacher**, Department of Psychiatry and Psychotherapy, University Hospital of Erlangen-Nuremberg, Friedrich-Alexander-University Erlangen, Germany; Computer Chemistry Center, Friedrich-Alexander-University Erlangen, Germany
- **Johannes Kornhuber**, Department of Psychiatry and Psychotherapy, University Hospital of Erlangen-Nuremberg, Friedrich-Alexander-University Erlangen, Germany
- **Radim Burget**, Brno University of Technology, Czech Republic
- **Václav Uher**, Brno University of Technology, Czech Republic
- **Jan Masek**, Brno University of Technology, Czech Republic
- **Marcin Blachnik**, Silesian University of Technology, Department of Management and Informatics, Poland
- **Miroslaw Kordos**, University of Bielsko-Biala, Department of Mathematics and Computer Science, Bielsko-Biala, Poland
- **Markus Goldstein**, German Research Center for Artificial Intelligence, Kaiserslautern, Germany

Acknowledgments

A lot of people have contributed to make this book and the underlying open source software solutions RapidMiner and RapidAnalytics happen. We are thankful to all of you.

We would like to thank the contributing authors of this book, who shared their experience in the chapters and who thereby enable others to have a quick and successful data mining start with RapidMiner providing successful application examples and blueprints for the readers to tackle their data mining tasks and benefit from the strength of using RapidMiner and RapidAnalytics.

Many thanks to Dr. Brian Nolan, Head of Department of Informatics, Institute of Technology Blanchardstwon (ITB), for continuously supporting the relationship between the institute and Rapid-I.

The entire team of the Taylor & Francis Group were very professional, responsive and always helpful in guiding us through this project. Should any of you readers consider publishing a book then we can highly recommend this publisher.

Before there could be any thought about a book like this one, there needed to be the open source data mining software RapidMiner to whose success many contributed.

A special thanks goes to Prof. Dr. Katharina Morik, Head of the Artificial Intelligence Unit at the Technical University of Dortmund, Germany, for providing an introduction to and deep insights into machine learning (ML), data mining, text mining, artificial intelligence (AI), and natural language processing (NLP), for providing an environment that enabled the initiation of an open source data mining software project named YALE (Yet Another Learning Environment), which was later improved and renamed to RapidMiner. She supports the open source project RapidMiner, the company Rapid-I, which is behind the project, and its founders until today, long after they left the university. We appreciate the good cooperation and exchange of ideas with her research and teaching unit.

Another big thank you goes to Dr. Ingo Mierswa and Dr. Simon Fischer who started the open source project YALE and later RapidMiner together with Ralf Klinkenberg and who took turns in supervising the international development team of the project. Without their ideas, passion, commitment, and enormous development work, we would not have such a powerful and flexible open source data mining framework and solution today, available for everyone to use and gain from.

Dr. Ingo Mierswa and Ralf Klinkenberg are also the co-founders of Rapid-I, the data mining and business analytics company behind the open source software RapidMiner.

We are grateful to all who joined the RapidMiner and Rapid-I teams, especially to Sebastian Land, Helge Homburg, and Tobias Malbrecht who joined the RapidMiner team in its early days and contributed a lot to its development and who are strong members of the team until today. We are also thankful to all contributors from the early days, like Michael Wurst, Martin Scholz, and Timm Euler, as well as to the newer team members like Marius Helf, Nils-Christian Whler, Marcin Skirzynski, Venkatesh Umaashankar, Marco Bck, Dominik Halfkann, and to those who support the team in other roles, like Nadja Mierswa, Simone Horstmann, Christian Brabandt, Edin Klapic, Balzs Brny, Dietrich Niederlintner, Caroline Hahne, Miguel Bscher, Thilo Kamradt, Jannik Zappe, Kyle Goslin, and Assumpta Harvey.

Open source projects grow strong with their community. We are thankful to all contributors to RapidMiner and RapidAnalyitcs and to all supporters of these open source projects. We are grateful not only for source code contributions, community support in the forum, bug reports and fixes, but also to those who spread the word with their blogs, videos, and words of mouth, especially to Thomas Ott,[3] Neil McGuigan,[4] Dr. Bala Deshpande,[5] Prof. Dr. Bonnie Holub,[6] Prof. Dr. Matthew North,[7] Sheamus McGovern,[8] Prof. Dr. David Wheismann, and many more.

Many bright minds have influenced our thoughts and inspired us with their ideas and valuable discussion. We would like to thank Prof. Dr. Thorsten Joachims, Prof. Dr. Hans-Ulrich Simon, Prof. Dr. Daniel St. Clair, Prof. Dr. Cihan Dagli, Prof. Dr. Tom Mitchell, and many others for widening our horizon and for many deep insights into the theoretical limits, enormous capabilities, and many practical applications of machine learning, data mining, data mining, text mining, statistical learning theory, and predictive analytics.

We would also like the many companies and organizations supporting the development of RapidMiner and RapidAnalytics by becoming customers of Rapid-I, including, for example, Sanofi, Daimler, Honda, Volkswagen, Miele, Siemens, Telekom Deutschland, T-Mobile International, mobilkom austria, Telenor, Nokia, Philips, Lufthansa, EADS, Salzgitter Mannesmann, ThyssenKrupp, Libri, Tchibo, KNV, PayPal, Intel, PepsiCo, GfK, Landesbank Berlin, E.ON, RWE, Axel Springer, 1&1, Schober, Schweizer Bundesbahn, FC Chelsea, The Cleveland Indians, and many more[9].

With best regards and appreciation to all contributors,

Dr. Markus Hofmann, Institute of Technology Blanchardstown, Dublin, Ireland

Ralf Klinkenberg, Co-Founder of RapidMiner and Rapid-I, CBDO, Rapid-I, Dortmund, Germany

[3]`http://www.NeuralMarketTrends.com/`

[4]`http://vancouverdata.blogspot.com/`

[5]`http://www.SimaFore.com/`

[6]`http://www.arclight.biz/uncategorized/rapidminer-introductory-tutorial-videos/`

[7]`http://docs.rapid-i.com/r/data-mining-for-the-masses`

[8]`http://www.cambridgecodeworks.com/wordpress/?author=2`, `http://www.meetup.com/Boston-Predictive-Analytics/`

[9]`http://rapid-i.com/content/view/8/56/`

List of Figures

List of Tables

Chapter 1

What This Book is About and What It is Not

Ingo Mierswa
Rapid-I, Dortmund, Germany

1.1 Introduction

Today, analytics is a very important topic and affects practically all levels in modern organizations. Analytics is also used by many data-driven researchers. Data is collected and analyzed, and the results of this analytical work either prove our hypothesis or delivers new insights.

When we talk about analytics in this book we are referring to what many call "advanced analytics". This field includes technologies known from statistics as well as from computer science. Isn't this just statistics done by computers then? By far not! Statistics often deal with the question if a hypothesis can be proven by a statistical test using a small but representative data sample. Although this is of biggest importance, it is even more useful to mix these ideas with algorithms from computer science to make sure that the methods we are talking about will be able to scale up and analyze even the largest datasets.

And I see another distinction: traditional statistics often requires that the analyst is creating a model or hypothesis right at the start of the analysis process. After creating such a model, parameters of the model are estimated or the applicability of this model is proven by means of a statistical test. Maybe this is because I am lazy, but I actually don't like this idea too much: Why should I manually do the work a computer is perfectly able to deal with itself? In this sense of manual analysis, statistical analysis is much more connected to online analytical processing (OLAP) than with "advanced analytics": In the OLAP world, people try to drill through their data to find the interesting patterns and the reasons in deeper data levels themselves. Fine, but again I think this is the wrong approach for solving the underlying problem for mainly two reasons: Firstly, people tend to see only what they are looking for. Most analysts have some expectations before they start and try to work as long as necessary on the data to prove their point. Secondly, OLAP is again a pretty tedious work I personally believe a computer is much better suited for. Did I already mention that I am pretty lazy in this respect? I prefer actually to describe myself as "unbiased" and "efficient".

Don't get me wrong: statistics and OLAP offer very important methods necessary for many day-to-day business cases and I myself am half computer scientist and half statistician. However, if you mix the methods described above with algorithms from computer science to scale those methods up to larger datasets and also throw in some ideas from artificial intelligence, especially from the field of machine learning, I personally think that much more interesting possibilities can arise. This is actually nothing new and has been an important field for research during the last decades. The methods and algorithms which have been developed over this time have actually formed complete new research fields under the names data mining, predictive analytics, or pattern detection. And one of the most amazing developments is a collection of methods which can not only be used on structured data, i.e., on tables, but also on unstructured data like texts or images. This has been the underlying motivation for fields like text mining, image mining, or audio mining.

Most recently a new buzzword has been used a lot: Big Data. Well, most recently means in the years 2012 and following, so if you are reading this book in 2092 you might want to use this as a historical lecture. OK, back to big data: What is so special about that? If you ask me, not that much from the point of view of a data miner. Big data is an umbrella term for many ideas and technologies, but the underlying point of all those things is that big data should be about infrastructure and methods for collecting, retrieving, and analyzing very large datasets which might be of structured, unstructured, or polystructured nature. Well, if you have read the paragraphs above you will certainly agree that this sounds actually like a perfect description of "data mining". As of 2013, the big data market is still in its early days and most people are worrying about data infrastructures. But this will change as soon as organizations do understand that the mere collection of data is worth nothing. It is the analysis of data which delivers new insights, explains underlying patterns, or creates models which can be extrapolated in order to predict the most likely future.

Reading this book might therefore be a very good idea for learning more about where and how data mining can be used to deliver those insights. It might also just serve you for your personal career—remember, the big data market will slowly but surely move into the direction of analytics in the future. Whatever the reason is, I hope that you will learn more about the use cases discussed in this book so that you are able to transfer them to your own business problems. RapidMiner is an excellent and very flexible tool for re-using those use cases and adapt them to your concrete problems. Have fun using it!

In the following, I would like to discuss what data mining actually is and to which types of problems it can be applied. At the same time I will give you an introduction to the most important terms. Whether you are already an experienced data mining expert or not, this chapter is worth reading in order for you to know and have a command of the terms used both here in this book as well as in RapidMiner.

1.2 Coincidence or Not?

Before we get properly started, let us try a small experiment:

1. Think of a number between 1 and 10.
2. Multiply this number by 9.
3. Work out the checksum of the result, i.e., the sum of the numbers.
4. Multiply the result by 4.
5. Divide the result by 3.
6. Deduct 10.

The result is 2.

Do you believe in coincidence? As an analyst you will probably learn to answer this question in the negative or even do so already. Let us take for example what is probably the simplest random event you could imagine, i.e., the toss of a coin. "Ah", you may think, "but that is a random event and nobody can predict which side of the coin will be showing after it is tossed." That may be correct, but the fact that nobody can predict it does in no way mean that it is impossible in principle. If all influence factors such as the throwing speed and rotation angle, material properties of the coin and those of the ground, mass distributions and even the strength and direction of the wind were all known exactly, then we would be quite able, with some time and effort, to predict the result of such a coin toss. The physical formulas for this are all known in any case.

We shall now look at another scenario, only this time we can predict the outcome of the situation: A glass will break if it falls from a certain height onto a certain type of ground. We even know in the fractions of a second when the glass is falling that there will be broken glass. How are we able to achieve this rather amazing feat? We have never seen the glass which is falling in this instant break before and the physical formulas that describe the breakage of glass are a complete mystery for most of us at least. Of course, the glass may stay intact "by chance" in individual cases, but this is not likely. For what it's worth, the glass not breaking would be just as non-coincidental, since this result also follows physical laws. For example, the energy of the impact is transferred to the ground better in this case. So how do we humans know what exactly will happen next in some cases and in other cases, for example that of the toss of a coin, what will not?

The most frequent explanation used by laymen in this case is the description of the one scenario as "coincidental" and the other as "non-coincidental". We shall not go into the interesting yet nonetheless rather philosophical discussions on this topic, but we are putting forward the following thesis:

The vast majority of processes in our perceptible environment are not a result of coincidences. The reason for our inability to describe and extrapolate the processes precisely is rather down to the fact that we are not able to recognize or measure the necessary influence factors or correlate them.

In the case of the falling glass, we quickly recognized the most important characteristics, such as the material, falling height, and nature of the ground, and can already estimate, in the shortest time, the probability of the glass breaking by analogy reasoning from similar experiences. However, it is just that we cannot do with the toss of a coin. We can watch as many tosses of a coin as we like; we will never manage to recognize the necessary factors fast enough and extrapolate them accordingly in the case of a random throw.

So what were we doing in our heads when we made the prediction for the state of the glass after the impact? We measured the characteristics of this event. You could also say that we collected data describing the fall of the glass. We then reasoned very quickly by analogy, i.e., we made a comparison with earlier falling glasses, cups, porcelain figurines, or similar articles based on a similarity measure. Two things are necessary for this: firstly, we need to also have the data of earlier events available and secondly, we need to be aware of how a similarity between the current and past data is defined at all. Ultimately we are able to make an estimation or prediction by having looked at the most similar events that have already taken place, for example. Did the falling article break in these cases or not? We must first find the events with the greatest similarity, which represents a kind of optimization. We use the term "optimization" here, since it is actually unimportant whether we are now maximizing a similarity or the sales figures of one enterprise or any other—the measurement concerned, in this case similarity, is always optimized. The analogy reasoning described then tells us that the majority of glasses we have already looked at broke and this very estimation then becomes our prediction. This may sound complicated, but this kind of analogy reasoning is basically the foundation for almost every human learning process and is done at a staggering speed.

The interesting thing about this is that we have just been acting as a human data mining method, since data analysis usually involves matters such as the representation of events or conditions and the data resulting from this, the definition of events' similarities and of the optimization of these similarities.

However, the described procedure of analogy reasoning is not possible with the toss of a coin: It is usually insufficient at the first step and the data for factors such as material properties or ground unevenness cannot be recorded. Therefore, we cannot have these ready for later analogy reasoning. This does in no way mean, however, that the event of a coin toss is coincidental, but merely shows that we humans are not able to measure these influence factors and describe the process. In other cases we may be quite able to measure the influence factors, but we are not able to correlate these purposefully, meaning that computing similarity or even describing the processes is impossible for us.

It is by no means the case that analogy reasoning is the only way of deducing forecasts for new situations from already known information. If the observer of a falling glass is asked how he knows that the glass will break, then the answer will often include things like "every time I have seen a glass fall from a height of more than 1.5 meters it has broken." There are two interesting points here: The relation to past experiences using the term "always" as well as the deduction of a rule from these experiences:

If the falling article is made of glass and the falling height is more than 1.5 meters, then the glass will break.

The introduction of a threshold value like 1.5 meters is a fascinating aspect of this rule formation. For although not every glass will break immediately if greater heights are used and will not necessarily remain intact in the case of lower heights, introducing this threshold value transforms the rule into a rule of thumb, which may not always, but will mostly lead to a correct estimate of the situation. Instead of therefore reasoning by analogy straight away, one could now use this rule of thumb and would soon reach a decision as to the most probable future of the falling article. Analogy reasoning and the creation of rules are two first examples of how humans, and also data mining methods, are able to anticipate the outcome of new and unknown situations.

Our description of what goes on in our brains and also in most data mining methods on the computer reveals yet another interesting insight: The analogy reasoning described does at no time require the knowledge of any physical formula to say why the glass will now break. The same applies for the rule of thumb described above. So even without knowing the complete (physical) description of a process, we and the data mining method are equally able to generate an estimation of situations or even predictions. Not only was the causal relationship itself not described here, but even the data acquisition was merely superficial and rough and only a few factors such as the material of the falling article (glass) and the falling height (approx. 2m) were indicated, and relatively inaccurately at that.

Causal chains therefore exist whether we know them or not. In the latter case, we are often inclined to refer to them as coincidental. And it is equally amazing that describing the further course is possible even for an unknown causal chain, and even in situations where the past facts are incomplete and only described inaccurately.

This section has given you an idea of the kind of problems we wish to address in this book. We will be dealing with numerous influence factors, some of which can only be measured insufficiently or not at all. At the same time, there are often so many of these factors that we risk losing track. In addition, we also have to deal with the events which have already taken place, which we wish to use for modeling and the number of which easily goes into millions or billions. Last but not least, we must ask ourselves whether describing the process is the goal or whether analogy reasoning is already sufficient to make a prediction. And in addition, this must all take place in a dynamic environment under constantly changing conditions—and preferably as soon as possible. Impossible for humans? Correct. But not impossible for data mining methods.

1.3 Applications of Data Mining

indexData Mining Applications Before we start to discuss the fundamental terms of data mining and predictive analytics, I would like to give you a feeling about the possible application fields for those techniques. There are literally hundreds of possible applications for data mining and predictive analytics across basically all verticals and horizontals you can think of. The following selection is therefore by no means complete; it should only give you some ideas where these technologies already have been successfully applied. The applications described in the following are grouped along selected verticals where they are most often used, but make sure to check them all since there is no reason why you should not use "fraud detection" also in retail instead of financial services.

1.3.1 Financial Services

- ***Fraud detection:*** Fraud detection is often used in the financial service industry but not only there. The basic idea here is to detect fraudulent activities among a very large set of transactions with methods from predictive analytics. Possible approaches include the detection of outliers or the modeling of "normal" cases against "fraudulent" cases and using this model for new transactions to check if they fall into the fraud segment.

- ***Churn prevention:*** Assume your insurance company has a contract with one of your customers. The optimal case for your insurance would that this contract stays and you keep an ongoing relationship with your customer. Sadly, some customers decide to quit the contract for a wide variety of reasons and you would like to know in advance for which customers this will happen with the highest likelihood in the near future. This is exactly the idea behind churn prevention: create predictive models calculating the probability that customers are likely to quit contracts soon so you can be pro-active, engage with them, offer incentives, etc. Besides the financial industry, churn prevention can also be often found in the retail industry, e-commerce, or telecommunications among others.

- ***Sentiment analysis:*** Sentiment analysis is not at all connected to the financial industry but we have seen it very often here lately. You can also find sentiment analysis in industries like consumer goods, retail, telecommunications, and life sciences a lot. The idea behind sentiment analysis is to connect to thousands of online sources in the web, collect statements about your brands or products, and analyze them by means of text analytics with respect to their tonality or sentiment. You can identify how sentiment changes over time and measure the success of marketing or PR by this or you can get new insights about how to improve your products. If you combine this with network analysis you can even detect key opinion leaders and see how they influence their peer groups.

- ***Trading analytics:*** If you are trading, building portfolios, or preparing deals, the natural idea would be to calculate the success rate of your decisions with help of predictive analytics. In general, you could analyze potential trade opportunities by looking at market data or inspect markets for trading activity which showing an emerging trend. Lately, many analysts combine more traditional methods like time series analysis with behavioral trading algorithms or even sentiment analysis.

- ***Risk management:*** This is another example done in the financial services industry which can also be applied to many other industries as well, especially when it comes to supply chain management, for example in manufacturing, or to logistics and transportation. Data mining and predictive analytics can be used for solving multiple problems connected to risk management, including error detection and quantification, unit reviews or internal audits, detecting fraud (see above), identifying suppliers with highest probability of failure, quantifying payment risks, and credit scoring.

1.3.2 Retail and Consumer Products

- ***Customer segmentation, channel selection, and next best action:*** In retail you can easily find most applications of data mining for sales and marketing. A typical approach is to use data about customers, including descriptions of the customers' purchase and transaction history, for creating segments of customers, for example based on classification but more often based on clustering techniques. Those clusters can build data-driven segments much more optimized than the gut-driven A-B-C segments which even today can often be seen in practice. The assignment of customers to segments is an important prerequisite for further analysis, for example, for selecting customers for specific sales or marketing channels or for predicting which is the optimal next best action to approach those customers or leads.

- ***Direct marketing:*** The bread-and-butter business for data mining and one of the stories how it all has begun. The idea behind direct marketing is to assign costs for different types of actions: If I contact a lead and he does not react, this costs A (for the contact efforts, etc.). If he does, I will get some gain B (gains are essentially negative costs). If I don't contact the lead and he would not have reacted anyway, this saves me cost A above. But if I decide to not contact the lead and he or she would have purchases, this might cause a huge loss C. So the whole point is to identify those leads with the highest probabilities for conversion for a certain marketing campaign so that you only would contact the most likely cases up to a limit where the contacting costs would no longer lead to revenue high enough to compensate the costs. Even with the advent of e-mail, direct marketing is necessary, the "cost" here might be that recipients are tired of getting too much spam and might opt out instead so we are using the lead.

- ***Recommendations, cross-selling and up-selling:*** Another one of the big success stories of data mining. Everybody who has already purchased a book at Amazon already came across the results of so-called recommender systems: "People who bought THIS book also purchased THAT one." At first sight, this problem type might not look too complicated: for each item, just search for those which have been purchased frequently together with the first one. The problem comes with the high number of available items and the high number of transactions typically available in retail and e-commerce. It is literally impossible to make those calculations for all combinations which might happen and so we need algorithms guiding our search to the most promising directions. We call those promising combinations "frequent item sets" and after we found those sets, we might want to recommend other items from those sets if the first one is added to the cart. This approach is called *cross-selling* and might complement or even replace traditional cross-selling approaches based on manual rules. Loosely connected to this is *up-selling* where we try to identify customers who are likely to purchase a higher-valued product or a larger quantity.

- ***Customer lifetime value***: Traditional systems for business intelligence based on OLAP approaches are great for answering questions like "who are the customers who bought most so far" or "what are the revenues created with my top 10 customers in the past". Although this is with no doubt important information, it unfortunately is also only reflecting the past. Previously good customers might change to another supplier or drop out for other reasons, and the fact that a customer has created much revenue so far does not mean this will continue also in the future. Instead of analyzing the historical customer value, many organizations now turn to predict how customers will develop in the future and what their total customer lifetime value will be to guide their sales and marketing efforts. Predictive analytics methods help to identify typical customer value cycles as well as to assign customers and leads to those cycles and determining at which state within their cycle they are.

1.3.3 Telecommunications and Media

- ***Network analysis:*** The telecommunications industry actually already is a driver for many of the application fields described above including churn prevention, channel selection, direct marketing, customer lifetime value, and sentiment analysis. One interesting finding is that the telecommunication industry in addition can make use of another data source describing social interactions between their customers. For example, the calls between customers might describe their social connections and describing those connections in addition to their usage behavior and transactional data can easily help to improve models for churn prevention and others. If a key opinion leader decides to change to another provider, other people influenced by this person might show a higher probability to also quit their contracts as well.

- ***Customer service process automation:*** Many companies spend a lot of efforts to improve customer support and they are right to do so. If a customer is happy with the support in case of a previously bad experience, this person might be turned to a more loyal customer than before. One of the most important influence factors for the happiness with customer service is the amount of time between when a request is sent and when the answer is delivered. Text analytics can help to either answer a question with an automatically selected text block or at least by assigning the request to the right employee without any further delay. This is another example not only applicable to the telecommunications industry but to all customer-centric businesses with many customer contacts.

1.3.4 Manufacturing, Construction, and Electronics

- ***Predictive maintenance:*** Predictive analytics can analyze sensor data right from the production processes and machines to determine if there is an upcoming problem with a machine which is likely to lead to a problem soon. Many problems manifest themselves at early stages already with certain types of error messages or changed behavior ultimately leading to changed sensor data. Analysts can create predictive models based on the failure events in the past and the historical data before those failures. Such a model can then be used to predict if a new failure is likely to happen before the next maintenance interval and should be addressed better now than later. Those models could also deliver insights into the reasons of those failures and hence deliver a root cause analysis.

- ***Patent text analysis:*** Another of those examples which is actually also applicable to other industries is the analysis of patent texts. This is most often done by methods derived from text analysis and text similarities for one of two reasons: either a company would like to detect emerging trends as soon as possible or it would like to be prepared for cases in which own patents are attacked.

- ***Supply chain management:*** There are multiple application fields for predictive analytics around supply chain management. We already have discussed risk management above, for supply chain management this could mean determining which suppliers have the highest risks for failure and what would be the expected impact in case of failure. Predictive analytics can also be used for demand forecasting and hence for improving logistics but also timely negotiations with suppliers. And finally those techniques can be used to predict prices and their changes in the supply chain, again allowing for pro-active and well-informed decisions.

- ***Optimizing throughput rates:*** The manufacturing industry even has started to connect data mining and predictive analytics with the control centers of their factories. Based on the sensor data describing the production process itself and also the input to this process, those models find the optimal settings for the process in real time in order to optimize for quality, or higher throughput rates, or even both at the same time. In some of those cases it is important not to leave certain parameter ranges in order to not damage the involved machines and even this is possible to do with the help of advanced analytics.

- ***Quality assurance:*** Another application field is the prediction of the quality of the outcome of a process even before the full process has been finished. Those predictive models use the data describing the process and combine it with sensor data describing the current state of an item in order to predict the quality of the final outcome. We even have seen cases where the item was taken out of further refinement, which would just have induced additional costs for a product which will not be sold due to quality restrictions anyway. Closely connected to this are questions like anomaly detection and why a certain item has a lower quality, hence a root cause analysis.

1.4 Fundamental Terms

We managed so far to get the general idea about what are data mining and predictive analytics about. Plus we have a good feeling that those technologies are very useful for many application fields across basically all industries. In the following, we will introduce some fundamental terms which will make dealing with the problems described later in this book easier for us. You will come across these terms again and again in the RapidMiner software too, meaning it is worth becoming acquainted with the terms used even if you are already an experienced data analyst.

First of all, we can see what the two examples looked at in the previous sections, namely the toss of a coin and the falling glass, have in common. In our discussion on whether we are able to predict the end of the respective situation, we realized that knowing the influence factors as accurately as possible, such as material properties or the nature of the ground, is important. And one can even try to find an answer to the question as to whether this book will help you by recording the characteristics of yourself, the reader, and aligning them with the results of a survey of some of the past readers. These measured reader characteristics could be, for example, the educational background of the person concerned, the liking of statistics, preferences for other, possibly similar books and further features, which we could also measure as part of our survey. If we now knew such characteristics of 100 readers and had the indication as to whether you like the book or not in addition, then the further process would be almost trivial. We would also ask you the questions from our survey and measure the same features in this way and then, for example using analogy reasoning as described above, generate a reliable prediction of your personal taste. "Customers who bought this book also bought...." This probably rings a bell.

1.4.1 Attributes and Target Attributes

Whether coins, customers, or production processes, there is, as previously mentioned, the question in all scenarios as to the characteristics or features of the respective situation. We will always speak of **attributes** in the following when we mean such describing factors of a scenario. This is also the term that is always used in the RapidMiner software when such describing features arise. There are many synonyms for this term and depending on your own background you might have already come across different terms instead of "attribute", for example

- characteristic,
- feature,
- influence factor (or just factor),
- indicator,
- variable, or
- signal.

We have seen that description by attributes is possible for processes and also for situations. This is necessary for the description of technical processes for example and the thought of the falling glass is not too far off here. If it is possible to predict the outcome of such a situation, then why not also the quality of a produced component? Or the imminent failure of a machine? Other processes or situations which have no technical reference can also be described in the same way. How can I predict the success of a sales or marketing promotion? Which article will a customer buy next? How many more accidents will an insurance company probably have to cover for a particular customer or customer group?

We shall use such a customer scenario in order to introduce the remaining important terms. Firstly, because humans are famously better at understanding examples about other humans and secondly, because each enterprise probably has information, i.e., attributes, regarding their customers and most readers can therefore relate to the examples immediately. The attributes available as a minimum, which just about every enterprise keeps about its customers, are for example address data and information as to which products or services the customer has already purchased. You would be surprised what forecasts can be made even from such a small number of attributes.

Let us look at an (admittedly somewhat contrived) example. Let us assume that you work in an enterprise that would like to offer its customers products in the future which are better tailored to their needs. Within a customer study of only 100 of your customers some needs became clear, which 62 of these 100 customers share all the same. Your research and development department got straight to work and developed a new product within the shortest time, which would satisfy these new needs better. Most of the 62 customers with the relevant needs profile are impressed by the prototype in any case, although most of the remaining participants of the study only show a small interest as expected. Still, a total of 54 of the 100 customers in the study said that they found the new product useful. The prototype is therefore evaluated as successful and goes into production—now only the question remains as to how, from your existing customers or even from other potential customers, you are going to pick out exactly the customers with whom the subsequent marketing and sales efforts promise the greatest success. You would therefore like to optimize your efficiency in this area, which means in particular ruling out such efforts from the beginning which are unlikely to lead to a purchase. But how can that be done? The need for alternative solutions and thus the interest in the new product arose within the customer study on a subset of your customers. Performing this study for all your customers is much too costly and so this option is closed to you. And this is exactly where data mining can help. Let us first look at a possible selection of attributes regarding your customers:

- Name
- Address
- Sector
- Subsector
- Number of employees
- Number of purchases in product group 1
- Number of purchases in product group 2

The number of purchases in the different product groups means the transactions in your product groups which you have already made with this customer in the past. There can of course be more or less or even entirely different attributes in your case, but this is irrelevant at this stage. Let us assume that you have the information available regarding these attributes for every one of your customers. Then there is another attribute which we can look at for our concrete scenario: Whether the customer likes the prototype or not. This attribute is of course only available for the 100 customers from the study; the information on this attribute is simply unknown for the others. Nevertheless, we also include the attribute in the list of our attributes:

- *Prototype positively received?*
- Name
- Address
- Sector
- Subsector
- Number of employees
- Number of purchases in product group 1
- Number of purchases in product group 2

If we assume you have thousands of customers in total, then you can only indicate whether 100 of these evaluated the prototype positively or not. You do not yet know what the others think, but you would like to! The attribute "prototype positively received" thus adopts a special role, since it identifies every one of your customers in relation to the current question. We therefore also call this special attribute a **label**, since it sticks to your customers and identifies them like a brand label on a shirt or even a note on a pin board. You will also find attributes which adopt this special role in RapidMiner under the name "label". The goal of our efforts is to fill out this particular attribute for the total quantity of all customers. We might therefore also speak of the **target attribute** instead of the term "label" since our target is to create a model which predicts this special attribute from the values of all others. You will also frequently discover the term **target variable** in the literature, which means the same thing.

1.4.2 Concepts and Examples

The structuring of your customers' characteristics by attributes, introduced above, already helps us to tackle the problem a bit more analytically. In this way we ensured that every one of your customers is represented in the same way. In a certain sense we defined the type or **concept** "customer", which differs considerably from other concepts such as "falling articles" in that customers will typically have no material properties and falling articles will only rarely buy in product group 1. It is important that, for each of the problems in this book (or even those in your own practice), you first define which concepts you are actually dealing with and which attributes these are defined by.

We implicitly defined above, by indicating the attributes name, address, sector, etc., and in particular the purchase transactions in the individual product groups, that objects of the concept "customer" are described by these attributes. Yet this concept has remained relatively abstract so far and no life has been injected into it yet. Although we now know in what way we can describe customers, we have not yet performed this for specific customers. Let us look at the attributes of the following customer, for example:

- Prototype positively received: yes
- Name: Miller Systems Inc.
- Address: 2210 Massachusetts Avenue, 02140 Cambridge, MA, USA

- Sector: Manufacturing
- Subsector: Pipe bending machines
- Number of employees: >1000
- Number of purchases in product group 1: 5
- Number of purchases in product group 2: 0

We say that this specific customer is an **example** for our concept "customer". Each example can be characterized by its attributes and has concrete values for these attributes which can be compared with those of other examples. In the case described above, Miller Systems Inc. is also an example of a customer who participated in our study. There is therefore a value available for our target attribute "prototype positively received?" Miller Systems Inc. was happy and has "yes" as an attribute value here, thus we also speak of a **positive example**. Logically, there are also negative examples and examples which do not allow us to make any statement about the target attribute.

1.4.3 Attribute Roles

We have now already become acquainted with two different kinds of attributes, i.e., those which simply describe the examples and those which identify the examples separately. Attributes can thus adopt different roles. We have already introduced the role "label" for attributes which identify the examples in any way and which must be predicted for new examples that are not yet characterized in such a manner. In our scenario described above, the label still describes (if present) the characteristic of whether the prototype was received positively.

Likewise, there are for example roles, the associated attribute of which serves for clearly identifying the example concerned. In this case the attribute adopts the role of an identifier and is called **ID** for short. You will find such attributes named ID in the RapidMiner software also. In our customer scenario, the attribute "name" could adopt the role of such an identifier.

There are even more roles, such as those with an attribute that designates the weight of the example with regard to the label. In this case the role has the name **Weight**. Attributes without a special role, i.e., those which simply describe the examples, are also called **regular** attributes and just leave out the role designation in most cases. Apart from that you have the option in RapidMiner of allocating your own roles and of therefore identifying your attributes separately in their meaning. Please note that for most data mining tasks in RapidMiner the regular attributes are used as input to the method, for example, to create a model predicting the attribute with role label.

1.4.4 Value Types

As well as the different roles of an attribute, there is also a second characteristic of attributes which is worth looking at more closely. The example of Miller Systems Inc. above defined the respective values for the different attributes, for example "Miller Systems Inc." for the attribute "Name" and the value "5" for the number of past purchases in product group 1. Regarding the attribute "Name", the concrete value for this example is therefore arbitrary free text to a certain extent; for the attribute "number of purchases in product group 1" on the other hand, the indication of a number must correspond. We call the indication whether the values of an attribute must be in text or numbers the **Value Type** of an attribute.

In later chapters we will become acquainted with many different value types and see how these can also be transformed into other types. For the moment we just need to know that there are different value types for attributes and that we speak of value type **text** in the case of free text, of the value type **numerical** in the case of numbers and of the value type **nominal** in the case of only few values being possible (like with the two possibilities "yes" and "no" for the target attribute).

Please note that in the above example the number of employees, although really of numerical type, would rather be defined as nominal, since a size class, i.e., ">1000" was used instead of an exact indication like 1250 employees.

1.4.5 Data and Meta Data

We want to summarize our initial situation one more time. We have a Concept "customer" available, which we will describe with a set of attributes:

- Prototype positively received? [Label; Nominal]
- Name [Text]
- Address [Text]
- Sector [Nominal]
- Subsector [Nominal]
- Number of employees [Nominal]
- Number of purchases in product group 1 [Numerical]
- Number of purchases in product group 2 [Numerical]

The attribute "Prototype positively received?" has a special **Role** among the attributes; it is our **Target Attribute** or **Label** here. The target attribute has the value type **Nominal**, which means that only relatively few characteristics (in this case "yes" and "no") can be accepted. Strictly speaking it is even binominal, since only two different characteristics are permitted. The remaining attributes all have no special role, i.e., they are **regular** and have either the value type **Numerical** or **Text**. The following definition is very important, since it plays a crucial role in a successful professional data analysis: This volume of information, which describes a concept, is also called meta data, since it represents data about the actual data.

Our fictitious enterprise has a number of Examples for our concept "customer", i.e., the information which the enterprise has stored for the individual attributes in its customer database. The goal is now to generate a prediction instruction from the examples for which information is available concerning the target attribute, which predicts for us whether the remaining customers would be more likely to receive the prototype positively or reject it. The search for such a prediction instruction is one of the tasks which can be performed with data mining.

However, it is important here that the information for the attributes of the individual examples is in an ordered form, so that the data mining method can access it by means of a computer. What would be more obvious here than a table? Each of the attributes defines a column and each example with the different attribute values corresponds to a row of this table. For our scenario this could look like Table 1.1 for example.

We call such a table an **Example Set**, since this table contains the data for all the attributes of our examples. In the following, and also within RapidMiner, we will use the

TABLE 1.1: A set of examples with values for all attributes together with label values.

Prototype positively received?	**Name**	**Address**	**Sector**	**Sub-sector**	**Number of employees**	**No of purchases group 1**	**No of purchases group 2**	...
Yes	Miller Systems Inc.	2210 Massachusetts Ave, Cambridge, MA, USA	Manufacturing	Pipe bending machines	>1000	5	0	...
?	Smith Paper	101 Huntington Ave, Boston, MA, USA	IT	Supplies	600-1000	3	7	...
No	Meyer Inc.	1400 Commerce St, Dallas, TX, USA	Retail	Textiles	<100	1	11	...
...	...	...	...	...	...	...	...	...

terms **Data**, **Dataset**, and **Example Set** synonymously. A table with the appropriate entries for the attribute values of the current examples is always meant in this case. It is also such data tables which have taken their name from data analysis or data mining. Note:

Data *describes the objects of a concept,* ***Meta Data*** *describes the characteristics of a concept (and therefore also of the data).*

Most data mining methods expect the examples to be given in such an attribute value table. Fortunately, this is the case here and we can spare ourselves any further data transformations. In practice, however, this is completely different and the majority of work during a data analysis is time spent transferring the data into a format suitable for data mining. These transformations are therefore dealt with in detail in later chapters when we are discussing the different use cases.

1.4.6 Modeling

Once we have the data regarding our customers available in a well-structured format, we can then finally replace the unknown values of our target attribute with the prediction of the most probable value by means of a data mining method. We have numerous methods available here, many of which, just like the analogy reasoning described at the beginning or the generating of rules of thumb, are based on human behavior. We call the use of a data mining method "**to model**" and the result of such a method, i.e., the prediction instruction, is a **model**. Just as data mining can be used for different issues, this also applies for models. They can be easy to understand and explain the underlying processes in a simple manner. Or they can be good to use for prediction in the case of unknown situations. Sometimes both apply, such as with the following model for example, which a data mining method could have supplied for our scenario:

If the customer comes from urban areas, has more than 500 employees and if at least 3 purchases were transacted in product group 1, then the probability of this customer being interested in the new product is high.

Such a model can be easily understood and may provide a deeper insight into the underlying data and decision processes of your customers. And in addition, it is an **operational model**, i.e., a model which can be used directly for making a prediction for further customers. The company "Smith Paper" for example satisfies the conditions of the rule above and is therefore bound to be interested in the new product—at least there is a high probability of this. Your goal would therefore have been reached and by using data mining you would have generated a model which you could use for increasing your marketing efficiency: Instead of just contacting all existing customers and other candidates without looking, you could now concentrate your marketing efforts on promising customers and would therefore have a substantially higher success rate with less time and effort. Or you could even go a step further and analyze which sales channels would probably produce the best results and for which customers.

In the rest of this book we will focus on further uses of data mining and at the same time practice transferring concepts such as customers, business processes, or products into attributes, examples, and datasets. This will train the eye to detect further possibilities of application tremendously and will make analyst life much easier for you later on. In the next chapter though, we would like to spend a little time on RapidMiner and give a small introduction to its use, so that you can implement the following examples immediately.

Chapter 2

Getting Used to RapidMiner

Ingo Mierswa

Rapid-I, Dortmund, Germany

2.1 Introduction

We will use the open source data mining solution RapidMiner for the practical exercises in this book and for demonstrating the use cases for data mining, predictive analytics, and text mining discussed here. Before we can work with RapidMiner, you of course need to download and install the software first. You will find RapidMiner in the download area of the Rapid-I website: `http://www.rapid-i.com`. If you want to follow the use cases in this book, it is highly recommended to download the appropriate installation package for your operating system and install RapidMiner according to the instructions on the website now. All usual Windows versions are supported as well as Macintosh, Linux or Unix systems. Please note that an up-to-date Java Runtime Environment (JRE) with at least version 7 is needed for all non-Windows systems. You can find such a JRE for example at `http://www.java.com/`.

2.2 First Start

If you are starting RapidMiner for the first time, you will be welcomed by the so-called Welcome Perspective. The lower section shows current news about RapidMiner, if you have an Internet connection. The list in the center shows the analysis processes recently worked on. This is useful if you wish to continue working on or execute one of these processes. You can open a process from this list to work on or execute it simply by double clicking. The upper section shows typical actions which you as an analyst will perform frequently after starting RapidMiner. Here are the details of these:

1. New Process: Starts a new analysis process. This will be the most often used selection

for you in the future. After selecting this, RapidMiner will automatically switch to the Design perspective (explained below).

2. Open Recent Process: Opens the process which is selected in the list below the actions. Alternatively, you can open this process by double-clicking inside the list. Either way, RapidMiner will then automatically switch to the Design Perspective.

3. Open Process: Opens the repository browser and allows you to select a process to be opened within the process Design Perspective.

4. Open Template: Shows a selection of different pre-defined analysis processes, which can be configured in a few clicks.

5. Online Tutorial: Starts a tutorial which can be used directly within RapidMiner and gives an introduction to some data mining concepts using a selection of analysis processes. Recommended if you have a basic knowledge of data mining and are already familiar with the fundamental operation of RapidMiner.

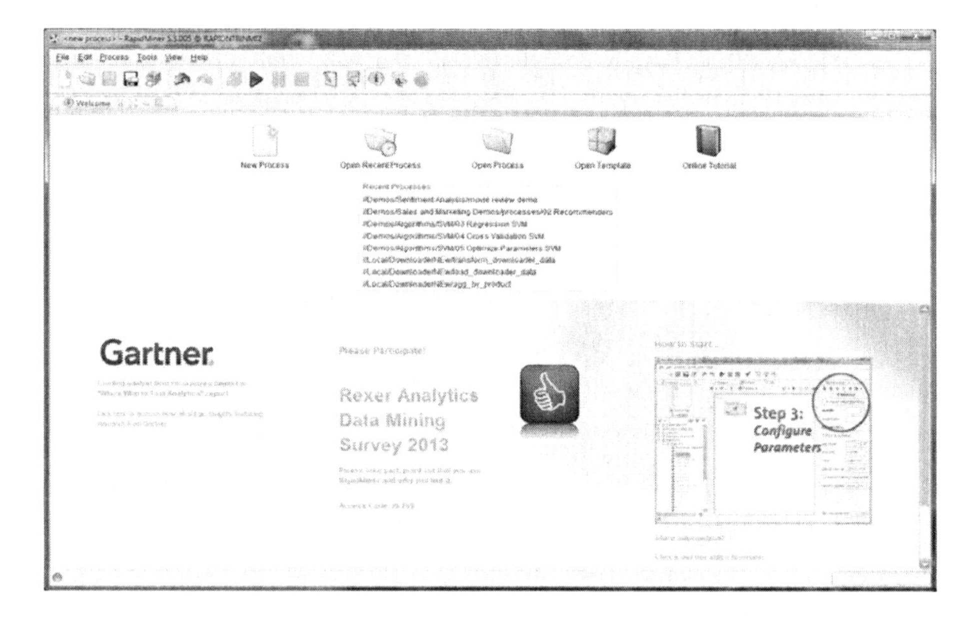

FIGURE 2.1: Welcome Perspective of RapidMiner.

At the right-hand side of the toolbar inside the upper section of RapidMiner, you will find three icons which switch between the individual RapidMiner perspectives. A **perspective** consists of a freely configurable selection of individual user interface elements, the so-called views. These can also be arranged however you like. In the Welcome Perspective there is only one view; one preset at least, namely the welcome screen, which you are looking at now. You can activate further views by accessing the "View" menu. Sometimes you may inadvertently delete a view or the perspective is unintentionally moved into particularly unfavorable positions. In this case the "View" menu can help, because apart from the possibility of reopening closed views via "Show View", the original state can also be recovered at any time via "Restore Default Perspective".

2.3 Design Perspective

You will find an icon for each (pre-defined) perspective within the right-hand area of the toolbar:

FIGURE 2.2: Toolbar icons for perspectives.

The icons shown here take you to the following perspectives:

1. *Design Perspective:* This is the central RapidMiner perspective where all analysis processes are created and managed.
2. *Result Perspective:* If a process supplies results in the form of data, models, or the like, then RapidMiner takes you to this Result Perspective, where you can look at several results at the same time as normal thanks to the views.
3. *Welcome Perspective:* The Welcome Perspective already described above, in which RapidMiner welcomes you with after starting the program.

You can switch to the desired perspective by clicking inside the toolbar or alternatively via the menu entry "View" – "Perspectives" followed by the selection of the target perspective. RapidMiner will eventually also ask you automatically if switching to another perspective seems a good idea, e.g., to the Result Perspective on completing an analysis process.

Now switch to the Design Perspective by clicking in the toolbar. This is the major working place for us while using RapidMiner. Since the Design Perspective is the central working environment of RapidMiner, we will discuss all parts of the Design Perspective separately in the following and discuss the fundamental functionalities of the associated views. In any case, you should now see the screen on the next page:

All work steps or building blocks for different data transformation or analysis tasks are called **operators** in RapidMiner. Those operators are presented in groups in the Operator View on the left side. You can navigate within the groups in a simple manner and browse in the operators provided to your heart's desire. If RapidMiner has been extended with one of the available extensions, then the additional operators can also be found here. Without extensions you will find at least the following groups of operators in the tree structure:

- *Process Control:* Operators such as loops or conditional branches which can control the process flow.
- *Utility:* Auxiliary operators which, alongside the operator "Subprocess" for grouping subprocesses, also contain the important macro-operators as well as the operators for logging.
- *Repository Access:* Contains the two operators for read and write access in repositories.
- *Import:* Contains a large number of operators in order to read data and objects from external formats such as files, databases, etc.

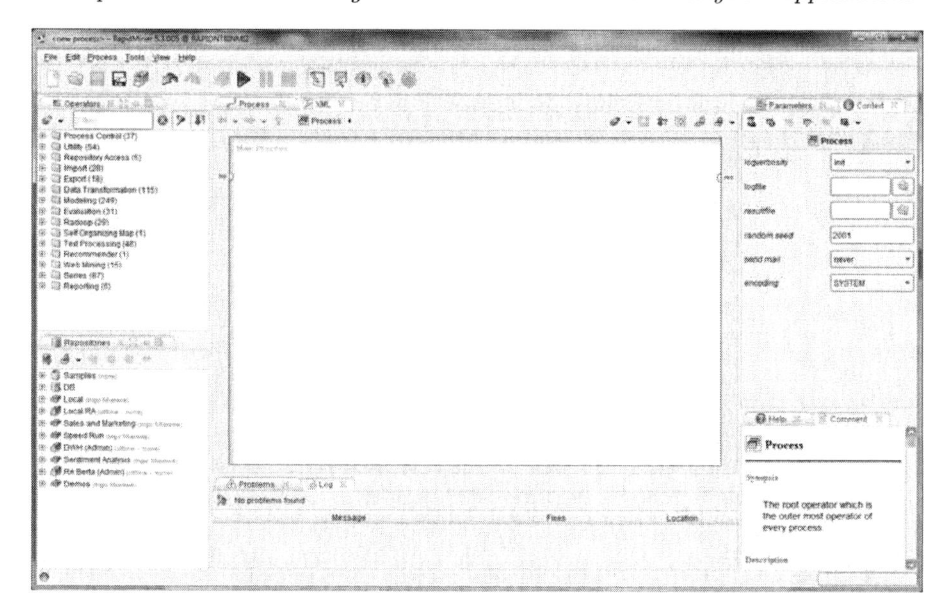

FIGURE 2.3: Design Perspective of RapidMiner.

- *Export:* Contains a large number of operators for writing data and objects into external formats such as files, databases, etc.

- *Data Transformation:* Probably the most important group in the analysis in terms of size and relevance. All operators are located here for transforming both data and meta data.

- *Modeling:* Contains the actual data mining process, such as classification methods, regression methods, clustering, weightings, methods for association rules, correlation and similarity analyses as well as operators, in order to apply the generated models to new datasets.

- *Evaluation:* Operators using which one can compute the quality of a modeling and thus for new data, e.g., cross-validations, bootstrapping, etc.

You can select operators within the Operators View and add them in the desired place in the process by simply dragging them from the Operators View and dropping them into the large white area in the center, the so-called Process View. Every analysis in RapidMiner is a **process**, and every process consists of one or several steps which are the operators.

Depending on your settings, those new operators might be directly connected with existing operators as suitably as possible on the basis of the available meta data information. If this is not happening or the automatically inserted connection is not desired, you can delete the connection by selecting them and pressing the Delete key or by pressing the Alt key while clicking on any of the connection **ports**. Ports are the round bubbles on the sides of the operators and they are used to define the data flow through your analytical processes. You can insert new connections by either clicking on the source port and then clicking again on a target port or by dragging a line between those ports.

Later on, when you have successfully defined your first RapidMiner processes, a typical result might look like in the image on the following page:

You could now simply try and select a few operators from the Operator View and drag them into the Process View. Connect their ports, even if this is probably not leading

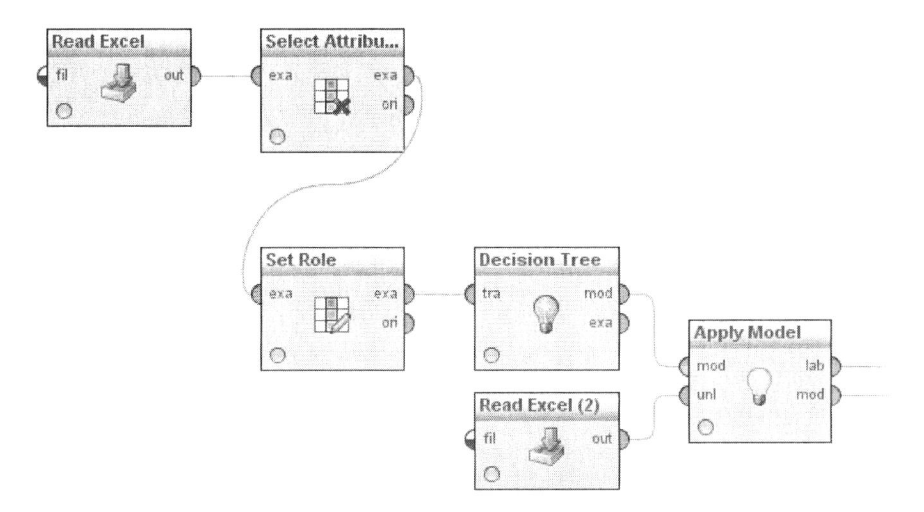

FIGURE 2.4: A typical process in RapidMiner consists of several operators. The data flow is defined by the connected ports, for example, the "Read Excel" operator loads a dataset from an Excel file and delivers it to the succeeding "Select Attributes" operator.

to working processes, and get familiar with the user interface of RapidMiner. In order to edit parameters you must select an individual operator. You will recognize the operator currently selected by its orange frame as well as its shadow. If you wish to perform an action for several operators at the same time, for example moving or deleting, please select the relevant operators by dragging a frame around these. In order to add individual operators to the current selection or exclude individual operators from the current selection, please hold the CTRL key down while you click on the relevant operators or add further operators by dragging a frame. You can also move operators around by selecting them and dragging them in the Process View. You will notice that the parameters in the **Parameter View** on the right side of RapidMiner changes sometimes if you select different operators. As you can see, most operators provide a set of parameters which control the actual working mode of the respective operator. You will find much information about RapidMiner and its user interface in the RapidMiner User Manual available at

`http://docs.rapid-i.com`

2.4 Building a First Process

One of the first steps in a process for data analysis is usually to load some data into the system. RapidMiner supports multiple methods for accessing datasets. It supports more than 40 different file types and of course all major database systems. If the data is not originally stored in a relational database system, the best approach is to import the data first into the RapidMiner repository. Please follow the instructions from the RapidMiner manual for more information or just try to import a dataset, for example an Excel file, with "File" – "Import Data". Later you will also realize that there are dozens of operators for data import in the operator group "Import", which can also be used directly as part of the process.

2.4.1 Loading Data

In the following we assume that you have managed to import the data into the Rapid-Miner repository and hence we will retrieve the data from there. If you are loading the data from a database or file, your following steps are at least similar to those described below.

It is always recommended to use the repository whenever this is possible instead of files. This will allow RapidMiner to get access to the describing meta data and will ease process design a lot. We will now create the beginning of a data analysis process and will add a first data mining technique using this data. The very first operation in our process should be to load the data from the repository again in order to make it available for the next analysis steps:

1. Go to the Repositories view and open the repository **Samples** delivered with Rapid-Miner. Click on the small plus sign in front of this repository. You should now see two folders named **data** and **processes**. Open the data folder and you will find a collection of datasets coming together with RapidMiner. Click on the dataset named **Iris** and drag it onto the large white view named **Process** in the center of your frame. After releasing the dataset somewhere on the white area, it should be transformed into an operator named **Retrieve** with a bluish output port on the right side. RapidMiner automatically has transformed the dataset into an operator loading the dataset. If you click on the operator, you can see a parameter in the **Parameters** view pointing to the data location. The Retrieve operator in general, well, retrieves objects from a repository and makes them available in your process.

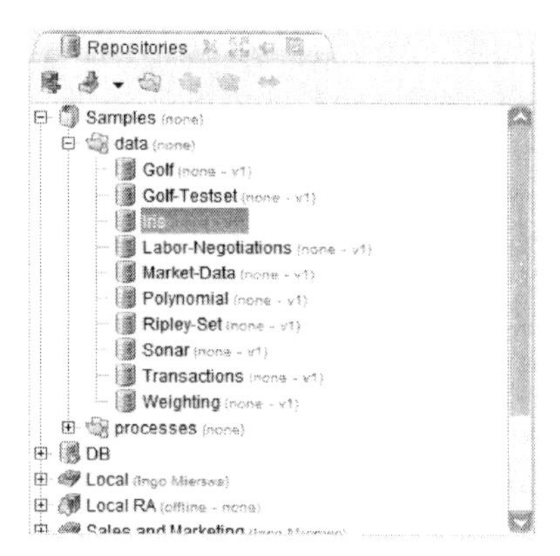

FIGURE 2.5: Drag the Iris dataset into the process view in order to create a new operator loading this dataset during process execution.

Background Information – Iris Data:
You might already be familiar with the "Iris" dataset since it is a well-known dataset among many data analysts. If not, here is the basic idea: the dataset describes 150 Iris plants with four attributes: sepal-length, sepal-width, petal-length, and petal-width. Since only experts for Iris plants actually understand the meaning of those attributes we will refer to those attributes with "a1" to "a4". And there is a fifth column describing to which class of Irises each of the 150 plants belong. There are three options: Iris-setosa, Iris-versicolor, and Iris-virginica. Each of those three classes is represented by 50 plants in the dataset. The goal is now to find a classification model using the measured lengths and widths (a1 to a4) to predict the class of the plant. This would

allow the classification of those plants even by non-experts like myself.

2. Maybe the output was automatically connected to one of the result ports named **res** on the right side of the Process area. If that is the case, you are done already. If not—and this depends on your current program settings—click on the output port of the Retrieve operator and then click on the first res port on the right side. Alternatively, you can also drag a connection line between the two ports. Your process should now look like the following:

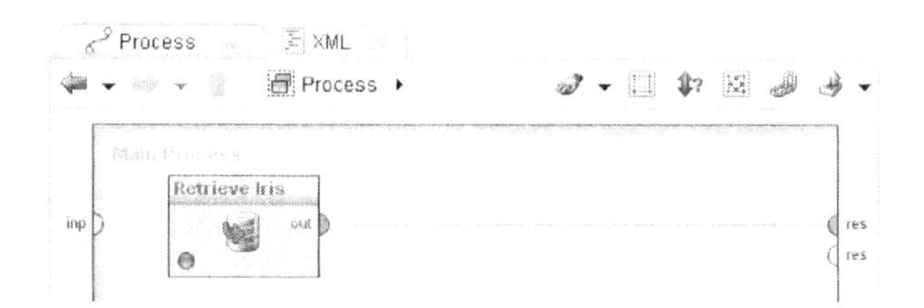

FIGURE 2.6: The probably most simple process which can be created with RapidMiner: It just retrieves data from a repository and delivers it as a result to the user to allow for inspection.

All results which are delivered at a result port named **res** will be delivered as a result to the user (or other systems if the process execution is integrated into other systems). The next step would be to create a decision tree on the Iris dataset and also deliver this model to the user.

2.4.2 Creating a Predictive Model

We have seen above how we create a new process just loading a dataset. The next step will be to create a predictive model using this dataset. This model predicts a categorical or nominal value, hence we could also say the model should describe rules which allow us to assign one of the three classes to new and unseen data describing new plants. We refer to this type of modeling as **classification**.

Adding a modeling technique to your process so that it calculated a predictive model is actually very easy. Just follow the following steps for creating such a model:

Go to the Operators view and open the operator group **Modeling, Classification, and Regression**, and then **Tree Induction**. You should now see an operator named **Decision Tree**. Click on it and drag it to your process, somewhere to the right of your initial retrieve operator.

You now only have to create the necessary connections. The dataset should be delivered to the modeling operator which is delivering a model then. However, you can also deliver the dataset itself to the user if you also connect the data port with one of the result ports. The complete process should look like the following figure.

In the next section we will learn how to execute this process and how to inspect the created results.

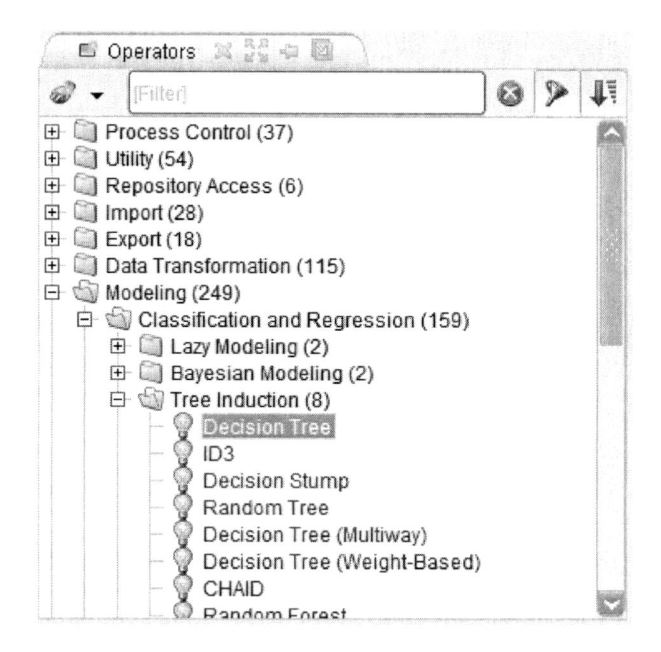

FIGURE 2.7: Drag the operator named "Decision Tree" into your process.

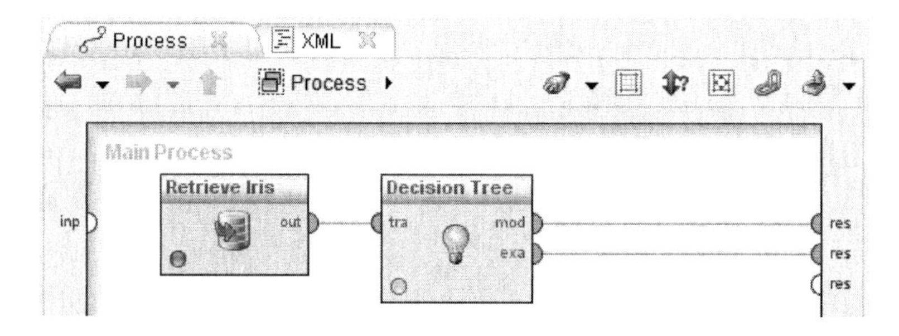

FIGURE 2.8: The complete process consisting of data loading and model creation.

Background Information: Decision Trees

Decision trees are probably one of the most widely used techniques in data mining. One of the biggest advantages is that they are easy to use and, maybe even more important, also easy to understand even by non-experts. The basic idea behind decision trees is a so-called divide-and-conquer approach. In each step the dataset is divided into different parts while each part should better represent one of the possible classes. The final result will be a tree-structure where each inner node represents a test for the value of a particular attribute and each leaf is representing the decision for a particular class. A new and unknown case is then routed down the tree until it reaches one of the leaves.

For each node we have two options depending on the value type of the attribute used at this node. For nominal attributes, the number of children is usually equal to the number of possible values for this attribute. If we are using a nominal attribute for a test in one of the inner nodes, this means that the dataset is at this stage basically divided according to the different values of this attribute. Hence, a nominal attribute will not get tested more than once since all examples further down the tree will have the same value for this particular attribute. This is different for numerical attributes: Here, we usually test if the attribute value is greater or less than a determined constant. The attribute may get tested several times for different constants.

The strategy for the construction of a decision tree is top-down in a recursive divide-and-conquer fashion. The first step is to select one of the attributes for the root node. Then we create a branch for each possible attribute value and split instances into subsets according to the possible values, i.e.,we will get one subset for each branch extending from the node. Finally we will repeat these steps recursively for each branch but only use instances that reach the branch now. We can stop this process if all instances have the same class.

The major question is: How should we select which attribute should be tested next? Our goal is to get the smallest tree possible since we believe that a small tree manages to explain the data best and in a general fashion leading to fewer errors for unseen data compared to a more complex tree. There is not a strict optimal solution for this task but a widely used heuristic: we should choose the attribute that produces the "purest" subsets of data with respect to the label attribute. A very popular choice for this type of measurement is the so-called ***information gain****. The information gain basically increases with the average purity of the subsets and hence we should choose the attribute which gives us the greatest information gain at each level.*

We will now discuss how we can measure information at all in order to be able to calculate which decision delivers the highest gain in information. We will see now that information can be easily measured in bits and that a concept exists for calculating this amount: the ***entropy****. The entropy measures the information required in bits (can also mean fractions of bits) to predict the outcome of an event if the probability distribution of this event is given. The formula is as follows:*

$$\textit{Entropy}\ (p_1,\ldots,p_n) = -p_1\ ld\ (p_1) - \ldots - p_n\ ld\ (p_n)$$

*p_1 to p_n are the possible probabilities for all outcomes while ld is the logarithm dualis, i.e.,the logarithm with base 2. The formula is easy to understand with a small example. Let's assume we are tossing an unbiased coin where each side can be shown with a probability of 0.5. In this casc thc cntropy is 0.5 * ld(0.5) 0.5 * ld(0.5) = 1. Thc ncccssary information to decide which side of a tossed coin is actually shown is hence 1 bit, which is the perfect amount for a binary decision with only two possible outcomes which are equally likely.*

*Now let's assume that the coin is biased and shows "head" with a probability of 75% and "tails" with 25%. In this case the entropy would be calculated as –0.75 * ld(0.75) – 0.75 ** *ld(0.75) = 0.81 and hence we need less information to decide. This is only natural since we already expected "head" with a higher probability.*

Algorithms for creating decision trees make use of the notion of entropy for selecting the optimal attribute for each split, i.e.,inner node. The entropy for the class distribution is first calculated for each possible value of a nominal attribute. Let's do this for an example and assume that we have an attribute with two possible values A and B. We have 10 examples with value A and 20 examples with value B. For those with value A we have 3 examples with class X and 7 examples with class Y. And for those with value B we have 15 examples with class X and 5 examples with class Y. We can now calculate the entropy values based on those class distributions in the subsets defined by those examples having values A and B respectively for this attribute:

1. ***Entropy A → entropy of class distribution (3/10, 7/10) = 0.881 bits***
2. ***Entropy B → entropy of class distribution (16/20, 4/20) = 0.322 bits***

For numerical attributes, each possible threshold for a numerical split is tested instead. In both cases the total information needed given a specific attribute is the weighted average of all entropy values; in this case this would be

(10 * Entropy A + 20 * Entropy B) / 30 = 0.508.

In order to calculate the ***information gain*** *by this split, we have to compare the entropy of the class distribution before the split with the entropy after the split, i.e.,in our example:*

gain = info (19/30,11/30) – weighted entropy = 0.948 – 0.508 = 0.44

The information gain for performing this split hence is 0.44. We can now calculate the gain for all possible attributes and compare all gains to find the one delivering the highest value. This attribute would then be used for the next split. The dataset is divided accordingly and the complete process is repeated recursively until all subsets only contain examples for a single class.

The section above has explained the basic ideas behind decision trees, but be assured that there is more about decision trees beyond this basic approach. Instead of the information gain, one could use different measurements including the so-called ***gain ratio*** *which prefers splits with less possible branches. Another important addition is* ***pruning*** *the tree after the growing phase in order to make it more robust for unseen data points. Please refer to the standard literature about data mining for getting more information about additions to the basic decision tree algorithm.*

2.4.3 Executing a Process

Now we are ready and want to execute the process we have just created for the first time. The status indicators of all used operators should now be yellow or green (the small traffic lights in each operator box) in the best case, and there should be no entries in the Problems View before you start executing a process. In such a case it should be possible to execute our process currently consisting of only one operator without any problems. However, the problems in the Problems view and also the red traffic lights only indicate that there might be a potential problem—you might be able to execute a process even if RapidMiner detects a potential problem. Just execute the process and check if it works despite the complaint as described below.

You have the following options for starting the process:

1. Press the large play button in the toolbar of RapidMiner.
2. Select the menu entry "Process" → "Run".
3. Press F11.

FIGURE 2.9: Press the Play icon in order to execute your process.

While a process is running, the status indicator of the operator being executed in each case transforms into a small green play icon. This way you can see what point the process is currently at. After an operator has been successfully executed, the status indicator then changes and stays green—until for example you change a parameter for this operator: Then the status indicator will be yellow. The same applies for all operators that follow. This means you can see very quickly which operators a change could have an effect on.

The process defined above only has a short runtime and so you will hardly have the opportunity to pause the running process. In principle, however, you can briefly stop a running process with the pause symbol, e.g., in order to see an intermediate result. The operator currently being executed is then finished and the process is then stopped. You can recognize a process that is still running but currently paused by the fact that the color of the play icon changes from blue to green.

Press the play button again to continue executing the process further. If you do not wish to merely pause the process but to abort it completely, then you can press the stop button. Just like when pausing, the operator currently being executed is finished and the process fully aborted immediately after. Please note that you can switch to the Design Perspective immediately after aborting the process and make changes to processes, even if the execution of the current operator is being finished in the background. You can even start further processes and do not need to wait for the first process to be completed.

Note: The operator being executed is always completed if you abort. This is necessary to ensure a sound execution of operators. However, completing an operator may need much more time in individual cases and also require other resources such as storage space. So if when aborting very complex operators you can see this taking hours and requiring additional resources, then your only option is to restart the application.

2.4.4 Looking at Results

After the process was terminated, RapidMiner should automatically have switched to the Result Perspective. If this was not the case, then you probably did not connect the output port of the last operator with one of the result ports of the process on the right-hand side. Check this and also check for other possible errors, taking the notes in the Problems View into consideration.

Feel free to spend a little time with the results. The process above should have delivered a dataset and a decision tree used to predict the label of the dataset based on the attributes' values. You can inspect the data itself as well as the meta data of this dataset and try out some of the visualizations in the plot view. You can also inspect the decision tree and try to understand if this makes sense to you. If you wish to return to the Design Perspective, then you can do this at any time using the switching icons at the right of the toolbar.

Tip: After some time you will want to switch frequently between the Design Perspective and the Result Perspective. Instead of using the icon or the menu entries, you can also use keyboard commands F8 to switch to the Design Perspective and F9 to switch to the Result Perspective.

What does that result mean to us? We now have managed to load a dataset from the RapidMiner repository and then we have built the first predictive model based on this data. Furthermore, we got a first feeling about how to build RapidMiner processes. You are now ready to learn more about the use cases for data mining and how to build corresponding processes with RapidMiner. Each of the following chapters will describe a use case together with the data which should be analyzed. At the same time each chapter will introduce new RapidMiner operators to you which are necessary to successfully solve the tasks at hand.

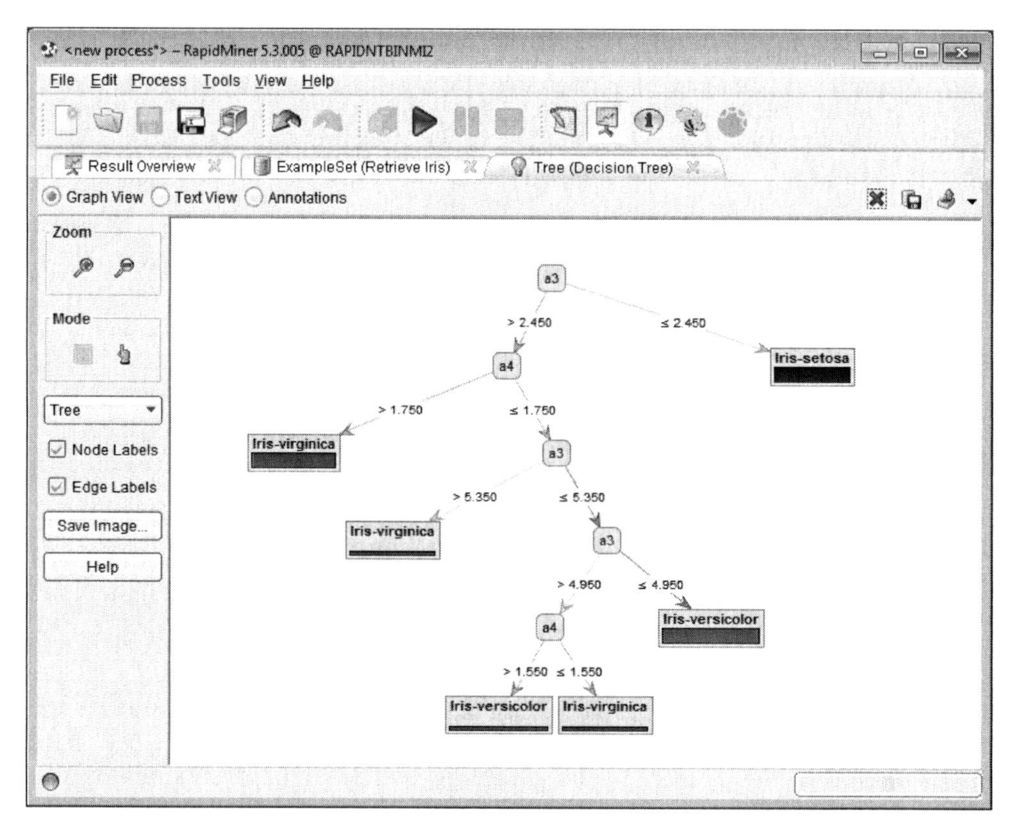

FIGURE 2.10: The decision tree described the rules how to assign the different classes to a new plant. First we check the value of a3 and if it small enough, we assign the blue class Iris-setosa to this plant. If not, we additionally check the value for a4 and if this is high we assign the red class Iris-virginica to it. Finally, we will check for attribute a3 again and decide for Iris-versicolor if the value is small enough.

Part II

Basic Classification Use Cases for Credit Approval and in Education

Chapter 3

k-Nearest Neighbor Classification I

M. Fareed Akhtar

Fastonish, Australia

3.1 Introduction

This chapter explains the k-NN classification algorithm and its operator in RapidMiner. The Use Case of this chapter applies the k-NN operator on the Teacher Evaluation dataset. The operators explained in this chapter are: Read URL, Rename, Numerical to Binominal, Numerical to Polynominal, Set Role, Split Validation, Apply Model, and Performance.

The k-Nearest Neighbor algorithm is based on learning by analogy, that is, by comparing a given test example with the training examples that are similar to it. The training examples are described by n attributes. Each example represents a point in an n-dimensional space. In this way, all of the training examples are stored in an n-dimensional pattern space. When given an unknown example, the k-nearest neighbor algorithm searches the pattern space for the k training examples that are closest to the unknown example. These k training examples are the k "nearest neighbors" of the unknown example. The "Closeness" is defined in terms of a distance metric, such as the Euclidean distance.

3.2 Algorithm

The neighbors are taken from a set of examples for which the correct classification or, in the case of regression, the value of the label is known. This can be thought of as the training set for the algorithm, though no explicit training step is required. The basic k-Nearest Neighbor algorithm is composed of two steps:

- Find the k training examples that are closest to the unseen example.
- Take the most commonly occurring classification for these k examples or, in the case of regression, take the average of these k label values and assign this value as the label of this unseen example.

These two steps are performed for all the unseen examples i.e., all examples of the training dataset.

3.3 The k-NN Operator in RapidMiner

The k-Nearest Neighbor algorithm is implemented by the k-NN operator in RapidMiner. This operator is located at "Modeling/ Classification and Regression/ Lazy Modeling" in the Operators Window. This operator expects an ExampleSet as input and it generates a k-Nearest Neighbor model from the given ExampleSet. This model can be a classification or regression model depending on the given ExampleSet. If the type of the label of the ExampleSet is polynominal or binominal, then this operator generates a classification model. If the type of the label is numerical, then this operator generates a regression model. Some important parameters of this operator are:

- **k**: This parameter specifies the number of nearest neighbors of the unseen example to look for. This parameter is equivalent to the k variable in the k-Nearest Neighbor algorithm. If the parameter k is set to 1, the example is simply assigned the class of its nearest neighbor.
- **weighted vote:** This parameter specifies if the votes should be weighted by similarity. If this parameter is set to true, the weight of examples is also taken into account. It can be useful to weight the contributions of the neighbors, so that the nearer neighbors contribute more than the more distant ones.
- **measure types:** This parameter is used for selecting the type of measure to be used for finding the nearest neighbors. In other words, this parameter specifies the type of measure to use for measuring closeness of examples. The following options are available: *mixed measures*, *nominal measures*, *numerical measures*, and *Bregman divergences*.

3.4 Da [illegible]

3.4.1 T [illegible]

This da [illegible] n be accessed through this link: `http:` [illegible] `ssistant+Evaluation`

3.4.2 B [illegible]

This d [illegible] nce of 151 teaching assistant (TA) [illegible] partment of the University of Wiscon [illegible] ghly equal-sized categories ("low", "n [illegible]

3.4.3 E [illegible]

This d [illegible] ues in this data set.

3.4.4 A [illegible]

This d [illegible]). The data set comes with some basi [illegible] t by the user of the dataset. Even the [illegible] xplanation of the attributes of this dat [illegible]

1. **Eng** [illegible] t the TA is a native English spea [illegible] the TA is a native English spea [illegible] hat the TA is not a native Eng [illegible] lues, its type should be set to binc [illegible] and 2 are not integer values here [illegible] nglish speakers, respectively.

2. **Ins** [illegible] r. To hide the identity of the inst [illegible] 1 to 25 (instead of instructor nan [illegible] ribute can have 25 possible valu [illegible] uld be set to polynominal in Ra [illegible] not integer values here; they are [illegible]

3. **Co** [illegible] identity of the courses, this attr [illegible] d of course name or id etc.). As [illegible] ble values, i.e., {1, 2, 3... 26}. The [illegible] in RapidMiner. It should be not [illegible] hey are used to represent 26 diff [illegible]

4. **Su** [illegible] ffered in summer or regular sen [illegible] hat the course was offered in the [illegible] t implies that the course was offe [illegible] two possible values, its type sho [illegible] e noted that values 1 and 2

are not integer values here; they are used to represent summer and regular semesters, respectively.

5. **Class Size:** This attribute specifies the number of students in the class. The type of this attribute should be set to integer in RapidMiner.

6. **Score Category** (label attribute): This attribute specifies the score category of performance evaluation. The values 1, 2, and 3 indicate that the score was low, medium or high, respectively. As this attribute has three possible values its type should be set to polynominal in RapidMiner. It should be noted that values 1, 2, and 3 are not integer values here; they are used to represent different score categories. The role of this attribute should be set to label because this is the target attribute or the attribute whose value will be predicted by the classification algorithms. The role of all other attributes should be set to regular.

More information about this dataset can be obtained from the UCI repositories.

3.5 Operators in This Use Case

3.5.1 Read URL Operator

The Read URL operator can be very useful if the required dataset is available on the Internet at a particular url. The url of the dataset should be given in the *url* parameter of the Read URL operator. Connectivity to the Internet is required for this operator to fetch data from the specified url. The *column separators* parameter is also important. It specifies the column separators in form of a regular expression. In most cases, the default value of the *column separators* parameter works well. It is important to note that the Read URL operator does not ask the user for meta data. It automatically guesses the types of attributes. The type conversion operators can be used after data import to change the type of the attributes (if required). The Set Role operator can be used to set the role of attributes after the dataset has been imported. Some role and type conversion operators are discussed later in this chapter.

3.5.2 Rename Operator

Pre-processing starts in RapidMiner as soon as the data has been loaded into RapidMiner. Some data import operators (e.g., Read CSV and Read Excel) import data in such a way that the attributes are assigned the correct name during the data import procedure. In this case, there may be no need for renaming the attributes. Names of attributes do not have any effect on the outcome of the classification model, but it is a good practice to assign meaningful names to attributes so that the attribute names reflect the sort of information stored in them. The most commonly used operators for renaming attributes in RapidMiner are the Rename and Rename by Replacing operators. The Rename operator can be used for renaming one or more attributes of the given dataset. The names of attributes in a dataset should be unique. The Rename operator has no impact on the type or role of an attribute.

3.5.3 Numerical to Binominal Operator

The Numerical to Binominal operator changes the type of the selected numeric attributes to binominal type. This operator not only changes the type of selected attributes but it may also change the data of those attribute in the dataset. Binominal attributes can have only two possible values e.g., "true" or "false". If the value of the selected attribute in an example is between the specified minimal and maximal value, this operator changes it to "false", otherwise it changes it to "true". Minimal and maximal values can be specified by the *min* and *max* parameters, respectively.

3.5.4 Numerical to Polynominal Operator

The Numerical to Polynominal operator changes the type of the selected numeric attributes to polynominal type. It simply changes the type of selected attributes i.e., every new numerical value is considered to be another possible value for the polynominal attribute. As numerical attributes can have a huge number of different values even in a small range, converting such a numerical attribute to polynominal type will generate a huge number of possible values for the new polynominal attribute. Such a polynominal attribute may not be a very useful one. This operator cannot be used for grouping numerical values into groups represented by a polynominal attribute. For grouping the numerical values discretization operators are used.

3.5.5 Set Role Operator

It is extremely important to assign the right role to the attributes in the dataset. Most classification operators will not work if there is no attribute with the label role in the dataset (or even if there are multiple attributes with the label role). It should be made sure that only one attribute (and the right one!) has the label role. In a very basic classification setup all other attributes will have regular roles. If an attribute uniquely identifies examples it can be assigned an id role. RapidMiner provides the Set Role operator for changing the role of the attributes. This operator is used for changing the role of one or more attributes. It is very simple to use. The user only needs to provide the name of the attribute and select the desired role for it. The name is provided in the *name* parameter and the desired role is selected by the *target role* parameter. Roles of multiple attributes can be set using the *set additional roles* parameter.

3.5.6 Split Validation Operator

The set of examples that the model is trained on is called the training data set. The trained model is applied on new examples to test the performance of the model; this is known as testing a model. The dataset on which the trained model is tested is called the testing dataset. Testing the model gives an idea of how the model will perform in practice. Testing the model validates the performance of the model.

Training and testing is very important part of a classification process. A classification model cannot be applied in practice without knowing the performance (e.g., prediction accuracy) of the model. Testing and training can be implemented in different ways. A very common method is to split the dataset into two portions. One portion is used for training the model and the other portion is used for testing the trained model. Usually the larger portion of the dataset is reserved for training the model.

The Split Validation operator is a nested operator. It has two subprocesses: a training sub-process and a testing sub-process. The training sub-process is used for learning or training a model. The classification operators are placed in this sub-process to train a classification model. The trained model is then applied in the testing sub-process. Mostly, the Apply Model operator is used for applying the trained model. The performance of the model is also measured during the testing phase. Performance measuring operators like the Performance, Performance (Classification), and Performance (Binominal Classification) operators are used for measuring the performance of the model. The given dataset is partitioned into two subsets. One subset is used as the training set and the other one is used as the testing set. The size of two subsets can be adjusted by parameters like *split, split ratio, training set size*, and *test set size* parameters. The *sampling type* parameter is used for selecting the type of sampling. The model is learned on the training set and is then applied on the testing set. This is done in a single iteration, as compared to the X-Validation operator that iterates a number of times using different subsets for testing and training purposes.

3.5.7 Apply Model Operator

The Apply Model operator applies a trained model on a dataset. Usually a model is trained on a training dataset and then tested on testing dataset. The Apply Model operator takes two inputs:

1. Trained model.
2. Testing dataset.

It is compulsory that both training and testing datasets should have exactly the same number, order, type and role of attributes. If these properties of meta data of the datasets are not consistent, it may lead to serious errors. The output of the Apply Model operator is the result of applying the trained model on the testing dataset. The resultant dataset has a new attribute with the *prediction* role. This attribute contains the values of the label predicted by the trained model.

3.5.8 Performance Operator

In contrast to the other performance evaluation operators like Performance (Classification) operator or Performance (Binominal Classification) operator, this operator can be used for performance evaluation of all types of learning tasks. It automatically determines the learning task type and calculates the most common criteria for that type. For more sophisticated performance calculations, task-specific performance operators should be used, for example, Performance (Binominal Classification) operator should be used for binominal classification tasks.

3.6 Use Case

The Teacher Evaluation dataset is used in this use case. The Read URL operator is used for importing the data. The name, type, and role conversion operators are used during the pre-processing. The Split Validation operator is used for assisting in training and testing of the model. The Performance operator is used for evaluating the performance of the model.

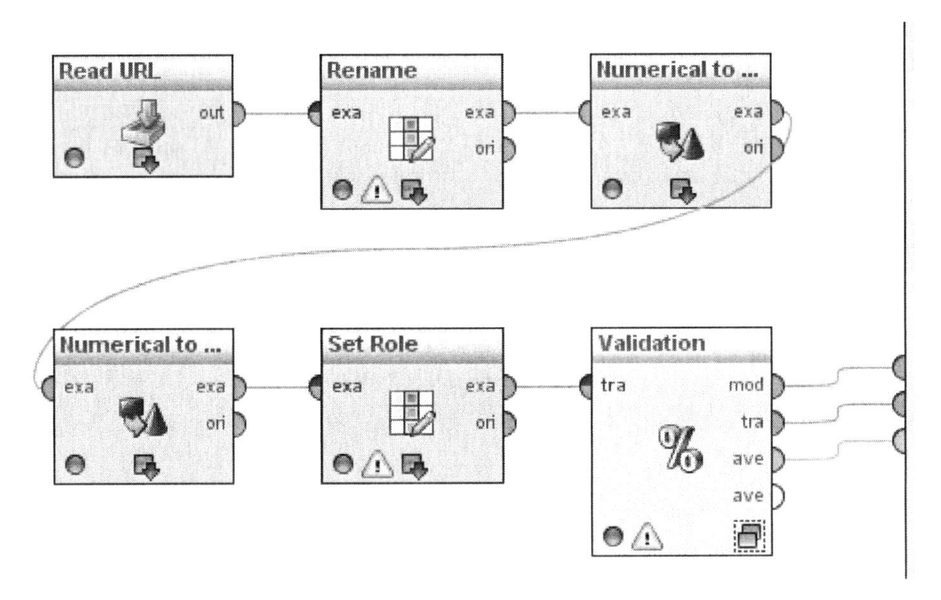

FIGURE 3.1: Workflow of the process.

3.6.1 Data Import

This process starts with the import of the Teacher Evaluation dataset. The Read URL operator is used for importing data. The *url* parameter of the Read URL operator is set to 'http://archive.ics.uci.edu/ml/machine-learning-databases/tae/tae.data' which is the URL of the Teacher Evaluation dataset. All other parameters of the Read URL operator are used with default values. The *column separators* parameter of the Read URL operator is set to ',\s*|;\s*|\s+' which is a regular expression. '|' stands for a disjunction (logical 'or'), thus, there are three statements separated by two disjunctions. The three statements are:

- **,\s*** : '\s' stands for a whitespace. '*' stands for zero or more occurrences of the previous element. Therefore, ',\s*' means zero or more whitespaces after a comma.
- **;\s*** : '\s' stands for a whitespace. '*' stands for zero or more occurrences of the previous element. Thus ';\s*' means zero or more whitespaces after a semi-colon.
- \s+ : '\s' stands for a whitespace. '+' stands for one or more occurrences of the previous element. Thus '\s+' means one or more whitespaces.

As the Teacher Evaluation dataset has commas as column separators there is no need to modify the default regular expression of the *column separators* parameter. A breakpoint is inserted after this operator so that the output of the Read URL operator can been viewed in the Results Workspace. A breakpoint is an interruption point in the process execution for inspection of intermediate results. These results are displayed in the Results View. To insert a breakpoint after an operator, right-click on that operator and select the "Breakpoint After" option. You can also select the "Breakpoint Before" option, in which case you will see the intermediate results obtained before execution of this operator.

3.6.2 Pre-processing

The Read URL operator fetches the data from the specified url but it does not specify the meta data of the dataset. In addition, the attribute names are not meaningful in the output of the Read URL operator. Therefore, some pre-processing is required to set the name, type, and role of the attributes.

3.6.3 Renaming Attributes

The Rename operator is used for renaming the attributes. The *old name*, *new name* and *rename additional attributes* parameters are used for renaming the attributes as shown in Figure 3.2. A breakpoint is inserted after this operator so that the effect of the Rename operator can be seen in the Results Workspace.

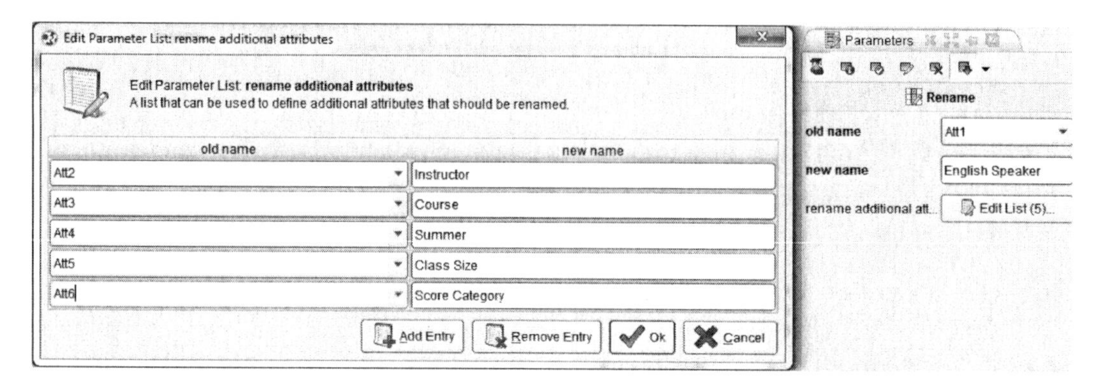

FIGURE 3.2: Parameters of the Rename operator.

3.6.4 Changing the Type of Attributes

The type of all attributes is integer when the dataset is fetched from the specified url which is not the correct type. As was explained in the description of the Teacher Evaluation dataset, the types of the Course, Instructor, and Score Category attributes should be polynominal. The Numerical to Polynominal operator is used for changing the type of these attributes from integer to polynominal. The parameters of the Numerical to Polynominal operator are shown in the Figure 3.3 . A breakpoint is inserted after this operator so that the effect of the Numerical to Polynominal operator can be seen in the Results Workspace.

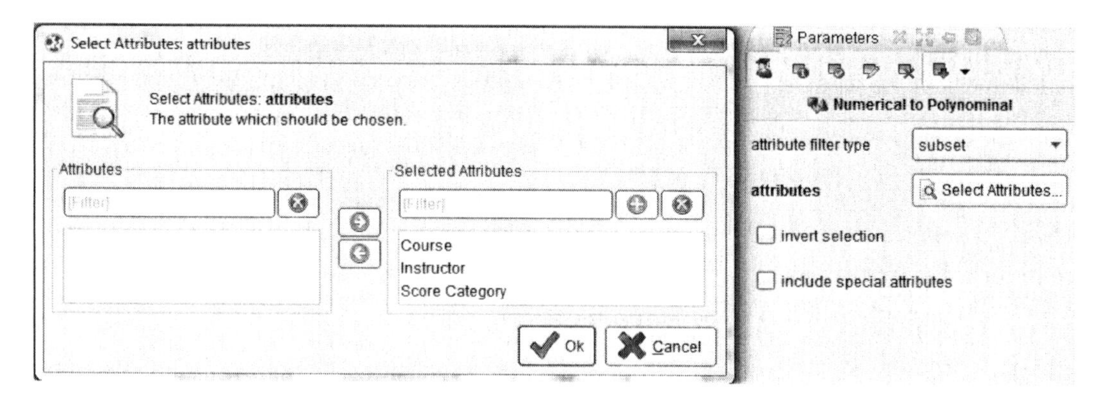

FIGURE 3.3: The parameters of the Numerical to Polynominal operator.

The type of the English Speaker and Summer attributes should be binominal. The

Numerical to Binominal operator is used for changing the type of these attributes. The parameters of the Numerical to Binominal operator are shown in the Figure 3.4.

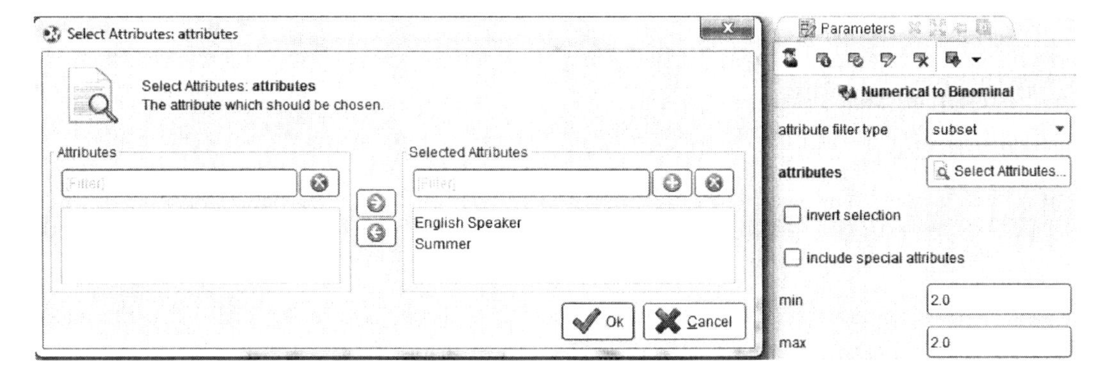

FIGURE 3.4: The parameters of the Numerical to Binominal operator.

The value '1' stands for 'true' and the value '2' stands for 'false' in the English Speaker and Summer attributes. To actually convert 1 and 2 to 'true' and 'false', respectively the *min* and *max* parameters of the Numerical to Binominal operator are set to 2. All the values in the range of *min* and *max* parameters are changed to false (in this case only one value i.e., 2 falls in this range) and all other values are changed to true (i.e., 1 is changed to true). A breakpoint is inserted after this operator so that the effect of the Numerical to Binominal operator can be seen in the Results Workspace.

3.6.5 Changing the Role of Attributes

The role of all the attributes is set to regular when the dataset is fetched from the specified url by the Read URL operator. As it was explained in the explanation of the Teacher Evaluation dataset, the role of the Score Category attribute should be *label*. The Set Role operator is used for changing the role. The parameters of the Set Role operator are shown in Figure 3.5. A breakpoint is inserted after this operator so that the effect of the Set Role operator can be seen in the Results Workspace.

3.6.6 Model Training, Testing, and Performance Evaluation

Now that the dataset has the correct name, type, and role for all attributes, the pre-processing has been completed for this process. Once the pre-processing of the dataset has

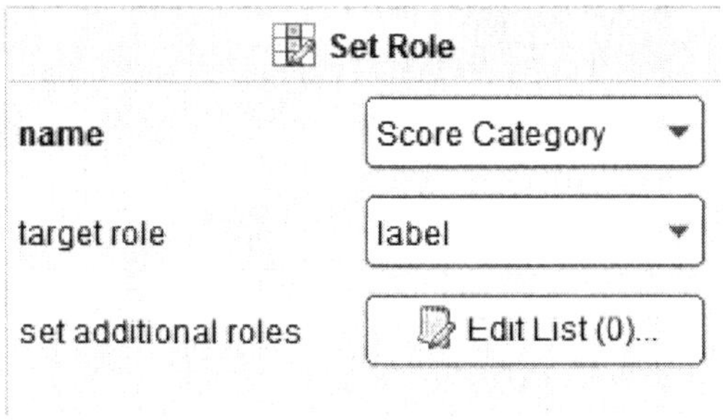

Set Role

name — Score Category

target role — label

set additional roles — Edit List (0)...

FIGURE 3.5: The parameters of the Set Role Operator.

been completed, a model can be trained on the training subset of the dataset. The trained model is then tested on the testing subset of the dataset. Finally, the performance of the model is measured. All these tasks can be implemented conveniently by using the Split Validation operator. The Split Validation operator splits the given dataset into training and testing datasets. The *split ratio* parameter specifies the ratio of the data to be used as training data (testing dataset is automatically assigned the ratio = 1 - split_ratio). This operator is applied on the pre-processed Teacher Evaluation dataset. The *split ratio* parameter is set to 0.7 and the *sampling type* parameter is set to "stratified sampling". The following is an explanation of what happens inside the Split Validation operator. You must double-click on the operator to get inside the operator for specifying the subprocesses.

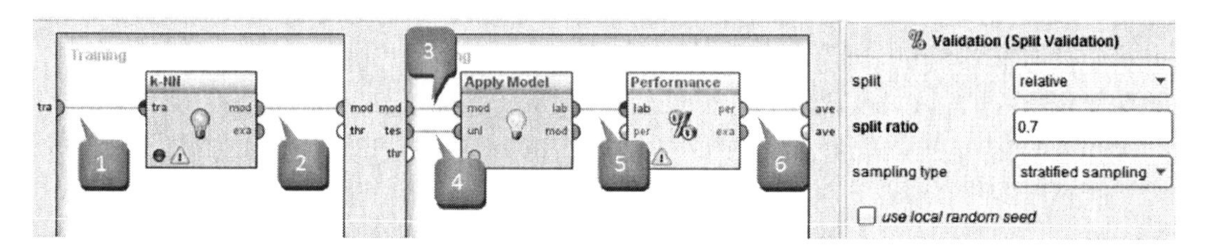

FIGURE 3.6: Subprocesses and Parameters of the Split Validation operator.

1. The Split Validation operator provides training dataset (composed of 70% of the examples) through the *training* port of the Training sub-process. This training dataset is used as input for the k-NN operator. Thus, the k-NN classification model is trained on this training dataset.

2. The k-NN operator provides the k-NN classification model as its output. This model is connected to the *model* port of the Training sub-process.

3. The k-NN classification model that was provided at the *model* port of the Training sub-process is delivered by the Split Validation operator at the *model* port of the Testing sub-process. This model is provided as input at the *model* port of the Apply Model operator.

4. The Split validation operator provides the testing dataset through the *test set* port of the Testing sub-process. As the *split ratio* parameter was set to 0.7, the testing dataset comprises 0.3 (i.e., 1.0 – 0.7) ratio of examples. This testing dataset is provided as input to the Apply Model operator (at the *unlabeled data* port).

5. The Apply Model operator applies the k-NN classification model (that was provided at its *model* input port) on the testing dataset (that was provided at its *unlabeled data* input port). The resultant labeled dataset is delivered as output by the Apply Model operator. This labeled dataset is provided as input to the Performance operator.

6. The Performance operator evaluates the statistical performance of the model through the given labeled dataset and generates a performance vector which holds information about various performance criteria.

If the k parameter of the k-NN operator is set to 1 (keeping everything else constant), the accuracy of the model turns out to be 41.30%. Figure 3.7 shows how the accuracy of the model varies with the change in value of parameter k.

It turns out that values 1 and 2 for parameter k produce the most accurate models, in this case. The accuracy of these models is not quite at par with the standards. This is because many steps were skipped in the process to keep it simple.

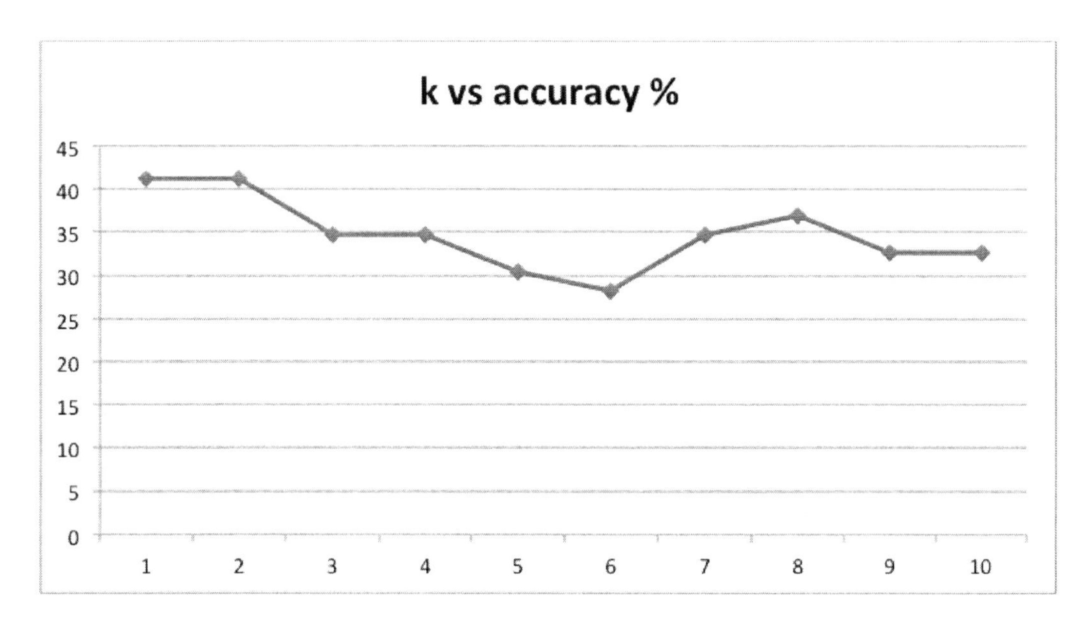

FIGURE 3.7: The change in accuracy of the model with the change in value of the parameter k.

Chapter 4

k-Nearest Neighbor Classification II

M. Fareed Akhtar
Fastonish, Australia

4.1 Introduction

The use case of this chapter applies the k-NN operator on the Glass Identification dataset (overview of the k-NN algorithm has been discussed in the previous chapter). The purpose of this use case is to predict the type of the glass depending on its components. The operators explained in this chapter are: Read CSV, PCA, Split Data, and Performance (Classification).

4.2 Dataset

Glass Identification Dataset This dataset has been taken from UCI repositories. This dataset can be accessed through this link: `http://archive.ics.uci.edu/ml/datasets/Glass+Identification`.

Basic Information: The aim of this dataset is to classify the glass into one of 7 forms depending on its oxide content (i.e., sodium, iron etc.).

Examples This dataset has 214 examples and there are no missing values in this data set.

Attributes This dataset has 11 attributes (including the label and Id attribute). The dataset comes with some basic information but the type and role of attributes is set by the user of the dataset. Even the attribute names are specified by the user. Except the label and id attributes, almost all other attributes are named according to short names (or symbols) of metals e.g. Na, Mg, and Al for Sodium, Magnesium, and Aluminum, respectively. The name, type, and range of attributes are explained below in this short format: '**name** (type): {range}'

1. **Id** (integer): {unique integer values}. This is the id attribute, i.e., attribute with id role for uniquely identifying examples.
2. **RI** (real): {real values}. RI stands for refractive index.
3. **Na** (real): {real values}
4. **Mg** (real): {real values}
5. **Al** (real): {real values}
6. **Si** (real): {real values}
7. **K** (real): {real values}
8. **Ca** (real): {real values}
9. **Ba** (real): {real values}
10. **Fe** (real): {real values}
11. **Type** (label attribute): This attribute has seven possible values: {1, 2, 3, 4, 5, 6, 7}. Each value represents a particular type of glass. As this attribute has seven possible values, its type should be set to polynominal in RapidMiner. The role of this attribute should be set to *label* because this is the target attribute or the attribute whose value will be predicted by the classification algorithms.

More information about this dataset can be obtained from UCI repositories.

4.3 Operators Used in This Use Case

4.3.1 Read CSV Operator

A comma-separated-value (CSV) file stores data (usually numbers and text) in simple text form. CSV files have all values of an example in one line. The values of different attributes are separated by a constant separator. It may have many rows but each row uses the same constant separator for separating the attribute values. The name 'CSV' suggests that the attribute values are separated by commas, but other separators can also be used.

The Read CSV operator imports CSV files into RapidMiner. The easiest and shortest way to import a CSV file is to use the *import configuration wizard* from the *Parameters* View. This configuration wizard is composed of a series of small steps which ensure that the file is read correctly. Moreover, the Read CSV operator provides the meta data with the data. The most important purpose of the configuration wizard is to set the meta data. It is important to make sure that the CSV file is read correctly before building a process using it. The file can also be read without using the wizard. The *csv file* parameter is used for specifying the path of the required CSV file. The constant separator can be specified in the form of a regular expression through the *column separators* parameter. The most important parameter is the *dataset meta data information* parameter. It is used for adjusting the meta data of the CSV file. The name, type, and role of the attributes can be specified through this parameter. The Read CSV operator tries to determine the appropriate type of the attributes by reading the first few lines and checking the occurring values. If all values are integers, then the attribute's type will be set to integer. Similarly, if all values are real numbers, then the attribute's type will be set to real. These automatically determined types can be overridden by the *dataset meta data information* parameter.

4.3.2 Principal Component Analysis Operator

Reducing the number of attributes in the dataset is a very important pre-processing step. Attribute set reduction is an advanced topic and will be discussed here, briefly. There are many reasons for reducing the number of attributes in a dataset; some of them are:

- Removing irrelevant attributes improves performance of algorithms
- Reducing the number of attributes reduces space requirement for algorithms
- Reducing the number of attributes reduces execution time of algorithms

All these benefits have more and more impact as the size of the dataset increases. The attributes that are not relevant to the problem in hand should be removed from the dataset. For example, suppose a dataset is being used for evaluation of performance of students and that dataset has information like address, father's name, and place of birth etc. in addition to the information about results of various courses. The attributes about the student's address are not relevant to the students' performance evaluation (unless one is trying to find the effect of the address of the student on his grades). Similarly, the father's name is not relevant (unless one believes that the name of the student's father has something to do with the student's grades). Such attributes should be removed from the dataset before training a model on the dataset because they do not add useful information into the classification model. In this example, it was very easy to find useless attributes because it was evident from their background that they will not add to the predictive accuracy of the model. In such cases, the Select Attributes operator can be used to select only the required attributes.

There are much more complex cases than the example just discussed. For example, if there is a dataset with five hundred attributes and the domain knowledge about the dataset and attributes is not given. In this case useless attributes cannot be removed manually. Advanced techniques like Principal Component Analysis, Singular Value Decomposition, and Iterative Component Analysis are used to reduce the number of attributes in such cases. RapidMiner provides numerous operators for automatically reducing the number of attributes, for example, the Remove Useless Attributes, Remove Correlated Attributes, Principal Component Analysis, Singular Value Decomposition and Iterative Component Analysis operators. These operators apply different algorithms to reduce the attribute set. The users can adjust various parameters of these algorithms.

Principal component analysis (PCA) is an attribute reduction procedure. It is useful especially for removing redundant attributes. In this case, redundancy means that some of the attributes are correlated with one another, maybe because they are measuring the same property. Because of this redundancy, it is possible to reduce the number of attributes into a smaller number of principal components. Principal components are artificial attributes that represent (or try to represent as closely as possible) the original attributes by accounting for most of the variance in the original attributes. It is important to note that this operator creates new attributes that represent the old attributes and old attributes are removed from the dataset.

4.3.3 Split Data Operator

This operator is different from other sampling and filtering operators as it is capable of delivering multiple partitions of the given dataset. The Split Data operator takes a dataset as its input and delivers the subsets of that dataset through its output ports. The number of subsets (or partitions) and the relative size of each partition are specified through the *partitions* parameter. The sum of the ratio of all partitions should be 1. The *sampling type* parameter decides how the examples should be shuffled in the resultant partitions. The *sampling type* parameter is present in many RapidMiner operators.

The *sampling type* parameter has the following possible values:

- **Linear sampling:** Linear sampling simply divides the dataset into partitions without changing the order of the examples, i.e., subsets with consecutive examples are created.
- **Shuffled sampling:** Shuffled sampling builds random subsets of the dataset, i.e., examples are chosen randomly for making subsets.
- **Stratified sampling:** Stratified sampling builds random subsets and ensures that the class distribution in the subsets is the same as in the whole dataset. For example, in the case of a binominal classification, Stratified sampling builds random subsets so that each subset contains roughly the same proportions of the two values of the label.

4.3.4 Performance (Classification) Operator

The entire goal of the classification process is to predict the value of the attribute with the label role. The performance of a model is measured by comparing the actual label values of the training dataset with the label values predicted by the model. Numerous operators are present in RapidMiner that measure the performance of a model. The performance operators require a labeled dataset in order to evaluate the performance. A labeled dataset has the label and prediction attributes. The Performance operators take the labeled dataset as input and deliver the performance vector. The performance vector has information about various performance criteria of the model that created the labeled dataset.

The Performance operator is used for statistical performance evaluation of classification tasks. The performance vector delivered by this operator has numerous classification-specific performance criteria, for example, accuracy, classification error, kappa, spearman rho, kendall tau, etc.

4.4 Data Import

The Read CSV operator is used for importing data in this process. The data at the following url is stored in the form of a text file.

`http://archive.ics.uci.edu/ml/machine-learning-databases/glass/glass.data`

Next, that text file is used as a data source in RapidMiner. The Read CSV operator is used for loading data from a csv file. In this process, the parameters of the Read CSV

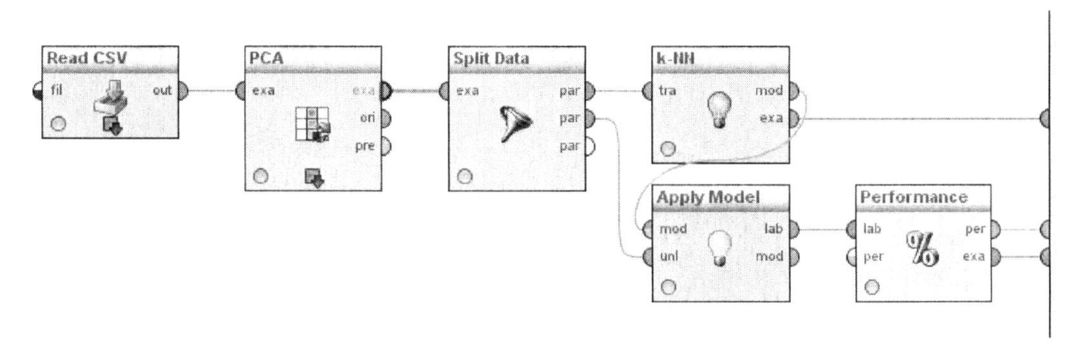

FIGURE 4.1: Workflow of the process.

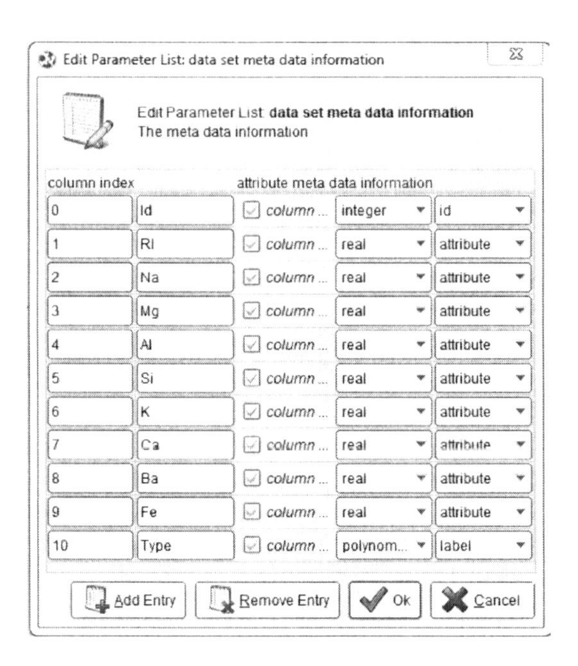

FIGURE 4.2: The dataset meta data information parameter of the Read CSV operator.

operator are used to configure the operator instead of the Import Configuration Wizard. The *csv file* parameter provides the path of the text file. The *column separators* parameter is set to ',' because the attribute values are separated by commas in the data file. The *dataset meta data information* parameter is used for setting the name, type and role of attributes. Figure 4.2 shows the values of the *dataset meta data information* parameter. All other parameters are used with default values. A breakpoint is inserted after this operator so that the output of the Read CSV operator can be seen in the Results Workspace.

4.5 Pre-processing

The PCA operator is introduced in this process. The PCA operator is used for attribute set reduction, i.e., reducing the number of attributes in a data set.

4.5.1 Principal Component Analysis

Principal component analysis is different from other attribute set reduction techniques because, instead of reducing the number of existing attributes, it creates new attributes altogether (called principal components) that cater to variance of the previous attributes. There are two ways of using the PCA operator:

1. Provide a variance threshold in the *variance threshold* parameter so that the PCA algorithm creates n number of principal components where n is the number of principal components required to produce the specified variance.
2. Provide the number of principal components required in the *number of components* parameter to make this number of principal components irrespective of the variance.

The first approach is used in this process and the variance threshold is set to 0.95. A breakpoint is inserted after this operator so that the results of this operator can be seen in the Results Workspace. The dataset had 9 regular attributes before application of this operator. After the execution of this operator, the dataset has 5 principal components that try to represent all the previous attributes.

4.6 Model Training, Testing, and Performance Evaluation

The Split Validation operator was used in the previous chapter for facilitating in model training, testing, and performance evaluation. In this process, an alternate scheme is used instead of the Split Validation operator which performs almost the same task.

The Split Data operator is used in this process for splitting the dataset into two partitions. One partition is used for training the model and the other partition is used for testing the model. The *partitions* parameter of the Split Data operator performs almost the same task that the *split ratio* parameter does in the Split Validation operator. The *partitions* parameter defines the number of partitions and ratio of each partition. The *sampling type* parameter of the Split Data operator behaves exactly the same as the *sampling type* parameter in the Split Validation operator. The parameters of the Split Data operator are shown in Figure 4.3.

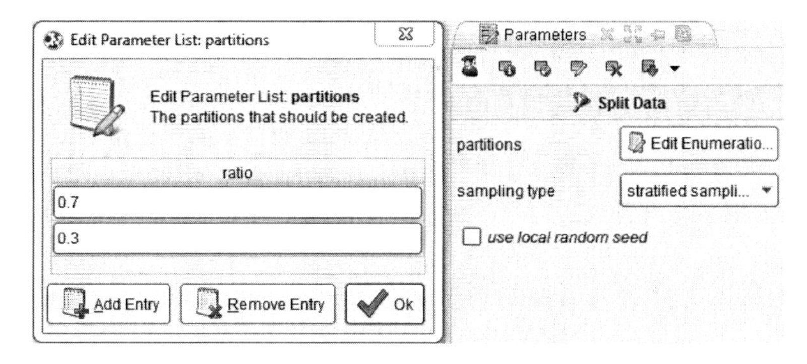

FIGURE 4.3: The parameters of the Split Data operator.

4.6.1 Training the Model

The Split Data operator provides the first partition (ratio = 0.7) through its first output port. This partition is used for training a k-NN classification model through the k-NN operator. Initially, all parameters of the k-NN operator are used with default values. The k-NN operator provides the k-NN classification model as its output. This model is given as input to the Apply Model operator.

4.6.2 Testing the Model

The Split Data operator provides the second partition (ratio = 0.3) through its second output port. This partition is used for testing the k-NN classification model through the Apply Model operator. The Apply Model operator applies the k-NN classification model (that was provided at its *model* input port) on the testing dataset (that was provided at its *unlabeled data* port). The resultant labeled dataset is delivered as output by the Apply Model operator. This labeled dataset is provided as input to the Performance (Classification) operator.

4.6.3 Performance Evaluation

The Performance operator was used in the previous chapter for evaluating the performance of the model. The Performance operator provides only limited criteria about the performance of the classification model. The Performance (Classification) operator should be used for performance evaluation of the classification models because it provides many different performance measures in its performance vector. The performance vector generated by the Performance (Classification) operator and the labeled dataset are connected to the *result* port of the process. They can be seen in the Results Workspace after the execution of the process.

The accuracy of this model turns out to be 70.77%. Right-click on the PCA operator and uncheck the "Enable operator" option to disable the PCA operator. Now execute the process again and note the accuracy. Without the PCA operator the accuracy of the model drops to 69.23%. This clearly shows that reducing the number of attributes (by using techniques like PCA) can significantly increase the accuracy of the resultant model.

Chapter 5

Naïve Bayes Classification I

M. Fareed Akhtar

Fastonish, Australia

5.1 Introduction

This chapter explains the Naïve Bayes classification algorithm and its operator in RapidMiner. The use case of this chapter applies the Naïve Bayes operator on the Credit Approval dataset. The operators explained in this chapter are: Rename by Replacing, Filter Examples, Discretize by Binning, X-Validation, and Performance (Binominal Classification) operator.

The Naïve Bayes algorithm is a simple probabilistic classifier based on applying Bayes' theorem with strong independence assumptions. In simple words, the Naïve Bayes algorithm assumes that the presence of a particular value of an attribute is unrelated to the presence of any other attribute value. For example, a ball may be classified as a tennis ball if it is green, round, and about 2 and half inches in diameter. Even if these properties depend on each other, the Naïve Bayes algorithm considers all of these features to independently contribute to the probability that this ball is a tennis ball.

Let **X** be an example that we want to classify. **X** is described by measurements made on a set of **n** attributes. Let **H** be some hypothesis, such as that the example **X** belongs to a specified class **C**. For classification problems, we want to determine $\mathbf{P(H|X)}$, the probability that the hypothesis **H** holds given the example **X**. In other words, we are looking for the probability that example **X** belongs to class **C**, given that we know the attribute description of **X**.

P(H|X) is the posterior probability of **H** conditioned on **X**. For example, suppose our dataset is about whether golf will be played or not in the given weather conditions like temperature and outlook. Thus, temperature and outlook are our attributes. The example **X** represents a particular scenario of weather conditions, for example, a weather condition with 80 degrees temperature and sunny outlook. Suppose that **H** is the hypothesis that golf will be played in the given weather conditions. Then **P(H|X)** reflects the probability that golf will be played in **X** weather condition given that we know the temperature and outlook of weather.

In contrast, **P(H)** is the prior probability, or a priori probability, of **H**. For our example, this is the probability that golf will be played in any given weather, regardless of temperature, outlook, or any other information. The posterior probability, **P(H|X)**, is based on more information, i.e., weather information, than the prior probability, **P(H)**, which is independent of **X**.

Similarly, **P(X|H)** is the posterior probability of **X** conditioned on **H**. That is, it is the probability that the weather, **X**, is sunny and 80 degrees hot, given that we know that golf will be played.

P(X) is the prior probability of **X**. Using our example, it is the probability that a weather condition will have temperature of 80 degrees and sunny outlook in our complete dataset.

P(H), **P(X|H)**, and **P(X)** can be estimated from the given data. Bayes' theorem is useful in that it provides a way of calculating the posterior probability, **P(H|X)**, from **P(H)**, **P(X|H)**, and **P(X)**. Bayes' theorem states that:

$$\mathbf{P(H|X) = P(X|H)\ .\ P(H)\ /\ P(X)}.$$

How Bayes' theorem is used in the Naïve Bayes classifier is not discussed in this chapter. The use case of the next chapter explains, in detail, the working of the Naïve Bayes classifier.

The Naïve Bayes operator in RapidMiner has only one parameter, i.e., Laplace correction. This expert parameter indicates if the Laplace correction should be used to prevent high influence of zero probabilities. To avoid zero probabilities, it can be assumed that our training dataset is so large that the addition of one to each count that we need would only make a negligible difference in the estimated probabilities. This technique is known as the Laplace correction.

5.2 Dataset

5.2.1 Credit Approval Dataset

This dataset has been taken from the UCI repositories. This dataset can be accessed through this link: `http://archive.ics.uci.edu/ml/datasets/Credit+Approval`.

This dataset concerns credit card applications, therefore, all attribute names and values have been changed to meaningless symbols to protect confidentiality of the data.

5.2.2 Examples

This dataset has 690 examples and there are some missing values in this dataset.

5.2.3 Attributes

This dataset has 16 attributes (including the label attribute). The data set comes with some basic information but the type and role of attributes is set by the user of the dataset. The name, type, and range of attributes are explained below in this short format: '**name** (type): {range}'

1. **A1** (binominal): {b, a}
2. **A2** (real): {real values}
3. **A3** (real): {real values}
4. **A4** (polynominal): {u, y, l, t }
5. **A5** (polynominal): {g, p, gg }
6. **A6** (polynominal): {c, d, cc, i, j, k, m, r, q, w, x, e, aa, ff }
7. **A7** (polynominal): {v, h, bb, j, n, z, dd, ff, o }
8. **A8** (real): {real values}
9. **A9** (binominal): {t, f }
10. **A10** (binominal): {t, f }
11. **A11** (integer): {integer values}
12. **A12** (binominal): {t, f }
13. **A13** (polynominal): {g, p, s }
14. **A14** (integer): {integer values}
15. **A15** (integer): {integer values}
16. **class** (label attribute): This is the class attribute. It has only two possible values: {+, −}. As this attribute has only two possible values, its type should be set to binominal in RapidMiner. The role of this attribute should be set to *label* because this is the target attribute or the attribute whose value will be predicted by the classification algorithms. The role of all other attributes should be set to regular.

As the label attribute of this dataset is of binominal type, the classification of this dataset will be binominal classification. More information about this dataset can be obtained from UCI repositories.

5.3 Operators in This Use Case

5.3.1 Rename by Replacing Operator

The Rename by Replacing operator replaces parts of the attribute names by the specified replacement. This operator is mostly used for removing unwanted parts of attribute names like whitespaces, parentheses, or other unwanted characters. This operator is very useful when many attribute names have some similar unwanted portion. The *replace what* parameter specifies the part of the attribute name that should be replaced. It can be defined as a regular expression which is a very powerful tool. The *replace by* parameter specifies the replacement.

5.3.2 Filter Examples Operator

Filtering can be defined as selecting a particular subset of examples from the dataset. The subset of examples can have some common properties, like having missing values. Filtering can be useful in a number of ways. It can be used to discard certain types of examples or it can be used to filter only the required type of examples. The Filter Examples operator in RapidMiner filters examples from the given dataset on the basis of some properties of the subset. The Filter Example Range operator filters examples within the specified range, for example, the first 100 examples.

The Filter Examples operator filters out only those examples that satisfy the specified condition. Several pre-defined conditions are provided in the *condition class* parameter. The users can also define their own conditions to filter examples. The *invert filter* parameter can be used to invert the condition specified in the *condition class* parameter. This operator can be used to filter examples with (or without) correct prediction, with (or without) missing values in attributes, and with (or without) missing values in the label attribute. The Filter Examples operator is frequently used to filter out examples that have missing values.

5.3.3 Discretize by Binning Operator

Dividing numerical values into different groups is known as discretization. This can be considered as a form of numerical-to-nominal type conversion. It is often required to group numerical values into certain number of groups. Many operators require all attributes to be of nominal form or many operators produce better results when applied on nominal attributes. In such cases, it is necessary to discretize numerical attributes to nominal type. The Discretization operators are located at 'Data Transformation / Type Conversion / Discretization'.

The Discretize by Binning operator discretizes the selected numerical attributes into a specified number of groups (or bins). The Values falling in the range of a group are named according to the name of that group. The range of all groups is equal. The number of the values in different groups may vary, i.e., the frequency of groups may vary. The resultant attribute is of nominal type. The groups are named automatically and the naming format can be changed by the *range name* parameter. Only a selected range of values can also be discretized by using the *define boundaries* parameter. The *min value* and *max value* parameters are used for defining the boundaries of the range. If there are values that are less than the *min value* parameter, a separate range is created for them. Similarly, if there are values that are greater than the *max value* parameter, a separate range is created for them. Then, the discretization by binning is performed only on the values that are within the specified boundaries.

5.3.4 X-Validation Operator

The X-Validation operator (also written as Cross-Validation) is a nested operator like the Split Validation operator. The major difference between Split Validation and X-Validation is that the Split Validation operator iterates just once. On the other hand, the X-Validation operator has multiple iterations. The given dataset is partitioned into k subsets of equal size. Of the k subsets, a single subset is retained as the testing data set, i.e., the input of the testing sub-process, and the remaining $k - 1$ subsets are used as the training dataset, i.e., the input of the training sub-process. The cross-validation process is then repeated k times, with each of the k subsets used exactly once as the testing data. The value k can be adjusted using the *number of validations* parameter.

5.3.5 Performance (Binominal Classification) Operator

This operator is used for statistical performance evaluation of binominal classification tasks, i.e., classification tasks where the label attribute is of binominal type. This operator delivers a list of performance criteria values of the binominal classification task. The performance vector delivered by this operator has numerous binominal classification-specific performance criteria, e.g., lift, fallout, f measure, false positive, false negative, true positive, true negative, sensitivity, specificity, positive predictive value, negative predictive value, etc.

5.4 Use Case

The Credit Approval dataset is loaded using the Read CSV operator. The Rename by Replacing, Filter Examples, and Discretize by Binning operators are used for pre-processing purposes. The X-Validation and Performance (Binominal Classification) operators are used during the testing phase. The effect of discretization and filtering on the accuracy of the model is also observed.

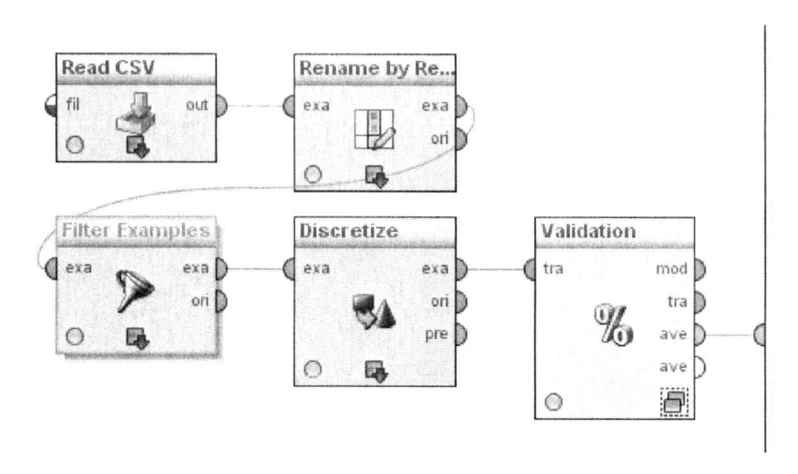

FIGURE 5.1: Workflow of the process.

5.4.1 Data Import

The Read CSV operator is used for importing data in this process. The data at the following url is stored in the form of a text file.

`http://archive.ics.uci.edu/ml/machine-learning-databases/credit-screening/crx.data`

That text file is used as data source in RapidMiner. The Read CSV operator is used for loading data from a csv file. The *Import Configuration Wizard* is used for loading the data from the file in this process. The steps of this configuration wizard are explained below.

Step 1: Select file.

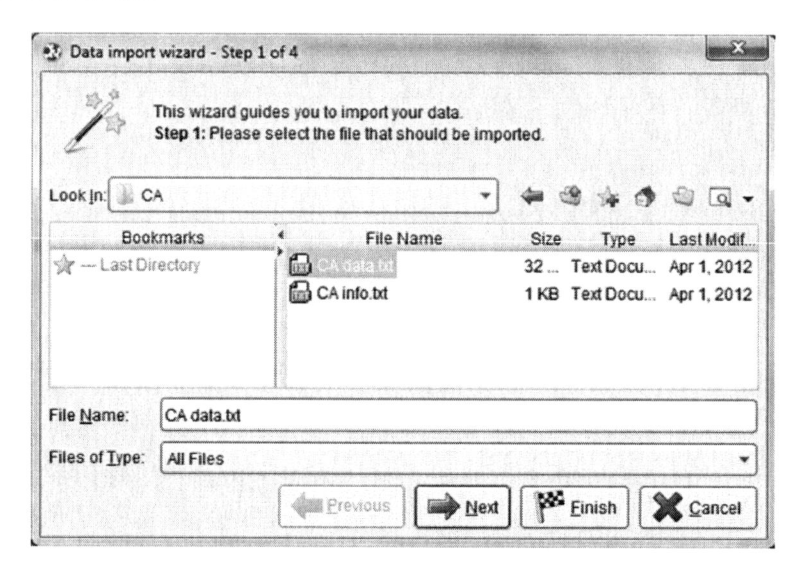

FIGURE 5.2: Step 1 of Import Configuration Wizard of the Read CSV operator.

Step 2: File reading and column separation.

RapidMiner fetches the data and displays it, in this step. The most important decision is selecting the column separator at this step. As the data in the Credit Approval dataset is comma separated, 'Comma' is selected as column separator.

Step 3: Annotations.

This step allows specifying the annotations. By default, the annotation of the first row is set to 'name'. If the first row of data does not contain the names of attributes, its annotation should be unchecked (as shown in Figure 5.4).

Step 4: Set the name, type, and role of attributes.

This is the most crucial step of the import wizard. The name, type, and role of attributes are specified in this step. Figure 5.5 shows the name, type, and role of attributes set for the Credit Approval dataset. A breakpoint is inserted after this operator so that the output of the Read CSV operator can be seen in the Results Workspace.

5.4.2 Pre-processing

Three pre-processing steps are discussed in this use case; renaming attributes, filtering missing values, and discretization.

Renaming attributes The name format of all the attributes of this dataset is of the form att1, att2 etc. This is the default naming format of the Read CSV operator. However, this naming format does not match the naming format that was mentioned in

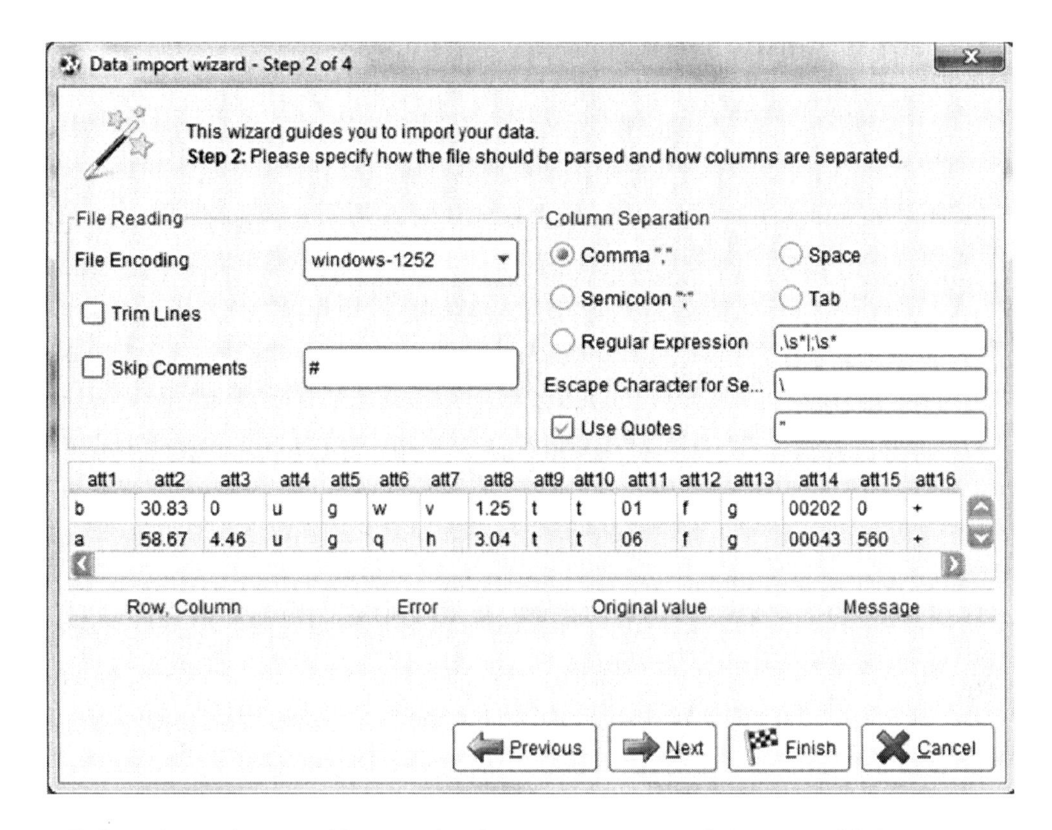

FIGURE 5.3: Step 2 of Import Configuration Wizard of the Read CSV operator.

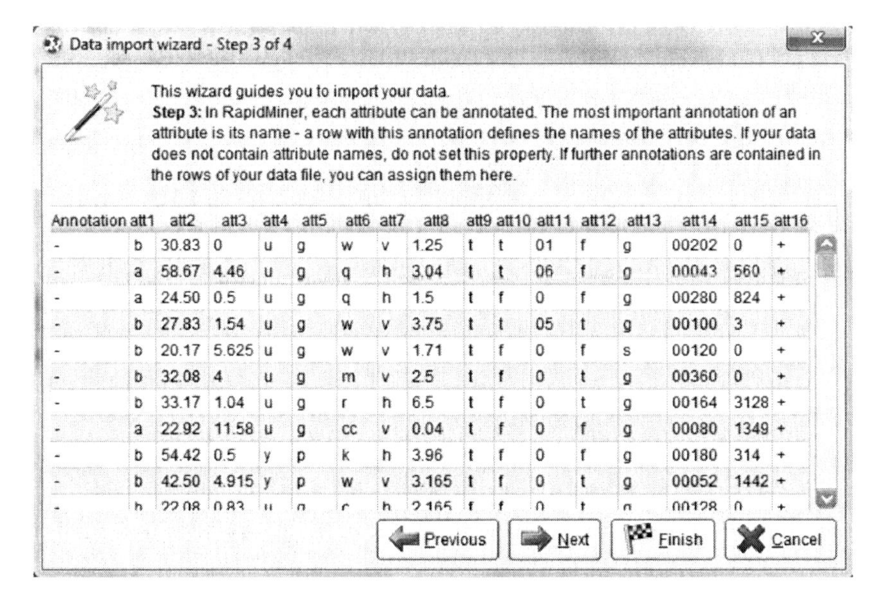

FIGURE 5.4: Step 3 of Import Configuration Wizard of the Read CSV operator.

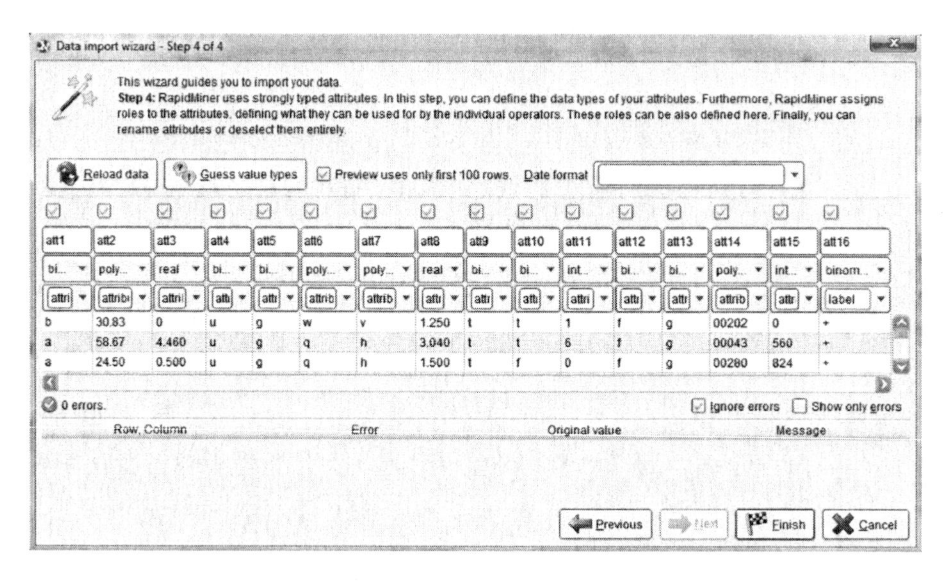

FIGURE 5.5: Step 4 of Import Configuration Wizard of the Read CSV operator.

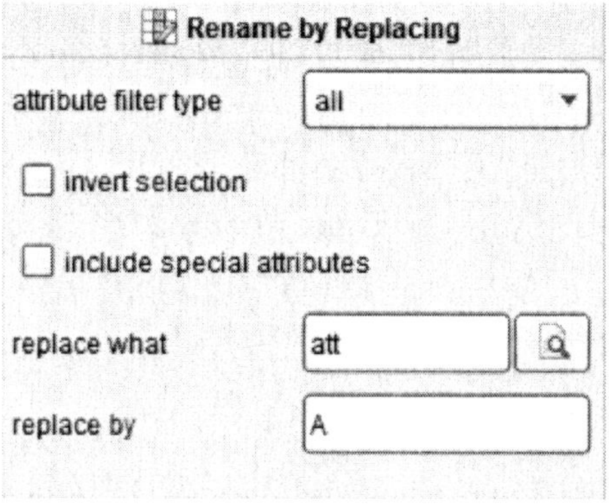

Rename by Replacing

attribute filter type: all

invert selection

include special attributes

replace what: att

replace by: A

FIGURE 5.6: Parameters of the Rename by Replacing operator.

the explanation of the Credit Approval dataset. The naming format should be of the form A1, A2, etc. To change the naming format of all the attributes the Rename by Replacing operator is used.

Figure 5.6 shows the parameters of the Rename by Replacing operator. The *attribute filter type* parameter is set to 'all' because all the attributes are to be renamed. The common pattern to replace is 'att' and it is to be replaced by 'A'; this explains the values of *replace what* and *replace by* parameters. A breakpoint is inserted after this operator so that the effect of this operator can be seen in the Results Workspace.

Filtering missing values The Credit Approval dataset has missing values in some attributes (A1, A2, and A14). This can be seen in the outputs of the previous two operators. The Filter Examples operator is used in this process to simply remove the examples with missing values. The *condition class* parameter is set to 'no missing values'. A breakpoint is inserted after this operator so that the results of this operator can be seen in the Results Workspace.

Discretize By Binning The Discretize By Binning operator is introduced in this process. The Discretize By Binning operator is used for converting the type of all the numeric attributes (A2, A3, A8, A11, A14, A15) to nominal. The parameters of this operator are shown in the Figure 5.7. A breakpoint is inserted after this operator so that the output of this operator can be seen in the Results Workspace. After application of this operator, all numeric attributes are changed to nominal type. All these attributes have three possible values, i.e., range1, range2, and range3. The *number of bins* parameter is set to 3, therefore, the ranges of all the attributes are divided into three equal partitions. All the values in a partition are named according to that partition. For example, the attribute A11 had range of values from 0 to 67. Three equal ranges of A11 are:

- range1: from 0 to 22.3
- range2: from 22.3 to 44.7
- range3: from 44.7 to 67

Most values fall in range1. Only the examples on Row No 45 and 47 fall in range2 (values 40 and 23, respectively). Only the example on Row No 117 falls in range3 (value 67).

5.4.3 Model Training, Testing, and Performance Evaluation

The X-Validation operator is used in this process for facilitating in model training, testing, and performance evaluation. All parameters are used with default values (default value of the number of validations parameter is 10). Here is an explanation of what happens inside the X-Validation operator.

1. The X-Validation operator provides the training dataset (composed of 9 out of 10 subsets) through the *training* port of the Training sub-process. This training dataset is used as input for the Naïve Bayes operator. Thus, the Naïve Bayes classification model is trained on this training dataset.
2. The Naïve Bayes operator provides the Naïve Bayes classification model as its output. This model is connected to the *model* port of the Training sub process.
3. The Naïve Bayes classification model that was provided at the *model* port of the Training sub-process is provided by the X-Validation operator at the *model* port of the Testing sub-process. This model is provided as input to the Apply model operator (at *model* port of the Apply Model operator).

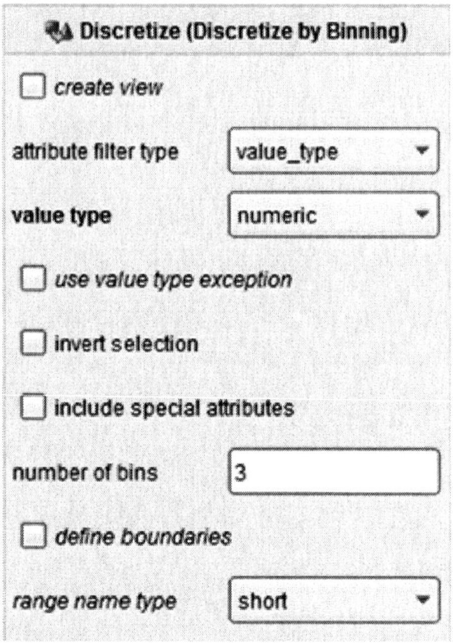

Discretize (Discretize by Binning)
create view
attribute filter type
value_type
value type
numeric
use value type exception
invert selection
include special attributes
number of bins
3
define boundaries
range name type
short

FIGURE 5.7: The parameters of the Discretize By Binning operator.

Training
Naive Bayes
tra
mod
exa
1
2
mod
mod
thr
tes
thr
3
4
Testing
Apply Model
mod
lab
unl
mod
5
Performance
lab
per
per
exa
6

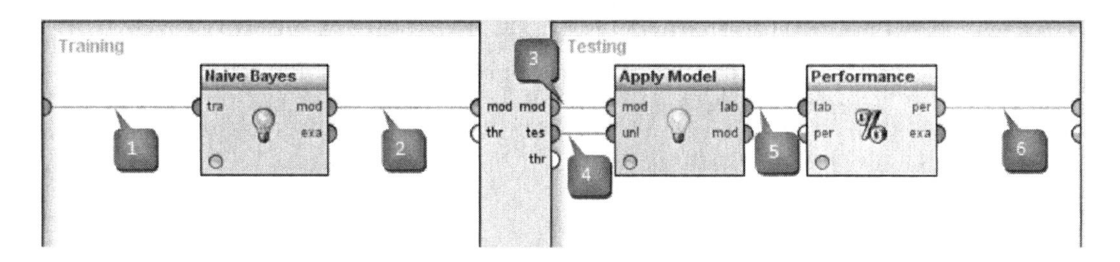

FIGURE 5.8: Training and Testing subprocesses of the X-Validation operator.

4. The X-Validation operator provides the testing dataset through the *test set* port of the Testing sub-process. The testing dataset is composed of the last remaining subset because 9 of the 10 subsets were part of the training dataset. This testing dataset is provided as input to the Apply Model operator (at *unlabeled data* port).

5. The Apply model operator applies the Naïve Bayes classification model (that was provided at its *model* input port) on the testing dataset (that was provided at its *unlabeled data* input port). The resultant labeled dataset is delivered as output by the Apply Model operator. This labeled dataset is provided as input to the Performance (Binominal Classification) operator.

6. The Performance (Binominal Classification) operator is used in this process instead of other performance operators because this operator is used specifically for evaluating the performance of binominal classification tasks. It has many binominal classification-related performance measures that are not present in other performance operators. The Performance (Binominal Classification) operator evaluates the statistical performance of the model through the given labeled dataset and generates a performance vector which holds information about various performance criteria.

7. All the steps from step-1 to step-6 are repeated 10 times; every time a different subset (of total 10 subsets) is used as the testing subset and the remaining 9 subsets as the training subset. Thus, in every iteration, a new model is trained and it is tested on a new testing subset, following which its performance is measured. Finally, at the end of all iterations, the X-Validation operator delivers the model with the best performance, together with its performance vector.

The accuracy of the model turns out to be 85.17%. Figure 5.9 shows how the accuracy of the model varies with discretization and filtering.

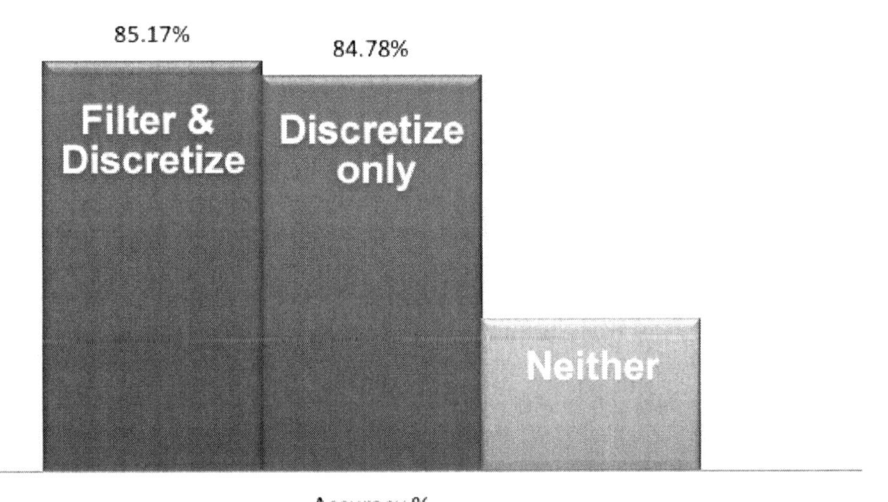

FIGURE 5.9: Effect of discretization and filtering on the accuracy of the Naïve Bayes model.

The figure shows that in this use case the most accurate model is obtained when both filtering and discretization are used.

Chapter 6

Naïve Bayes Classificaton II

M. Fareed Akhtar
Fastonish, Australia

The use case of this chapter applies the Naïve Bayes operator on the Nursery dataset (overview of the Naïve Bayes algorithm has been discussed in the previous chapter). The working of the Naïve Bayes operator is also discussed in detail. The purpose of this use case is to rank applications for nursery schools using the Naïve Bayes algorithm. The operators explained in this chapter are: Read Excel and Select Attributes operators.

6.1 Dataset

6.1.1 Nursery Dataset

This dataset has been taken from the UCI repositories. This dataset can be accessed through this link: `http://archive.ics.uci.edu/ml/datasets/Nursery`.

6.1.2 Basic Information

Nursery Database was derived from a hierarchical decision model originally developed to rank applications for nursery schools. It was used during several years in the 1980s when there was excessive enrollment in these schools in Ljubljana, Slovenia, and the rejected applications frequently needed an objective explanation.

6.1.3 Examples

This dataset has 12,960 examples and there are no missing values in this dataset.

6.1.4 Attributes

This dataset has 9 attributes (including the label attribute). The dataset comes with some basic information, but the type and role of attributes is set by the user of the dataset. Even the attribute names are specified by the user. Here is an explanation of the attributes of this dataset:

1. **Parents:** This attribute has information about the parents of the child. It has three possible values: usual, pretentious, and great_pret. As this attribute can have three possible values, the type of this attribute should be set to polynominal in RapidMiner.

2. **Has_nur:** This attribute has information about the nursery of the child. It has five possible values: proper, less_proper, improper, critical, and very_crit. As this attribute can have five possible values, the type of this attribute should be set to polynominal in RapidMiner.

3. **Form:** This attribute has information about the form filled out by the applicants. It has four possible values: complete, completed, incomplete, and foster. As this attribute can have four possible values, the type of this attribute should be set to polynominal in RapidMiner.

4. **Children:** This attribute has information about the number of children of the applicant. It has four possible values: {1, 2, 3, more}. As this attribute can have four possible values, the type of this attribute should be set to polynominal in RapidMiner.

5. **Housing**: This attribute has information about the housing standard of the applicant. It has three possible values: convenient, less_conv, and critical. As this attribute can have three possible values, the type of this attribute should be set to polynominal in RapidMiner.

6. **Finance:** This attribute has information about the financial standing of the applicant. It has two possible values: convenient, inconv. As this attribute can have only two possible values, the type of this attribute should be set to binominal in RapidMiner.

7. **Social:** This attribute has information about the social structure of the family. It has three possible values: nonprob, slightly_prob, and problematic. As this attribute can have three possible values, the type of this attribute should be set to polynominal in RapidMiner.

8. **Health:** This attribute has information about the health picture of the family. It has three possible values: recommended, priority, and not_recom. As this attribute can have three possible values, the type of this attribute should be set to polynominal in RapidMiner.

9. **Rank** (label attribute): This attribute specifies the rank of the application. It has five possible values: not_recom, recommend, very_recom, priority, and spec_prior. As this attribute has five possible values its type should be set to polynominal in RapidMiner. The role of this attribute should be set to label because this is the target attribute or the attribute whose value will be predicted by the classification algorithms. The role of all other attributes should be set to regular.

More information about this dataset can be obtained from UCI repositories.

6.2 Operators in this Use Case

6.2.1 Read Excel Operator

For applying any algorithm on a dataset the first step is to import the dataset. Importing the dataset means loading the dataset into RapidMiner. RapidMiner provides numerous operators for loading datasets. These operators can be found at the 'Import/Data' section in the Operators window. Data is available in different formats. Therefore, RapidMiner provides different operators for importing data in these different formats.

Mostly data is stored in CSV files, Excel files, or in databases. To access such datasets, RapidMiner provides operators like Read CSV, Read Excel, and Read Database.

The Read Excel operator imports data from Microsoft Excel spreadsheets. The user has to define which of the spreadsheets in the workbook should be used as data table. The table must have a format so that each row is an example and each column represents an attribute. The data table can be placed anywhere on the sheet.

The easiest and shortest way to import an Excel file is to use *import configuration wizard* from the *Parameters* View. The import configuration wizard, Excel file, first row as names, and dataset meta data information parameters of the Read Excel operator are very similar to corresponding parameters in the Read CSV operator. Other important parameters of the Read Excel operator include the *sheet number* parameter which specifies the number of the sheet which contains the required dataset. The *imported cell range* parameter specifies the range of cells to be imported. It is specified in "xm:yn" format where "x" is the column of the first cell of range, "m" is the row of the first cell of range, "y" is the column of the last cell of range, "n" is the row of the last cell of range. For example, "A1:E10" will select all cells of the first five columns from row 1 to 10.

6.2.2 Select Attributes Operator

This is a very powerful operator for selecting required attributes of the given dataset because it gives many different options for selecting attributes. Attributes can be selected by simply mentioning their names. Attributes of a particular type or block type can also be selected. Attributes can be selected on the basis of missing values or even attribute value filters. The most powerful option for selecting attributes is through regular expression. Moreover, exceptions can be provided for any method of attribute selection.

6.3 Use Case

The main goal of this process is to have a deeper look at the working of the Naïve Bayes algorithm. The Read Excel operator is used for importing the Nursery dataset. The Split Data operator is used for splitting the data into training and testing datasets. Finally, the Performance (Classification) operator is used for evaluating the performance of the model.

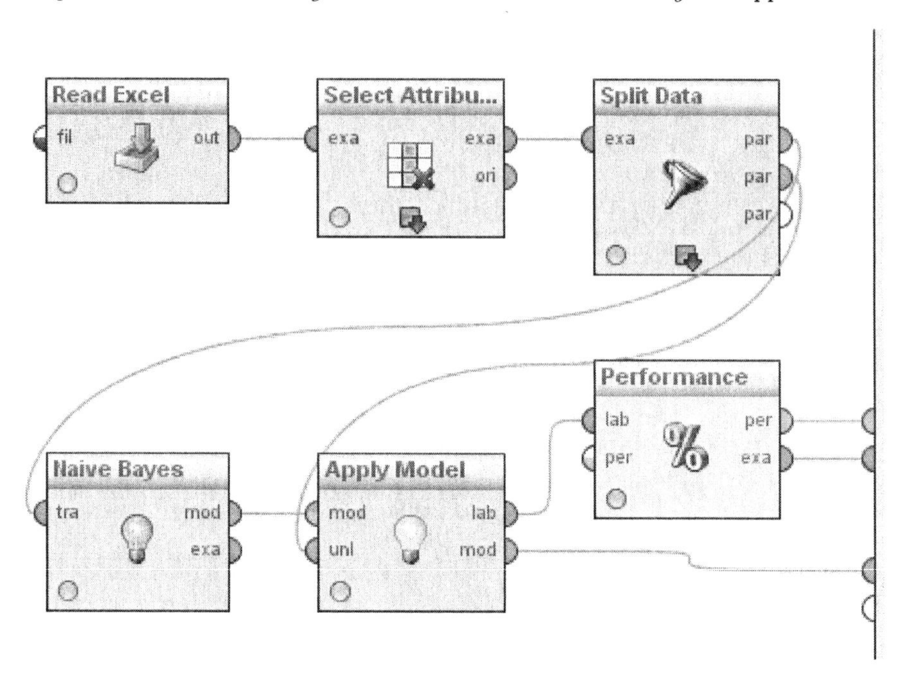

FIGURE 6.1: Workflow of the process.

6.3.1 Data Import

The Read Excel operator is used for importing data in this process. The data at the following url is stored in an Excel file. `http://archive.ics.uci.edu/ml/machine-learning-databases/nursery/nursery.data`.

This Excel file is then used as the data source in RapidMiner. The Read Excel operator is used for loading data from an Excel file. The *Import Configuration Wizard* is used for loading the data in this process. The steps of this configuration wizard are explained below.

Step 1: Select Excel file.

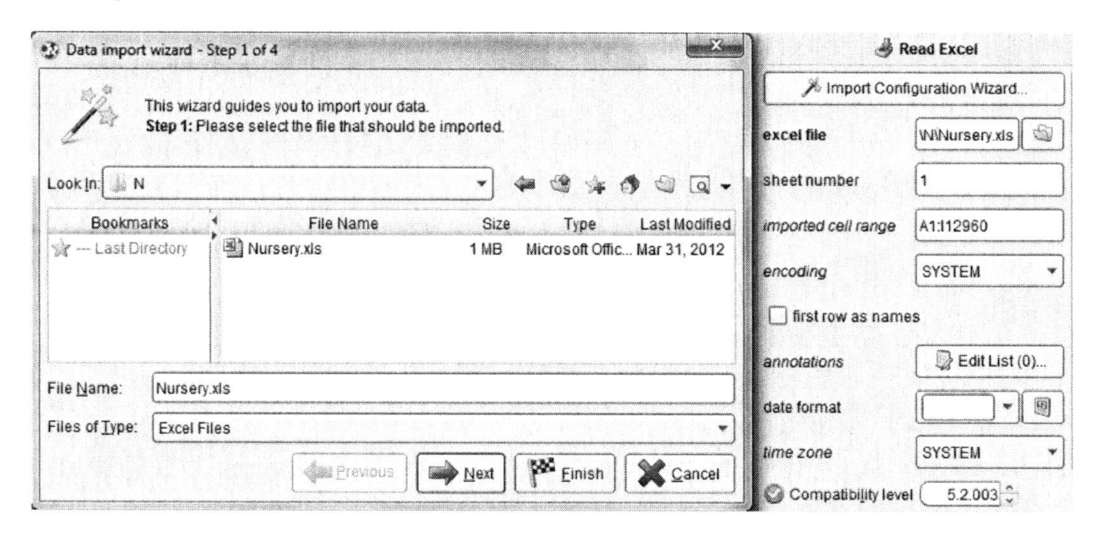

FIGURE 6.2: Step 1 of import Configuration Wizard of the Read Excel operator.

The first step is choosing the Excel file that contains the required data.

Step 2: Select the required sheet and cell range.

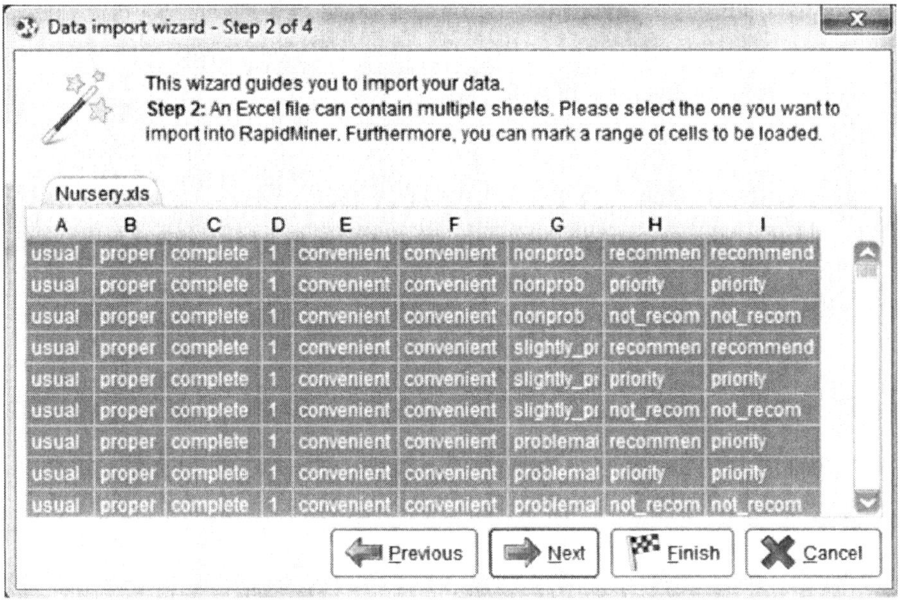

FIGURE 6.3: Step 2 of import Configuration Wizard of the Read Excel operator.

RapidMiner retrieves the Excel file and displays it in this step. The desired Excel sheet and cells are selected. By default, all cells are selected. In this process, there is no need to change anything at this step because there is only one sheet and all the required cells are already selected.

Step 3: Annotations.

This step enables the annotations to be specified. By default, the annotation of the first row is set to 'name'. If the first row of data does not contain the names of attributes, its annotation should be unchecked (as shown in Figure 6.4).

Step 4: Set the name, type, and role of attributes.

This is the most crucial step of the import wizard. The name, type, and role of attributes are specified in this step. Figure 6.5 shows the name, type, and role of attributes that are specified for the Nursery dataset. A breakpoint is inserted after this operator so that the output of the Read Excel operator can be seen in the Results Workspace.

6.3.2 Pre-processing

The Select attributes operator is used for selecting only the Form and Parents attributes. Only two attributes are selected to simplify this process (in order to make the explanation of working of Naïve Bayes operator simple).

6.3.3 Model Training, Testing, and Performance Evaluation

The Split Data operator is used in this process for splitting the dataset into two partitions. One partition is used for training the model and the other partition is used for testing the model. The *partitions* parameter of the Split Data operator performs almost the same task that the *split ratio* parameter does in the Split Validation operator. The *partitions*

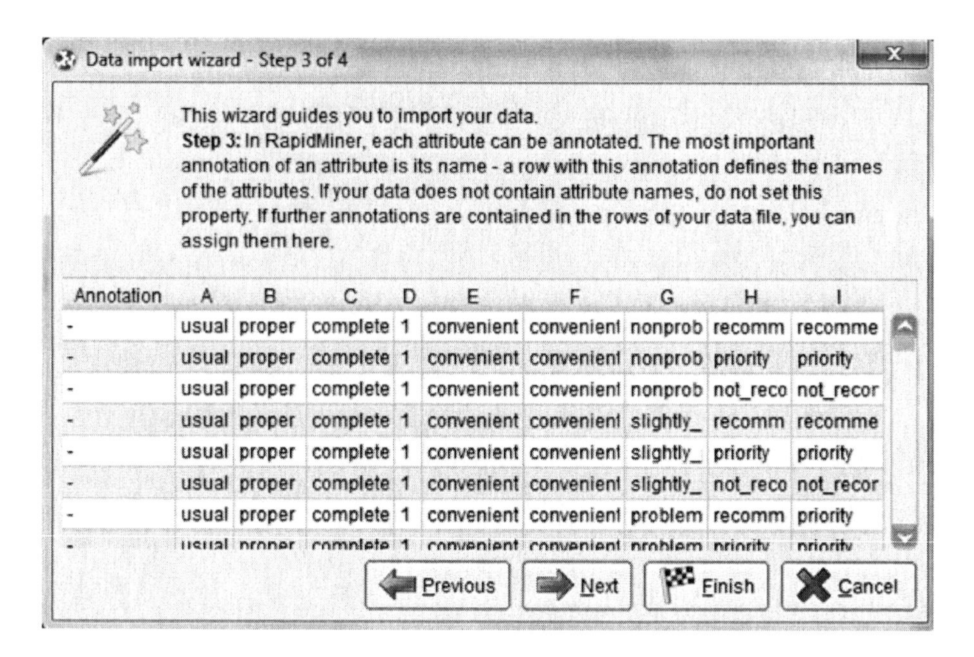

FIGURE 6.4: Step 3 of Import Configuration Wizard of the Read Excel operator.

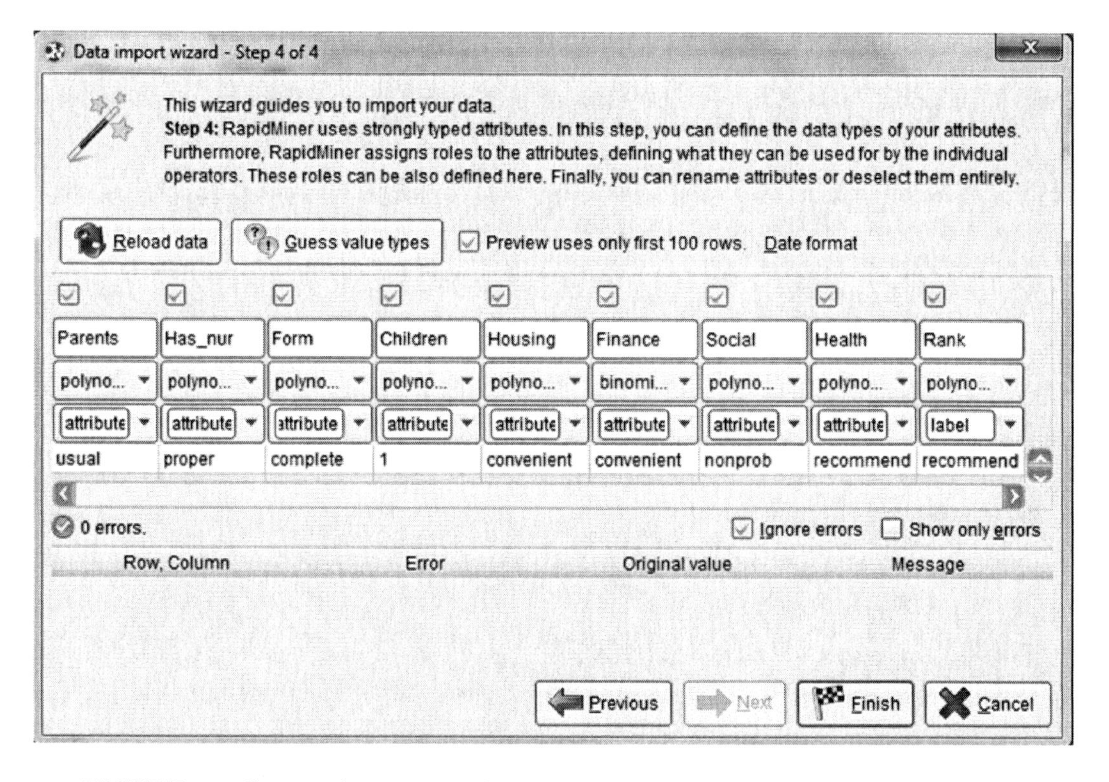

FIGURE 6.5: Step 4 of import configuration Wizard of the Read Excel operator.

parameter defines the number of partitions and ratio of each partition. The *sampling type* parameter of the Split Data operator behaves exactly the same as the *sampling type* parameter in the Split Validation operator. The parameters of the Split Data operator are shown in Figure 6.6.

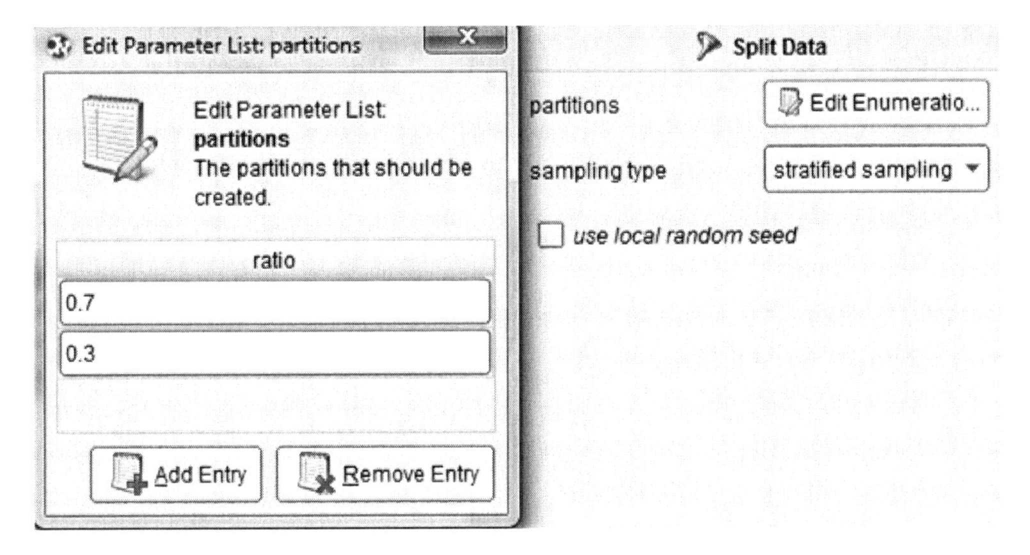

FIGURE 6.6: The parameters of the Split Data operator.

Training the Model The Split Data operator provides the first partition (ratio = 0.7) through its first output port. This partition is used for training a Naïve Bayes classification model through the Naïve Bayes operator. Initially, all parameters of the Naïve Bayes operator are used with default values. The Naïve Bayes operator provides the Naïve Bayes classification model as its output. This model is given as input to the Apply Model operator.

Testing the Model The Split Data operator provides the second partition (ratio = 0.3) through its second output port. This partition is used for testing the Naïve Bayes classification model through the Apply Model operator. The Apply Model operator applies the Naïve Bayes classification model (that was provided at its *model* input port) on the testing dataset (that was provided at its *unlabeled data* port). The resultant labeled dataset is delivered as output by the Apply Model operator. This labeled dataset is provided as input to the Performance (Classification) operator.

Performance Evaluation The Performance (Classification) operator should be used for performance evaluation of classification models because it provides many different performance measures in its performance vector. The performance vector generated by the Performance (Classification) operator and the labeled data set are connected to the *result* port of the process.

6.3.4 A Deeper Look into the Naïve Bayes Algorithm

The results of this process show three outputs:

1. Performance Vector
2. Labeled dataset (obtained by application of model on testing dataset)
3. Naïve Bayes classification model

Table 6.1 shows the posterior probabilities of the label values. Posterior probability is calculated by dividing the number of examples with that label value by the total number of examples.

TABLE 6.1: Posterior probabilities.

Rank (label)	No. of Examples	Posterior Probability
recommend	1	1/9072
priority	2986	2986/9072
not_recom	3024	3024/9072
very_recom	230	230/9072
spec_prior	2831	2831/9072

Figure 6.7 shows the first row of the labeled dataset. There is a confidence attribute for each possible value of the label. The label value with the highest confidence is assigned as predicted value for the example. In this example, the highest confidence (i.e., 0.450) is for label value = priority. Therefore, this example is predicted as priority.

The following steps explain how the confidences for each label value are calculated. They are calculated using the posterior probabilities and the distribution table (shown in Figure 6.8). The following calculations are done for finding the confidence of a label value:

1. Multiply the probabilities of all the attributes for that label value.
2. Multiply this product with the posterior probability of that label value.
3. Divide the resultant value by sum of all the confidences.

Here is an explanation of how the confidences for the first example were calculated. The rows where "Parents = usual" and "Form = complete" are highlighted in Figure 6.8 because these are the only rows that will be used for calculations for confidences in this example (because in this example, Parents and Form attributes have values "usual" and "complete" respectively). Firstly, confidences of all the label values are calculated without normalization. These values will not match the values in Figure 6.7. After calculating all the confidences, these values will be normalized. These normalized values will be the same as the values in Figure 6.7.

Rank	confidence(recommend)	confidence(priority)	confidence(not_recom)	confidence(very_recom)	confidence(spec_prior)	prediction(Rank)	Parents	Form
priority	0.001	0.450	0.339	0.062	0.148	priority	usual	complete

FIGURE 6.7: First row of labeled ExampleSet.

confidence (recommend) confidence (recommend) = P(Parents = usual |Rank = recommend) * P(Form = complete |Rank = recommend) * Posterior(recommend)

$$= 1.000 \quad * \quad 1.000 \quad * \quad 1/9072 = \quad 0.0001$$

confidence (priority) confidence (priority) = P(Parents = usual |Rank = priority) * P(Form = complete |Rank = priority) * Posterior(priority)

$$= 0.449 \quad * \quad 0.262 \quad * \quad 2986/9072 = \quad 0.0387$$

confidence (not_recom) confidence (not_recom) = P(Parents = usual |Rank = not_recom) * P(Form = complete |Rank = not_recom) * Posterior(not_recom)

$$= 0.344 \quad * \quad 0.254 \quad * \quad 3024/9072 = \quad 0.0291$$

Text View | Plot View | Distribution Table | Annotations

Attribute	Parameter	recommend	priority	not_recom	very_recom	spec_prior
Parents	value=usual	1.000	0.449	0.344	0.587	0.191
Parents	value=pretentious	0.000	0.348	0.329	0.413	0.314
Parents	value=great_pret	0.000	0.203	0.326	0.000	0.495
Parents	value=unknown	0.000	0.000	0.000	0.000	0.000
Form	value=complete	1.000	0.262	0.254	0.357	0.214
Form	value=completed	0.000	0.256	0.244	0.300	0.245
Form	value=incomplete	0.000	0.247	0.252	0.230	0.267
Form	value=foster	0.000	0.235	0.250	0.113	0.274
Form	value=unknown	0.000	0.000	0.000	0.000	0.000

FIGURE 6.8: Distribution table.

confidence (very_recom) confidence (very_recom) = P(Parents = usual |Rank = very_recom) * P(Form = complete |Rank = very_recom) * Posterior(very_recom)

$$= 0.587 \quad * \quad 0.357 \quad * \quad 230/9072 = \quad 0.0053$$

confidence (spec_prior) confidence (spec_prior) = P(Parents = usual |Rank = spec_prior) * P(Form = complete |Rank = spec_prior) * Posterior(spec_prior)

$$= 0.191 \quad * \quad 0.214 \quad * \quad 2831/9072 = \quad 0.0127$$

Normalization All the confidences are divided by the sum of all the confidence values to get normalized confidence values. The sum of confidence values is 0.0859. The normalized confidence values are shown in the table below.

TABLE 6.2: Normalized confidences.

Rank (label)	**Confidence**
Recommend	0.0001/0.0859 = 0.001
Priority	0.0387/0.0859 = 0.450
not_recom	0.0291/0.0859 = 0.339
very_recom	0.0053/0.0859 = 0.062
spec_prior	0.0127/0.0859 = 0.148

As the confidence of the "priority" label value is the highest, the rank of this example is predicted as "priority".

Part III

Marketing, Cross-Selling, and Recommender System Use Cases

Chapter 7

Who Wants My Product? Affinity-Based Marketing

Euler Timm

viadee IT-Consultancy, Münster/Köln (Cologne), Germany

Acronyms

CRISP-DM - Cross-Industry Standard Process for Data Mining
DWH - Data Warehouse
BI - Business Intelligence

7.1 Introduction

A bank introduces a new financial product: a type of current (checking) account with certain fees and interest rates that differ from those in other types of current accounts offered by the same bank. Sometime after the product is released to the market, a number of customers have opened accounts of the new type, but many others have not yet done so. The bank's marketing department wants to push sales of the new account by sending direct mail to customers who have not yet opted for it. However, in order not to waste efforts on customers who are unlikely to buy, they would like to address only those 20 percent of customers with the highest affinity for the new product.

This chapter will explain how to address the business task sketched above using data mining. We will follow a simple methodology: the Cross-Industry Standard Process for data mining (CRISP-DM) [1], whose six stages will be applied to our task in the following subsections in their natural order (although you would typically switch back and forth between the stages when developing your own application). We will walk step by step through the fictitious sample data, which is based on real data structures in a standard data warehouse design, and the RapidMiner processes provided with this chapter to explain our solution.

Thus, how can we determine whether a customer has a high affinity for our new product? We can only use an indirect way of reasoning. We assume those customers who have already bought the product (the buyers) to be representative of those who have a high affinity toward the product. Therefore, we search for customers who have not yet bought it (the non-buyers) but who are similar to the buyers in other respects. Our hope is that the more similar they are, the higher their affinity.

Our main challenge, therefore, is to identify customer properties that can help us to find similarities and that are available in the bank's data. Section 7.4 discusses what properties might be useful, and how to build data that reflect them. Assuming that we have good data, we can use a standard data mining method, namely binary classification, to try to differentiate between buyers and non-buyers. Trying to keep buyers and non-buyers apart in order to find their similarities may sound paradoxical; however, "difference" and "similarity" belong to the same scale. Therefore, it is crucial that our data mining algorithm be able to provide that scale. Fortunately, most algorithms can do that by delivering a *ranking* of customers, with higher-ranked customers being predicted to be buyers with higher confidence or probability than lower-ranked ones.

Thus, in what follows, a number of mining models will be developed that each deliver a ranking of non-buyers in which the top-ranked customers are those for which the model is most confident that they ought, in fact, to be buyers (if only they knew it!). We will also see how to decide which model is most useful. We can then serve the marketing department by delivering the top 20 percent of non-buyers from our final ranking. Finally, in the last part of this chapter, we will discuss how to apply the same methodology to other business tasks.

7.2 Business Understanding

The purpose of the CRISP-DM Business Understanding phase is to thoroughly understand the task at hand from the perspective of the business users: what terminology do they use, what do they want to achieve, and how can we assess after the project whether its goals have been met?

Thus we conduct a number of interviews with our bank's business experts and learn the following:

- The bank now has four types of current (checking) accounts on offer, differentiated by their internal names CH01 through CH04. CH04 is the new type that is to be pushed by our marketing action. Basically, each type of account comes with certain fixed monthly fees and interest rates for credit and debit amounts, but some customers can have deviating rates or can be freed from the monthly fee due to VIP status or other particularities.

- A customer can have any number of current accounts, including zero, and any mix of types.
- Accounts have an opening and an ending date. When an account ends, its balance is zero and no further money transaction can occur. An open account whose ending date has not yet passed is called *active* and a customer with at least one active account is also called active.
- Each money transaction for every account is automatically classified by an internal text analysis system, based on an optional free-text form field that can be filled by the initiator of the transaction. There are many categories such as "cash withdrawal", "salary", "insurance premium", and so on, including an "unknown" category.
- For most customers, personal information such as date of birth or family status is known, but may not always be up to date.
- Customers can buy many other products from the bank, including savings accounts, credit cards, loans, or insurance. While data from these business branches is a valuable source of information, in our simplified example application, we do not include such data.

From this we quickly develop the idea of using information about the customers' *behavior*, derived from the money transaction data, to characterize our customers. We delve into this in Section 7.4 after taking a look at the available data sources in Section 7.3. For now we decide to exclude inactive customers from the analysis because their behavior data might be out of date. Thus, all active customers who have ever had a CH04 account are buyers, and all other active customers are non-buyers.

As stated above, eventually we want to send mail to about 20% of the non-buyers. In order to evaluate whether the mailing was successful, we can look at the sales rates of the CH04 account among recipients of the mail and other current non-buyers some time after the mailing. While this is an important part of our project, we will not discuss it further in this chapter as it involves no data mining-related techniques.

7.3 Data Understanding

Continuing our interviews with the bank's staff, we now turn to the IT department to learn about the available data. This phase of the CRISP-DM process is central to planning the project details, as we should never mine data whose structure and contents we have not fully understood. Otherwise, we run the risk of misinterpreting the results.

Not surprisingly, the bank has implemented a central data warehouse (DWH), that is, a data source that is separate from the operational information systems and provides an integrated view of their data, built specifically for analysis purposes. While data warehouses are not typically aimed at allowing data mining directly, data mining projects can benefit greatly from a well-designed data warehouse, because a lot of issues concerning data quality and integration need to be solved for both. From the perspective of a data miner, data warehouses can be seen as an intermediate step on the way from heterogeneous operational data to a single, integrated analysis table as required for data mining.

It is best practice [2] to use a dimensional design for data warehouses, in which a number of central fact tables collect measurements, such as *number of articles sold* or *temperature*, that are given context by dimension tables, such as *point of sale* or *calendar date*. Fact tables are large because many measurements need to be collected, so they do not store any context information other than the required reference to the dimension tables. This is the well-known star schema design.

We consider a particular star schema, depicted in Figure 7.1, for our example application; the data are provided as RapidMiner example sets in the material for this book. The central fact table holds money transactions, linked to three dimensions: calendar date, customer, and account.

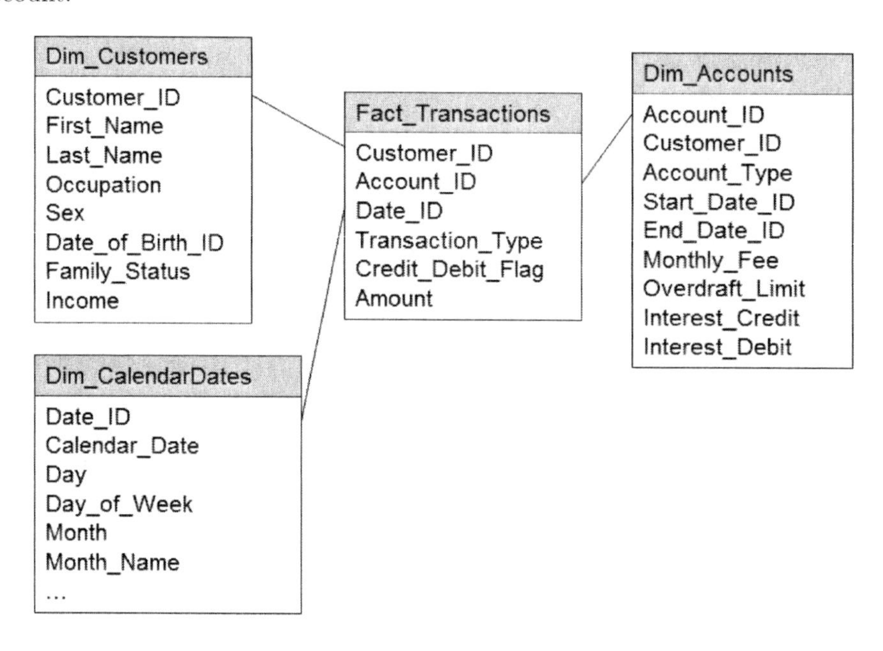

FIGURE 7.1: The source data schema.

Let us briefly examine the contents of these tables:

- Dim_Customers: One row for each customer. Includes personal data: first and last name, date of birth (as a reference to the Dim_CalendarDates table), occupation (one of ten internal codes, 00 through 09, whose interpretation is irrelevant for our application), sex (M or F), family status (single, married, or divorced, with a few missing values), and income. The income attribute is set to zero for young customers, but has many missing values for adults.

- Dim_Accounts: One row for each account, with a reference to the account owner (customer). Each account belongs to one of the types CH01 through CH04 and has a start date and an end date (both are references to the Dim_CalendarDates table). The monthly fee, the overdraft limit, and the interest rates for credit and debit amounts are stored individually with each account because, as we saw in the previous section, for some customers individual fees or rates may apply.

- Dim_CalendarDates: As recommended by [3], our data warehouse includes an explicit calendar date dimension rather than using date fields. In our case, it holds one row for each of the 50,000 days from January 1st, 1900 through November 22nd, 2036, plus one row with ID 100,000 that represents "infinity" (December 31st, 9999). This last row is used for the end date of accounts whose end date is not yet fixed (active accounts). Because the `Date_ID` attribute numbers the days consecutively, we can use it to compute the number of days between two given dates directly without needing to join to this table, as long as we do not use the special date "infinity". The columns of this table can be used in reporting or other applications to qualify a given calendar date; we won't need most of them for our use case.

- Fact_Transactions: For each account, all money transactions that have ever been made are recorded in this table. A reference to the account owner (customer) and the transaction date is stored with each fact (transaction). The type of transaction holds categories like "salary" that are found automatically as explained in Section 7.2 (our sample data includes only a rather small number of categories). The credit/debit flag (values *CR* and *DR*) tells us whether money was taken from the account or booked into it, while the amount attribute gives the amount of money transferred, which is always a positive value.

We should be aware that the data for this example application is simplified, compared to real data, in at least the following ways:

- Our star schema does not accommodate "slowly changing dimensions" [2], a mechanism to keep old values of changed data together with a reference to the period of time during which these old values were current. For predictive data mining projects, such a mechanism can be valuable because it may be necessary to restore data values from past points in time, in order to learn from relationships between those values and subsequent developments. This is not needed for the application discussed here, however.
- Customers belonging to the same household are grouped together and analysed at the household level in many financial applications. In addition, a customer may not represent a natural person but a married couple or a corporate entity. These issues would need to be considered to find an appropriate definition of a mining example in a real application. However, our sample data includes only natural persons.
- Unlike a real schema, our star schema includes only a few, simple attributes without many distinct values.

These simplifications allow us to focus on the data mining issues of our task, yet still include some standard data preparation subtasks and a RapidMiner solution to them in the following section. You should easily be able to extend the methods that we discuss in this chapter to your real data.

7.4 Data Preparation

Virtually all data mining algorithms use a single table with integrated and cleaned data as input. Data Preparation is the CRISP-DM phase in which this table is created. The table should provide as much information about the examples (here, the customers) as is feasible, so that the modelling algorithm can choose what is relevant. Therefore, it is a good idea to bring background knowledge about the business into the data.

In this section we examine two RapidMiner processes; one to assemble the data from our warehouse schema into a single example set (Section 7.4.1), and the second (Section 7.4.2) to perform some tasks that are specific to enabling data mining on the result.

7.4.1 Assembling the Data

Bringing data from the warehouse into a form suitable for data mining is a standard task that can be solved in many ways. Let us take a look at how to do it with RapidMiner, as exemplified by the first sample process from the material for this chapter, Ch7_01_Create-MiningTable. After opening the process in RapidMiner, you can click on Process, Operator Execution Order, Order Execution to display the order of operator execution; we will use these numbers for reference in the following.

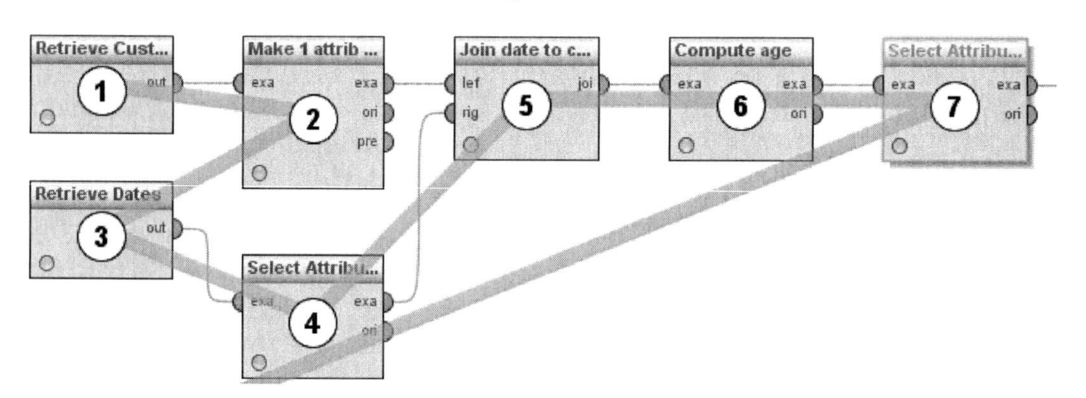

FIGURE 7.2: The substream in Ch7_01_CreateMiningTable for processing customer data.

Each of our four tables is used as one of the four inputs for the process. You could replace the four *Retrieve* operators (numbers 1, 3, 8, and 9) that read our sample data with operators like *Read Database* if you were using a real warehouse as your source. Each input is fed into a substream of our process consisting of a number of operators, and we will discuss the substreams in turn below. Figure 7.2 shows the customer substream. Figure 7.3 shows the last steps in this process, where our substreams are eventually combined into a single resulting example set.

One area of consideration in this phase is data types. RapidMiner associates each attribute with one of several types: real, integer, date, binominal, or nominal. The attributes in our four source datasets have an automatically determined type which may or may not be appropriate; ensuring that our resulting example set is correctly typed makes it easier for RapidMiner to support our data mining issues later on.

Substream: customer data: For example, the first operator we apply to our customer data (number 2 in the execution order, compare Figure 7.2) changes the type of the sex attribute to binominal, because we know (or can see from the example set statistics) that this attribute has exactly two distinct values, M and F. The operator *Nominal to Binominal*, which we use here, normally creates new, binominal attributes for each value of the input attribute, but in this case it recognizes that the input attribute already is binominal, so it just changes the type. (We have set the parameter *attribute filter type* to *single* because we only want to change one attribute.)

Next, we want to include a customer's current age into our data mining basis. In our application we assume that the data is current for December 31st, 2011, so we compute the current age as of this date. Because the customer table refers to the calendar date table in the birth date attribute, rather than using a date attribute, we join the calendar date table to the customer data using the *Join* operator (number 5) with the join type "left".

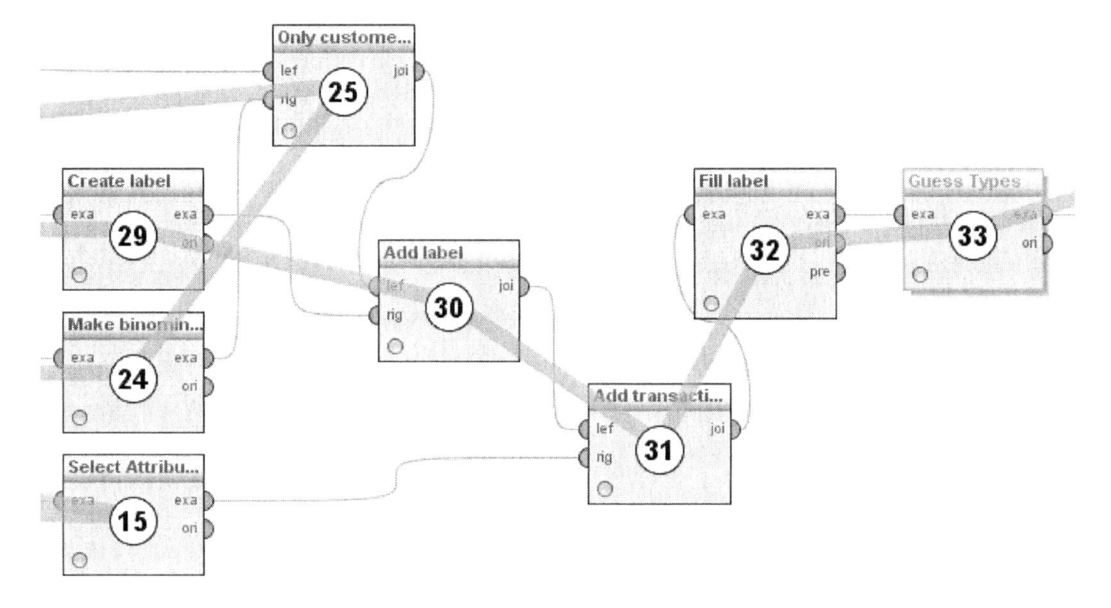

FIGURE 7.3: Last steps in the process Ch7_01_CreateMiningTable.

Our example sets have no attribute with the RapidMiner ID role yet, so we uncheck the option *use id attribute as key* in the *Join* operator; instead we choose the join key attributes explicitly. Note that in order to avoid taking over all the attributes from Dim_Calendar-Dates, we use the *Select Attributes* operator (number 4) before the join. After the join, we can compute a customer's current age (operator 6) by subtracting their year of birth from 2011: `year` is an attribute from the calendar data, so we set up the simple function expression `2011 - year` as a definition for the new attribute `current_age` in the *Generate Attributes* operator. Finally, we remove three attributes from the customer data (7), as they are useless for data mining because they nearly take on individual values for every customer: first and last name, and the birth date ID.

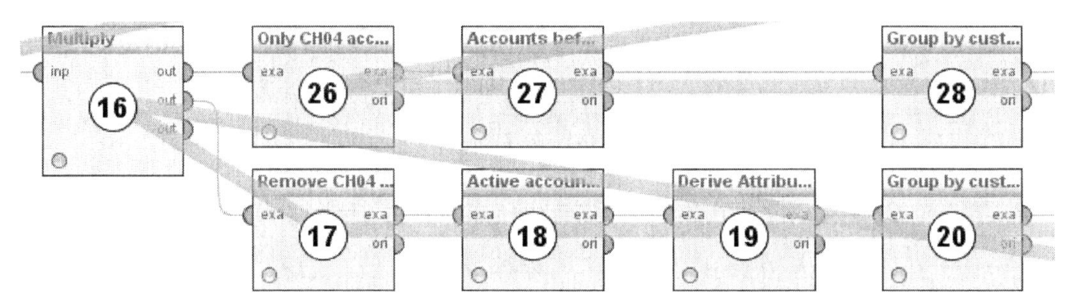

FIGURE 7.4: The first part of the substream in Ch7_01_CreateMiningTable for processing account data.

Substream: account data: Now we turn to the account data, see Figures 7.4 and 7.5. One important goal of this process is to label each customer as buyer or non-buyer; all owners of a CH04 account (as of December 31st, 2011) are buyers. We use the operator *Multiply* (number 16) to copy the input data into two streams, keeping only CH04 accounts (26) in one stream and dealing with all other accounts in the second stream, by applying the *attribute_value_filter*s `account_type = CH04` and `account_type != CH04`, respectively, in two *Filter Examples* operators. Operator 27 leaves only accounts that have opened on or before 12/31/2011 in the stream (for simplicity, we use the ID of this date from our calendar dimension, which is 40907, directly for filtering). Note that we include inactive CH04 accounts as long as they were opened before 12/31/2011, because their owners have taken the decision to open a CH04 account, so they are buyers even if they have closed the account again in the meantime.

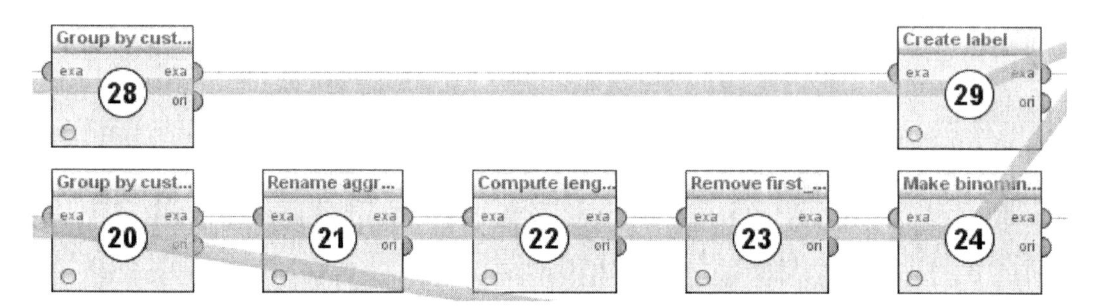

FIGURE 7.5: The second part of the substream in Ch7_01_CreateMiningTable for processing account data.

A customer can have more than one account, even more than one CH04 account, so we group by customer ID with the *Aggregate* operator (28). The resulting example set has only one attribute, the customer ID; we do not need any aggregation attributes since all customers in this substream are buyers. Thus we can now add the label attribute, `is_buyer`, by setting its value to *true* for all customers in this stream (operator 29, another *Generate Attributes* operator).

In the second stream that processes the account data, we handle all other account types (CH01 through CH03, operator 17), and here we keep only active accounts (18). The aim of this substream, as well as of the substream that deals with the transaction data, is to gather information about the customer's behavior (see Section 7.2).

Excursion: Creating the label: There is an important point to be discussed here. Why do we not include (active) CH04 accounts in the data from which we derive our behavior indicators? After all, customers display their behavior in all four types of accounts. But the CH04 accounts make a customer a buyer, so there may be some behavior information in them that betrays the label. Our overall goal is to predict which non-buyers could be most easily persuaded to become buyers, so when we apply our data mining model to the non-buyers, no CH04-based information will be available. Therefore, our model should be trained on data that does not include CH04-based information either (other than the simple buyer/non-buyer distinction itself). Otherwise it might discover biases in the data that result from the fact that the buyers already have bought a CH04 account, and such biases cannot hold for the non-buyers. Indeed, if you are going to implement a similar project using real data, you should be extremely careful to avoid including information in your mining base that is only available because the label is already known for the training data; it can be quite difficult to discover the hidden links between the label and other data sources due to the complexity of your business. The problem is that even evaluating your model on test data does not necessarily tell you that it is too good to be true, if these effects are small.

Indicators for customer behavior: We now return to the question of how to represent customer behavior in our mining base. Candidate attributes are the number of active accounts, the average duration that accounts have been active, whether a customer seems to be willing to pay a fee for their accounts, and, of course, the type and volume of money transactions that have been shifting money from and to their accounts. In a real business, you should be able and willing to think of many more such "derived" attributes that may indicate, however indirectly, what kind of customer someone is. You can let the data mining algorithm decide whether an attribute is useful, but if you do not construct the attribute in the first place, you may lose an opportunity to improve your model. In fact, including more information in your data mining basis is probably the single most promising approach if you do not get good results in your first data mining runs.

ExampleSet (2663 examples, 0 special attributes, 14 regular attributes)

account_id	customer_id	fee_indicator	ch01_indicator	ch02_indicator	ch03_indicator
63	59	0	0	0	1
65	61	0	0	0	1
66	61	1	0	1	0
67	62	1	0	1	0
68	63	1	1	0	0
69	63	1	1	0	0
70	64	0	0	0	1

FIGURE 7.6: Indicator attributes at the account level.

The next step in our example process, therefore, is to compute a few attributes at the account level (operator 19) before aggregating them to the customer level (20). In particular, we compute binominal indicator attributes (values 0 and 1) to determine whether an account is associated with a monthly fee or if it is of type CH01, CH02 or CH03; see Figure 7.6. We use the *Generate Attributes* operator (19) with if-expressions (whose second argument is returned if the first argument, a Boolean condition, is true, while the third argument is returned otherwise) to do so. For example, the function expression

```
if(account_type == "CH01", 1, 0)
```

creates the binominal attribute `ch01_indicator` shown in Figure 7.6.

ExampleSet (2385 examples, 0 special attributes, 8 regular attributes)

customer_id	maximum(fee_indicator)	sum(ch01_indicator)	sum(ch02_indicator)	sum(ch03_indicator)
59	0	0	0	2
61	1	0	1	1
62	1	0	1	0
63	1	2	0	0
64	0	0	0	1

FIGURE 7.7: Indicator attributes at the customer level (after applying the *Aggregate* operator).

Summing up these indicators to the customer level (Figure 7.7) gives us, for instance, the counts of each type of account, while computing the maximum preserves the binominal character of the indicator, like in our fee indicator. The *Aggregate* operator is used for this, with `customer_id` as the only *group by attribute*. (The RapidMiner types of these generated attributes must be explicitly set, for which the operators *Numerical to Binominal* (24) and *Guess Types* (33) are used).

To compute the account duration (19), we exploit the fact that the date IDs in our calendar dimension are consecutively numbered, so that the difference between two IDs yields the number of days between the corresponding dates. The number of years a customer has been with our bank now—the duration of the customer relationship—is computed similarly (22), using the earliest start date of any account as the start of the relationship.

Substream: transaction data: Our last data source is the transaction data (see Figure 7.8). The first operator applied to it (10) implements an application-specific decision to consider only transactions from the last year (2011 in our example) as a basis for behavior-indicating attributes, for fear of older data being no longer representative of a customer's behavior. The second operator (11) sorts the transaction amounts into new attributes according to their transaction type. For example, the function expression

```
if(transaction_type=="CreditCard", amount, 0)
```

is used to compute the attribute `creditcard_last_year`, which holds the value of the `amount` attribute for all rows in which the transaction type is *CreditCard*, and 0 for other rows. This step is simply a preparation for the aggregation (operator 12) to the customer level, because we want to have separate attributes with the number of transactions, or the (total and average) amount of money involved in such transactions, for our relevant transaction types. From a business perspective, these attributes are very important because we hope that they reflect how the customers handle their money. As mentioned above, many more such attributes can and should be invented in a real application.

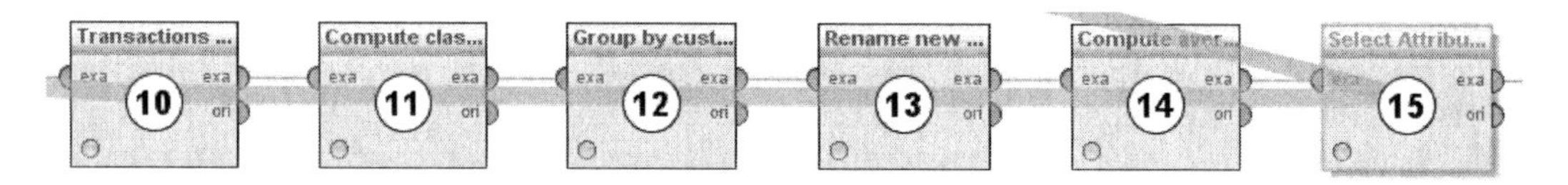

FIGURE 7.8: The substream in Ch7_01_CreateMiningTable for processing transaction data.

Joining the substreams: See also Figure 7.3. Joining the active accounts to the customer example set by an inner join (25) leaves only active customers in our stream. The next join is a left join to add the label to our data (30); it results in missing values in the label attribute `is_buyer` for all non-buyers, which is remedied by the operator *ReplaceMissingValues* (operator 32, called "FillLabel" in our process), with *false* as the replenishment value. By left joining our active customers with the transaction-based data (31), we complete our assembly of a single example set with diverse, information-rich attributes, so we store the result in our RapidMiner repository (34).

7.4.2 Preparing for Data Mining

Having illustrated how RapidMiner can be used for data assembly, we now turn to data mining-specific issues in our next example process, Ch7_02_PrepareDataForMining. It uses the output of the previous process and applies a simple chain of operators to it that make the data ready for modelling.

Most importantly, using the operator *Set Role* we tell RapidMiner that our attribute `is_buyer` contains the label that is to be predicted. We also declare `customer_id` to be an ID attribute, making RapidMiner ignore it during modelling, since no similarities or patterns can be based on an attribute that takes an individual value for each example.

Data cleaning: Modelling operators usually cannot handle missing values, so we must decide on a strategy to deal with the missing values in the `income` and `family_status` attributes. We choose to fill them with a standard value, which is the new value *unknown* for the family status and simply zero for `income` (Operator *Replace Missing Values*). A simple strategy like this is sufficient in many applications, but also provides one of the many options we have for trying something different if our modelling results are unsatisfactory. Other strategies include deleting customers with missing values, filling up with the most frequent or average value, and training extra models that predict how to fill, using the RapidMiner operator *ImputeMissingValues*.

We include the operator *Remove Useless Attributes* in our process; it is a useful cleaning operator that can remove attributes whose values are the same for almost all examples. Its parameters must be adjusted to the data at hand. In our case it would remove any nominal attribute whose most frequent value occurs in less than 1% or more than 99% of examples, because we set the parameters *nominal useless above* and *nominal useless below* to 0.99 and 0.01, respectively, but no such attributes exist in our sample data.

Discretization: The remaining steps discretize some of the numeric attributes. They are optional in the sense that modelling can be done without them, as long as the modelling algorithm can deal with numeric data, but they are useful for bringing some background knowledge into the data, which can improve modelling results.

A classic example that we also use here is the discretization of the age attribute into age groups like kids, adults, or pensioners. The age intervals for these groups are based on real-world knowledge or business rules, rather than statistic criteria as might be employed by an automatic discretization method. Similarly, in our example process we discretize income and the duration of the customer relationship by user-specified intervals. The operator *Discretize by User Specification* accepts an arbitrary number of intervals, each specified by its upper limit, which is included in the interval, implicitly including "minus infinity" as the lower limit for the first class.

In contrast, the operator *Discretize by Binning* that we apply to our monetary attributes automatically creates a number of intervals that are specified by a parameter, by equally dividing the range of values. You can easily follow this behavior, as well as that of other discretization operators, by looking at the meta data view of our attributes before and after the discretization.

In general, however, it would be difficult to give recommendations as to whether discretization should be used at all, or which type of discretization would be most helpful in a particular application. It is worth tinkering with discretization if the modelling results do not meet your expectations, but there is no guarantee that this will lead to improvement.

7.5 Modelling and Evaluation

Our data is prepared for data mining now, so let us prepare ourselves too, by briefly discussing a couple of methodological issues.

7.5.1 Continuous Evaluation and Cross Validation

The CRISP-DM process mentioned in the introduction formally distinguishes between the process stages *modelling* (other terms are learning, training, mining, building a model) and *evaluation* of the model. However, in practice an unevaluated model is useless, so the two stages are intertwined. We will evaluate each model that we build until we have found a model with satisfactory quality. Of course, the datasets used for modelling on the one hand and evaluation on the other must always be distinct.

One standard way of evaluating/validating a model is cross validation, which is directly supported by the RapidMiner operator *X-Validation*. It takes the available data and randomly divides it into a parameter-specified number of distinct subsets. Each subset then serves as a test set for a model that has been trained using the remaining subsets. Thus, as many models are created as there are subsets, and each is evaluated on one subset. Afterward, the evaluation results are averaged. This scheme ensures that we do not overestimate the quality of our model, which might happen if we accidentally choose a single test set with favourable data.

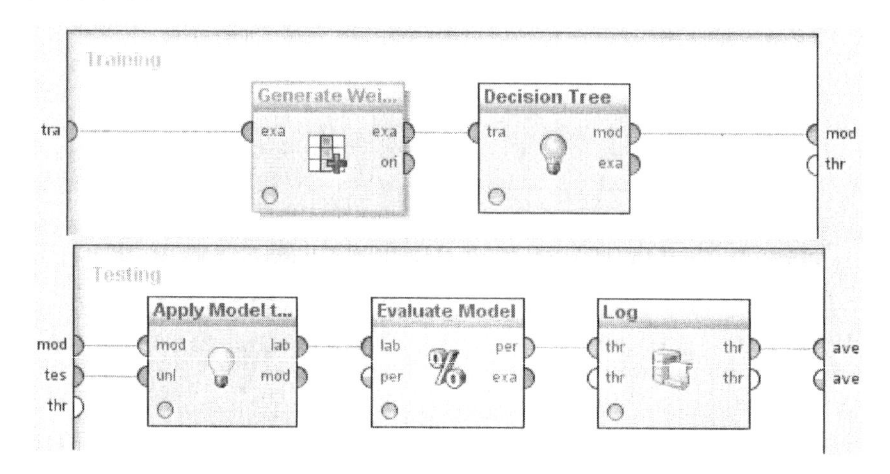

FIGURE 7.9: Inside the RapidMiner *X-Validation* operator.

The *X-Validation* operator is a nesting operator as two RapidMiner sub-processes can be embedded into it, one for training the model and one for testing/evaluating it. Double-clicking on the operator shows the two sub-processes and their expected in- and output (see Figure 7.9): labelled data as input, a model as output of the training sub-process and input of the testing sub-process, and a performance evaluation result as the testing output. Each sub-process is executed by the operator as many times as is specified by the parameter *number of validations*. More details will follow below.

7.5.2 Class Imbalance

Our application displays a rather common characteristic. The distribution of the values of our label attribute `is_buyer` is unbalanced: there are many more non-buyers than buyers; only around 7% of our customers are buyers. This can mislead many data mining algorithms into disregarding the buyers, because a high accuracy (ratio of correct predictions) can be achieved by simply predicting all customers to be non-buyers. There are, at least, two possible remedies for this: weighting and re-balancing.

Weighting introduces a special attribute into the data that RapidMiner recognizes by its role. It is a numeric attribute that associates each example, in this case the customer, with a weight factor, allowing more weight to be given to the minority label, in this instance the buyers. Some, but not all modelling operators can account for this weight factor and change their prediction behavior, accordingly. The RapidMiner operator *Generate Weight (Stratification)* can be used for the automatic creation of a weight attribute, which is demonstrated in the example process Ch7_03_XValidateWeighted (Figure 7.9). It distributes a given total weight over the examples in the input, such that for each label, the sum of the weights of all examples with this label is equal. In our example, both the buyers and the non-buyers receive a total weight of 5,000 because we set the parameter *total weight* to 10,000 (have a look at the new weight attribute in the output of this operator).

Re-balancing creates a special subset of the data in which both labels (buyers and non-buyers) occur with roughly equal frequency. This subset is used only for training/creating the model, but not for testing or deploying it. This approach is shown in the example process Ch7_04_XValidateBalanced. The only change is that the *Sample* operator replaces *Generate Weight (Stratification)* in the training sub-process. We use it to randomly draw 10% from the non-buyers in the training data (i.e., a fraction of 0.1, to be set as ratio for the class *false*), but all buyers (a fraction of 1.0, the ratio for *true*), resulting in a roughly even distribution of buyers and non-buyers.

Good results have been achieved in real applications with both approaches, so it is usually worth trying which works best. On our fictitious, randomly created sample data, re-balancing gives better evaluation results with a decision tree model while the author has frequently found the weighting scheme to be more successful in his real projects.

7.5.3 Simple Model Evaluation

Let us now take a first look at model evaluation: inside our *X-Validation* operators, the models that are returned by the training sub-processes are used to make predictions on the test data by applying the operator *Apply Model* (compare Figure 7.9). This operator creates (among others which are discussed below) an attribute with the predicted label, called `prediction(is_buyer)` in our case (insert a breakpoint after *Apply Model* if you want to see it). In order to see how well the predicted label matches the true label in our test data, we include the operator *Performance (Binominal Classification)*, which is specialized for evaluating applications with a binominal label like ours. It automatically creates a contingency table and several useful metrics, depending on the selected parameters.

Multiclass Classification Performance   Annotations

Table View   Plot View

accuracy: 72.70% +/- 10.62% (mikro: 72.70%)

	true false	true true	class precision
pred. false	1661	111	93.74%
pred. true	540	73	11.91%
class recall	75.47%	39.67%	

FIGURE 7.10: A RapidMiner contingency table.

Compare the contingency tables created by the two example processes Ch7_03_XValidate-Weighted and Ch7_04_XValidateBalanced (the latter is shown in Figure 7.10); each is the result of adding up all contingency tables created by the *X-Validation* operator in each process. Similarly, other performance metrics yielded by *X-Validation* are averages of its individual runs. Inspecting these individual runs can be done with the help of a *Log* operator as shown in our processes (Figure 7.9). The *Log* operator is executed in each testing sub-process; it adds the values listed in the *log* parameter to the end of the specified file in each run. The elements of the contingency table have to be specified individually within the *Log* operator using the pre-defined categories *false_positives*, *false_negatives*, and so on—note that these categories are only available in the *Log* operator if they have been checked in the *Performance* operator.

Two important evaluation metrics for applications like ours are recall and precision of the prediction of buyers (i.e., of the positive class). They can be read directly from the contingency table (column "true true" and line "pred. true"). Typically, we cannot get high values for both of them. Instead, as we find models that increase one of these values, the other one is usually decreased. So which metric is more important? The answer depends on your application. In our marketing application, the goal is to find non-buyers that are similar to the buyers. So the set of customers predicted to be buyers should include a lot of real buyers, and must not be too small, which it is likely to become if we aim for high precision on real data. Therefore, we should put more emphasis on achieving a high recall value; at the same time, the precision value must remain significantly higher than it would be if our prediction was random, which means that it must remain higher than the overall ratio of positive labels (7% in our sample data). With real data, we might not be able to achieve precision values higher than two or three times this ratio, depending on the application, at hand. This would still mean that our predictions are two or three times as good as random predictions, and would not prevent a useful deployment of the model.

7.5.4 Confidence Values, ROC, and Lift Charts

But we can do deeper evaluations of our models by looking beyond recall and precision, to obtain metrics that can directly be translated into business terms like costs and expected returns of our deployment. To see how, insert a breakpoint after the *Apply Model* operator and look at the special attributes `confidence(true)` and `confidence(false)` that it creates. They reflect a numeric confidence or probability (depending on the type of model) with which our model makes its predictions. You see that our prediction in `prediction(is_buyer)` is *true* whenever `confidence(true)` takes a value of 0.5 or higher.

You might have the idea of changing this threshold from 0.5 to a lower or higher value. You can manually do so using the RapidMiner operators *Create Threshold* and *Apply Threshold*, which we will discuss below in Section 7.6. As you increase the threshold, the set of customers predicted to be buyers becomes smaller and vice versa. If our model is any good, a higher threshold should also result in a higher precision.

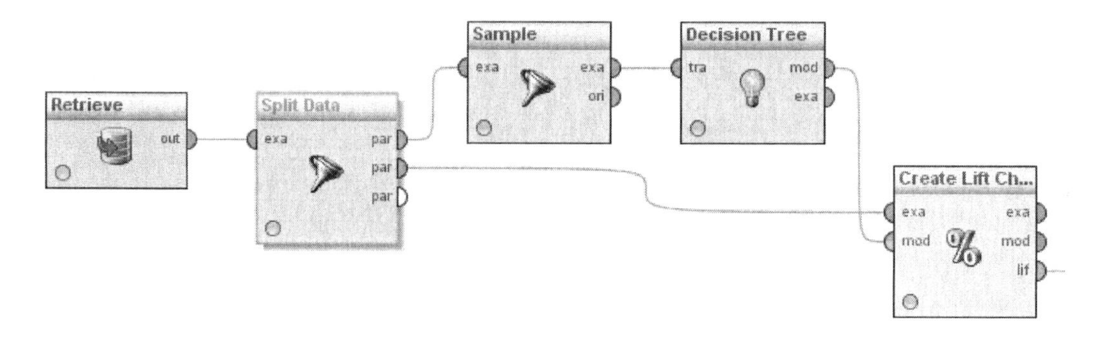

FIGURE 7.11: A RapidMiner process for creating lift charts.

Lift charts: Lift charts quantify this relation by showing the increase in the number of customers that are predicted to be buyers, and also the increase in the real buyers among them, as the threshold is decreased—see Figure 7.12. From the material for this book, check the process Ch7_05_LiftChart, shown in Figure 7.11, to see how to apply the RapidMiner operator *CreateLiftChart*, which was used to create Figure 7.12. In this process, the data is split evenly into a training set and a distinct test set to which *CreateLiftChart* is applied using a model that is built on the training set (the *Sample* operator is used to make the training data balanced, see Section 7.5.2). The chart is drawn by creating a number of bins for the confidence values associated with the specified *target class* (label), so we enter the value *true* for this parameter (we want to examine the lift for buyers). The bars indicate for each bin the number of examples whose confidence value is higher than the threshold for this bin *and* which belong to the target class. The count to the right of the slash gives the full number of examples whose confidence exceeds the value for the bin, while the thin line shows the cumulated counts.

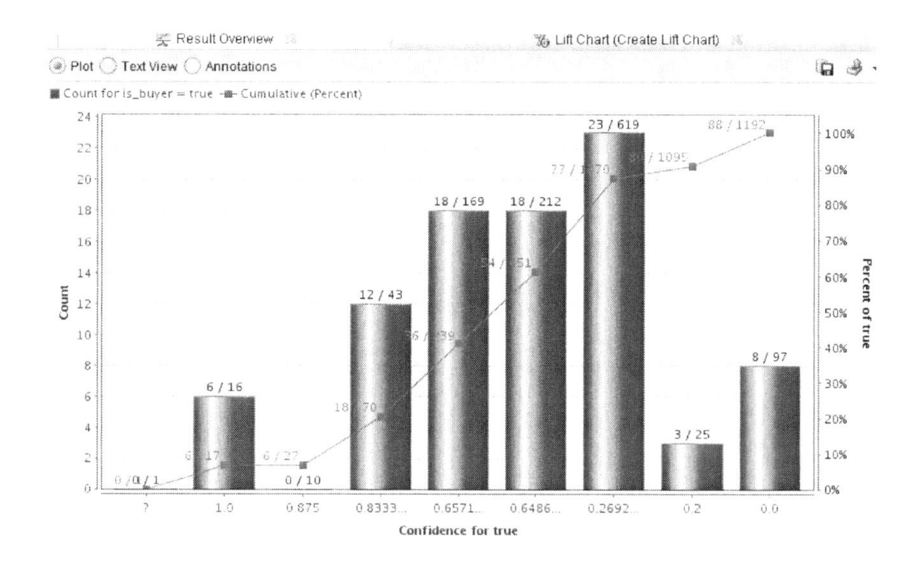

FIGURE 7.12: A lift chart created with RapidMiner.

The lift chart is useful for response-oriented marketing applications because the cumulated counts give us an idea of how many customers must be targeted in a campaign in order to receive a certain number of responses. For example, based on the chart in Figure 7.12, 451 customers would have to be addressed in order to find 54 buyers among them—note that this ratio of buyers, thanks to our model, is higher than the 7% we could easily get without data mining, by randomly choosing 451 customers (though the increase in this ratio is not great on our fictitious data). However, in this scenario, we wish to address only non-buyers, so we use the lift chart mainly to confirm that our model behaves as expected. In contrast, if we were to target both buyers and non-buyers because we wanted to sell a different, brand-new product that nobody has bought yet, and if we hoped that the buyers would tend to buy more readily again, the lift chart would be more useful. This corresponds to the classic scenario where previous responses to campaigns are used to predict future responses [4].

ROC charts: A similar argument holds for ROC charts [5], which are provided by the RapidMiner operator *Performance (Binominal Classification)* if the *AUC* parameter is selected. Figure 7.13 shows a ROC chart obtained on real data. The horizontal axis shows the ratio of non-buyers wrongly predicted as buyers to all non-buyers (*false positive rate*). So, as you go from left to right, more and more non-buyers are predicted to be buyers, which can only happen if the confidence threshold is lowered, which is indeed shown by the blue line (named "(Thresholds)") for which the vertical axis indicates the threshold value. The red line uses a different vertical axis (but the same horizontal one), namely, the recall for the positive class (buyers), also called *true positive rate*. As the threshold is lowered, more and more true buyers are in the set of customers predicted as buyers, so the recall increases; but the earlier it does so, the better the model. A random model would lead to a red line that follows the diagonal from lower left to upper right, as you can indeed (almost)

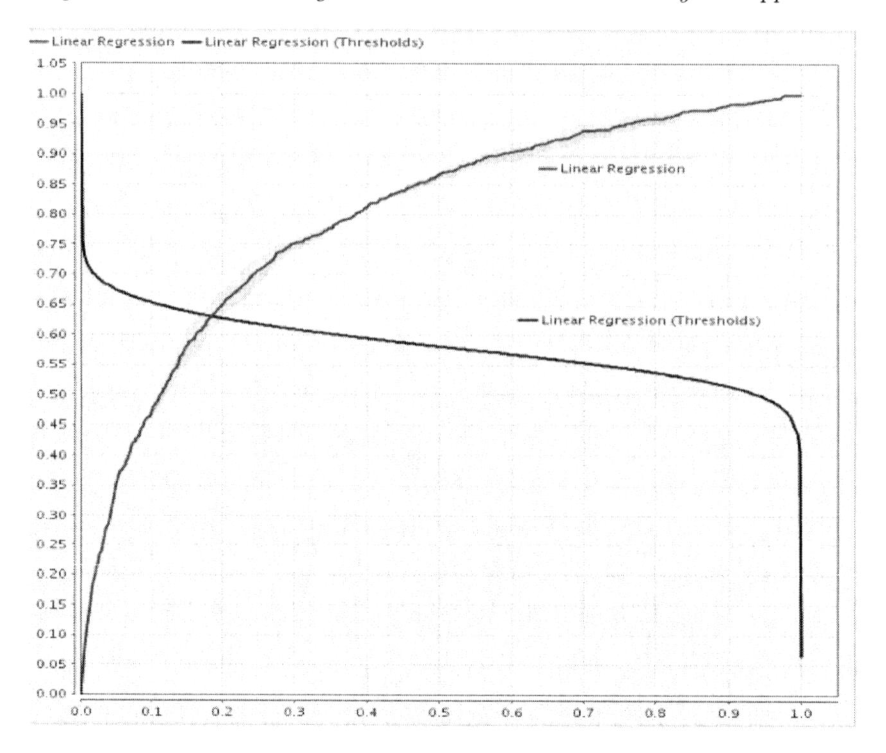

FIGURE 7.13: A ROC chart created with RapidMiner on real data.

see in ROC charts created on our fictitious, randomly created data. So the more the red line deviates from the diagonal, the more powerful our model. This deviation is measured by the area between the line and the diagonal, a metric called area-under-the-curve (AUC). It is a useful metric for comparing the qualities of different models, a task to which we turn next.

7.5.5 Trying Different Models

So far we have left the core of data mining aside in our discussion—the actual creation of a model. You noticed, of course, the *Decision Tree* operator in the last few example processes that we discussed. It was applied to the training datasets, and created a decision tree that we used as input for *Apply Model* in order to get actual predictions and confidence values. This operator may well be used in a real application, but for us it serves mainly as a placeholder for the large number of data mining operators that RapidMiner provides, many of which could have been used instead. For more information about decision trees and other data mining algorithms see, for example, [6].

Type compatibility: Some mining operators come with restrictions regarding the type of data they can be applied to. For example, most of the function fitting and neural net operators need numeric data, i.e., they cannot be applied to data with nominal attributes. You can see this in RapidMiner in the operator information under "capabilities". In addition, RapidMiner warns you if you attempt to apply an operator to incompatible data, which is why we ensured correct data types in our data preparation process (see Section 7.4). But it is usually not difficult to transform the data so that its attributes have the right types, and in fact, frequently RapidMiner can automatically determine what kind of transformation to make and suggest it as a quick fix.

To see an example for type conversion, take a look at the example process Ch7_06_CompareMiningMethods. Its only top-level operator, *Compare ROCs*, performs a cross validation on each of its inner data mining operators and draws a ROC curve for each of them into the same chart. Thus it serves to compare any number of mining methods on the same data at one glance. However, if some of these methods are incompatible with the given data, type conversions have to be done. In our example, the operator *Nominal to Numerical* is applied inside *Compare ROCs*, to prepare the data for *Linear Regression* and *Logistic Regression*, two mining operators that need numeric data. With default settings, *Nominal to Numerical* replaces a nominal attribute with an integer attribute, using integers to code the nominal values. For example, a nominal attribute `colour` with values *red*, *green*, and *blue* is replaced by a like-named numeric attribute with values 0, 1, and 2. You can change this behavior to dummy coding by setting the parameter *coding type* accordingly. Dummy coding creates new attributes with the values 0 and 1 for each value of each nominal input attribute, so it would create the attributes `colour=red`, `colour=green`, and `colour=blue` in our example. Their values are 1 for all examples where the input attribute `colour` takes the value corresponding to the respective new attribute, and 0 otherwise. Effect coding uses -1 instead of 0.

There are several other types of conversion operators available in RapidMiner for various purposes, providing the flexibility for setting up mining schemes on rather diverse data.

Data mining parameters: Almost all data mining operators have parameters that specify details of their learning behavior. For example, the decision trees that the operator *Decision Tree* produces can be constrained to a maximum depth (parameter *maximal depth*), or to have a minimum number of examples in any leaf (parameter *minimal leaf size*). Further, how a tree is built can be influenced by parameters like *minimal size for split* or *minimal gain*. It is the main goal during modelling and evaluation to decide not only on a suitable data mining operator, but also to find settings of its parameters that yield good results. This is supported by the *Optimize Parameters* operators, which automatically search through different settings, as described elsewhere in this book.

7.6 Deployment

In the previous sections we have described how to prepare our data and how to try and evaluate different preparation methods, data mining models, and parameter settings. In a real project, probably quite a few variants of our processes would have been tested until a decision on a final preparation process, mining method, and parameter settings was made, based on the recall and precision or AUC values that could be achieved. Now we finally want to find the 20% of non-buyers that are the most promising target for our marketing campaign. This is the last CRISP-DM phase, called deployment.

Our example process Ch7_07_Deployment assumes that we decided on the balancing scheme to deal with class imbalance and on the *Decision Tree* operator with certain parameter settings as our mining algorithm. You can see that we do not take one of the decision trees constructed during our evaluation experiments, but that we build a final decision tree for deployment, using all the available data for training it, rather than keeping some test data apart for evaluation. Note that while our balancing scheme reduces the amount of data used for training, in general we can use all data for the final model. The tree is stored in the RapidMiner repository for reference and documentation purposes, using the *Write Model* operator. It is then applied to all our non-buyers (and only the non-buyers). The *Performance* operator shows us that in this way, some 40% of non-buyers are predicted to be buyers.

Those 40% are more than we want, but remember that behind these predictions there is a default threshold value of 0.5 that is applied internally to the ranking of customers that is induced by the confidence values produced by *Apply Model* (Section 7.5.4). We can easily choose a different confidence threshold. The process Ch7_08_Deployment_Thresholds shows a variant of the previous process in which the threshold is set manually. In this variant, the parameterless operator *Apply Threshold* is applied to the output of *Apply Model*, using the threshold created by a *Create Threshold* operator. We need to tell the *Create Threshold* operator which values of our label correspond to the confidence values 0 (parameter *first class*, which must be set to *false* in our process) and 1 (*second class*, to be set to *true*).

We can try different thresholds to find one that comes as close to our desired result as possible, given the model. The final example set returned by this process lists all non-buyers, with a ranking of who should be most likely to open a new CH04 account in the `confidence(true)` attribute.

7.7 Conclusions

In this chapter, we have developed a marketing application for data mining with the aim of advertising a new product to those customers with a high affinity for it. We have seen how to find and assemble suitable data, how to test and evaluate different mining algorithms on it, and how to deploy a final model to our customer base. We have introduced several RapidMiner operators and processes that support all of these stages and provide a reference as well as starting points for your own applications and experiments. In addition we have clarified a number of methodological issues in the hope of enabling you to confidently realize your own data mining projects in due course.

We have used the buyers of our product of interest as "model customers" for finding similar customers among the non-buyers. This is a common approach in other use cases as well. For example, in churn prediction, customers who have churned—discontinued their contract with our organisation—serves to find other would-be churners in a similar way. Analogous to our use case, churners and non-churners form two groups for binary classification, based on which ranking of non-churners can be created that reflects their propensity to churn. In fact, one would be able to directly re-use several of the processes presented in this chapter for churn prediction, which illustrates the generality and flexibility of our methods and the RapidMiner software.

Glossary

Accuracy In supervised learning, the ratio of the number of correctly predicted examples to the number of all examples in a test set.

Area Under the Curve (AUC) The area between a ROC curve and the diagonal line in a ROC chart (see ROC).

Contingency Table A table that displays the number of correctly and incorrectly predicted examples, for each label, from a single model application to a test dataset. For binary classification, this results in four cells with the numbers of true positives, false positives, true negatives, and false negatives, respectively.

Cross Validation An evaluation scheme for supervised learning that uses all of the available labelled data. The data are divided into a given number of subsets and a model is tested on each subset after training it on the remaining subsets. In the extreme case of using as many subsets as there are examples, this is called leave-one-out. The individual evaluation results are averaged to get an overall result that is statistically more reliable than would have been a single result gathered from a single test set.

Data Cleaning Good data quality is a prerequisite for data mining, but even the best real-world data usually contain some omissions or errors. During Data Cleaning, which is one stage in the Data Preparation phase (see separate entry), erroneous data values are repaired where possible, and missing values are filled (unless it is decided to simply delete any examples with erroneous or missing values). For filling missing values, several methods exist, like inserting a distinct value or the most frequent or average value. As a more refined approach, a separate model can be learned for each attribute with missing values; such a model predicts how to fill the missing value based on the present values of other attributes, for each example.

Data Preparation Preparing data for mining involves a number of technical tasks: bringing the relevant data into a single table that has one row for each example, creating the label attribute (for supervised learning tasks), cleaning the data (see Data Cleaning), and converting the data types to make them suitable for the chosen data mining method. For some mining methods, for instance association rule mining, quite specific input formats are needed, whose creation is also part of Data Preparation. But apart from these technical tasks, there is also the challenge of finding or creating attributes that give as much information about the examples as possible, perhaps including some background knowledge from the business domain. For example, it may help to explicitly create attributes that reflect certain trends in the data, like increase/decrease in revenue, rather than hoping that the mining algorithm will discover or use such trends by itself.

Deployment The final phase in a data mining project when a trained model is applied to some dataset as part of a business process.

Example The representation in data of a real-world object that exemplifies the phenomenon of interest for a data mining application. A collection of examples is needed in order to find patterns or relationships among them.

False Negatives In binary classification, the number of positive examples in a test dataset that have been predicted to be negative.

False Positives In binary classification, the number of negative examples in a test dataset that have been predicted to be positive.

False Positive Rate In binary classification, the ratio of the number of negative examples predicted to be positive (false positives) to the number of all negative examples.

Precision The ratio of the number of *correct* predictions of a given label to the number of predictions of that label in a test dataset. In binary classification, there are two precision values, one for the positive and one for the negative label. Precision for the positive label is thus defined as the ratio of true positives to all examples predicted to be positive in a test dataset.

Recall The ratio of the number of *correct* predictions of a given label to the number of all examples with that label in a test dataset. In binary classification, there are two recall values, one for the positive and one for the negative label. Recall for the positive label is thus defined as the ratio of true positives to all positive examples in a test dataset.

Receiver Operating Characteristic (ROC) In binary classification, a chart that plots model evaluations (one point in the chart for each evaluation), using the false positive rate for the horizontal axis and the true positive rate for the vertical axis (usually both expressed as values between 0 and 1). It is common to evaluate the same model based on different thresholds for the minimum confidence value that leads to a positive prediction, so that each threshold leads to one point on the ROC curve. Random predictions would lead to a ROC curve that is identical to the diagonal line from (0,0) to (1,1). The area between the ROC curve for a given model and this diagonal line is a measure for the model's quality called area under the curve (AUC).

Training The application of a data mining algorithm to a dataset in order to create (learn) a data mining model, like a decision tree, a neural net, or a regression function.

True Negatives In binary classification, the number of negative examples in a test dataset that have been predicted to be negative.

True Positives In binary classification, the number of positive examples in a test dataset that have been predicted to be positive.

True Positive Rate The same as recall for the positive label: the ratio of true positives to all positive examples in a test dataset (for binary classification).

Bibliography

[1] P Chapman, J Clinton, R Kerber, T Khabaza, T Reinartz, C Shearer, and R Wirth. CRISP-DM 1.0. Technical Report, The CRISP-DM Consortium, 2000.

[2] R Kimball, M Ross, W Thornthwaite, J Mundy, and B Becker. *The Data Warehouse Lifecycle Toolkit,*. 2nd edition. John Wiley & Sons, 2008.

[3] R Kimball and M Ross. *The Data Warehouse Toolkit: The Complete Guide to Dimensional Modeling,*. 2nd edition. John Wiley & Sons, 2002.

[4] G Linoff and M Berry. *Data Mining Techniques: For Marketing, Sales, And Customer Relationship Management,*. 3rd edition. Wiley Publishing, 2011.

[5] T Fawcett. ROC Graphs: Notes and Practical Considerations for Data Mining Researchers. Technical Report, HP Laboratories. 2003.

[6] M North. *Data Mining for the Masses.* A Global Text Project. Global Text Project, 2012.

Chapter 8

Basic Association Rule Mining in RapidMiner

Matthew A. North

The College of Idaho, Caldwell, Idaho, USA

8.1 Data Mining Case Study

Consider the hundreds, even thousands of products you can buy at your local grocery store. These items are often life-sustaining; almost every person on Earth nowadays relies on grocery stores for food and other household necessities. Because of this reliance, we have developed a certain expectation that our grocery stores will have what we need, when we need it, and for the most part, grocery stores around the world are successful in meeting this expectation. But, how do our grocery stores know which items to stock on their shelves, which items to direct our attention to in their marketing, which items to put together in packages or promotions? Increasingly, the answer to these types of questions is "data mining". More specifically, Association Rule modeling within the larger discipline of data mining has become a valuable tool for not only grocery stores, but many types of organizations, in determining patterns of behavior and relationships in large sets of data. This chapter will present a case example of how a supermarket can use simple modeling and data manipulation tools in RapidMiner to create meaningful Association Rules.

Supermarkets are retail businesses, and modern retail businesses are predominantly managed through digital systems. Each transaction generates data, and as these data are aggregated, the resulting body of facts becomes extremely large. Just think about your own supermarket shopping behavior. When did you last go to the grocery store? How often do you go? How many products do you usually buy? How much do you spend each time you go? The answer to each of these questions leads to the generation of data—data which can be aggregated by product, product type or category, day of week, time of day—the list of possibilities is virtually endless. If we consider then not just your behavior, but the behavior of hundreds or thousands of customers who shop in a certain grocery store on a given day, we can see how quickly a dataset for a single store location can grow, not to mention a dataset for a chain or large conglomerate with multiple store locations. As items are passed across bar code scanners, money is collected, and receipts are handed to customers, the data continue to grow. These data can become a valuable asset to manage the store, but only if they can be analyzed quickly and accurately to inform store managers about customer preferences and choices.

Since inventory databases, universal product bar codes and scanners, and other such supply chain management technologies have been around for years, the idea of using data to help manage retail operations is not new. However, more recently, the use of data mining to more thoroughly understand patterns of consumer behavior that affect retail operations has become more prevalent. In order to truly understand consumer behavior though, it is beneficial to understand both *what they buy* and *who they are*. Thus, in the past decade or so, we have seen an increase in the implementation of customer loyalty programs. You have probably seen these programs, and may even participate in them yourself. Generally, if you participate in such programs, you are given some form of reward, either a lower price on items in the store, or 'points' redeemable toward some future good or service. Airlines have been in the business of using such programs to encourage customer loyalty for many years, with grocery and other retail establishments adapting the concept to their operations more recently. But consider what you give when signing up for these programs. In order to receive the card which you subsequently use to gain the added benefit, you fill out a form. On this form, you give your name, gender, address, phone number, birth date, and perhaps any number of other personal characteristics. With this information, your grocer can go beyond traditional inventory management, and craft a much more personalized shopping experience for you. As we begin to examine how this might be accomplished, a comment about ethical behavior is in order. All organizations that collect, store, and analyze data have a responsibility to protect privacy, to guard against misuse and abuse, and to share data only within the constraints of fairly developed and disclosed policies. Aggregation of data, with the removal of personally identifiable information, is one way to ensure that peoples' privacy is protected. As a data miner, you bear a great responsibility to guard privacy and to behave ethically with the data you collect.

Figure 8.1, below, depicts a simplified relational model which might realistically be used by a supermarket to gather and store information about customers and the products they buy. It is simplified in that the attributes represented in each of the tables would likely be more numerous in an actual grocery store's database. However, to ensure that complexity of the related entities does not confound the explanation of Association Rules in this chapter, the tables have been simplified.

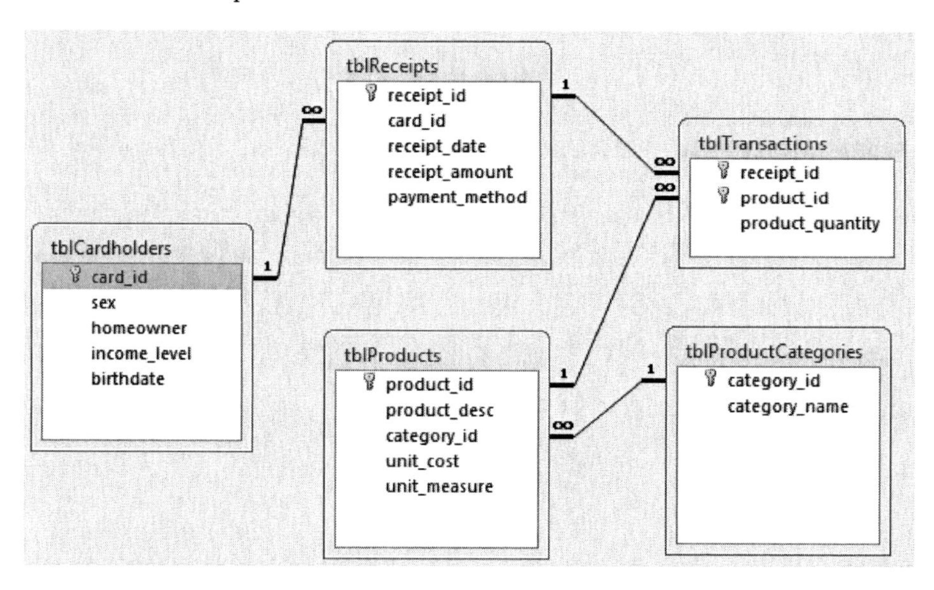

FIGURE 8.1: A simplified relational model of supermarket's database.

In this model we see five entities:

1. tblCardholders is made possible by the aforementioned customer loyalty programs. As customers fill out and submit forms to participate in the program and receive

their loyalty card, the supermarket logs birth date, sex, whether or not the person is a homeowner, etc. into the database. While most retailers would likely include the person's full name, address, phone number, email address, etc., these types of attributes are omitted from this example, as an illustration of ways to protect privacy through anonymized data. Of course, omitting these data will also preclude me from being able to mail personalized discounts or offers to each person's home, to send them an electronic advertisement via email, or even to greet them by name on a personalized communication. These are choices that must be made during data collection in any data mining project. In reality, most merchants would choose to collect, retain, and (hopefully) protect personally identifiable information such as name, address, etc. An example of a record (known as an *example* in RapidMiner, or simply a *tuple* in database terminology) is shown in Figure 8.2.

card_id	sex	homeowner	income_level	birthdate
8047	0	1	3	12/19/1981

FIGURE 8.2: An example customer record from a supermarket's database.

2. tblProducts serves as a repository for all of the products the supermarket will sell. In addition to a barcode identifier, a product description, and other related information is stored in this table. The category identifier serves as a foreign key in this table. A sample record for tblProducts is shown in Figure 8.3.

product_id	product_desc	category_id	unit_cost	unit_measure
H43845RQ	Campell's Vegetable Soup	1	$1.19	12

FIGURE 8.3: An example product record from a supermarket's database.

3. tblProductCategories gives the merchant a way to group products into similar bins or containers. Some data mining models might look very specifically at differences or similarities between brands or types of products. For example, vegetable soup might be compared to noodle soup. This example however will look at similarities between *categories* of products. Each product in our database is connected to a category through the previously mentioned foreign key in tblProducts. Retailers will likely set up categories differently from one another, perhaps even cross-listing some products in more than one category, but Figure 8.4 offers examples of the very simple categorization we will use for this chapter's case.

category_id	category_name
1	canned goods
2	dairy
3	meats

FIGURE 8.4: Example of product categories in a supermarket's database.

4. tblReceipts begins to connect customers with the products they buy. Because each receipt is connected with a loyalty card holder, we can begin to connect personal information with shopping behavior. This, of course, presumes that all customers participate in the loyalty program—a situation which is admittedly not always true.

For those customers who do not have a card_id, a generic placeholder identifier could be used, or the card_id foreign key could be left null. Note that again, additional attributes, such as a timestamp or applicable sales tax (if any), might also be present in an actual retailer's database. Figure 8.5 shows examples of receipt records both with and without connections back to tblCardholders.

receipt_id	card_id	receipt_date	receipt_amount	payment_method
1	8047	1/26/2012	$24.57	Cash
2		2/4/2012	$12.39	Ccard

FIGURE 8.5: Examples of receipt data in a supermarket's database.

5. Finally, tblTransactions brings the other four entities together. This table is an associative entity with a composite primary key, linking receipts to individual products that were purchased in the shopping basket. This enables the diverse range of relationships found in our example database, for example, many customers generate many receipts and a great deal of products are represented on many receipts, etc. Figure 8.6 depicts records connecting individual receipts with individual products.

receipt_id	product_id	product_quantity
1	H43845RQ	6
1	NSR14IP	1
2	NSR14IP	2

FIGURE 8.6: Products connected to their corresponding receipts in a supermarket's database.

With these five entities in place, we can construct a dataset for use in mining for Association Rules. To begin, a unified dataset must be generated by extracting data using an SQL statement. As a first illustration of Association Rules, we can attempt to answer the question: "Are any of our product categories strongly associated with one another, and if so, which ones?" In order to answer this question, we can extract from our database each receipt, and an indication of whether or not products in each of our categories were found on each receipt. Although the SQL approach will vary depending upon the database management system used, an example of how this would be accomplished in a Microsoft Access or SQL Server database is offered:

```
TRANSFORM Nz(Count(tblTransactions.product_quantity),0) AS Product_Count
SELECT tblReceipts.receipt_id
FROM tblProductCategories INNER JOIN (tblReceipts INNER JOIN (tblProducts IN-
NER JOIN tblTransactions ON tblProducts.product_id = tblTransactions.product_id) ON
tblReceipts.receipt_id = tblTransactions.receipt_id) ON tblProductCategories.category_id
= tblProducts.category_id
GROUP BY tblReceipts.receipt_id
PIVOT tblProductCategories.category_name;
```

This query fetches each category from our category table where at least one product was purchased. It then uses each of those category names as attribute names. Receipt identifiers are grouped together so that each record in the resulting dataset represents a single receipt. For each category, if a product assigned to that category is present on the receipt, a '1' is listed, since we are using the COUNT aggregate function. There will either be a value in the product_quantity field of the tblTransactions table, resulting in a count of '1' for the above query, or, in the absence of a value because that receipt did not contain any products from the category, the value will be 0. Figure 8.7 depicts the results of the above query where only two receipts exist in the system. Receipt 1 had both a 'canned goods' product and a 'dairy' product on it (at least one of each), while Receipt 2 had only a 'dairy' product (at least one item for this category) on it.

receipt_id	canned goods	dairy
1	1	1
2	0	1

FIGURE 8.7: Query results indicating which product categories are found on each receipt.

Thus far, this chapter has focused on explaining how data from a transactional and relational database might be structured and manipulated in preparation for Association Rule mining. This has deliberately been done using limited numbers of attributes and records in order to more clearly explain basic concepts. However, at this time we are reaching a point where we are ready to connect to a larger and more comprehensive dataset for mining in RapidMiner. Figure 8.8 depicts the first four rows of our previously discussed query, however this query was run on tables containing 108,131 receipts from 10,001 different loyalty card holders.

receipt_id	desserts	meats	juices	paper_goods	frozen_foods	snack_foods	canned_goo	beer_wine_spirits	dairy	breads	produce
1	0	1	1	0	1	0	0	0	0	0	1
2	1	0	1	1	0	0	0	0	1	0	0
3	0	0	0	0	0	0	0	0	0	0	0
4	1	1	0	1	1	0	0	0	0	0	1

FIGURE 8.8: Query results from an expanded dataset, ready for further analysis in RapidMiner.

The purchases on each receipt, denoted as before with '1', now represent dozens of products across eleven categories define by the grocer. These query results can be accessed by RapidMiner in a number of different ways. They can be exported as a Comma Separated Values (CSV) file. They can be put into a spreadsheet. Alternatively, using Java Database Connectivity (JDBC), RapidMiner can read datasets straight from the database. If JDBC is used, creating a named view that RapidMiner can access would make mining the data easier. For this chapter example, we will export the dataset from Access as a CSV file and then connect to it in RapidMiner. Figure 8.9 shows the attribute names and the first few rows of data, exported from our database in CSV format.

```
receipt_id,desserts,meats,juices,paper_goods,frozen_foods,snack_foods,canned goods,beer_wine_spirits,dairy,breads,produce
1,0,1,1,0,1,0,0,0,0,0,1
2,1,0,1,1,0,0,0,0,1,0,0
3,1,1,1,1,1,0,1,1,1,1,1
4,1,1,0,1,1,0,0,0,0,0,1
5,0,0,0,0,0,1,0,1,0,0,0
6,1,0,1,0,0,0,0,0,0,0,0
```

FIGURE 8.9: Supermarket data, extracted from the relational database in CSV format.

The actual file contains all 108,131 receipts, numbered starting from '1' in this dataset. The receipt identifier is arbitrary, and so in RapidMiner we will exclude this from the

Association Rule model. However, first, we must connect to the dataset, and then do a bit more pre-processing as part of building our data mining model. As with all dataset connections in RapidMiner, we will use the Repository to add and maintain our data.

Given that we have previously extracted our data from our relational database and saved it to a CSV file, connecting to the data is relatively simple. Using the down arrow to the right of the Data Import icon, we can use the CSV file utility and follow the prompts to connect to our data (Figure 8.10).

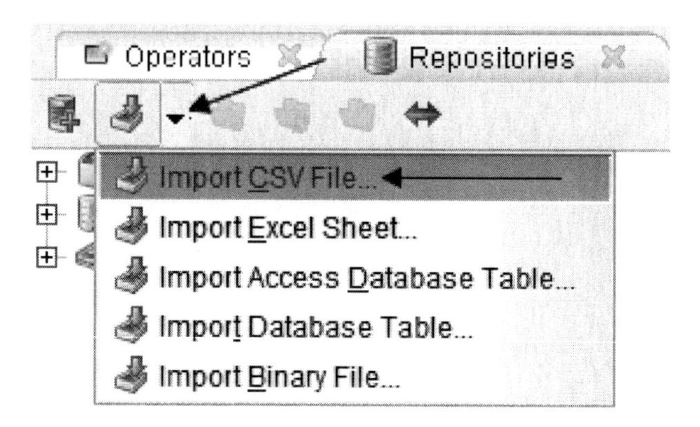

FIGURE 8.10: Connecting to a CSV file in RapidMiner.

Note that there are a number of options for importing data in addition to the CSV option. It is also worth noting that although we will not do it in this case, using the New Repository icon, just to the left of the Data Import icon, it is possible to connect to remote datasets as well. If we had created a view in our relational database, rather than exporting our data to CSV, we could connect to that view to mine and model data as well. However, in this instance we will follow the five steps of the Data Import wizard to bring our data into RapidMiner. While the wizard is relatively self-explanatory, there are a few areas that merit some explanation.

Upon clicking the Import CSV File option, the Data Import Wizard will open. Begin by browsing to the file location where you have saved your data. In RapidMiner it is possible to navigate to a network location that is not mapped to a drive letter, however it is simpler if the networked directory is mapped. If, however, the location is not mapped, you may reach it by typing the path into the File Name field and pressing 'Enter'. This will open the network directory's contents in the Data Import window, allowing you to select the file to which you want to connect. This is depicted in Figure 8.11.

Once the file has been selected, we can use the Next button to move through the ensuing screens of the Data Import Wizard to complete our connection to the dataset. On step two, it is important to ensure that we indicate the correct delimiter character. In this example, we will need to choose the "Comma" option as the indicator for column separation. Remember that if your data contain naturally occurring commas, you might want to export your data as something other than CSV, perhaps as a tab, semicolon, or pipe-delimited text file. RapidMiner can handle all of these, so think about what format makes the most sense for your data. For our example data in this case, we have only numbers (mostly ones and zeros), so a comma-delimited format poses no problem for us. Figure 8.12 shows the columns and rows previewed from the CSV file. You can see that our attribute names in this screen image are being treated as a row of data, but we will correct this in step three of the Data Import wizard.

Step three gives us the opportunity to identify our first row as our attribute names. This

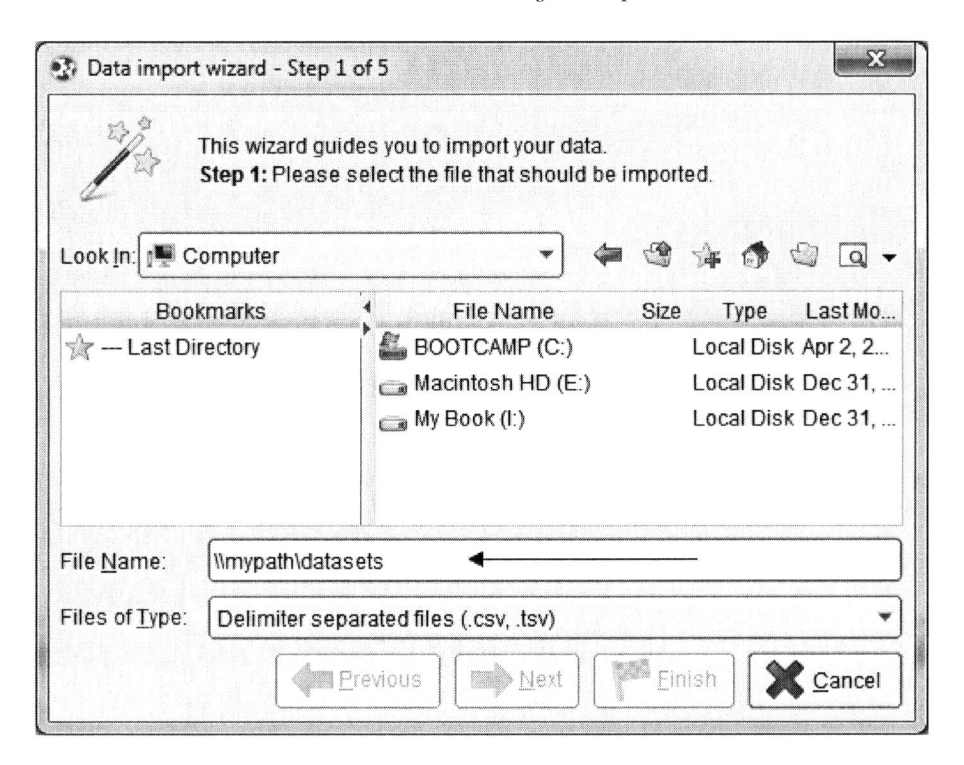

FIGURE 8.11: Connecting to an unmapped network location.

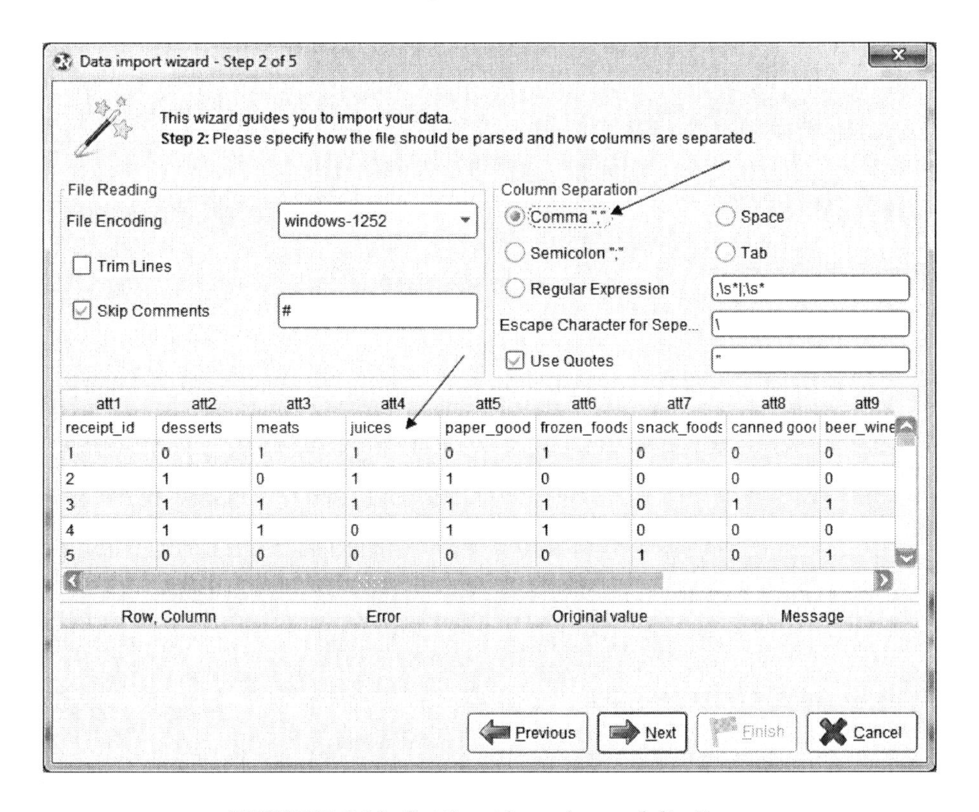

FIGURE 8.12: Setting the column delimiters.

is done by clicking in the cell just below "Annotation" and selecting "Name", as noted in Figure 8.13.

Data import wizard - Step 3 of 5

This wizard guides you to import your data.
Step 3: In RapidMiner, each attribute can be annotated. The most important annotation of an attribute is its name - a row with this annotation defines the names of the attributes. If your data does not contain attribute names, do not set this property. If further annotations are contained in the rows of your data file, you can assign them here.

Annotation	att1	att2	att3	att4	att5	att6	att7	att8
Name	receipt_id	desserts	meats	juices	paper_good	frozen_food:	snack_food:	canned go
-	1	0	1	1	0	1	0	0
-	2	1	0	1	1	0	0	0
-	3	1	1	1	1	1	0	1
-	4	1	1	0	1	1	0	0
-	5	0	0	0	0	0	1	0

FIGURE 8.13: Setting the first row in our CSV import as our attribute names.

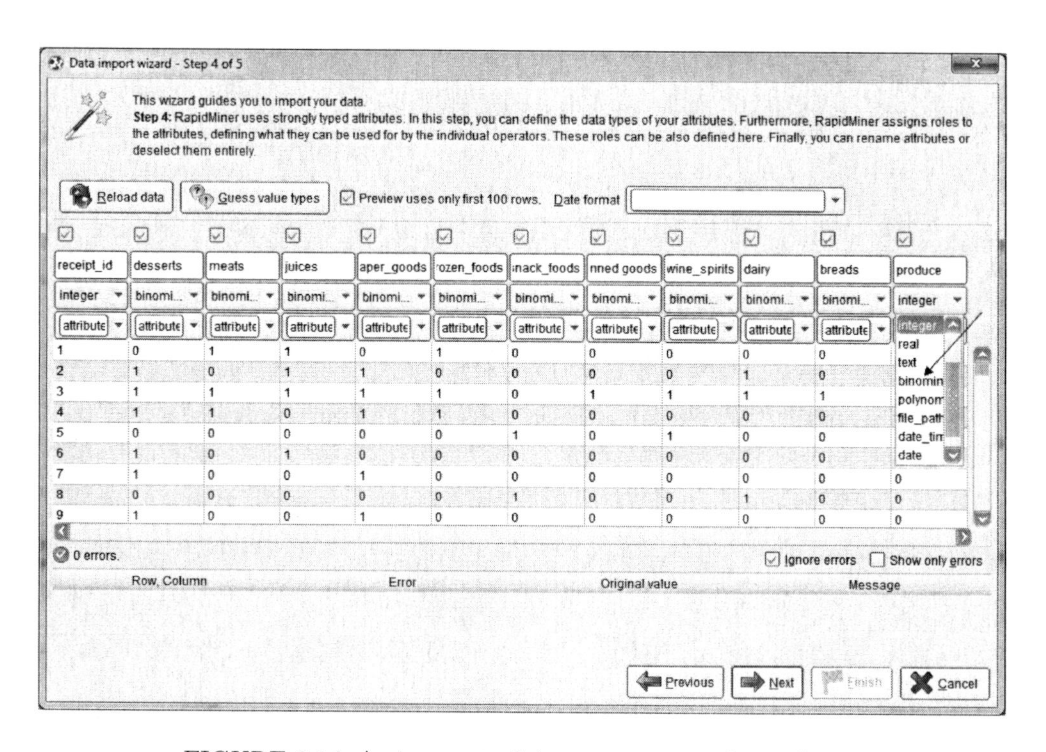

FIGURE 8.14: Assignment of data types to each attribute.

On the step four screen (Figure 8.14), it is important to ensure that the data type of each attribute is assigned correctly. Though RapidMiner does have operators which allow you to change an attribute's data type after it has been imported, it is usually preferable to set the data type during import. RapidMiner will guess the data type for each attribute based on the first 100 observations in the data set, so it is important to double check the default data types for accuracy and relevance to your data mining activities. For Association Rules, it is best for binary attributes, such as our grocery store product categories, to be set to "binominal", as has been done in Figure 8.14.

The final step in importing data is simply to choose the repository where the dataset will be stored and give it a name. In Figure 8.15, we have selected the repository named RapidMiner by clicking on it, then added GroceryStore as the name of the dataset to be saved. Clicking finish will add your dataset to the selected repository in RapidMiner.

FIGURE 8.15: Saving the imported dataset.

It is important to consider when importing a dataset into RapidMiner that the data are actually being copied into an XML file that RapidMiner stores and uses. One drawback to this approach is that if the data are updated or added to, the dataset would need to be re-imported into RapidMiner in order to reflect the changed or added data. One way to avoid this problem is to use a Read operator instead of importing. Here we will depart from the present example to briefly illustrate two ways to read data in RapidMiner.

The first way is simply to use a Read CSV operator. This operator is depicted in Figure 8.16.

The Read CSV operator is found under Import >Data in the Operators tab. When this operator is dropped into a Main Process window, as has been done in Figure 8.16, the Parameters pane on the right side of the RapidMiner window offers options to connect to a CSV file directly. Using the folder button, users can browse for the location of the desired CSV file. The other settings can be accepted as defaults, or modified as needed to match the format of the data in any CSV file. For example, in our example dataset depicted in Figure 8.9, our column separator is a comma, rather than the defaulted semicolon, so we would need to modify that parameter. We could also un-check the 'use quotes' parameter, since our dataset does not contain quotation marks. Once our parameters are set, running our main process would read the CSV dataset into our data mining process, making it available for other operators and processes, while still enabling us to make changes to the source data. Any such changes would be brought into RapidMiner automatically, since the Read CSV operator goes back and checks the source CSV each time our data mining process is run.

Similarly, the Read Database operator can be used to incorporate datasets into a Rapid-

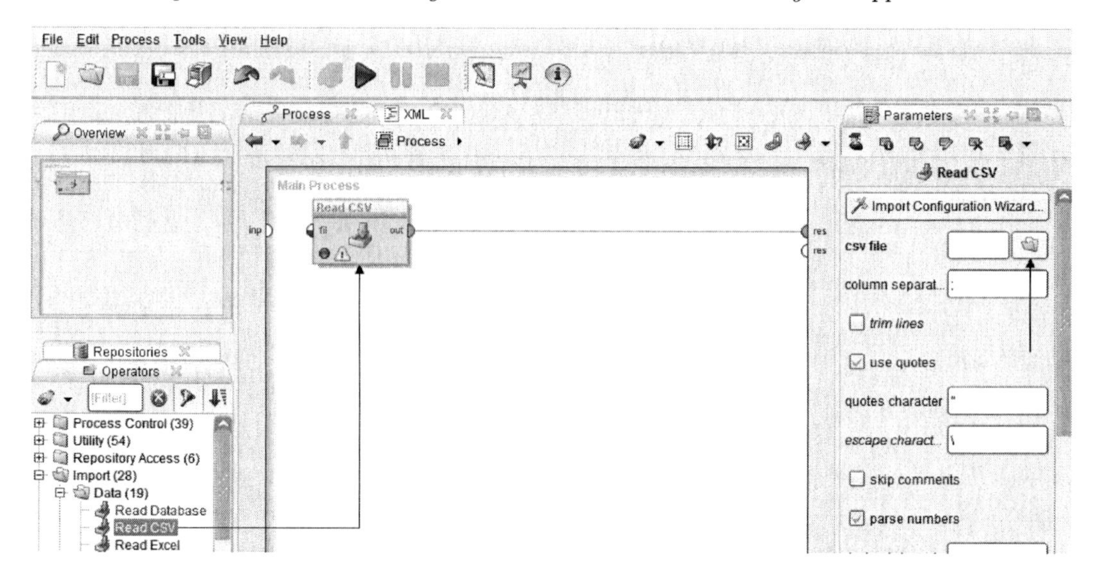

FIGURE 8.16: Using the Read CSV operator.

Miner process without importing. In this instance, assume that the dataset was loaded into a MySQL database table called grocery_receipts. A SQL SELECT call to that table would yield something similar to Figure 8.17.

SQL Query*

```
1 • select * from grocery_receipts;
```

Overview Output Snippets Result (1)

Fetched 14 records

ReceiptID	desserts	meats	juices	paper_goods	frozen_foods	snack_foods	canned_goods	beer_wine_spirits	dairy	breads	produce
1	0	0	1	1	0	1	0	1	0	1	1
2	1	1	0	0	1	1	0	1	0	1	1
3	1	1	0	0	1	0	1	0	0	0	0
4	0	0	1	1	1	1	0	1	0	1	1
5	1	0	0	0	0	0	0	0	1	0	0
6	0	1	1	0	0	1	0	0	0	1	1

FIGURE 8.17: An example of binominal receipt data in a MySQL database table.

With the data in this format, RapidMiner can easily connect to the dataset using the Read Database operator. This operator is also found under Import >Data, just as the Read CSV operator was.

This operator will allow the construction of data mining models on datasets that may change—anytime an update or insert is performed on the database table, RapidMiner will pick up and incorporate the new data when the process is run. In order to connect to a database table, parameters must be set. The black arrow on the right side of Figure 8.18 indicates the button to click to set up database connection and query parameters.

Figure 8.19 shows the parameter configuration window for the Read Database operator. Database connections can be given a descriptive name in the 'Name' field. The 'Database system' drop-down menu includes many major database management systems. MySQL is selected by default, however Oracle, Microsoft SQL Server, and numerous others are also available. Selecting your database type will also auto-select the default connection port for that database, however this can be changed if your database server is configured to

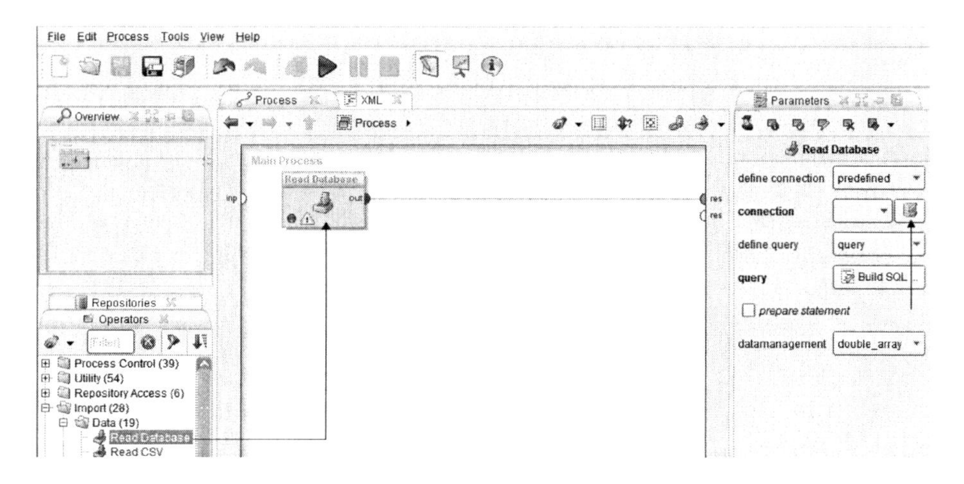

FIGURE 8.18: Using the Read Database operator.

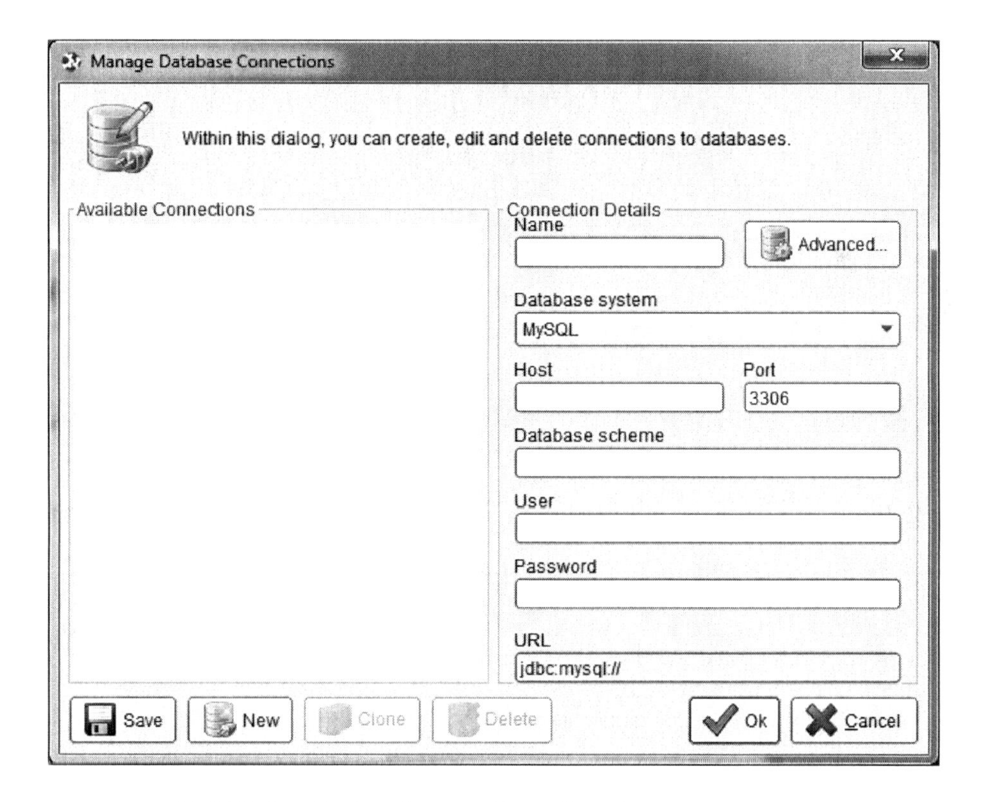

FIGURE 8.19: Configuring Read Database parameters.

communicate on a port other than the default. The 'Host' can be entered as either an IP address or machine/domain (e.g., doc.RapidMiner.com). The 'Database scheme' is the name of the schema within the database where your tables are located. The 'User' and 'Password' fields are used to provide the login credentials for the database. A nice benefit of using the Read Database operator is that RapidMiner will automatically inherit the table attributes' data types, so you do not need to specify data types separately within RapidMiner, although you can use RapidMiner operators to change data types within your data mining stream when necessary. Once you have configured your database connection parameters, you can save them and return to the Main Process window. With the Read Database operator still selected, your Parameters pane on the right-hand side will allow you to build an SQL query, load an SQL query from a file, or directly connect to a table and include all of that table's data as your dataset. In the example introduced in Figure 8.18, we would change the 'define query' parameter to 'table name', and then we would select grocery_receipts from the resultant list of table names in the schema we set up in our database connection parameters. For this chapter's example, however, we began by simply importing our data from a CSV file, so although we have just discussed the Read CSV and Read Database operators, we will now proceed based upon the assumption that Import CSV option has been used.

The default workspace in RapidMiner shows a Main Process in the center of the screen. To begin building an Association Rule data mining model, we locate the dataset we have just imported and drag it into the Main Process window. The software will add a "Retrieve" object to our process window, and automatically connect a spline between the output (out) port and a result (res) port. Since this is the only dataset we will be working with in this case example, we will switch back to our list of operators by clicking on the Operators tab as indicated in Figure 8.20.

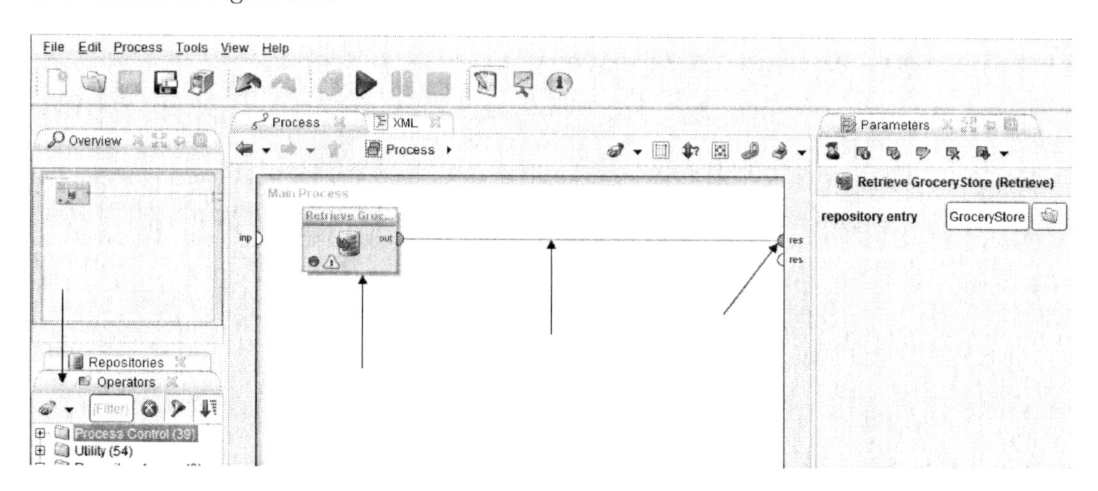

FIGURE 8.20: Setting up the process window to build a model.

With the dataset connected, we can run the model by clicking on the blue "Play" button. This will not yield much at this point because the dataset is the only thing we have added to our process. RapidMiner will switch from Design Perspective to Results Perspective. You can toggle between these two views by clicking on the Notepad-and-Pencil and Graph Display icons in the toolbar. These are noted in Figure 8.21. In addition, RapidMiner will likely ask you if you would like to save your process, and if you would like to switch to Results Perspective. These questions will be posed to you in a series of pop-up dialog boxes, where you can indicate your response, and whether or not RapidMiner should remember your response. Once you have navigated these dialogs and switched to Results Perspective, your attributes, together with their roles and data types will be displayed to you. You may also wish to examine your dataset, which can be done by selecting the Data View radio button, also indicated in Figure 8.21.

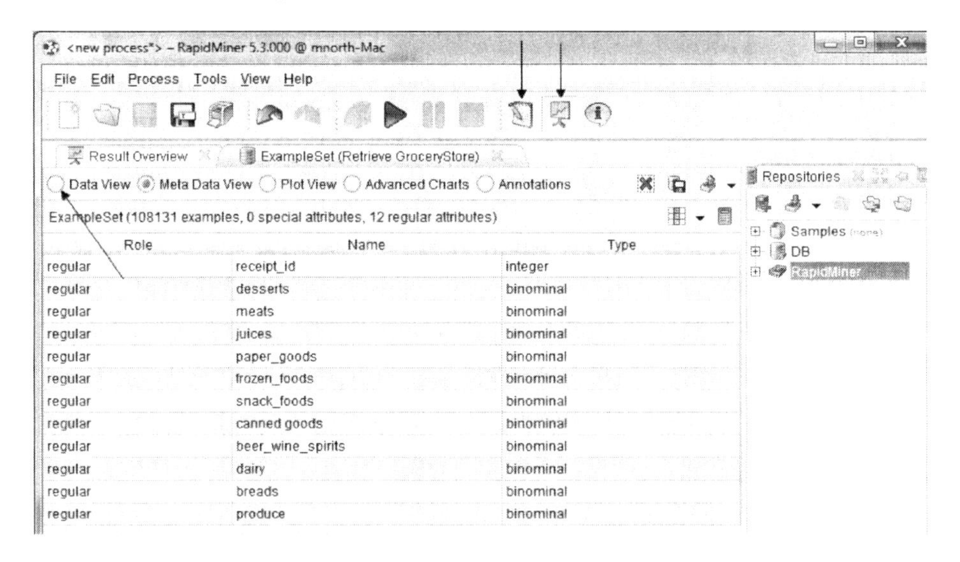

FIGURE 8.21: Results Perspective after running a model with only a Retrieve object in it.

Thus, we have successfully connected to the dataset, but we have not yet *mined* it. To enhance our model and add actual association rule operators to the process flow, we need to switch back to the Design Perspective by clicking on the notepad-and-pencil icon in the application toolbar. When connecting to this dataset, we brought in one attribute that is actually not relevant to our data mining model. The receipt id is arbitrarily assigned to each transaction, but does not actually indicate anything of interest in association with the products purchased. Fortunately, RapidMiner offers an operator that allows users to remove attributes from consideration in a model. The Select Attributes operator, depicted in Figure 8.22, enables us to exclude any attributes, in this case just one, that we do not wish to mine.

Note that the Select Attributes operator is found within the Data Transformation folder as depicted in the Operators menu tree on the lower left side of Figure 8.22. The Select Attributes operator has been dragged and dropped onto the spline between the Retrieve object and the Resultset port (res). If the ports do not connect, you can click on the out port and connect it to the exa (example) port to include it in the process flow. The parameters for the Select Attributes operator are found on the right side of the screen, depicted in Figure 8.23.

To filter attributes that we do not want to include in our model, we can select only a subset of the attributes in our dataset. After selecting 'subset' as our filter type, we can click the Select Attributes button and in the resultant dialog window, we can use the left and right green arrows to include or exclude attributes. When finished, we can simply click OK to return to our main process window. At this point there are a number of other operators that we could use to prepare our dataset for mining. Some likely candidates include:

- Sample: Found within the Filtering folder in the Data Transformation operators folder, this operator allows you to take various types of samples to use in your model. This can be a nice way to test and prove a model on a small subset of your dataset before

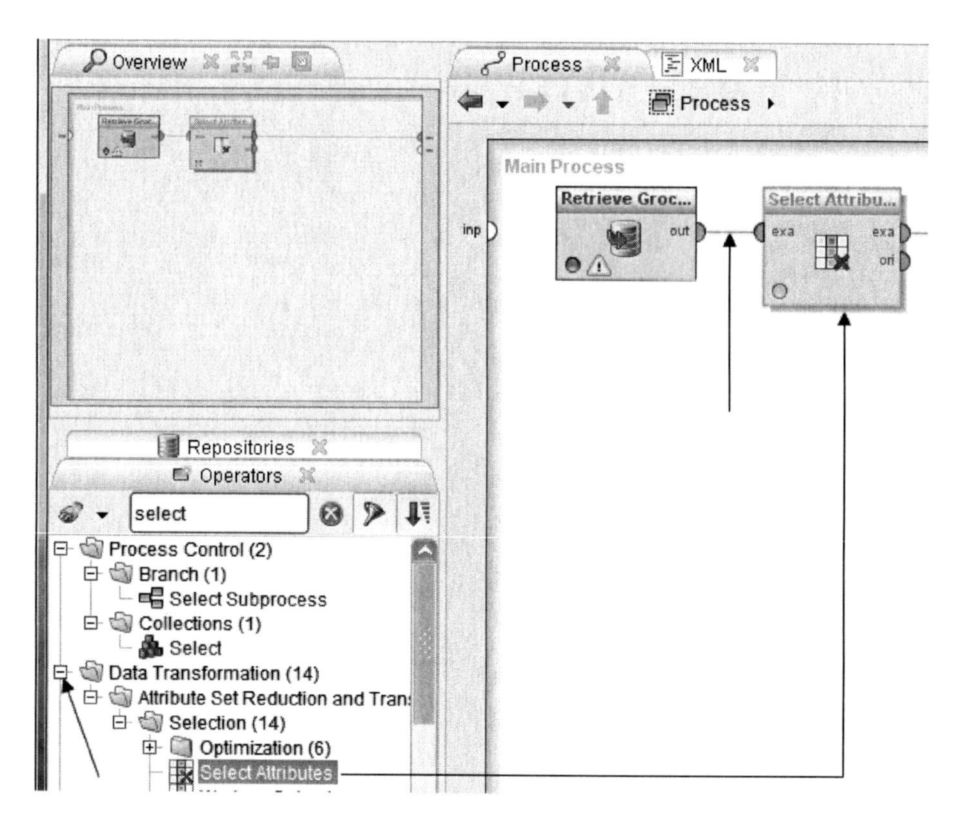

FIGURE 8.22: Addition of a Select Attributes operator to the model.

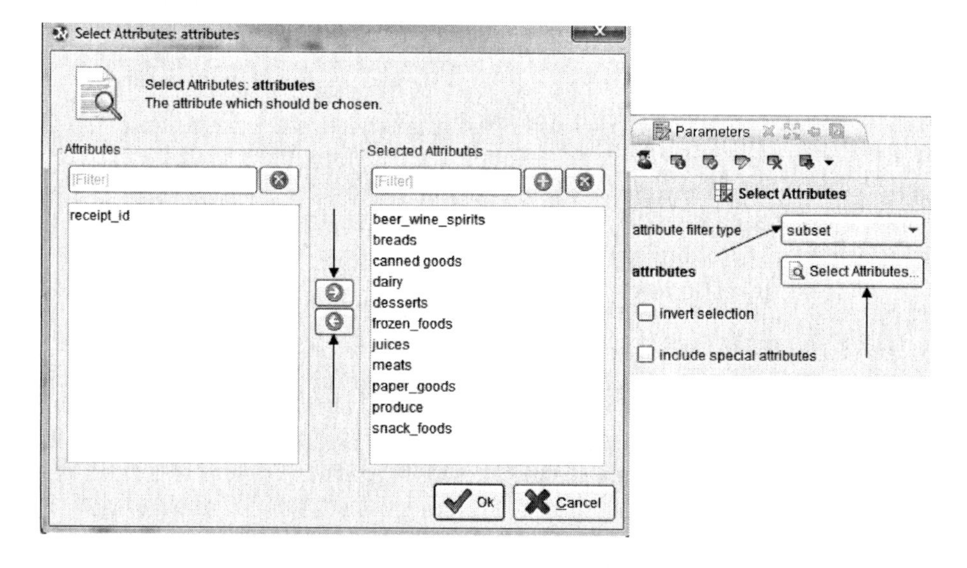

FIGURE 8.23: Configuring the Select Attributes operator.

running the model on the entire dataset. If there are hundreds of thousands or millions of observations in your dataset, the model may take some time to run. Tuning the model on a smaller sample can save time during development, then once you are satisfied with your model, you can remove the sample operator and run the model on the entire dataset.

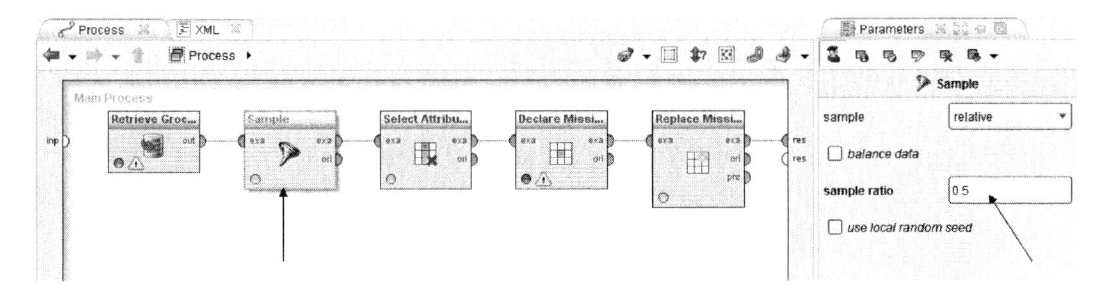

FIGURE 8.24: A Sample operator configured to take a 50% sample of the data.

- Nominal to Binominal (and other similar operators): Sometimes, variables are not data typed as Binominal, which is the data type required for mining for association rules. This operator and a wide array of other data type conversion operators, are found under Type Conversion in the Data Transformation folder.
- Filter Examples: This operator, located in the Data Transformation >Filtering folder, enables you to remove some observations from your dataset if they contain things like missing values. If a record has a value missing in one of more attributes, you may choose to exclude it from your data mining model, rather than guess what the value should be.

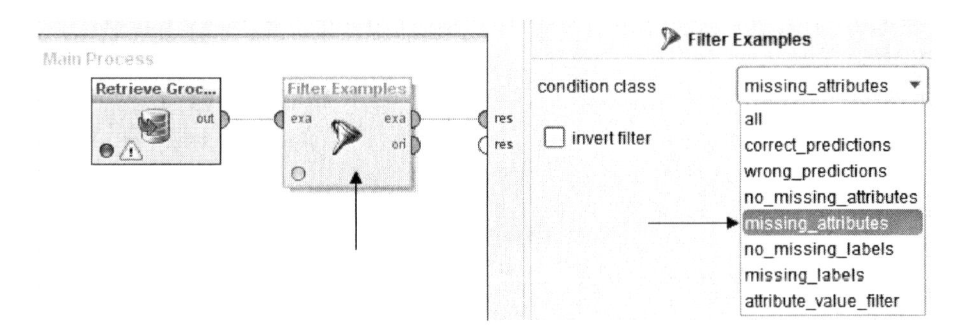

FIGURE 8.25: Using the Filter Examples operator to remove any records containing missing values in any of the attributes in the dataset.

- Declare Missing Value/Replace Missing Value: These two operators help you to handle null values that may exist in your data. Both are found in the Data Transformation folder. The former, found in Value Modification can find undesired values and set those to missing (null) in your dataset; while the latter, found in Data Cleansing, can locate the missing values in an attribute, or series of attributes, and change them from missing to a desired value.

An example of how some of these operators could be used in the current case might be in handling backordered items. Suppose that instead of 0 (item not purchased) and 1 (item purchased), we have a third value: 2 (item purchased but backordered). We could use these two operators to set all '2' values to missing and then all missing values to '1', since the item was in fact purchased. Figures 8.26 through 8.29 depict the use of the Replace operator to address inconsistent data which has been introduced into in our dataset for illustration purposes.

Result Overview | ExampleSet (Select Attributes)

Data View | Meta Data View | Plot View | Advanced Charts | Annotations

ExampleSet (108131 examples, 0 special attributes, 11 regular attributes) View Filter (108131 / 108131): all

Row No.	desserts	meats	juices	paper_goods	frozen_foods	snack_foods	canned goo...	beer_wine_...	dairy	breads	produce
1	0	1	1	0	1	0	0	2	0	0	1
2	1	0	1	1	0	0	0	0	1	0	0
3	1	1	1	1	1	2	1	1	1	1	1
4	1	1	0	1	1	0	0	0	0	0	1

FIGURE 8.26: Inconsistent data that will complicate an Association Rule model.

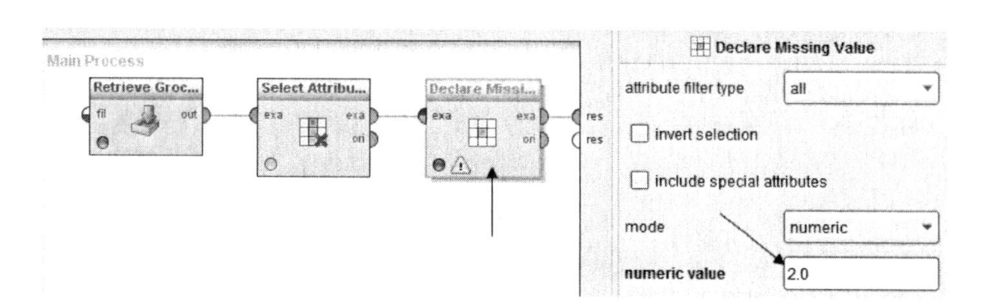

FIGURE 8.27: Inclusion of a Declare Missing Value operator, removing any '2' values.

Result Overview | ExampleSet (Declare Missing Value)

Data View | Meta Data View | Plot View | Advanced Charts | Annotations

ExampleSet (108131 examples, 0 special attributes, 11 regular attributes) View Filter (108131 / 108131): all

Row No.	desserts	meats	juices	paper_goods	frozen_foods	snack_foods	canned goo...	beer_wine_...	dairy	breads	produce
1	0	1	1	0	1	0	0	?	0	0	1
2	1	0	1	1	0	0	0	0	1	0	0
3	1	1	1	1	1	?	1	1	1	1	1
4	1	1	0	1	1	0	0	0	0	0	1
5	0	0	0	0	0	1	0	1	0	0	0

FIGURE 8.28: In Results View, the '2' values have been changed to 'missing' (?).

Running the model as depicted in Figure 8.29 would show that any values of '2' in the dataset are now changed to '1'. Be sure to pay close attention to which values are being modified in your dataset when doing this type of data preparation. Setting '2' values to be missing, and then setting *all* missing values to '1' might inadvertently indicate an item was purchased, when in fact it was not. If your dataset has other missing values, it would be crucial to handle those missing values first, and then set inconsistent values, such as '2' to missing and then to '1'. Also be sure to check the data types of your attributes in the Meta Data View of Results Perspective after each operator is added. It may be necessary to change some of your attributes' data types to binominal after handling missing values. In this chapter's example, because we did most of our data cleansing and preparation prior to connecting to the dataset in RapidMiner, our process in this case will not require the

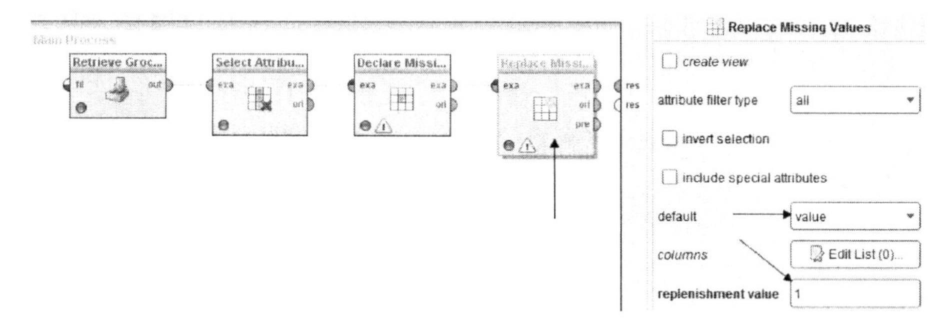

FIGURE 8.29: Addition of a Replace Missing Values operator, changing missing values to '1'.

use of additional data handling operators. Although we have modified the data a bit for Figures 8.26 through 8.29 to illustrate some of the operators available in RapidMiner, for simplicity's sake, we will now return to our original dataset as it was imported in Figure 8.15. We have retained the Select Attributes operator in order to remove the Receipt_ID attribute from our data mining model, as was discussed previously in Figure 8.22.

Once any inconsistencies or other required transformations have been handled, we can move on to applying modeling operators to our data. The first modeling operator needed for association rules is FP-Growth (found in the Modeling folder). This operator, depicted in Figure 8.21, calculates the frequent item sets found in the data. Effectively, it goes through and identifies the frequency of all possible combinations of products that were purchased. These might be pairs, triplets, or even larger combinations of items. The thresholds used to determine whether or not items are matches can be modified using the tools on the right-hand side of the screen.

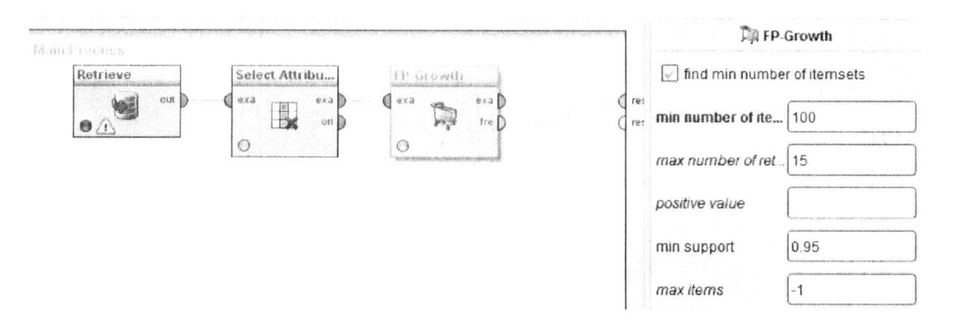

FIGURE 8.30: The addition of FP-Growth to our data mining process.

At this point, we can run our model and look at the item sets that were found, but we cannot necessarily see the strength of the associations. Figure 8.31 shows the results of running the model with just the FP-Growth operator in place. Note that the frequency (fre) port is connected to the result set port.

As we can see, the operator found frequencies for most items individually, and began to find frequencies between items as well. Although the screen capture does not show all 32 item sets that were found, it if did, you would be able to see that the final set found contains four products that appear to be associated with one another: juices, meats, frozen foods, and produce. There are a number of three-product combinations, and even more two-product sets. The Support attribute seen in Figure 8.31 indicates the number of observations in the dataset where the single or paired attributes were found; in other words, out of the 108,131

Result Overview | FrequentItemSets (FP-Growth)

Table View | Annotations

No. of Sets: 32
Total Max. Size: 4
Min. Size: 1
Max. Size: 4
Contains Item:
Update View

Size	Support	Item 1	Item 2
1	0.780	juices	
1	0.673	meats	
1	0.671	frozen_foods	
1	0.665	produce	
1	0.339	beer_wine_s	
1	0.338	snack_foods	
1	0.322	paper_good	
1	0.312	desserts	
1	0.219	breads	
2	0.522	juices	meats
2	0.514	juices	frozen_foods
2	0.519	juices	produce

FIGURE 8.31: Item sets generated by FP-Growth.

receipts in our dataset, what percent of the time was that attribute or pairing found on a receipt. While this begins to unveil potential connections between products in our data set, we can examine the relationships even further.

By switching back to Design Perspective, there is one final operator that can be added to our process. The Create Association Rules operator, also found in the Modeling folder, takes the product pairings that were frequently found by the FP-Growth operator and organizes them into rules according to certain user-configurable parameters. Figure 8.32 shows the addition of the Create Association Rules operator in our process stream.

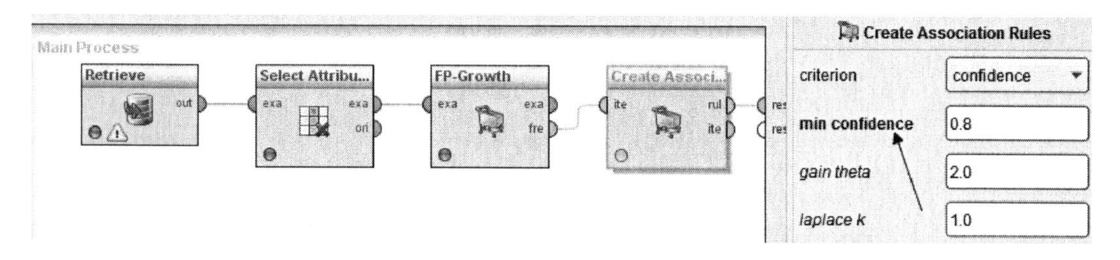

FIGURE 8.32: Create Association Rules in the main process of our data mining model.

RapidMiner is programmed with default parameters for Association Rules. The confidence percentage is the main parameter used when creating rules. This setting tells the software how often an association must be found in a dataset in order for it to be considered a rule. Association rules are laid out using premises (sometimes called antecedents) and conclusions (sometimes called consequents). The premise is the first attribute in the association, and the conclusions are all attributes associated with the given premise. So, for example, assume ten grocery store receipts. Assume further that Milk was purchased on six of those receipts and that cookies were purchased on five. If Milk and Cookies appeared together on five of the ten receipts, then the confidence percentage with Cookies as the premise and Milk as the conclusion would be 100%—every time Cookies were purchased, so was Milk. However, the reciprocal does not yield the same confidence percentage: Milk was purchased six times and Cookies only five, so the confidence percent with Milk as the premise is only 83%—that is, the rule only held five times out of the six times Milk was purchased ($5/6 = .8333$). In each of these example scenarios, the rule, with the attributes in either order, would satisfy the default confidence percentage in RapidMiner (.8, or 80%). If we extend the example just a bit further, we could find that it would be possible that Milk and Cookies where only purchased together one time in the ten receipts. If Milk were purchased on receipt numbers one through six, and Cookies were purchased on receipt numbers six through ten, then the two products, though still purchased six and five times respectively, would only be found together on a single receipt. The confidence percentages in this scenario would then become Milk→Cookies: 17% (.1667); and Cookies→Milk: 20% (.2). These confidence levels would be far too low to convince us that the relationship between cookies and milk, in any order, should be considered a rule.

As we mine for association rules, the confidence percentage threshold can be changed on the right side of the process window to find rules of higher or lower confidences in the dataset. An association rules model can be further tuned to look for and evaluate rule pairings that might be interesting by modifying the differential criteria (Gain and Laplace). The field to modify the confidence percentage is indicated with an arrow on the right side of Figure 8.32, and the differential criteria fields are seen just below this. For this chapter's case example, we will accept all defaults offered by RapidMiner and run the model.

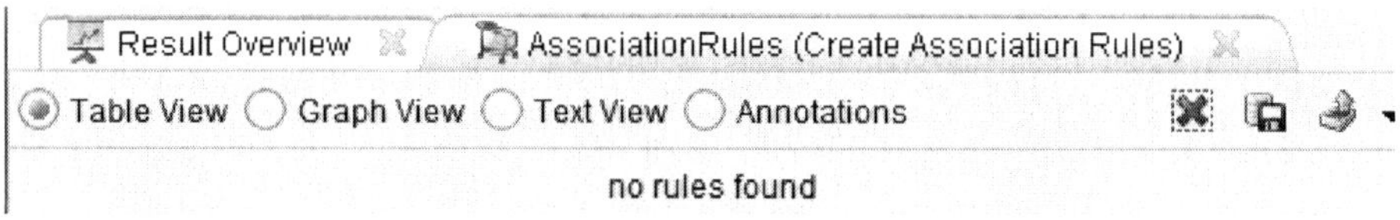

Result Overview | AssociationRules (Create Association Rules)

Table View | Graph View | Text View | Annotations

no rules found

FIGURE 8.33: Results of the Associate Rules model at 80% confidence.

We can see in Figure 8.33 that no rules were found at the 80% threshold in this dataset. That is not to say that no associations were found. We can refer back to Figure 8.31 and confirm that indeed, some products in our data set do appear together frequently, but perhaps not frequently enough to be considered rules. If we lower our expectations of what constitutes a rule of association, it is possible we will find some pairings to evaluate. Returning to Design Perspective, we will set the confidence threshold to 75% (.75) and run the model again.

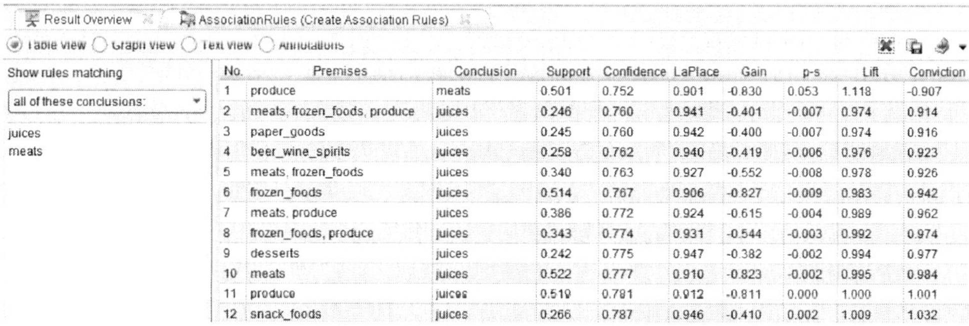

Result Overview | AssociationRules (Create Association Rules)

Table View | Graph View | Text View | Annotations

Show rules matching: all of these conclusions: juices, meats

No.	Premises	Conclusion	Support	Confidence	LaPlace	Gain	p-s	Lift	Conviction
1	produce	meats	0.501	0.752	0.901	-0.830	0.053	1.118	-0.907
2	meats, frozen_foods, produce	juices	0.246	0.760	0.941	-0.401	-0.007	0.974	0.914
3	paper_goods	juices	0.245	0.760	0.942	-0.400	-0.007	0.974	0.916
4	beer_wine_spirits	juices	0.258	0.762	0.940	-0.419	-0.006	0.976	0.923
5	meats, frozen_foods	juices	0.340	0.763	0.927	-0.552	-0.008	0.978	0.926
6	frozen_foods	juices	0.514	0.767	0.906	-0.827	-0.009	0.983	0.942
7	meats, produce	juices	0.386	0.772	0.924	-0.615	-0.004	0.989	0.962
8	frozen_foods, produce	juices	0.343	0.774	0.931	-0.544	-0.003	0.992	0.974
9	desserts	juices	0.242	0.775	0.947	-0.382	-0.002	0.994	0.977
10	meats	juices	0.522	0.777	0.910	-0.823	-0.002	0.995	0.984
11	produce	juices	0.519	0.781	0.912	-0.811	0.000	1.000	1.001
12	snack_foods	juices	0.266	0.787	0.946	-0.410	0.002	1.009	1.032

FIGURE 8.34: The Association Rule model with a 75% confidence percentage.

Simply by changing the confidence percentage downward by 5 points enabled our model to yield 12 potential rules of association. Since we combined a number of products into product categories for this case example, we cannot tell exactly what items are associated with one another, but we can tell what *kinds* of products are frequently purchased together. Our first rule Produce→Meats, is supported by 50% of our 108,131 observations, meaning that the association was found in our dataset roughly 54,174 times. Our confidence that when a customer buys produce products (the premise) they will also buy meat products (the consequent) is 75.2%. The other differential indicators for this rule (the five columns on the right) also generally indicate that this rule is fairly reliable. We can similarly analyze the other 11 rules in this way to find interesting patterns or connections. Rule number two is interesting in that it associates four product categories. While this rule is supported by just less than 25% of the records, we are 76% confident that if products from all three of the premise categories are purchased, juice products will also be purchased. In fact, in this particular example, it appears that juice products are quite popular with our customers.

Another excellent way to analyze association rules in RapidMiner is to use the visualization tools available in Results Perspective. In Figure 8.34, there is a radio button option labeled Graph View. Clicking this button will yield results similar to those shown in Figure 8.35. The arrow in Figure 8.35 highlights the Graph View radio button.

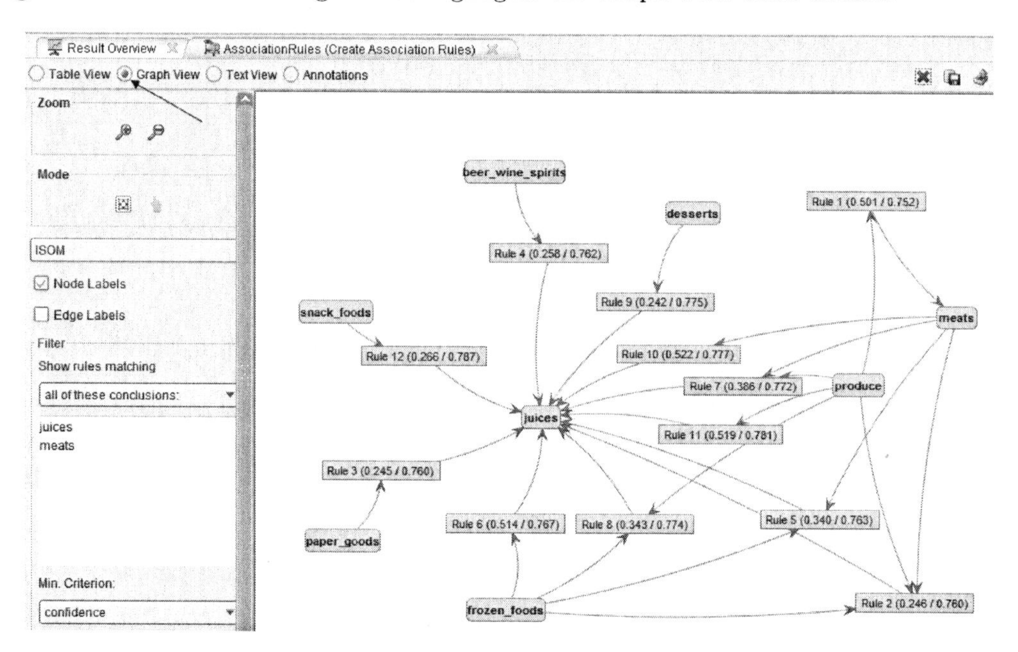

FIGURE 8.35: Graph View of our Association Rules in Results Perspective.

This visualization provides not only the support and confidence percentages for each rule found, it offers a map of how our different attributes relate to one another relative to each rule. This paints a much clearer picture of our associations, showing that in this example, the 'juices' category is a primary hub for associations with many other grocery categories. Using the 'pointing hand' icon under Mode on the left hand side will allow you to click and drag the various diagram objects if the default view is too cluttered or compact to allow you to read the associations clearly. With the objects spread apart from one another a bit so that the categories, rules, and associations can be more clearly seen, we can see fairly clearly that 'juices' are the products that are most frequently associated with other items in our grocery store.

This observation can help us draw this chapter's case to a close. Association Rules can be sought out in almost any kind of data; they are simply measurements of the strength of attribute connections in a dataset. In this chapter's example, we have found that juice products are relatively strongly connected to essentially every other product category in our grocery store, but what can we do with this information? Perhaps we already know, through daily experience, that we sell a lot of juice products, in which case this particular data mining model is of little help to us. But perhaps we might not have realized, without this model, just how pervasive juice products are in our product sales. As grocery store managers, we may begin to design product promotions which pair juice products with other strongly associated products in order to boost sales. We may go back and lower our confidence percentage a bit more, to see if other product categories emerge as the next most common conclusions (e.g., frozen foods and produce both have associations above 70% confidence). Or we may decide that we need more detail about specifically what juice products are frequently sold with other items, so we may choose to go back to the data extraction and preparation phase to group juice products into more specific attributes across our 108,131 receipts, to see if we might find even better clarity about what products our customers are most frequently buying together.

The power of Association Rules is limited only by your ability to gather and prepare data, and your creativity in seeking out interesting pairings in the results. RapidMiner makes it easy to import, manipulate, and mine your data for associations, and to evaluate the results for real-world applicability. While this example has used the classic scenario of supermarket basket analysis, you can adapt the principles and techniques illustrated in this chapter to examine relationships between attributes in essentially any dataset.

Chapter 9

Constructing Recommender Systems in RapidMiner

Matej Mihelčić

Electrical Engineering, Mathematics and Computer Science, University of Twente, Netherlands; Rudjer Boskovic Institute, Zagreb, Croatia

Matko Bošnjak

University of Porto, Porto, Portugal; Rudjer Boskovic Institute, Zagreb, Croatia

Nino Antulov-Fantulin

Rudjer Boskovic Institute, Zagreb, Croatia

Tomislav Šmuc

Rudjer Boskovic Institute, Zagreb, Croatia

Acronyms

k-NN - k Nearest Neighbours
ROC curve - Receiver Operating Characteristic curve
AUC - Area Under the Curve
Prec@k - Precision at k
MAP - Mean Average Precision
NDCG - Normalized Discounted Cumulative Gain

9.1 Introduction

Making choices is an integral part of everyday life, especially today when users are overwhelmed with information, from the Internet and television, to shelves in local stores and bookshops. We cope with this information overload by relying on daily recommendations from family, friends, authoritative users, or users who are simply willing to offer such recommendations. This is especially important when we lack information to make a rational decision, for example, choosing a hotel for vacations in a new city, selecting a new movie to watch, or choosing which new restaurant to visit.

Recommender systems [1] facilitate decision-making processes through informed assistance and enhanced user experience. To aid in the decision-making process, recommender systems use the available data on the items themselves, such as item taxonomies, descriptions and other, and/or data on user experience with items of interest, for example, user choices, rankings, scores, tags, etc. Personalized recommender systems subsequently use this input data, and convert it to an output in the form of ordered lists or scores of items in which a user might be interested. These lists or scores are the final result the user will be presented with, and their goal is to assist the user in the decision-making process.

Recommender systems represent a broad and very active [2, 3, 4, 5] field of research and were, from their origins in the 1990s [6, 7, 8, 9], somewhat detached from the data mining field. This was mostly due to the specific form of data that recommender systems used. There are three basic types of recommender systems: collaborative filtering, content-based, and hybrid systems. Collaborative filtering recommender systems use social information, preferences, and experiences of other users in order to find users with similar taste. The assumption of these systems is that users of similar tastes might enjoy and consume similar items. Content filtering recommender systems use the available structured, and unstructured information on users or items to recommend further items of interest. They assume that the user might be interested in items similar to the ones in which an interest has already been displayed. Both of these systems have their advantages, which are additionally reinforced, or disadvantages, which are diminished with the usage of hybrid systems—systems which combine both the collaborative and the content-based recommendation approach.

Recommender systems are ubiquitous, and an average Internet user has almost certainly had experiences with them, intentionally or not. For example, the well-known Internet commerce, Amazon.com employs a recommender system that recommends products its users might be interested in, based on the shopping habits of other users. Social networking sites like Facebook or LinkedIn use recommender systems for recommending new friends to users based on their social network structure. The music website Last.fm uses a recommender system to recommend new music to a user, based on the listening habits of users with similar music taste. The Internet Movie Database (IMDb) recommends similar movies, based on the content and style of the movies user previously browsed. Streaming provider Netflix tries to predict new movies a user might be interested in based on his watching habits and movie ratings, compared to other users. These, and numerous other examples like StumbleUpon, Google AdSense, YouTube, etc., which differ in services provided, like audio, video, general item, social network, other Internet content, books, etc., demonstrate the importance of these systems.

Recommender systems facilitate making choices, improve user experience, and increase revenue, therefore should be easily accessible for deployment to interested parties. This led us to write a chapter on recommender systems in a clearly understood and easily applied way through RapidMiner. We believe that RapidMiner's workflow approach entices systematic research and facilitates its implementation in combination with RapidAnalytics. The combination of research and production environment renders itself as an excellent environment for understanding recommender systems through practice. Throughout this chapter, you will learn the basics of theory related to recommender systems, with a strong emphasis on practical implementation. This practical work will introduce you to all the necessary knowledge for rapid prototyping of recommender systems, thus enabling you to master them through application of your data. The implementation of recommender systems in RapidMiner has been additionally simplified through the Recommender Extension.

9.2 The Recommender Extension

The Recommender Extension [10, 11] is a RapidMiner extension, which contains various recommendation operators based on the MyMedia library [12]. The main purpose of the extension is to provide a simple and an efficient way for users to create recommender systems. Although recommender systems can be implemented using the existing RapidMiner operators [13], the Recommender Extension enables utilizing dedicated and optimized recommender operators for the recommendation task. Using these operators results in achieving better performance, lower execution time, and a smaller memory footprint, depending on the operator and the users' needs. The resulting system is therefore run-time efficient, since it is specialized for the task, and its production-time cost is significantly shorter. In addition, every recommendation operator in the extension contains various advanced user-set parameters, which enable greater freedom in experimenting, and achieving better performance.

To install the extension, simply download it from the marketplace integrated in RapidMiner, if you are using RapidMiner 5.3 or above. In case you are using an older version of RapidMiner, download the extension from the Rapid-I marketplace[1] and place it in your RapidMiner library folder (`"<RapidMiner folder>/lib/"`).

The Recommender Extension brings a new set of operators to RapidMiner, specifically designed to model recommender systems. We describe these operators in the following text.

9.2.1 Recommendation Operators

In version 5.0.10, the Recommender Extension has a total 26 recommendation operators. These operators are grouped in the following categories: Item Recommendation, Item Rating

[1] `http://rapidupdate.de/`

Prediction, and Recommender Performance. The overview of operators supported in the Recommender Extension is given in the Table 9.1.

TABLE 9.1: Recommendation operators supported by the Recommender Extension.

Item Recommendation (IR)	Collaborative Filtering-based Item Recommendation	Item k-NN, Most Popular, Bayesian Personalized Ranking Matrix Factorization, User k-NN, Weighted Regularized Matrix Factorization, Random
	Attribute-based Item Recommendation	User Attribute k-NN, Item Attribute k-NN
Item Rating Prediction (RP)	Collaborative Filtering-based Rating Prediction	Random, Global Average, User Item Baseline, User k-NN, Item k-NN, Slope One, Bi-Polar Slope One, Matrix Factorization, Biased Matrix Factorization, Factor Wise Matrix Factorization
	Attribute-based Rating Prediction	User Attribute k-NN, Item Attribute k-NN
Recommender performance	Model Application	Apply Model (IR), Apply Model (RP), Model Combiner (IR), Model Combiner (RP)
	Performance Evaluation	Performance (IR), Performance (RP)

In this chapter we describe only the Item Recommendation (IR) operators. The Item Rating Prediction (RP) operators are referred to and categorized in the Table 9.1. The reasons behind this decision lie in the desire to simplify and shorten the content of the chapter, and to focus on IR operators which, in our opinion, should be more interesting to the end user. In addition, workflows, input data, and operator usage are almost identical between the IR and the RP tasks. Therefore, by understanding the usage of the IR operators, readers will easily construct the corresponding RP workflows. More details on the Item Rating Prediction capabilities of the Extension can be found in [10].

We outline the summary of all the Item Recommendation operators grouped by their requirements and effects in Table 9.2. More details on the usage of Recommender Extension operators, as well as some utility workflows, can be found in the Recommender Extension user guide [11].

Item recommendation operators operate over large matrices that contain information about which user consumed which item. These input matrices often contain large numbers of empty entries, and can thus be used in a more space-efficient way. We describe the appropriate format used for efficient data loading and storage in the next subsection.

9.2.2 Data Format

Typical recommender systems, operating on user usage data, are built on top of large matrices called utility matrices. These matrices usually contain elements from a limited set of numbers, for example from $0, 1, 2, 3, 4, 5$, where 0 typically denotes that the user had no interaction with the item, and the rest describes the level of that interaction in a form of rating. Due to a large number of both users and items, these matrices are typically very large. In addition, since users mostly consume a very small portion of items out of the total number of items, these matrices tend to contain a lot of zeros—they tend to be very sparsely populated. This is why special data structures for handling sparse data need to be

TABLE 9.2: Summary of the requirements and effects of the recommendation operators.

Operator group	Input	Output	Short description
Collaborative Filtering-based Item Recommendation	1. Example set containing user ID, and item ID roles	1. Unchanged training data 2. Recommendation model	Operators take the input and train a collaborative filtering model.
Attribute-based Item Recommendation	1. Example set containing user ID, and item ID roles 2. Example set containing user ID or item ID, and attribute ID roles.	1. Unchanged training data 2. Recommendation model	Operators take the input and learn a model using attribute information describing each user or item.
Model Application	1. Example set containing user ID role 2. Recommendation model	1. Example set containing user ID and item ID roles, and rank attribute 2. Recommendation model	Operator takes the input and returns it together with the first n ranked items. The online update option updates the recommendation model with the data from the query dataset.
Performance	1. Example set containing user ID and item ID roles 2. Example set containing user ID and item ID roles 3. Recommendation model	1. Performance vector containing error measures for recommendation 2. Example set containing error measures for recommendation	Operator takes the input and returns the standard error measures on testing data as both the Performance vector and the Example set.
Model Combiner	1. Multiple recommendation models	1. Combined model	Operator does the weighted combination of input models into a single model. Importance weights for each input model are user defined.

implemented. In RapidMiner, we can use AML reader operators to read such datasets. Input datasets used to learn a recommender system model must be formatted in two columns; for example, the first column can contain user IDs, while the second can contain item IDs. Attributes names, and their positioning can be arbitrary. Prior to applying recommendation operators to input datasets, proper roles have to be set for these attributes, as seen in Table 9.3. Any additional attributes will not be considered. An example of an AML, and a related DAT file for item recommendation operators is given in Table 9.3.

TABLE 9.3: An example of an AML and a related DAT file for item recommendation operators.

AML file (item_recommendation.aml)

```
<?xml version="1.0" encoding="windows-1252"?>
<attributeset default_source="sample.dat">

  <attribute
    name        = "user_id"
    sourcecol   = "1"
    valuetype   = "integer"/>

  <attribute
    name        = "item_id"
    sourcecol   = "2"
    valuetype   = "integer"/>

</attributeset>
```

DAT file (sample.dat)

```
1 71
1 169
2 211
2 562
3 670
4 576
```

After loading the AML file, it is advisable to save it as a regular RapidMiner example set and use it afterward from the Repository since this format keeps the roles required for further usage. Note that when using other reading operators and other, non-sparse formats, enforcing the *datamanagement* property to a variant of sparse array and sparse map, when handling data is advisable, and rather necessary when dealing with large datasets.

We explained the data format needed for using Recommender Extension operators. Before going further into the details of various recommendation approaches, and corresponding operators, we briefly explain the performance measures used for the recommendation task.

9.2.3 Performance Measures

There are different measures used for measuring performance of recommender systems on various recommendation tasks. Some of these measures are calculated and outputted by the Performance operator.

First, we need to introduce some formal notation and definitions from [14]. Then we continue by giving an example of how to calculate each of them by hand for better understanding.

By $X \subset \mathbb{R}^d$ we denote a training set, $|X| = n$. Y denotes ranking of items for different users. For a query q, representing a user to whom we want to recommend items, we let X_q^+ denote the subset of relevant points of the training se Xt. For a ranking $y \in Y$ and two points $i, j \in X$, we use $i \prec_y j$ to indicate that i is placed before j in y. Now we can define error measures we used to evaluate performance of our operators.

Precision at k (Prec@k): Prec@k is the fraction of relevant results out of the first k returned. It is a localized evaluation criterion which captures the quality of rankings of the top k items.

Average Precision (AP): AP is the precision-at-k score of a ranking y, averaged over all positions k of relevant items:

$$AP(q, y) = \frac{1}{|X_q^+|} \sum_{k=1}^{n} Prec@k(y) \mathbb{1}[k \in X_q^+]$$

$\mathbb{1}[k \in X_q^+]$ is the indicator function equalling 1 if the item at rank k is a relevant item, and zero otherwise.

Mean Average Precision (MAP): MAP is the mean average precision score over all queries in the query set Q.

$$MAP(Q, y) = \sum_{q=1}^{|Q|} \frac{AP(q, y)}{|Q|}$$

Normalized Discounted Cumulative Gain (NDCG): NDCG is a measure that scores all of the top k items at a decaying discount factor.

$$NDCG(1, y; k) = \frac{\sum_{i=1}^{k} D(i) \mathbb{1}[i \in X_q^+]}{\sum_{i=1}^{k} D(i)}$$

$$D(i) = \begin{cases} 1 & i = 1 \\ 1/log_2(i) & 2 \leq i \leq k \\ 0 & i > k \end{cases}$$

We will explain how to compute the described measures on one example:

Let us consider the case in which we have $X = \{1, 2, 3, 5, 6, 7, 8, 9, 10, 11\}$, and $|X| = 10$. Suppose our recommendation algorithm gives us the following output:
$1 \rightarrow \{7, 5, 9, 11, 8, 6, 3, 10\}$, $2 \rightarrow \{2, 9, 8, 11, 6, 1, 10, 3\}$, $3 \rightarrow \{1, 9, 2, 10, 3, 7, 11, 5\}$, $4 \rightarrow \{11, 6, 5, 3, 1, 2, 8, 7\}$, $5 \rightarrow \{1, 8, 9, 2, 6, 7, 5, 10\}$.

$1 \rightarrow \{7, 5, 9, 11, 8, 6, 3, 10\}$ denotes that for user with ID 1, our algorithm returned item with ID 7 as the first ranked, item with ID 5 as the second, and so on. Suppose we have the following user-item pairs in our test set: $(1, 5)$, $(2, 6)$, $(3, 7)$, $(4, 5)$, $(5, 6)$. Let us calculate the measure values in the order we defined them above:

1. Prec@k:

   Prec@5: For each user we count the number of positive items among top 5 recommended and calculate the average over all users.

   $$\text{Prec@5} = \frac{\frac{1}{5} + \frac{1}{5} + 0 + \frac{1}{5} + \frac{1}{5}}{5} = 0.16.$$

2. MAP:

   Per definition, we calculate Prec@k averaged over all possible positions k of relevant items for each user and then take the average sum.

We have only one relevant item for every user so we have to calculate Prec@2 for user with ID 1, Prec@5 for user with ID 2 etc.

$$\text{MAP} = \frac{\frac{\frac{1}{2}}{1} + \frac{\frac{1}{5}}{1} + \frac{\frac{1}{6}}{1} + \frac{\frac{1}{3}}{1} + \frac{\frac{1}{5}}{1}}{5} = 0.28.$$

3. NDCG:

We calculate $NDCG(1, y; k)$ for all users y and take the average sum as a final $NDCG$ result. Since we have only one relevant item per user, we have $k = 1$.

$$\text{NDCG} = \frac{\frac{D(2)}{D(1)} + \frac{D(5)}{D(1)} + \frac{D(6)}{D(1)} + \frac{D(3)}{D(1)} + \frac{D(5)}{D(1)}}{5} =$$

$$\frac{0.5 + 0.43068 + 0.38685 + 0.63093 + 0.43068}{5} = 0.476.$$

9.3 The VideoLectures.net Dataset

The recommender system datasets used throughout this chapter consists of content and collaborative data. Content data was taken from the "VideoLectures.Net Recommender System Challeng" [15] public dataset. However, due to privacy-preserving policies, collaborative data was synthesized using a variant of the memory biased random walk model [16].

Though the synthesized data is not real, it shares important statistical properties with the real data, and is therefore useful for model building, and most importantly, it is freely available. We generated synthetic data by using only the public datasets while the content data originates from the open access multimedia repository of video lectures VideoLecture.Net.[2] In the following text we use lectures and items interchangeably, when discussing items of recommendation.

The content dataset described contains the following content attributes for each item:

- item ID: a unique integer that represents a lecture
- item name: a text string containing a name of a particular lecture
- item description: a text string denoting a description of a particular lecture

A particular item identifier is denoted by the small letter i and the set of all items is denoted by the capital letter I. Collaborative data contains synthetic clickstreams of users, where each clickstream is a sequence of items viewed by a particular user in some time interval. In the following text, we refer to the synthetic users as users. A particular user identifier is denoted by the small letter u and the set of all users is denoted by the capital letter U. Clickstreams are transformed into the sparse matrix A, which is called the usage matrix. The non-zero elements of the usage matrix $(A(i, u))$ tell us that the item i was consumed by the user u.

Using this dataset, we construct collaborative and content recommender systems in the following sections. The collaborative recommender systems rely on the usage matrix A while the content recommender systems rely on items textual descriptions.

[2] http://videolectures.net/

9.4 Collaborative-based Systems

The main idea of collaborative recommendation approaches is to use information about the past behaviour of existing users of the system for predicting which item the current user will most probably like and thus might consume. Collaborative approaches take a matrix of given user-item ratings or viewings as an input and produce a numerical prediction indicating to what degree the current user will like or dislike a certain item, or a list of n recommended items. The created list should not contain items the current user has already consumed.

We can divide collaborative-based systems in two categories:

1. Neighbourhood-based recommender systems
2. Matrix factorization-based recommender systems

9.4.1 Neighbourhood-based Recommender Systems

Neighbourhood-based recommender systems work by counting common items two users have viewed for every pair of users in the system, or the number of common users that viewed the same pair of items. Using this count, similarity between two users or items is calculated. Neighbourhood systems use intuition that two users who have viewed a large number of common items have similar tastes. That information can be used to recommend items that one user consumed and the other one did not. We are interested in finding pairs of users having the most similar taste, or pairs of items having the most users that viewed both items. Those pairs of users/items are called "the closest neighbours".

We describe two main approaches of the neighbourhood-based recommender systems: user and item-based nearest neighbour recommendation.

User-based nearest neighbour recommendation
Given a user-item viewing matrix and the ID of the current user as input, identify other users having similar past preferences to those of the active user. Subsequently, for every product the active user has not yet consumed, compute aprediction based on the product usage of the selected user subset. These methods assume that users, who have previously shared similar tastes, will share similar tastes in the future, and that user preferences remain stable and constant over time.

To calculate similarity between users, two typical similarity measures are used: the Pearson correlation and the Cosine correlation [17]. In our item recommendation problem we used cosine correlation as a similarity measure. Typically, we do not consider all users in the database when calculating user similarity, rather the k most similar ones.

Item-based nearest neighbour recommendation
When dealing with large problems, consisting of millions of users, user-based collaborative filtering approaches lead to increased memory usage and execution time. Since the system is required to calculate a large volume of potential neighbours, it becomes impossible to compute predictions in real time. In some cases, the number of users dominates the number of items in the system so it would be natural to try to use items for making recommendations. That is the reason for creating a second neighbourhood-based recommender system based on items instead of users.

As opposed to the user-based approach, the item-based recommendation approach computes prediction using the similarity between items. We use a cosine similarity measure, as we did in the user-based approach. Likewise, as in the user-based approach, we use k-nearest neighbours, i.e. the k most similar items, for prediction.

9.4.2 Factorization-based Recommender Systems

Factorization-based recommender systems use low-rank matrix approximation techniques, like the singular value decomposition, non-negative matrix factorizations, regularized matrix factorizations, etc., in order to predict item recommendations [18, 19]. We demonstrate how factorization-based techniques work on a small movie recommendation example. Let us assume a case of a movie recommender system with only five users and six items. In addition, we assume that we obtained information about users' past behaviour on the system (see Table 9.4).

TABLE 9.4: Implicit feedbacks in movie recommender system. "1″ denotes user consumed an item, "0″ denotes he/she did not.

	UserA	UserB	UserC	UserD	UserE
The Hangover	1	1	0	0	0
Snakes on a Plane	0	1	1	0	0
Little Miss Sunshine	1	0	0	1	1
The Descendants	1	0	0	0	1
Schindler's List	0	0	0	1	0
The Pianist	0	0	0	1	1

We construct the user-item usage matrix A to be:

$$A = \begin{pmatrix} 1 & 1 & 0 & 0 & 0 \\ 0 & 1 & 1 & 0 & 0 \\ 1 & 0 & 0 & 1 & 1 \\ 1 & 0 & 0 & 0 & 1 \\ 0 & 0 & 0 & 1 & 1 \\ 0 & 0 & 0 & 1 & 1 \end{pmatrix}.$$

Let us assume that we want to predict the affiliation $\hat{A}(i,j)$ of a user j toward an item i by linear regression in the following form:

$$\hat{A}(i,j) = \theta_j(1) * x_i(1) + \theta_j(2) * x_i(2) + ... + \theta_j(k) * x_i(k),$$

where $(x_i(1), ..., x_i(k))$ represent attributes of an item i and $(\theta_j(1), ..., \theta_j(k))$ represent weights for a user j. This can also be written as a scalar product of two vectors: $\hat{A}(i,j) = \mathbf{\theta_j}^T \mathbf{x_i}$, where $\mathbf{x_i} \in R^k$ denotes the attribute vector and $\mathbf{\theta_j} \in R^k$ denotes the weight vector. Prediction of affiliation for all users and items can be written as a matrix product:

$$\hat{\mathbf{A}} = \begin{pmatrix} \theta_1^T x_1 & \theta_2^T x_1 & \theta_3^T x_1 & \theta_4^T x_1 & \theta_5^T x_1 \\ \theta_1^T x_2 & \theta_2^T x_2 & \theta_3^T x_2 & \theta_4^T x_2 & \theta_5^T x_2 \\ \theta_1^T x_3 & \theta_2^T x_3 & \theta_3^T x_3 & \theta_4^T x_3 & \theta_5^T x_3 \\ \theta_1^T x_4 & \theta_2^T x_4 & \theta_3^T x_4 & \theta_4^T x_4 & \theta_5^T x_4 \\ \theta_1^T x_5 & \theta_2^T x_5 & \theta_3^T x_5 & \theta_4^T x_5 & \theta_5^T x_5 \\ \theta_1^T x_6 & \theta_2^T x_6 & \theta_3^T x_6 & \theta_4^T x_6 & \theta_5^T x_6 \end{pmatrix}.$$

In matrix notation this can be written as a product of two matrices $\mathbf{X}$ and $\boldsymbol{\Theta}$, which represent item and user attributes in latent space, respectively.

$$\hat{\mathbf{A}} = \begin{pmatrix} x_1(1) & x_1(2) \\ x_2(1) & x_2(2) \\ x_3(1) & x_3(2) \\ \dots & \\ x_6(1) & x_6(2) \end{pmatrix} \begin{pmatrix} \theta_1(1) & \theta_2(1) & \theta_3(1) & \dots & \theta_5(1) \\ \theta_1(2) & \theta_2(2) & \theta_3(2) & \dots & \theta_5(2) \end{pmatrix} = \mathbf{X\Theta}.$$

In order to factorize the original matrix $\mathbf{A}$ to two matrices $\mathbf{X}$ and $\boldsymbol{\Theta}$, we need to find a solution to the following optimization problem:

$$\underset{\mathbf{X\Theta}}{\arg\min} \sqrt{\sum_i \sum_j (A(i,j) - (X\Theta(i,j))}.$$

The solution of this optimization problem is a pair of matrices $\mathbf{X}$ and $\boldsymbol{\Theta}$ that represent the best low-rank approximation of the original matrix $\mathbf{A}$. To obtain the solution, we can use an alternating least squares optimization procedure [20, 21] among others. Regularization techniques are often employed in order to prevent data overfitting.

One possible solution to our problem are the following $\mathbf{X}$ and $\boldsymbol{\Theta}$ matrices:

$$\mathbf{X} = \begin{pmatrix} 0.2367 & 1.1682 \\ 0 & 1.2783 \\ 1.6907 & 0.1011 \\ 1.0714 & 0.2091 \\ 0.6192 & 0 \\ 1.3077 & 0 \end{pmatrix}, \boldsymbol{\Theta} = \begin{pmatrix} 0.4404 & 0 & 0 & 0.6004 & 0.6675 \\ 0.3960 & 0.8135 & 0.4260 & 0 & 0 \end{pmatrix}.$$

We can observe that each movie item is described with two latent features (matrix $\mathbf{X}$). In this case, the first latent feature could describe "drama movie genre" and the second latent feature could describe "comedy movie genre". The first row of matrix $\mathbf{X}$ describes the movie "*The Hangover*" with the value of 0.2367 for "drama genre" and the value of 1.1682 for "comedy genre". The last row of matrix $\mathbf{X}$ describes the movie "*The Pianist*" with the value of 1.3077 for "drama genre" and the value of 0 for "comedy genre".

In addition, users are described with the same latent features within the matrix $\boldsymbol{\Theta}$. For example, the second column in the matrix $\boldsymbol{\Theta}$ describes "UserB" with the value of 0 affiliation to "drama genre" and the value of 0.8135 affiliation to "comedy genre".

If we multiply these two matrices $(\mathbf{X}, \boldsymbol{\Theta})$, we obtain the recommendation matrix $\hat{\mathbf{A}}$ in the user and item space.

TABLE 9.5: Recommendation table obtained as a product of two matrices **X** and **Θ**. Recommendation table represents 2-rank matrix approximation of the original user item matrix obtained by a non-negative matrix factorization.

	User A	User B	User C	User D	User E
The Hangover	0.5668	0.9503	0.4977	0.1421	0.1580
Snakes on a Plane	0.5062	1.0399	0.5446	0	0
Little Miss Sunshine	0.7846	0.0822	0.0431	1.0151	1.1285
The Descendants	0.5546	0.1701	0.0891	0.6433	0.7152
Schindler's List	0.2727	0	0	0.3718	0.4133
The Pianist	0.5759	0	0	0.7851	0.8729

We are now able to make a recommendation from the 2-rank matrix $\hat{\mathbf{A}}$ by selecting the unseen movies with the highest score. From our example, we would recommend unseen movie "*The Descendants*" to "user B" and "*Schindler's List*" to "user E".

9.4.3 Collaborative Recommender Workflows

Collaborative recommender operators use the user-item matrix to build a recommendation model. This user-item matrix is presented as an example set of user-item pairs describing user consumption history. The recommendation model built with this matrix is used to recommend items to users from a query set. The query set is an example set containing identification numbers of users for which we want to make recommendations. For each user in the query set we recommend only the items not consumed by this user. Figure 9.1 depicts a basic collaborative recommender operator workflow.

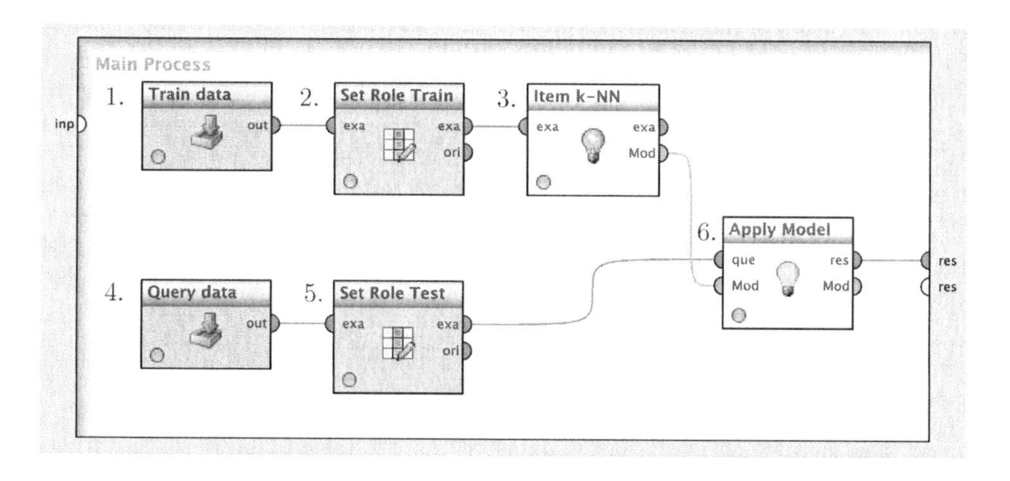

FIGURE 9.1: An example of an item recommendation workflow.

In the item recommendation workflow, the first two operators read the train and the query example sets using the *Read AML* operators (1,4). Following, the appropriate roles are set to attributes using the *Set Role* operator (2). The `user identification` role was set to `user_id` attribute and `item identification` role to `item_id` attribute. Data attributes can have arbitrary names but roles for those attributes must be set. Next, we use the train data with the appropriately set roles to train an *Item k-NN* model (3). At this point we can use our trained model to recommend new items to users in the query set using the *Apply Model* operator (6). Prior to model application, the user identification role was set for the query set (5). The *Apply Model* operator (6) returns an example set containing the first n ranked recommendations for every user in a query set.

In Figure 9.1 we have seen how to make recommendations for particular users. In the following figure, Figure 9.2, we show how to measure performance of a recommendation model.

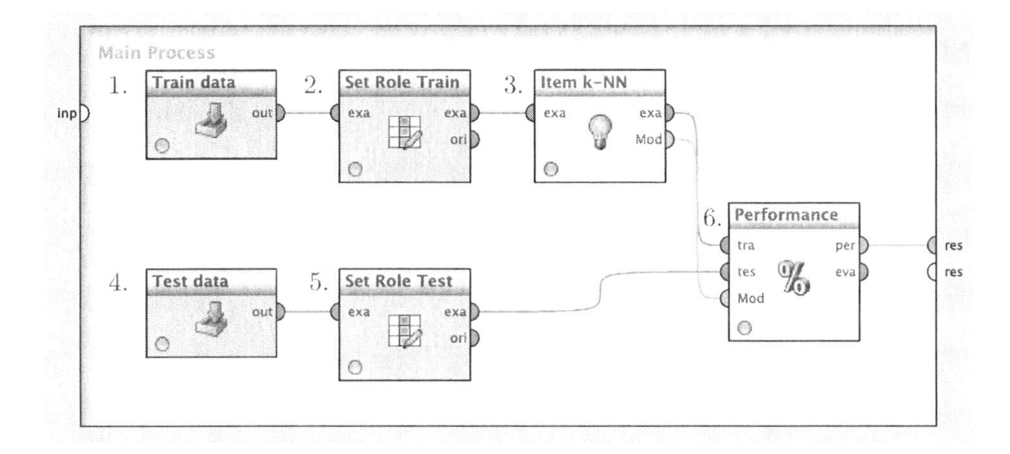

FIGURE 9.2: Measuring performance of a recommendation model.

The data management part of the workflow for measuring recommender model performance in Figure 9.2 is the same as in Figure 9.1. We use the *Read AML* operators (1,4) to load the data input, and the *Set Role* operators (2,5) to set the appropriate roles. In this workflow we use the test data (4) containing two attributes, the `user_id` and the `item_id` attribute and we set `user identification` and `item identification` roles to those attributes, respectively. The difference from the previous workflow is the need to calculate the performance of our built recommendation model (3). We use the *Performance* operator (6) to measure standard recommendation error measures we previously defined: AUC, Prec@k, NDCG, and MAP. The *Performance* operator (6) returns a performance vector and an example set containing performance measures. This enables a user to choose which format suits his or her needs.

9.4.4 Iterative Online Updates

The majority of the supported item recommendation operators has a built-in online update option. This option allows users of the extension to update recommendation models with new data, without retraining. An updated model is an approximation of the retrained model, but its execution time of online updates is much shorter than model retrain execution time. The time difference increases with the size of the problem. We can see a typical Item Recommendation online update workflow in Figure 9.3.

The data management of the online update workflows is the same as in the item recommendation workflow; there is a train (1), a test (7) dataset, and the Item k-NN model (3) trained on the train data (1), and applied on the test data (7). The main difference between these two workflows is that in the online update workflow there is an additional dataset, the query dataset (4), which is being used for model update after the model building, and

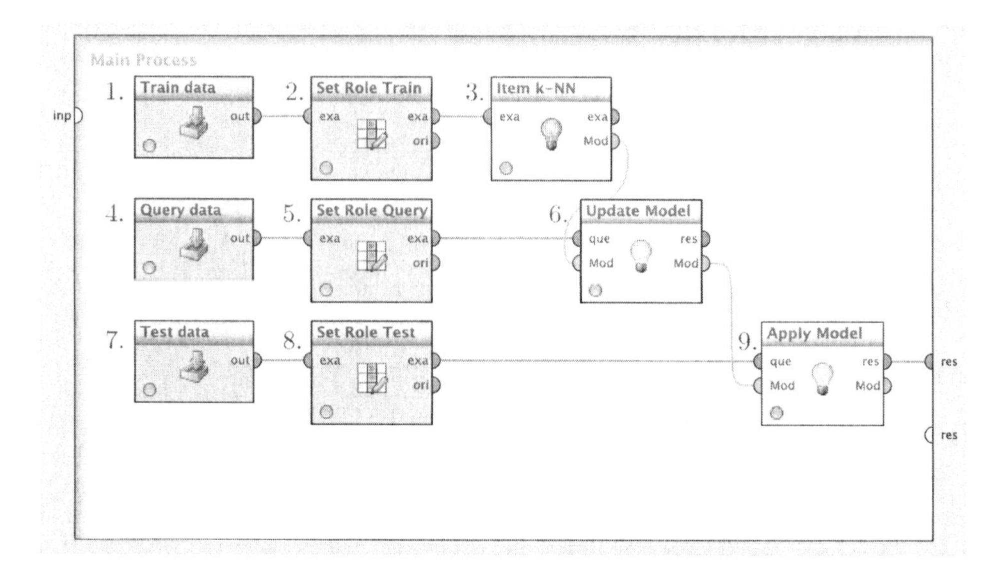

FIGURE 9.3: Item recommendation online update workflow.

before the model application. In order to use the query data for model updating without retraining the whole model, the query dataset is inputted to the Apply Model operator (6) with the online update parameter checked. Note that the Apply Model operator (6) that performs the online updates, instantaneously applies the new updated model and calculates the best n items for all the users in the query set. To apply the newly updated model to a test set, we pass the model and the appropriate test set to a new Apply Model operator (9).

9.5 Content-based Recommendation

Whereas collaborative filtering uses existing users' usage information, content-based recommendation [22] uses item description, together with a user profile for recommending new items, similar to ones used before. For example, in a restaurant, web page, or news articles recommendation setting, textual content of users' reviews, web pages, or articles can be used for recommendation, respectively. In general, any item description such as free-text description, taxonomy, or structured description of the item, can be used for recommendation purposes. Other than obtaining sufficient information about the item, maintaining a descriptive user profile that describes his past usage and/or his preferences is required. Such user profile is used together with the item description to infer new recommendations.

We can already observe that content-based recommendation is domain dependent and relies on the type of data we can use. The need for the sufficient amount of information for all the items, and the inability to produce new interesting recommendations, that are not similar to previously used items, are the weaknesses of the content-based system recommendation approach. However, content-based systems are widely used when there is no user tracking, and are very useful for further recommendation filtering of collaborative approaches.

Based on the type of data usage and user profile creation, we can divide content-based recommender systems into two methods: attribute based, also called model-based, and similarity based, also called case-based, content recommender systems. We go further into these types, along with examples of their construction.

9.5.1 Attribute-based Content Recommendation

Attribute-based content recommendation is a recommendation technique based on learning a function, i.e., a model, over a set of attributes describing items or users. In this case, the user profile is presented by the model itself. The model contains user's preferences in an applicable form and the model is used directly for recommendation. Machine learning-based models like rule induction models or a decision tree model, applied on item data, are examples of such systems. Both of the aforementioned models contain a set of rules or decisions describing a user's past preferences. These rules are directly applied to new items in order to obtain information for classification; does the item correspond to user's preferences, and possibly a numerical value describing the strength of the correspondence?

In the Recommender Extension, we implemented a simple attribute-based recommender operator for item recommendation which uses binomial user/item attributes for finding the nearest neighbours. Attribute-based recommender operators take two input datasets: the attribute dataset and the same training set used for collaborative recommender operators. The user-item matrix is not used for anything other than checking new items for which we can make a recommendation. The attribute dataset consists of pairs in which we have user or item identification as one component and attribute identification as the other component. If we have items with IDs $1 \ldots n$ in our system, and we have some item attributes with IDs $1 \ldots k$, then one pair (i, j) in our attribute dataset denotes that item i satisfies/has attribute with ID j. Every user/item in our train data must have at least one attribute in the attribute set. The trained recommendation model is used to recommend new items to each user in the query set. Figure 9.4 depicts a basic attribute-based recommender operator workflow.

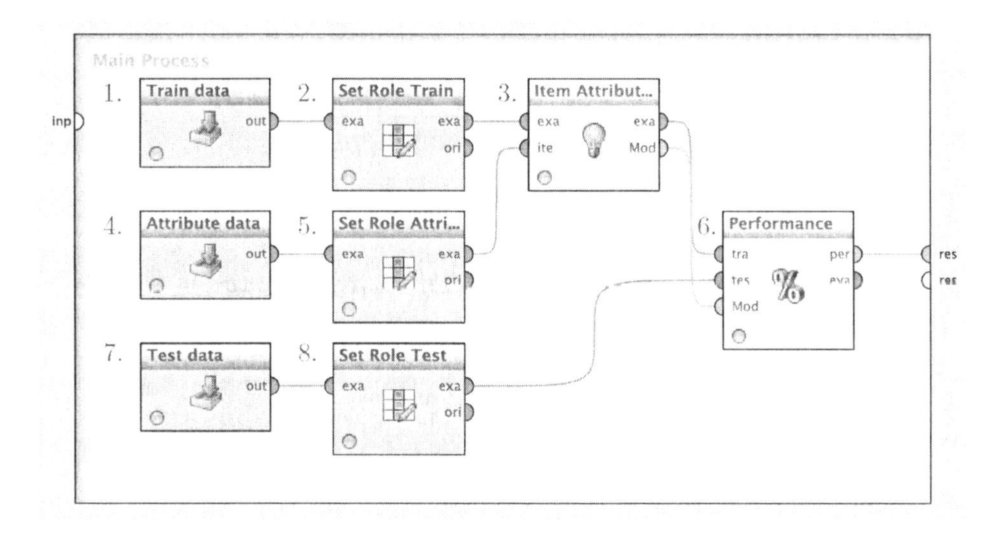

FIGURE 9.4: Item attribute recommendation workflow.

In Figure 9.4 we use three input datasets: Train data (1), Attribute data (4), and Test data (7) with their respective roles set using *Set Role* operators (2,5,8). The attribute set (4) has two attributes, item_id and attribute_id in our case. We have to set the user/item identification and the attribute identification role for those attributes before training our Attribute-based model (3). We train the Item Attribute model (3) using the Train data (1) and the Attribute data (4). Finally, the performance of the model is calculated using the Train data (1), the Test data (7), and the trained model (3) inputted into the *Performance* operator (6).

9.5.2 Similarity-based Content Recommendation

Another way of obtaining a recommendation is by using the concept of item or user similarity. New items similar to previously consumed ones might be useful to a user. Therefore, a similarity metric for obtaining a ranked list of items can be directly used for recommendation. For example, if we know a user's preference in movies (he might like old thrillers with a specific set of actors), we can find other movies with asimilar description (thrillers with some of the actors) and recommend them as the most similar ones.

Since most of the available information comes in a free textual form, we will show an example of such a recommender on our VideoLectures dataset, using RapidMiner's Text Mining operators. These operators have been previously described in this book; therefore, we will not go into further detail.

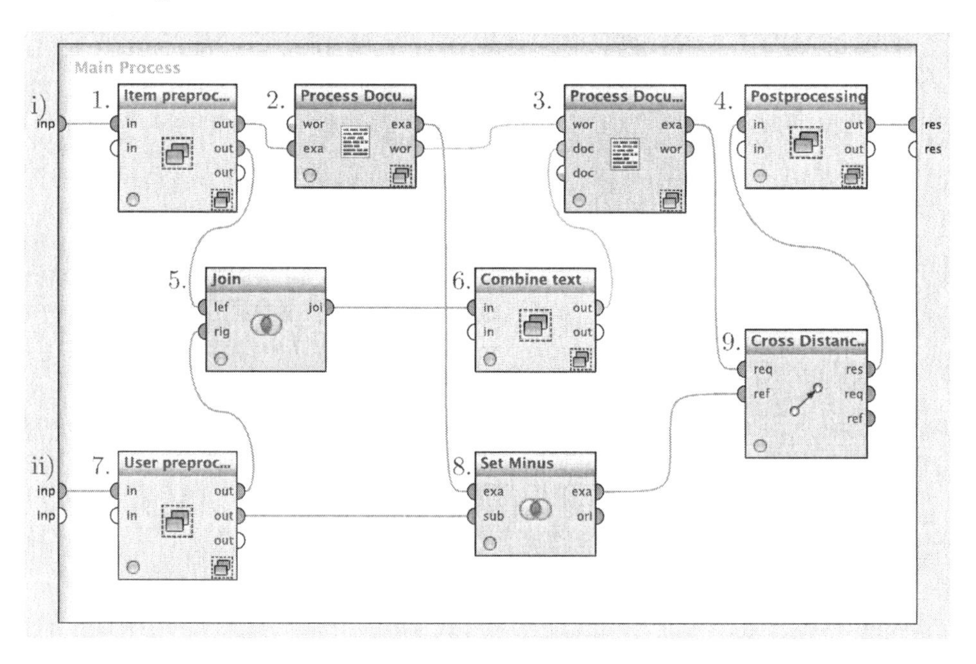

FIGURE 9.5: Similarity-based content recommendation workflow. Similarity is based on textual similarity between consumed and new items.

Figure 9.5 depicts the similarity-based content recommender operating on textual descriptions of items. The workflow takes two distinct example sets as input, defined in the workflow context: i) a dataset consisting of item IDs and appropriate textual attributes, and ii) a set of IDs of items our user consumed, also called the user profile. By defining a value of the recommendation_no macro, we limit the number of outputting recommendations, in our case 10.

The *Item pre-processing* sub-process (1) contains basic operators to set the ID role on the id attribute, to select name and description attributes (select the important text attributes) and to pass multiple copies of the resulting example set using the *Multiply* operator. The *User pre-processing* sub-process (7) sets the ID role to the `id` attribute and simply passes multiple copies of the example set. The *Join* operator's (5) role is to extract the data on the viewed lectures, following which the *Combine text* sub-process (6) essentially converts this example set to a set of documents and concatenates them all. This essentially concatenates the text of all the viewed lectures, creating a textual user profile. The *Process Documents from Data* operator (2) analyses all the video lectures text by lowering the case, tokenizing over non-letter characters, filtering out short and long tokens, removing English stopwords, and applying the Porter stemmer algorithm. The resulting example set is a TFIDF word-vector representation containing only words present in at least 1 and at most 30% of documents. The *Process Documents* (3) operator analyses the textual user profile while applying the word list produced by the previous operator. The viewed lectures are removed from the total set of the lectures by utilizing *Set Minus* operator (8). The *Cross Distance* operator (9) calculates the cosine similarity and returns the `recommendation_no` most similar items. Finally, the *Post-processing* sub-process (4) removes the `request` variable and renames the resulting variables into `recommendation` and `score`. The workflow results in the `recommendation_no` textually most similar video-lectures to the video-lectures in the user's profile, and their similarity score.

The user profile dataset is set to contain mostly random lectures, but with three lectures connected to logic, knowledge representation, and ontologies, the resulting recommendations are mostly logic and knowledge representation-based. This is due to the fact that those lectures were additionally described while the remainder were not, therefore, these lectures dominated textual analysis. This demonstrates the importance of balanced descriptions or the need for different textual analysis (i.e., extract only the top N words from each lecture). We leave further experimentation and improvements of this recommender to the interested reader.

This workflow can be easily used over RapidAnalytics for real-time usage on smaller datasets. We can radically speed it up by doing the textual analysis offline, periodically, and using only its saved results (word-vectors and word list). The problem of big data remains due to the fact that the entire workflow is unable to scale to a large number of items for real-time purpose. However, pre-calculation and further optimization may reduce the execution time.

9.6 Hybrid Recommender Systems

The hybrid approach to recommender systems consists of combining different recommendation techniques [23, 24, 25] in order to achieve better performance and to circumvent limitations of single techniques [2]. Although there are several views on the types of hybrid systems one can create stated in the literature [2, 25], we reason that, regardless of the type of the hybrid system, most of the hybrid methods can be produced in RapidMiner with the Recommender Extension. This is because RapidMiner is a powerful environment for experimentation with various types of models and data transformation methods. We will not go into the full extent of implementing all of the possible hybrid methods. Nevertheless, we will implement one simple hybrid method, namely the weighted hybrid recommender system [25], and give several tips on constructing other hybridization methods.

The construction of the simple weighted hybrid recommender system, which uses both content and collaborative data, as described in Section 9.3, is straightforward using the Model Combiner operator. This construction is performed in several steps.

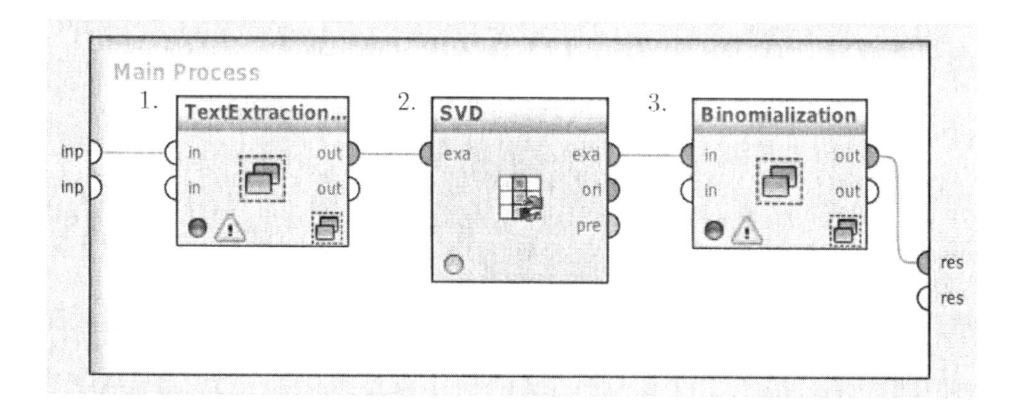

FIGURE 9.6: Text analysis workflow for VideoLectures Case study.

The first thing we need to do is to download the content data for Videolectures.net in RapidMiner format from the myExperiment repository.[3] This can be directly imported into the RapidMiner repository. If we construct a dataset which contains item textual attributes in binomial form, then we can use the `attribute-based item recommendation` operator from the Recommender Extension. We will use the standard text analysis workflow, as used in Section 9.5.2, to represent each lecture item as word vectors.

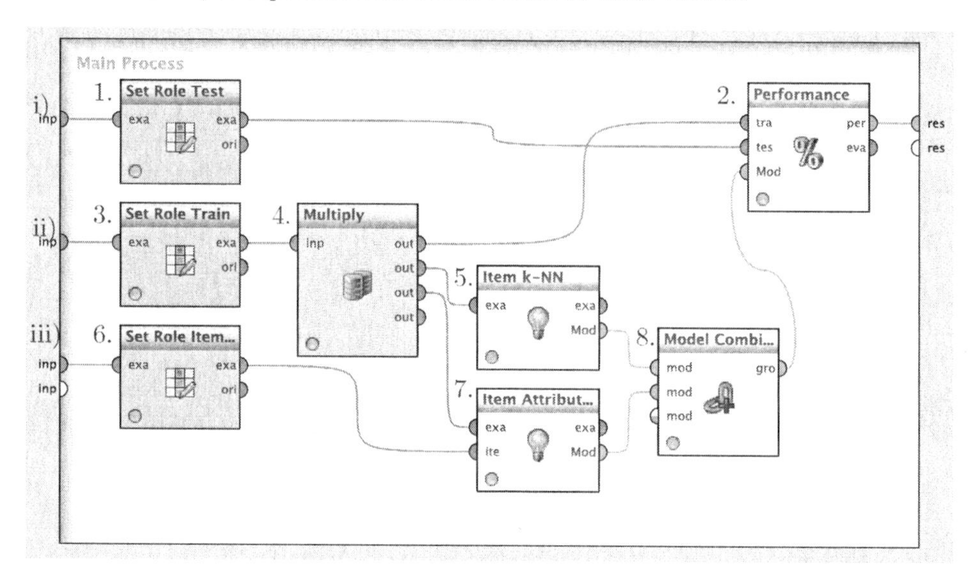

FIGURE 9.7: Hybrid recommender system for VideoLectures Case study.

Next, we need to download the workflow[4] for the text analysis from the myExperiment repository (see Figure 9.6). This workflow takes all textual attributes for each lecture item and performs textual processing. The text is analysed (1) by first transforming it to the

[3] `http://www.myexperiment.org/files/765.html`
[4] `http://www.myexperiment.org/workflows/2932.html`

lower case, then tokenizing it by splitting on all non-letter characters, and then filtering the resulting tokens with respect to token length, filtering out tokens shorter than 4 and longer than 10 letters. The last steps are stopword filtering (using English stopwords) and stemming using Porter stemmer algorithm. In the end, tokens are discarded if they show up in too few (less than 2%) or too many (more than 10%) documents. The result of the analysis is a word-vector example set A of binary-term occurrence vectors for each lecture item.

After the text processing part, we use Singular Value Decomposition (2) $A_k = U_k * S_k * V_k^T$ to capture latent semantics between lecture items and words and to obtain low-dimensional representation of lecture items. The low-dimensional representation of lecture items U_k was obtained by calculating $k = 50$ greatest left singular vectors of matrix A with the SVD RapidMiner operator. The items are represented as word-vectors in the original space, where each row in matrix A represents a word-vector of a particular item. Matrix U_k, on the other hand, represents items in the k dimensional concept space. After the SVD reduction procedure, the item-concept example set was transformed to binomial format (3), in order to be directly applicable with the `attribute-based item recommendation` operator.

After content data pre-processing, we are ready to create a full hybrid recommender system (see Figure 9.7). This hybrid recommender will use the synthetic collaborative data and the previously constructed binominal attribute-based content data. This collaborative and processed content data can also be downloaded[5] from the myExperiment portal. The resulting workflow[6] is now simple. We first set the appropriate roles (1,3,6) (user id, item id) to the dataset (i), ii), iii)) attributes. Then we feed the training set to the *Item k-NN* operator (5) and the *Item Attribute k-NN* operator (7). The *Item Attribute k-NN* operator (7) additionally receives the previously constructed binominal attribute-based content data (iii)). The models resulting from both recommender operators are combined via the *Model Combiner* operator (8) with the same weights applied to both models. This constitutes our hybrid recommender, which is then subjected to performance evaluation via the *Performance* operator (2).

Other weighted combinations of various recommendation models are possible too, as well as the other types of hybrid systems, mentioned in the literature [2, 25]. By utilizing the *Branch* or *Select Subprocess* over various conditions on data, or even clustering results, one can create switching hybrid recommender systems [25]. A switching system could, for example, detect if there is enough data on the user for the collaborative filtering method, and if there is not, direct the recommender system to execute a content-based recommendation. A mixed hybrid system [25] can combine lists resulting from different recommender systems into a single one, by utilizing RapidMiner's *Data Transformation* operators for concatenating, sorting, and filtering the results. The cascade hybrid system [25] can be built by employing attribute and item recommendation models working in a cascade. In the end, the plethora of available operators in RapidMiner enables easy construction of almost every imaginable hybrid combination, therefore solving a wide range of recommender system problems.

[5]`http://www.myexperiment.org/files/674.html`

[6]`http://www.myexperiment.org/workflows/2684.html`

9.7 Providing RapidMiner Recommender System Workflows as Web Services Using RapidAnalytics

The most suitable place for recommender systems is the world wide web, therefore, we explain how to construct web services from our recommender system workflows by using the RapidAnalytics server. Firstly, we explain how to transform a simple item recommendation workflow into a web service. Subsequently, we will explain how to construct more complex recommender system web engine.

9.7.1 Simple Recommender System Web Service

In this section, we will create one simple item recommendation web service. The simple recommendation web service has a requirement to return top-n recommendations for a specific user request. The simple recommendation web service workflow (see Figure 9.8) takes three inputs:

1. Macro value `user`—identification number of user request
2. Train set from RM repository
3. Learned Model from RM repository

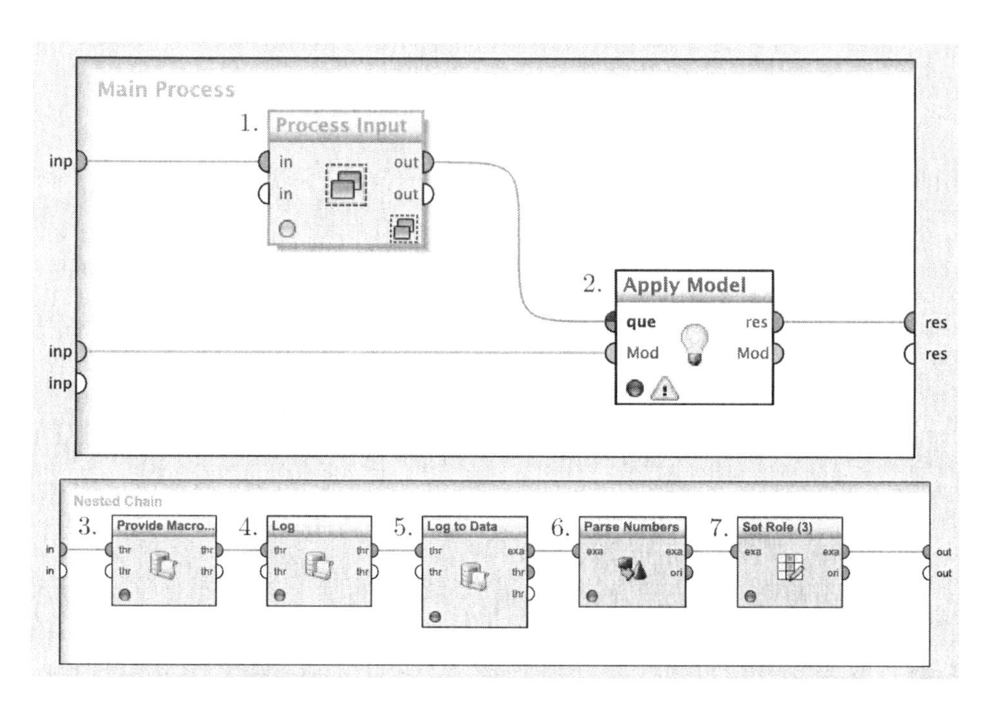

FIGURE 9.8: Simple recommendation web service workflow (upper image) and the content of the ProcessInput operator (lower image).

This workflow[7] takes the macro value (defined in the process context) of a user request and computes the top-n recommendations from an already learned model on a train set. The

[7] `http://www.myexperiment.org/workflows/2959.html`

macro value is transformed to an example set (operators 3, 4 and 5), the attribute *user_id* is parsed to a number (operator 6), and the appropriate *user identification* role is set (operator 7) within the *ProcessInput* sub-process operator (1). The next step is to save this workflow and the input datasets (Train and Model) to the RapidAnalytics server repository. We will assume access (Repository URL, User name, and Password) to some RapidAnalytics repository. If there is none, then installation and configuration of RapidAnalytics is required.[8] The next step is to access the workflow from the RapidAnalytics web interface and choose the option *Expose as service* from the process *Execution* options.[9] In order to speed-up the web service response time, *Cache input* option should be checked. Finally, we can *submit* the service (see Figure 9.9). Now, we can test this simple recommender system web service from a web browser by typing the public[10] url link from the web service, together with the user request parameter.[11]

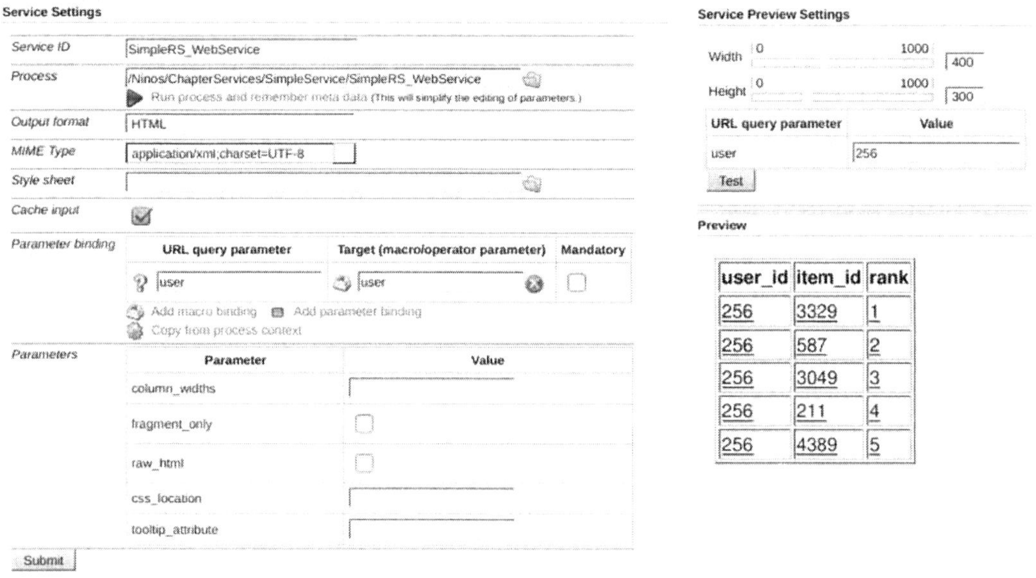

FIGURE 9.9: Settings and test of the simple recommender web service.

9.7.2 Guidelines for Optimizing Workflows for Service Usage

In the previous subsection we demonstrated how to construct a simple recommendation web service. However, in order to build a recommendation engine web service that needs to serve a website in real time, we need to assure response time in milliseconds. Therefore, we need to devise a more complex architecture of recommender web services. In this section, we will provide guidelines for constructing optimized web services in RapidAnalytics. Our recommender system web engine architecture (Figure 9.8) consists of the following parts in RapidAnalytics server:

1. Front-end recommendation web service

[8] `http://rapid-i.com/`

[9] `http://tinyurl.com/RapidAnalyticsTutorial`

[10] Public url link in RapidAnalytics (1.1.014): `http://RapidAnalyticsServerURL/RA/public_process/ServiceName`

[11] Web service public url link + string "?user=25"

2. Write activity web service
3. Online update recommendation web service
4. Offline update recommendation web service

These services communicate with the RapidMiner repository on RapidAnalytics and with the SQL database. In order to have low recommendation response time, we save the top-n recommendations for each user in the SQL database.

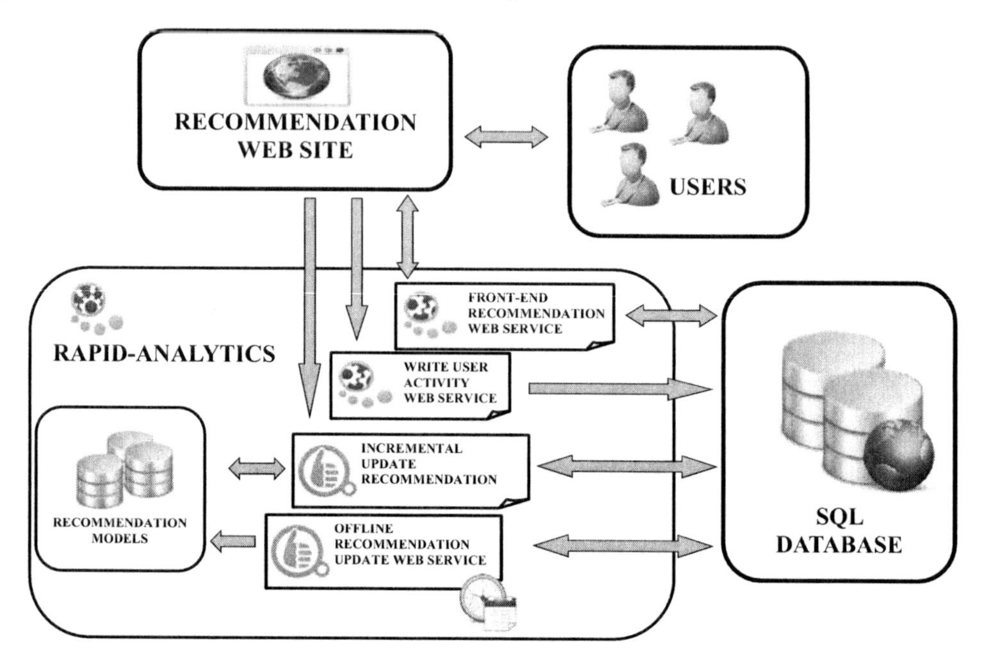

FIGURE 9.10: Architecture of recommendation web engine.

Our SQL database has only two tables:

- Item recommendation table with the following columns; (`user_id, item_id, rank`)
- Train set table with the following columns; (`user_id, item_id`)

Note: It is very important to put indexes on columns: (`user_id, item_id`) in both tables to reduce query response time.

The front-end recommendation web service[12] has a simple job to query the cached recommendations from the *item recommendation table*. When a web page requires a recommendation for a specific user, it calls the front-end recommendation web service. When the specific user i consumes certain item j, the write activity web service[13] is called, which writes activity (i, j) to the *train set table* and removes recommendation j for user i from the *item recommendation table* in the SQL database.

After an arbitrary number of recommendations to specific users, the system has to update the recommendations in the *item recommendation table*. This is accomplished by calling the online update recommendation web service,[14] which updates the recommendation model in RapidAnalytics repository and updates the recommendations for specific

[12]http://www.myexperiment.org/workflows/2904.html
[13]http://www.myexperiment.org/workflows/2906.html
[14]http://www.myexperiment.org/workflows/2955.html

users in the *item recommendation table*. Note that the online update procedure updates recommendations only for users for which some recommendations and feedback have been obtained. Therefore, periodically we have to do a full retraining on the whole train dataset by the offline update recommendation web service.[15]

9.8 Summary

Recommender systems became essential in an information- and decision-overloaded world. They changed the way users make decisions, and helped their creators to increase revenue at the same time. Bringing recommender systems to a broader audience is essential in order to popularize them beyond the limits of scientific research and high technology entrepreneurship. The goal of the Recommender Extension for RapidMiner and this chapter was to bring recommenders to a broad audience, in a theoretical, practical, and above all, applicational way.

In this chapter we presented recommender systems and their different techniques: collaborative filtering, content-based recommender systems, and hybrid systems. We presented the advantages and disadvantages of each of those systems and demonstrated how they could be implemented easily in RapidMiner. The application of recommender systems outlined was just a small introduction to the possibilities of the extension. We hope you will use the knowledge obtained through this chapter in your own applications, problems, and businesses, and that recommender systems will assist you in reaching quality, informed decisions.

Glossary

Recommender System Recommender System is an information-filtering system that predicts ratings or preferences for particular items to specific users.

User-Item Matrix The user-item matrix is a matrix which contains the interaction features between users and items, such as ratings or preferences.

Collaborative Filtering Collaborative Filtering is a group of recommender system techniques for predicting user ratings or preferences based on ratings or preferences of many other users.

Content-Based Recommendation Content-Based Recommendation is a group of recommender system techniques which make predictions about user ratings or preferences by collecting content data about items and particular users.

Hybrid Recommender System Hybrid systems are recommender systems which make predictions about user ratings or preferences by combining the collaborative and the content-based recommendation techniques.

Matrix Factorization - Matrix factorization is a decomposition of a matrix into a product of two or more matrices.

[15]`http://www.myexperiment.org/workflows/2954.html`

Bibliography

[1] Paul Resnick and Hal R. Varian. Recommender systems. *Communications of the ACM*, 40(3):56–58, March 1997.

[2] Gediminas Adomavicius and Alexander Tuzhilin. Toward the Next Generation of Recommender Systems: A Survey of the State-of-the-Art and Possible Extensions. *IEEE Transactions on Knowledge and Data Engineering*, 17(6):734–749, June 2005.

[3] Linyuan Lü, Matúš Medo, Yi-Cheng Zhang Chi H. Yeung, Zi-Ke Zhang, and Tao Zhou. Recommender systems. *Physics Reports*, February 2012.

[4] Miquel Montaner, Beatriz López, and Josep Lluís de la Rosa. A taxonomy of recommender agents on the Internet. *Artificial Intelligence Review*, 19:285–330, June, 2003.

[5] Steffen Rendle, Karen Tso-Sutter, Wolf Huijsen, Christoph Freudenthaler, Zeno Gantner, Christian Wartena, Rogier Brussee, and Martin Wibbels. Report on state of the art recommender algorithms (update), MyMedia public deliverable D4.1.2. Technical report, 2011.

[6] David Goldberg, David Nichols, Brian M. Oki, and Douglas Terry. Using collaborative filtering to weave an information tapestry. *Communications of the ACM*, 35(12):61–70, December 1992.

[7] Will Hill, Larry Stead, Mark Rosenstein, and George Furnas. Recommending and evaluating choices in a virtual community of use. In *Proceedings of the SIGCHI Conference on Human factors in computing systems*, CHI '95, pages 194–201, May 1995.

[8] Paul Resnick, Neophytos Iacovou, Mitesh Suchak, Peter Bergstrom, and John Riedl. Grouplens: an open architecture for collaborative filtering of netnews. In *Proceedings of the ACM Conference on Computer Supported Cooperative Work*, CSCW '94, pages 175–186, October 1994.

[9] Upendra Shardanand and Pattie Maes. Social information filtering: algorithms for automating "word of mouth". In *Proceedings of the SIGCHI Conference on Human Factors in Computing Systems*, CHI '95, pages 210–217, May 1995.

[10] Matej Mihelčić, Nino Antulov-Fantulin, Matko Bošnjak, and Tomislav Šmuc. Extending RapidMiner with recommender systems algorithms. *Proceedings of the 3rd RapidMiner Community Meeting and Conference*, pages 63–74, August 2012.

[11] Matej Mihelčić, Nino Antulov-Fantulin, and Tomislav Šmuc. *Rapid Miner Recommender Extension - User Guide*, 2011.

[12] Zeno Gantner, Steffen Rendle, Christoph Freudenthaler, and Lars Schmidt-Thieme. MyMediaLite: a free recommender system library. In *Proceedings of the 5th ACM Conference on Recommender Systems*, RecSys '11, pages 305–308, October 2011.

[13] Matko Bošnjak, Nino Antulov-Fantulin, Tomislav Šmuc, and Dragan Gamberger. Constructing recommender systems workflow templates in rapidminer. *Proceedings of the 2nd RapidMiner Community Meeting and Conference*, pages 101–112, June 2011.

[14] Brian McFee and Gert Lanckriet. Metric learning to rank. In *Proceedings of the 27th International Conference on Machine Learning (ICML'10)*, June 2010.

[15] Nino Antulov-Fantulin, Matko Bošnjak, Martin Žnidaršič, Miha Grčar, and Tomislav Šmuc. ECML/PKDD 2011 Discovery Challenge Overview. In *Proceedings of ECML-PKDD 2011 Discovery Challenge Workshop*, pages 7–20, September 2011.

[16] Nino Antulov-Fantulin, Matko Bošnjak, Vinko Zlatić, Miha Grčar, and Tomislav Šmuc. Memory biased random walk approach to synthetic clickstream generation. *arxiv.org/abs/1201.6134*, 2012.

[17] Asela Gunawardana and Guy Shani. A survey of accuracy evaluation metrics of recommendation tasks. *Journal of Machine Learning Research*, 10:2935–2962, December, 2009.

[18] Steffen Rendle, Christoph Freudenthaler, Zeno Gantner, and Lars Schmidt-Thieme. BPR: Bayesian personalized ranking from implicit feedback. In *Proceedings of the Twenty-Fifth Conference on Uncertainty in Artificial Intelligence*, UAI '09, pages 452–461, 2009.

[19] Badrul Sarwar, George Karypis, Joseph Konstan, and John Riedl. Incremental singular value decomposition algorithms for highly scalable recommender systems. In *Fifth International Conference on Computer and Information Science*, pages 27–28, 2002.

[20] Dave Zachariah, Martin Sundin, Magnus Jansson, and Saikat Chatterjee. Alternating least-squares for low-rank matrix reconstruction. *IEEE Signal Process. Lett.*, 19(4):231–234, 2012.

[21] Yunhong Zhou, Dennis Wilkinson, Robert Schreiber, and Rong Pan. Large-scale parallel collaborative filtering for the Netflix prize. In *Proceedings of the 4th International Conference on Algorithmic Aspects in Information and Management*, AAIM '08, pages 337–348, 2008.

[22] Michael J. Pazzani and Daniel Billsus. The adaptive web. pages 325–341. Springer, 2007.

[23] Marko Balabanović and Yoav Shoham. Fab: Content-based, collaborative recommendation. *Communications of the ACM*, 40(3):66–72, March 1997.

[24] Justin Basilico and Thomas Hofmann. Unifying collaborative and content-based filtering. In *Proceedings of the 21st International Conference on Machine learning*, ICML '04, pages 65–72, August 2004.

[25] Robin Burke. The Adaptive Web. pages 377–408. Springer, 2007.

Chapter 10

Recommender System for Selection of the Right Study Program for Higher Education Students

Milan Vukićević

Faculty of Organizational Sciences, University of Belgrade, Belgrade, Serbia

Miloš Jovanović

Faculty of Organizational Sciences, University of Belgrade, Belgrade, Serbia

Boris Delibašić

Faculty of Organizational Sciences, University of Belgrade, Belgrade, Serbia

Milija Suknović

Faculty of Organizational Sciences, University of Belgrade, Belgrade, Serbia

Abstract

Educational Data Mining (EDM) is a new emerging field that discovers knowledge from data originating from educational environments. It has a focus on different aspects of educational objects, for example, students, teachers, teaching materials, organization of classes, etc. In this research, we used RapidMiner for development of student success prediction models for a recommender system for selection of the correct study program for higher education students.

10.1 Introduction

Educational data mining (EDM) is an emerging field that attempts to discover knowledge using data originating from educational (traditional or distance learning) environments in order to improve the learning process from various aspects [1]. Increasing research interests in using data mining in education is recorded in the last decade [2] [3] [4] [5] with a focus on different aspects of the educational process, i.e., students, teachers, teaching materials, organization of classes, etc.

In this chapter, we present a recommender system designed in RapidMiner [6] for selection of the correct study program for higher education students. The main idea is early analysis of students' success on each study program and recommendation of a study program where a student will likely succeed.

Operators used are Select Attributes, Sub-process, Multiply, Filter examples, Loop, Wrapper Split Validation, X-Validation, Set Macros, Provide Macro As Log Value, Log, Optimize Selection, Select Sub-process, Write Weights, Naïve, Decision Tree, W-LMT, W-RandomForest, and W-SimpleCart.

10.2 Literature Review

RapidMiner is often successfully used in the application of classification algorithms [7]. Furthermore, it provides a support for Meta learning for classification [8] and constructing of recommender system workflow templates [9]. In this paper, we focus on building recommender system for higher education students.

[10] conducted a systematic survey on EDM from 1995 to 2005. Due to the increasing popularity and volume of research conducted in this area, these authors produced another survey about the state-of-the-art in this area, with over 300 references cited [10]. In this paper, we will focus solely on research that is closest to our work. A study [11] was conducted in order to investigate how data mining techniques can be successfully used for adaptive learning. [1] described how different data mining techniques can be used in e-learning environment to improve the course and the students' learning.

Applications or tasks that have been resolved through data mining techniques are classified in eleven categories: Analysis and visualization of data, providing feedback for supporting instructors, recommendations for students, Predict students' performance, student modeling, detecting undesirable student behaviors, grouping students, social network analysis, developing concept maps, constructing courseware, planning and scheduling [1].

One of the most frequent research topics in the area of EDM (also investigated in this research) is the prediction of student performance [1] [11] [12]. The main idea behind this research direction is that, based on student activity, one can predict the future outcome of student performance. For the purpose of predicting students' final outcome on a course, researchers used various techniques and algorithms. [13] proposed an incremental ensemble of classifiers as a technique for predicting students' performance in distance education. Neuro-fuzzy system rules are used for student knowledge diagnosis through a game learning environment [14]. In addition, [15] proposed a prototype version of decision support system for prediction of students' performance based on students' demographic characteristics and their marks in a small number of written assignments.[16] used neural networks (multilayer perception), and [17] utilized a combination of Artificial Neural Networks and Evolutionary

Computation models to predict students' performance. Similarly, [18] used a genetic algorithm for optimization of multiple classifier performance. Delgado et al., [19] used neural networks to predict success of students defined with binary classes (pass or fail).

10.3 Automatic Classification of Students using RapidMiner

In this section, we will describe data, RapidMiner processes, and the evaluation results of the processes for automatic classification of students.

10.3.1 Data

The data represents real-life data from the Faculty of Organizational Sciences, University of Belgrade, Serbia. In the Faculty, there are two study programs, Information Systems (IS) and Management (MN). Students select the study program in the second year of studies as the first year is common to all students. The data involves 366 records regarding graduated students and their success. Predictors include the success on the first-year exams (11 attributes), together with the number of points on entrance examination, sex, and the student's region of the country. The output variable is the average grade of the student, following graduation. In this chapter, we are using classification algorithms for prediction of the student's success, therefore, the average grade is discretized (this will be explained in next sub-section) on "Bad", "Good", and "Excellent" as class. Based on the predicted membership of students in one of the classes on study programs, a recommendation of the study program is made.

10.3.2 Processes

For the purpose of implementation and evaluation of the system, we designed two RapidMiner processes. The first process, "simple", automatically evaluates several classification algorithms. The second process, "complex", together with automatic evaluation of the algorithms, also includes automatic feature selection for every algorithm. This is conducted in order to find the feature subset that best suits the classification algorithm judging by classification accuracy. The main process, which includes, data preparation phase, is the same for both of the processes ("simple and complex") and it is demonstrated in Figure 10.1.

"Read AML" (operator 1) is used for loading the data. "Select attributes" (operator 2) is used for selection of predictors (described in text above) from the original dataset. In this operator, we removed the attribute "Years_Of_Studies", because this attribute does not help us with the result of the process (because large majority of students finish studies after four years).

In this study we evaluated classification algorithms, and so, we had to discretize the output variable (operator 3). Students' average grade is discretized into 3 classes, "Bad", "Good", and "Excellent". All students that had the average grade below 8 are classified as "Bad", between 8 and 9, "Good", and between 9 and 10 "Excellent". Setting of the "Discretize" operator is showed on Figure 10.2.

After discretization, the "label" role (operator 4) is assigned the output attribute "Average_Grade".

Further, the "Subprocess" (operator 5) operator is used. This operator introduces a process within a process. Whenever a "Sub-process" operator is reached during a process

TABLE 10.1: Description of I/O variables used in this study.

Variables		**Variable Description**
Input variables	*Sex*	Student gender Value: male/female (1/0) type: binominal
	Score on enter qualification exam	Points achieved in the entrance qualification exam for Faculty Values: 40–100 type: real
	Study program	Choice of study program in time when student is enrolled in Faculty Values: Information system and technology program and Management program type: nominal
	Grades at the first year of studies	Grades: marks of each of 11 subjects: examined at the first year of studies Values: 6–10 type: integer
Output variables	*Students' success*	Students' success at the end of studies Values:6–10 type: continuous

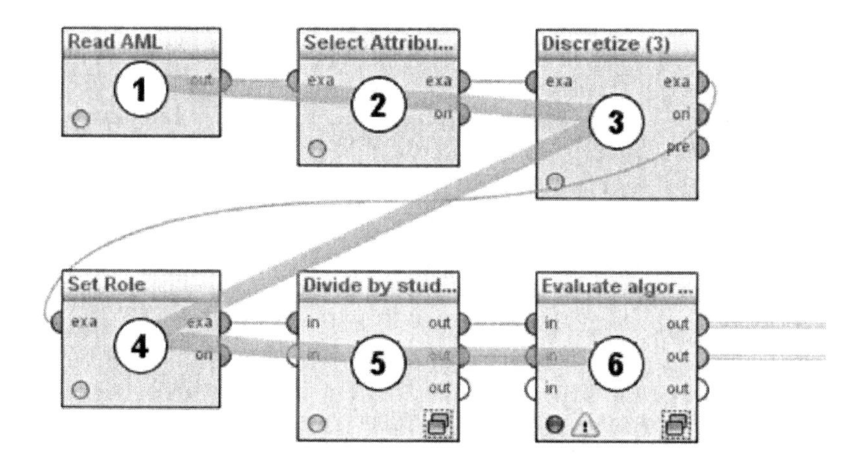

FIGURE 10.1: Process for automatic evaluation of classification algorithms.

Edit Parameter List: **classes**
Defines the classes and the upper limits of each class.

class names	upper limit
Bad	8.0
Good	9.0
Excelent	10.0

Add Entry | Remove Entry | Ok | Cancel

FIGURE 10.2: Discretization of the output attribute.

Set Role

name	Average_Grade
target role	label
set additional roles	Edit List (0)...

FIGURE 10.3: Assigning of "label" Role to the "Average_Grade" attribute.

execution, the entire sub-process is executed and then the flow is returned to the parent process. We used this operator to encapsulate the process for division of the data based on study program ("Divide by study program" on Figure 10.1). This way we make the main process easier to follow.

Separation of student's data based on different study programs is done because the data can have different distributions. Therefore, for every study program different classification models will be created and evaluated, separately. This way the solution becomes more robust when adding new data. For the separation of the data, we used "Multiply" (operator 1) to make two copies of the original dataset and then "Filter examples" (operators 2 and 3) to separate the data by study programs (Figure 10.4).

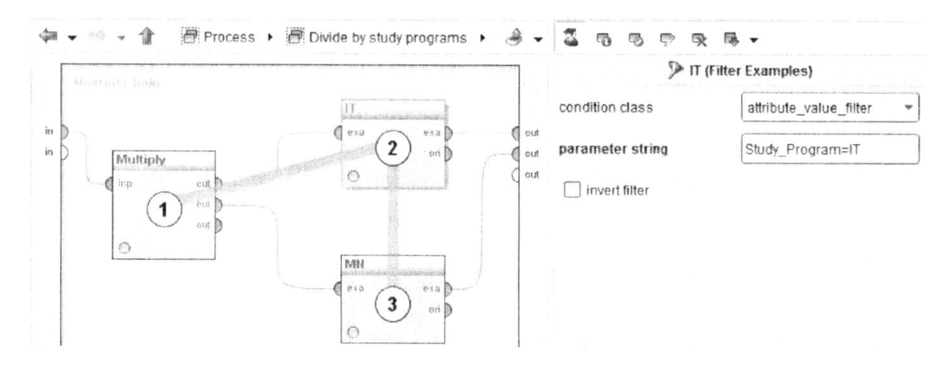

FIGURE 10.4: Separation of the data by study programs.

Data, separated by study programs, is forwarded to the "Evaluate algorithms" sub-process (operator 6 on Figure 10.1).

For every study program, the "Loop" operator is defined (nested in "Evaluate algorithms" sub-process) for iteration over the algorithms in the experiment. We used the "Loop" operator for both study programs which are separated in "Filter examples" operators (Figure 10.4). Because inner operators for both "Loop" (operators 1 and 2 on Figure 10.5) operators are the same, the following text will focus solely on the one that deals with the IT study program (operator 1 on Figure 10.5). "Loop" contains the "iteration" macro (in our case its start value is set to 1 and the number of iterations to 5 since we are evaluating 5 algorithm) which is used as a parameter for the "Select Sub-process" operator for selection of the algorithm in every iteration (this will be more thoroughly explained in the following sub-section).

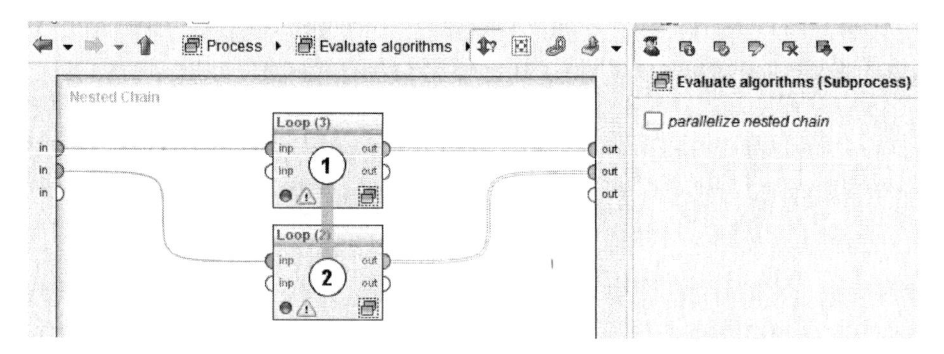

FIGURE 10.5: Separation of the data by study programs.

10.3.2.1 Simple Evaluation Process

Simple evaluation process (Figure 10.6), as an inner operator of "Loop" (described in previous sub-section), uses "Simple Validation" (operator 1) for separation of training and testing data, creating and evaluation of models. For splitting the data in all experiments we used stratified sampling with split ratio of 0.9. In order to enable reproducibility of the results, local random seed is used with a value of 1992.

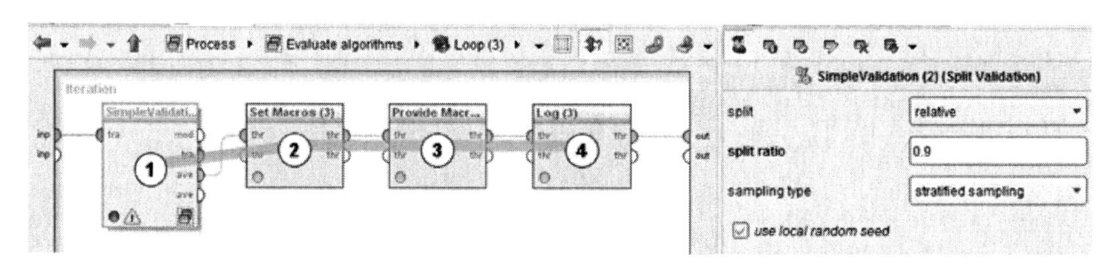

FIGURE 10.6: "Simple Validation" and logging operators.

The remaining three operators in Figure 10.6 are used for logging of evaluation results. "Set Macros" defines a macro "Study program" (with values "IT" and "MN") and "Provide Macro" provides these values to "Log". "Log" writes the results (Study program, Algorithm and the performance) in the ".csv" file.

The first inner operator of "Simple Validation" ("Training") is "Select sub-process", which contains five different classification algorithms. The second inner operator ("Testing") contains the "Apply Model" and "Performance (Classification)" operators (Figure 10.7).

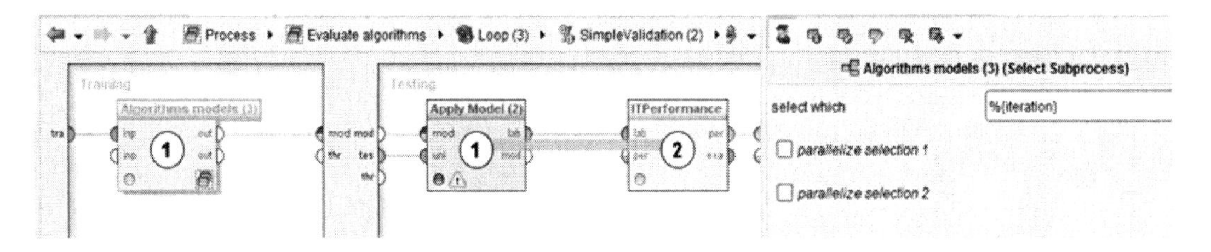

FIGURE 10.7: "Simple Validation" inner operators.

In order to automatically evaluate all defined algorithms, we used "Select Sub-process" (operator 1 in "Training" part on Figure 10.7). This operator consists of multiple sub-processes but it executes only one sub-process at a time. In our case "Select Sub-process" uses "iteration macro" provided by "Loop" (as described in previous sub-section). In each iteration of the "Loop" one algorithm is executed and the model created is passed (Figure 10.8) to the "testing" sub-process (Figure 10.7), where the model is applied and evaluated. Evaluation results are written in a log file (Figure 10.6). For all of our experiments, algorithms are used with default parameters.

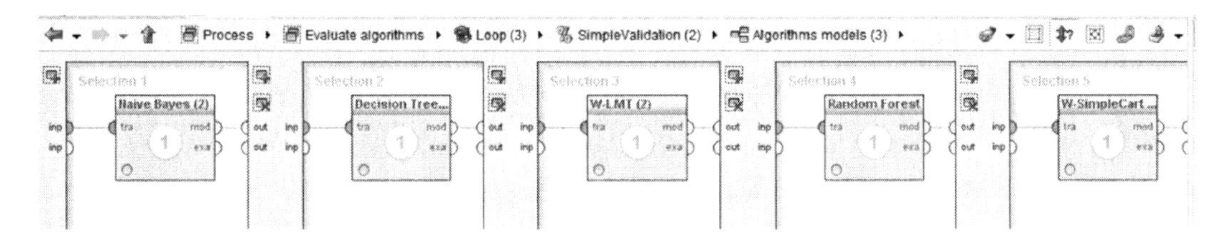

FIGURE 10.8: Classification algorithms as inner operators of the "Select Sub-process".

The evaluation process is the same for every study program, so we do not describe the evaluation process for the "MN" study program.

Finally, after execution of the process, classification performance is shown in RapidMiner for every algorithm and every study program (Figure 10.9).

Table / Plot View · Text View · Annotations

Criterion Selector: accuracy

Multiclass Classification Performance · Annotations

Table View · Plot View

accuracy: 73.33%

	true Bad	true Good	true Excelent	class precision
pred. Bad	3	1	0	75.00%
pred. Good	0	7	2	77.78%
pred. Excelent	0	1	1	50.00%
class recall	100.00%	77.78%	33.33%	

FIGURE 10.9: Classification algorithms as inner operators of the "Select Sub-process".

Note that algorithms used in this case can be replaced with any other classification algorithms. So in further text we just briefly explain the types of algorithms used in this study.

Probabilistic classifier (Naïve Bayes) assumes that the presence (or absence) of a particular feature of a class (i.e., attribute) is unrelated to the presence (or absence) of any other feature. The advantage of the Naive Bayes classifier is that it only requires a small amount of training data to estimate the means and variances of the variables necessary for classification.

A decision tree classifiers, recursively partition, example set and produces, tree-like graph in order to predict the "label" attribute. This representation of the data has the advantage, compared with other approaches, of being meaningful and easy to interpret. Algorithms of this class used in our case are: Decision tree (standard RapidMiner operator), and two algorithms from the Weka extension for RapidMiner: LMT (Logistic Model Tree) and Simple Cart.

The ensemble classifier (Random forest) consists of many random decision trees and creates a voting model of all the random trees. Ensemble classifiers often produce highly accurate models.

10.3.2.2 Complex Process (with Feature Selection)

The idea of the second process is to enable automatic feature and algorithm selection that is best suited for the data. Pre-processing of the data is done in the same way as described for the first process (Figure 10.1). Therefore, the only difference between the "simple" and "complex" process is in the evaluation phase.

In this process, besides algorithm selection, we are optimizing feature selection for every algorithm. Because of the optimization in the learning process, the learner may adjust to very specific random features of the training data, that have no causal relation to the "label" attribute ("over-fitting"). If this happens, the performance on the training examples increases, while the performance on unseen data becomes worse.

In order to avoid the "over-fitting" effect, instead of "Simple Validation" we used the "Wrapper Split Validation" operator (operator 1 on Figure 10.10). This operator evaluates the performance of feature weighting algorithms. It splits the data in training and test sets. Training set is used for optimization of feature weighting algorithms (based on classification accuracy in our case), while the test set is used for algorithm evaluation on the unknown data. This way the "over-fitting" effect is prevented. The remaining three operators in Figure 10.10 are used for logging of evaluation results the same way as described in Section 10.3.2.1.

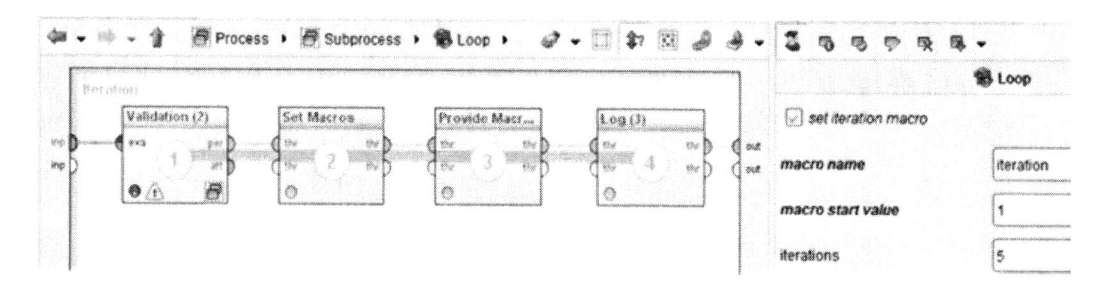

FIGURE 10.10: "Wrapper Split Validation" operator.

The parameters of "Wrapper Split Validation" (operator 1 on Figure 10.10) are set the same way as for "Simple Validation" described in the previous sub-section (split ratio of 0.9, local random seed with value of 1992).

"Wrapper Split Validation" has three inner operators (Figure 10.11). The first inner operator is the feature weighting algorithm. It returns an attribute weights vector which is applied on the data. The second inner operator defines an algorithm that produces a model based on features retrieved from the first inner operator. Performance is retrieved using the third inner operator.

For the feature selection we used the "Optimize Selection" operator. As an input, it uses an example set with all features, and allows two deterministic greedy feature selection algorithms (forward selection and backward elimination) to optimize selection of features based on some performance vector (in our experiments we tried both of these feature selection strategies, for every study program). All other parameters are set to their default values.

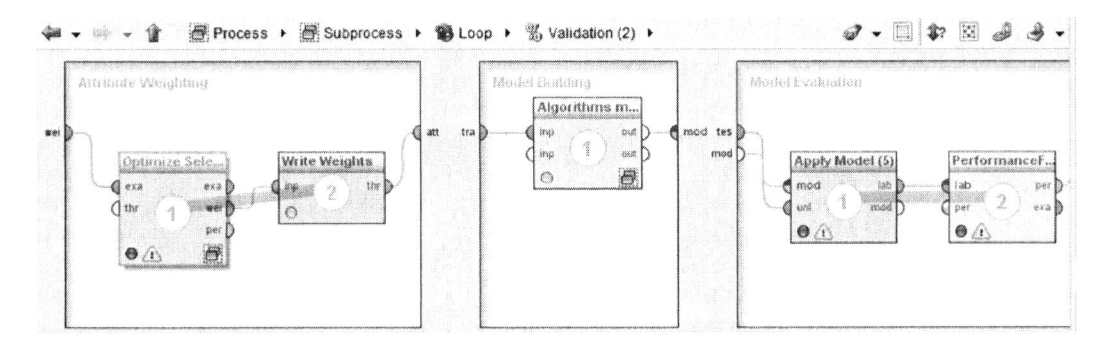

FIGURE 10.11: Inner operators of "Wrapper Split Validation".

As an inner operator of "Optimize Selection", "X-Validation" is used to estimate the performance of 5 algorithms (we used 10-fold x-validation with stratified sampling, Figure 10.12) with different selections of features. Training sub-process of "X-Validation" operator contains the "Select Subprocess" operator, which is defined exactly the same as in the "simple process" (in every iteration it selects one out of five algorithms based on "iteration" macro provided by "Loop"). So for every algorithm, feature selection is optimized based on cross-validation estimation of the accuracy.

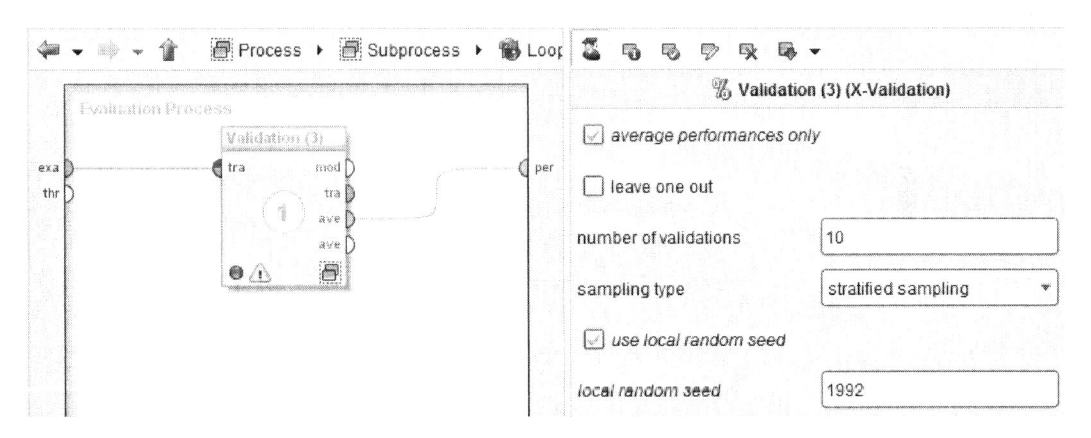

FIGURE 10.12: "X-Validation" for the estimation of algorithm performance with feature selection.

As a result of this sub-process, attribute weights are gathered for every algorithm. These weights are written in files using "Write Weights" (because they can be used with other parameter settings of the algorithms in other processes) and forwarded to the "Model Building" sub-process of the "Wrapper Split Validation" operator (Figure 10.11).

"Model building" uses the same "Select sub-process" operator (as used for accuracy estimation described for "simple process") for iterating over the 5 algorithms with the same "iteration" macro value. Every model is built on "optimal" selection of attributes passed from "Optimize Selection".

Finally, the model is passed to the "Model evaluation" part of "Wrapper Split Validation" in order to evaluate the model with selected features on the test data.

10.4 Results

When both processes are executed, users can compare the results in order to evaluate if automatic feature selection really improved the classification accuracy of the algorithms. Table 10.1 shows classification accuracy of the evaluated models for the "IT" study program. Results are demonstrated for the "simple process" (without feature selection) and for the "complex" process (with forward selection and backward elimination). The best values are displayed in bold.

TABLE 10.2: Classification accuracy for "IT" study program.

IT	**Simple**	**Complex (Forward selection)**	**Complex (Backward elimination)**
Naïve Bayes	60.00	**66.67**	60
Decision tree	**53.33(60)**	**53.33**	**53.33**
W – Linear model tree	73.33	**80**	66.67
Random forest	60	**73.33**	60
W – Simple CART	**60**	**60**	**60**

It can be seen from Table 10.1 that classification accuracy is improved (except on W-Simple CART) when forward selection is used. Backward elimination did not give any improvement in accuracy when compared with algorithms without feature selection.

Table 10.2 shows the classification accuracy of the evaluated models for the "MN" study program. Here, backward elimination shows an improvement in accuracy for 3 algorithms, while the remainder had the same accuracy as algorithms without feature selection. Forward selection did not yield any accuracy improvement, and for some algorithms this feature selection gave worse results than algorithms without feature selection.

TABLE 10.3: Classification accuracy for the "MN" study program.

MN	**Simple**	**Complex (Forward selection)**	**Complex (Backward elimination)**
Naïve Bayes	**86.36**	**77.27**	**90.91**
Decision tree	**77.27**	**72.72**	**77.72**
W – Linear model tree	**90.91**	**77.27**	**90.91**
Random forest	**63.64**	**63.63**	**72.73**
W – Simple CART	**72.73**	**63.63**	**72.73**

10.5 Conclusion

In this chapter, we used RapidMiner's possibilities for designing of complex processes to enable automatic feature and algorithm selection for classification of students' performance. Five classification algorithms were automatically evaluated with or without feature selection. Optimization of feature selection was done for every algorithm separately, meaning that different algorithms had different sets of features. It was shown that optimization of feature selection often yields better results, however, it also demands more processing time. For the purpose of classification of students' success, this is not a big problem because design and selection of the models should be done at the end of every semester.

The processes defined in this chapter could be adapted for the application in any area, just by changing the data preparation process. Adding new algorithms in the automatic evaluation process is also fairly simple. Users just have to expand the "Select Sub-process" operator with new algorithms and to set the "iterations" parameter of the "Loop" macro on the number of algorithms defined.

Bibliography

[1] Cristóbal Romero, Sebastián Ventura, and Enrique García. Data mining in course management systems: Moodle case study and tutorial. *Computers & Education*, 51(1):368–384, 2008.

[2] Varun Kumar and Anupama Chadha. An empirical study of the applications of data mining techniques in higher education. *IJACSA) International Journal of Advanced Computer Science and Applications*, 2(3):80–84, 2011.

[3] Mohammad Hassan Falakmasir and Jafar Habibi. Using educational data mining methods to study the impact of virtual classroom in e-learning. In *3rd International Conference on Educational Data Mining, Pittsburgh*, pages 241–248, 2010.

[4] Félix Castro, Alfredo Vellido, Àngela Nebot, and Francisco Mugica. Applying data mining techniques to e-learning problems. *Evolution of teaching and learning paradigms in intelligent environment*, pages 183–221, 2007.

[5] Cristobal Romero and Sebastian Ventura. Educational data mining: A survey from 1995 to 2005. *Expert Systems with Applications*, 33(1):135–146, 2007.

[6] Ingo Mierswa, Michael Wurst, Ralf Klinkenberg, Martin Scholz, and Timm Euler. Yale: Rapid prototyping for complex data mining tasks. In *Proceedings of the 12th ACM SIGKDD International Conference on Knowledge Discovery and Data Mining*, pages 935–940. ACM, 2006.

[7] M Jovanović, B Delibašić, M Vukićević, and M Suknović. Optimizing performance of decision tree component-based algorithms using evolutionary algorithms in rapid miner. In *Proc. of the 2nd RapidMiner Community Meeting and Conference, Dublin, Ireland*, 2011.

[8] Faisal Shafait, Matthias Reif, Christian Kofler, and Thomas M Breuel. Pattern recognition engineering. In *RapidMiner Community Meeting and Conference*, 2010.

[9] Nino Antulov-Fantulin Matko Bosnjak. Constructing Recommender Systems Workflow Templates in RapidMiner. In *RapidMiner Community Meeting and Conference*, 2011.

[10] C. Romero and S. Ventura. Educational data mining: A review of the state of the art. *Systems, Man, and Cybernetics, Part C: Applications and Reviews, IEEE Transactions on*, 40(6):601–618, November 2010.

[11] Ya-huei Wang and Hung-Chang Liao. Data mining for adaptive learning in a TESL-based e-learning system. *Expert Systems with Applications*, 38(6):6480–6485, 2011.

[12] César Vialardi, Jorge Chue, Juan Pablo Peche, Gustavo Alvarado, Bruno Vinatea, Jhonny Estrella, and Álvaro Ortigosa. A data mining approach to guide students through the enrollment process based on academic performance. *User Modeling and User-Adapted Interaction*, 21(1):217–248, 2011.

[13] S Kotsiantis, K Patriarcheas, and M Xenos. A combinational incremental ensemble of classifiers as a technique for predicting students performance in distance education. *Knowledge-Based Systems*, 23(6):529–535, 2010.

[14] Sinisa Ilic Kristijan Kuk, Petar Spalevic. A model for student knowledge diagnosis through game learning environment. *Technics Technologies Education Management*, 7(1):103–110, 2012.

[15] S Kotsiantis. Use of machine learning techniques for educational proposes: A decision support system for forecasting students grades. *Artificial Intelligence Review*, 37:331–344, 2012.

[16] Osmar R Zaïane and J Luo. Web usage mining for a better web-based learning environment. In *Conference on Advanced Technology for Education*, pages 60–64, 2001.

[17] Des Traynor and J Paul Gibson. Synthesis and analysis of automatic assessment methods in CS1: generating intelligent MCQs. In *ACM SIGCSE Bulletin*, volume 37, pages 495–499. ACM, 2005.

[18] Behrouz Minaei-Bidgoli, Deborah A Kashy, G Kortmeyer, and William F Punch. Predicting student performance: An application of data mining methods with an educational web-based system. In *Frontiers in Education, 2003. FIE 2003 33rd Annual*, volume 1, pages T2A–13. IEEE, 2003.

[19] M Delgado Calvo-Flores, E Gibaja Galindo, MC Pegalajar Jiménez, and O Pérez Piñeiro. Predicting students marks from moodle logs using neural network models. *Current Developments in Technology-Assisted Education*, 1:586–590, 2006.

Part IV

Clustering in Medical and Educational Domains

Chapter 11

Visualising Clustering Validity Measures

Andrew Chisholm

Institute of Technology, Blanchardstown, Dublin, Ireland

Acronyms

ARI - Adjusted Rand Index

FM - Fowlkes Mallow Index

RI - Rand Index

11.1 Overview

This chapter is about clustering, i.e., the task of automatically grouping objects (examples) into groups of similar objects. This is an unsupervised technique; no human is involved in the production of the answer. This, of course, means that the answers could be meaningless. Because there are many possible answers, all potentially equally meaningless and certainly difficult to tell apart, it is difficult for a human to understand what clustering is best. It is vital, therefore, that to allow a human to be involved and be helped to identify the best answers, methods should be provided to highlight those to focus on.

The focus of this chapter is therefore to give an example of a clustering algorithm to put data into clusters, measure how well these work, and crucially present information to a human to help drive investigations, increase understanding, and avoid meaningless answers. This crucial part involves calculating cluster validity measures and presenting the results to users and, thus gives this chapter its title.

The aim of clustering is twofold. Firstly, it seeks to separate data items into a number of groups so that items within one group are more similar to other items in the same group. Secondly, it aims to arrange so that items in one group are different from items in other groups. Visualizing how the clusters are formed and what data points are in what cluster is difficult. Fortunately, a number of cluster validity measures exist and these can be used to give guidance about potential clustering. (It is important to note that these validity measures are themselves not guaranteed to give the right answer). Instead, they show how the measure varies for different clusterings and allow a person to determine which clustering leads to something of interest. In essence, the clustering validity measures give guidance about which clusterings are interesting, thus enabling a person to focus on these specific ones.

This chapter gives a detailed walk-through of a RapidMiner process that clusters some data and produces various cluster validity measures.

11.2 Clustering

Many clustering techniques are available, and it is beyond the scope of this chapter to review these, in detail. For a more detailed review, see, for example, [1], [2], and [3].

At a high level, however, clustering techniques can be divided into partitional and hierarchical approaches. Partitional approaches attempt to divide or partition the data into regions or groupings where points in one region or group are more similar to points in the same region or group whilst being different from points in others. One partitioning approach divides the space into regions and calculates how far the points are from the centre of the region. By moving the centres using techniques to minimise the distances from centre to points, an optimum partitioning can be found. Examples of this type of algorithm are k-means [4] and k-medoids. These approaches are strongest when the points are arranged in spherically shaped clusters each of similar density. Another partitioning approach is density based. These algorithms attempt to find connected regions in the space of points where the number of points per unit of volume is similar. These approaches are stronger when the clusters are of an arbitrary shape. DB-Scan [5] is an example of an algorithm of this type.

Hierarchical clustering techniques have two main approaches. Firstly, the divisive approach starts with all points in one cluster and splits this into smaller and smaller clusters.

Secondly, the agglomerative approach starts with all points in separate clusters and builds larger clusters from these. Distance measures determine which is the furthest cluster to split or closest to merge, and stopping criteria are used to determine when to terminate. Hierarchical approaches are more suited to some problems, including situations where the data points are not arranged in spherical clusters. They can also give better results but can be more computationally expensive and suffer from the disadvantage that once a merge or split operation has happened, it cannot be reversed and this can cause problems where data are noisy.

11.2.1 A Brief Explanation of k-Means.

For the purposes of this chapter, the k-means algorithm [4] is used. This algorithm is simple and is best suited to finding clusters that are spherical in shape and are sufficiently distant from one another to be distinct. The artificial data in Section 11.4.1 fits this requirement. It is not known whether the *E-coli* data in Section 11.4.2 has similar characteristics but there is still value in using k-means with appropriate validity measures because the results are instructive as a use case for determining the value of these measures.

The k-means algorithm is a simple algorithm that finds the best location for k centroids, each corresponding to the centre of a cluster. The algorithm starts with k random locations and examines the data points in turn and assigns each to the closest centroid. For each cluster, the points within it are considered to determine a new centroid. This new centroid is then used to determine the closest points potentially causing points to move into different clusters. This process repeats until the centroid stops moving or until an error function is minimised. The strength of the algorithm is its simplicity and relatively quick run-time. Its weaknesses are that it is limited to spherical clusters of similar density and it sometimes gets stuck in a local minimum where the centroids do not represent the best clustering. One other point is that the number of clusters, k, is not determined automatically but this does allow different runs to be performed to allow cluster validity measures to be compared.

Nonetheless, the algorithm is valuable from the point of view of this chapter to demonstrate the use of a clustering algorithm and some validity measures.

11.3 Cluster Validity Measures

As with clustering, there are many validity techniques that aim to find the best clustering possible. See, for example [6], where details of various measures are given as well as an assessment of how they perform with k-means on a selection of datasets.

The validity measures are classified as internal, external, and relative and these terms will be explained below and used throughout this chapter. These techniques use a mathematical definition of *best* that may or may not translate into something meaningful in the real world. Nonetheless, they can give invaluable guidance to help give focus to what can be an overwhelming amount of data. The interpretation of validity measures is often not straightforward and comes with practice. This chapter aims to help the reader on this road.

11.3.1 Internal Validity Measures

Internal validity measures concern themselves with information contained within the clusters and how data points are arranged with respect to them. A good clustering is one where all the points in a cluster are close to one another: compactness measures characterise this aspect. In addition, clusters themselves should be well separated: separability measures characterise these. Validity measures use variations on these themes and in this chapter, the following validity measures will be used (the mathematical details are avoided where possible).

Average within centroid distance
This considers each cluster one by one and calculates the average distance of each point in the cluster from the centroid. The average of these gives the final measure. As clusters get more compact, this measure reduces. Of course, as the number of clusters increases, the average distance will decrease naturally anyway and so this measure can be difficult to interpret. Typically, what is looked for is an obvious outlier from a smooth progression.

Cluster density
This measure considers each cluster in turn and finds the average of the distances between all the pairs of points in the cluster and multiplies by the number of points in the cluster. This results in a measure that is equivalent to a distance per point within the cluster and which is, therefore, similar to a density. This measure tends to zero as the number of clusters increases, but smaller values indicate more compact clusters. The interpretation of this can be difficult, but again as with average within centroid distance, an outlier from a smooth progression is what is being sought.

Sum of squares item distribution
This measure considers each cluster in turn and divides the number of points in the cluster by the number of data points in all clusters. This is squared and the values for all clusters are summed. Where there is one big cluster, this measure tends to 1; where the clusters are all of the same size, the measure equals $1/n$ where n is the number of clusters. This measure gives an indication of how well particular clusters are changed as different starting parameters are used to produce the clustering. If the measure decreases more slowly than expected as the number of clusters increases, this gives evidence that there are large stable clusters that are remaining intact.

Davies Bouldin
This considers each cluster one by one. For each cluster, it determines which other cluster has the maximum ratio of average intracluster distance for points in the two clusters to the distance between the clusters. Having found these maxima, average them for all the clusters. The resulting value is lower if clusters are compact and far apart. This measure considers both separability and compactness and often gives clear indications of good clusterings.

11.3.2 External Validity Measures

External validity measures compare clusters that are previously known with the clusters produced by the clustering algorithm. Clearly, in real-world situations, the clusters will not be known and would have to be provided by experts as part of model development and testing. Throughout this chapter, the term *ground truth* will be used to refer to the known clusters.

The measures used in this chapter are described below. All are variations on a theme of calculating how close the ground truth is to the clusters that have been produced and so they tend to 1 if the fit is close and tend to 0 if not.

Rand Index

This measure [7] is the ratio of the sum of true positive and true negatives to the sum of true positive, false positive, true negative, and false negative.

Adjusted Rand Index

This measure [8] modifies the Rand index and makes a correction to allow for chance occurrences.

Jaccard Index

This measure is the ratio of the size of the intersection of the two clusterings to the size of their union.

Fowlkes-Mallow Index

This measure [9] gives a better indication for unrelated data and is less susceptible to noise.

Wallace Indices

These measures [10] give an indication of the probability that a pair of points are in one cluster and the other to which it is being compared.

11.3.3 Relative Validity Measures

Relative validity measures compare clusterings produced by two different algorithms to determine how different the clusters are. In this chapter, an approach is taken to use different clusterings produced by repeated applications of the k-means algorithm with different values of k to allow pairs of clusterings to be compared. One of the pairs is regarded as the ground truth to allow external validity measures to be used. This approach gives insight into how resilient clusters are as different algorithm starting conditions are used.

11.4 The Data

Two sources of data are considered. Firstly, some artificial data that is easy to visualise so clustering results and validity measures can be interpreted with the data in mind.

The second dataset contains data relating to protein sites based on experimental data for the *E-coli* organism [11], [12]. It is less likely that the reader of this book will have expert knowledge in this domain. Nonetheless, the clustering results and the guidance the validity measures give could allow a data mining practitioner to start a dialogue with such an expert.

11.4.1 Artificial Data

The artificial data consists of 100 examples generated using the *Generate Data* operator. This is best seen by looking at the graph in Figure 11.1.

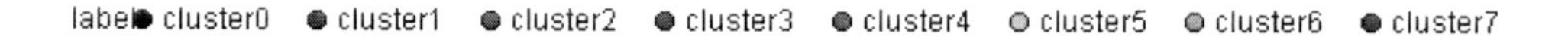

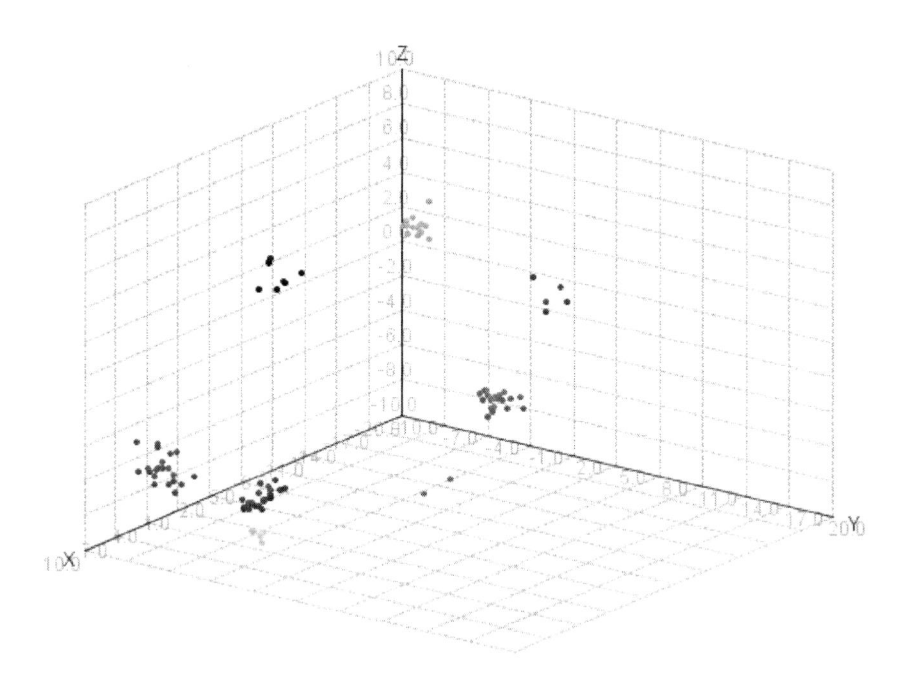

FIGURE 11.1 **(see color insert)**: 3-D scatter plot of artificial data showing 8 clusters.

The data is generated using the *Gaussian Mixture Clusters* option. With three attributes this leads to 8 clusters (2^3). More detail is given in Section 11.6.1

11.4.2 *E-coli* Data

This data [11] contains information on the sites where protein localisation occurs in *E-coli* bacteria; this is the label for the examples. In addition, there are 7 other attributes that are to be used to predict the label. In total there are 8 different sites and 336 examples. The data contains no missing values. Figure 11.2 shows some of the raw *E-coli* data. More detail is given in Section 11.6.1

The dataset contains 8 groups and previous researchers have used this data to test various classification and clustering investigations. For example, [12] achieved an 81% classification accuracy using various techniques, [13] performed clustering to achieve 75% accuracy on the data and [14] shows a 96% accuracy using an enhanced naïve Bayes algorithm.

The objective of this chapter is not to replicate the accuracy of the previous studies but to show how clustering validity measures can help focus attention to areas of interest. The

ExampleSet (336 examples, 2 special attributes, 7 regular attributes)

Row No.	site	sequenceName	mcg	gvh	lip	chg	aac	alm1	alm2
1	cp	AAT_ECOLI	0.490	0.290	0.480	0.500	0.560	0.240	0.350
2	cp	ACEA_ECOLI	0.070	0.400	0.480	0.500	0.540	0.350	0.440
3	cp	ACEK_ECOLI	0.560	0.400	0.480	0.500	0.490	0.370	0.460
4	cp	ACKA_ECOLI	0.590	0.490	0.480	0.500	0.520	0.450	0.360
5	cp	ADI_ECOLI	0.230	0.320	0.480	0.500	0.550	0.250	0.350
6	cp	ALKH_ECOLI	0.670	0.390	0.480	0.500	0.360	0.380	0.460
7	cp	AMPD_ECOLI	0.290	0.280	0.480	0.500	0.440	0.230	0.340
8	cp	AMY2_ECOLI	0.210	0.340	0.480	0.500	0.510	0.280	0.390
9	cp	APT_ECOLI	0.200	0.440	0.480	0.500	0.460	0.510	0.570
10	cp	ARAC_ECOLI	0.420	0.400	0.480	0.500	0.560	0.180	0.300
11	cp	ASG1_ECOLI	0.420	0.240	0.480	0.500	0.570	0.270	0.370
12	cp	BTUR_ECOLI	0.250	0.480	0.480	0.500	0.440	0.170	0.290
13	cp	CAFA_ECOLI	0.390	0.320	0.480	0.500	0.460	0.240	0.350
14	cp	CAIB_ECOLI	0.510	0.500	0.480	0.500	0.460	0.320	0.350
15	cp	CFA_ECOLI	0.220	0.430	0.480	0.500	0.480	0.160	0.280
16	cp	CHEA_ECOLI	0.250	0.400	0.480	0.500	0.460	0.440	0.520
17	cp	CHEB_ECOLI	0.340	0.450	0.480	0.500	0.380	0.240	0.350
18	cp	CHEW_ECOLI	0.440	0.270	0.480	0.500	0.550	0.520	0.580
19	cp	CHEY_ECOLI	0.230	0.400	0.480	0.500	0.390	0.280	0.380
20	cp	CHEZ_ECOLI	0.410	0.570	0.480	0.500	0.390	0.210	0.320
21	cp	CRL_ECOLI	0.400	0.450	0.480	0.500	0.380	0.220	0

FIGURE 11.2: Raw *E-coli* data.

data was chosen because it is relatively simple and has a previously known ground truth grouping.

11.5 Setup

Before the process is run, it is assumed that RapidMiner has been installed. The version used to produce this chapter was version 5.3.000.

R itself must also be installed and it is very important to match the architecture of the R executable to that of RapidMiner. The following R commands will display the version and architecture from the R console.

```
> R.Version()$version.string
[1] "R version 2.15.1 (2012-06-22)"
>

> R.Version()$platform
[1] "x86_64-pc-mingw32"
>
```

The version used in the production of this chapter is 2.15.1 as shown above.

In addition, the RapidMiner R extension must be downloaded and installed. This can lead to some difficulties arising from different operating systems and platforms. Some common problems and their resolution are given in the next section, although this is by no means complete and pointers to help forums are also given.

11.5.1 Download and Install R Extension

The R extension is available from the Rapid-I update site. The version used for this chapter was 5.1.4. When installing the extension for the first time, a number of steps must be followed. These are displayed during the installation. These are not repeated here but the most important things to get right are as follows.

- On Windows machines, Java, R, and RapidMiner must all have the same 32- or 64-bit operating system architecture. Any mismatch will cause problems.
- The R package rJava must be downloaded and installed in the R environment. This allows RapidMiner to connect to R.
- The location of the rJava library must be entered into the RapidMiner GUI.
- Environment variables to allow RapidMiner to find R and Java and for R to find Java must be set up correctly. These are PATH, R_HOME, and JAVA_HOME. It is most important to get the architecture correct in the Windows case.
- On Linux and Mac machines, there are additional differences.

There is no one simple set of instructions but persistence will pay off. There are many sources of help on the public Internet with the best starting point being the support forum at Rapid-I `http://forum.rapid-i.com` with a video example being available from `http://rapid-i.com/content/view/239/`.

11.5.2 Processes and Data

The processes and the data are all available from the website for this book. Simply download them and import the two processes. The exact path to the processes will need to be modified depending on the specifics of the environment where the processes are run. The processes themselves give an indication when they are loaded about what to do. The *readAndProcessEcoliData* process reads the *E-coli* data from a data file *ecoli.data*. This process must be edited to point to the data. The *clusterVisualisation* process calls the *readAndProcessEcoliData* and clearly the location of the latter process must be entered into the *Execute Process* operator.

First time run

An important point is the first time the R script is run, various R packages will be downloaded. The script detects the presence of the library after the first run and so should not need to do the download again.

11.6 The Process in Detail

The RapidMiner process is shown in Figure 11.3. The process has been divided into

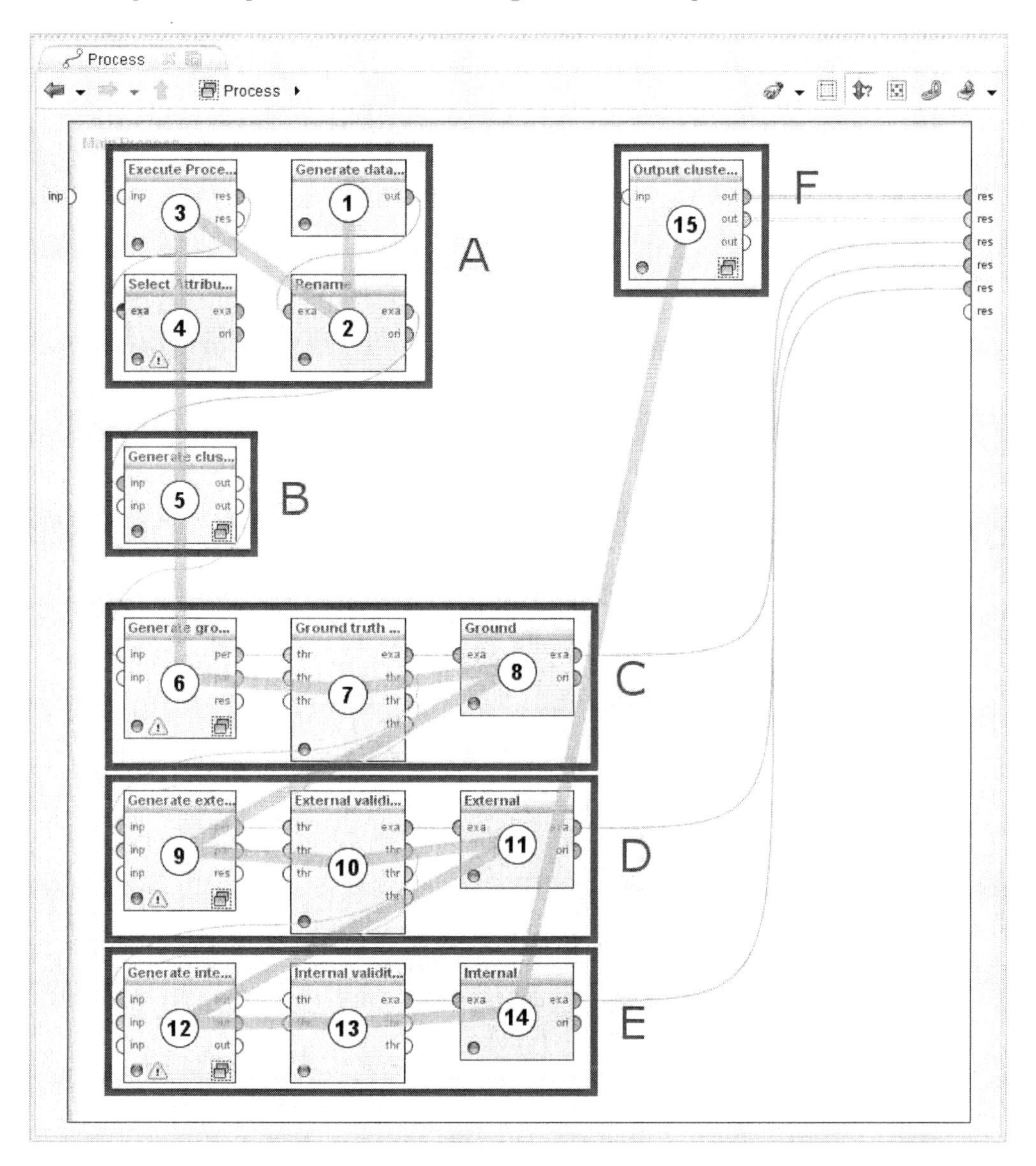

FIGURE 11.3: Overview of the RapidMiner process.

process sections labelled A to F. Each process section contains a number of operators labelled from 1 to 15. Each of the section names and contents are summarised in Table 11.1 and references are given to where in the rest of this chapter more information can be found.

TABLE 11.1: Process sections.

Process Section	Description (Section Reference)	Contains Operators	Page Reference
A	Import Data (11.6.1)	1	168
		2	168
		3	168
		4	169
B	Generate Clusters (11.6.2)	5	169
C	Generate Ground Truth Validity Measures (11.6.3)	6	170
		7	171
		8	172
D	Generate External Validity Measures (11.6.4)	9	173
		10	173
		11	173
E	Generate Internal Validity Measures (11.6.5)	12	173
		13	174
		14	174
F	Output results (11.6.6)	15	174

11.6.1 Import Data (A)

This process section creates or imports the data to be processed. Operators 1 and 2 generate the artificial data whilst 3 and 4 read in the *E-coli* data. It is straightforward to connect the output of 2 or 4 to process 5 in order to choose between artificial or *E-coli* data.

Generate data for testing (1)

This process uses the *Generate Data* operator to create 100 examples. Each example consists of 3 regular attributes and a label. The target function is *Gaussian mixture clusters* and this has the effect of generating 8 clusters from the 3 attributes. Each label takes on a nominal value from cluster0 to cluster7. With the initial random seed set to 2, the output from this operator is shown in Figure 11.1.

This plot view can be re-created by setting a breakpoint after the operator and switching to the results perspective. Selecting the Scatter 3D Color plotter with att1, att2, and att3 as x, y, and z, respectively, and Color set to label will give the same result. As can be seen, there are 8 distinct clusters in the data. Any clustering algorithm and guidance given by a validity measure can be easily compared with reality.

This process simply renames the label attribute to the value *site*. This corresponds to the name of the label attribute in the *E-coli* data. This is done simply to make subsequent processing easier.

Rename (2)

Execute Process (3)

This process reads in the *E-coli* data. It does this by making use of the *Execute Process* operator rather than using the *Read CSV* operator directly. The reason for this is simply that during development of the process, a number of different data sources were investigated and it was more convenient at the time to separate the reading of the data from its processing. When the *E-coli* dataset was finally chosen, it was felt that retaining the *Execute Process* method was valuable since this is an operator that deserves to be better known.

In this case, the external process that is executed is *readAndProcessE-ColiData.* This is a simple process that uses the *Read CSV* operator to read the *E-coli* data file downloaded from [11]. The data file contains data records with spaces separating the records. This means that the column separators parameter must be set to something like the following regular expression [ ]*\s in order to work properly.

The meta data information for the read CSV operation creates the names and types of the attributes. The end result is shown in Table 11.6.1.

TABLE 11.2: *E-coli* data.

Attribute name	Role	Type
site	Label	polynominal
mcg	regular	real
gvh	regular	real
lip	regular	real
chg	regular	real
aac	regular	real
alm1	regular	real
alm2	regular	real

The interested reader is encouraged to refer to [11] and [12] for more details. For the purpose of this chapter, the details of the data are not relevant. As long as the data contains regular attributes that are all real, a polynominal label, and no missing values, the clustering activity can proceed.

Select Attributes (4)
This operator selects the attributes that are required for the clustering activity. In this case, the SequenceName id is removed.

11.6.2 Generate Clusters (B)

Generate clusters is a loop operator that makes multiple partitioned clusters with k = 2 up to a maximum number of clusters as defined by the numberOfClusterIterations macro (which is set in the process context). The cluster models and example sets are stored using the *Remember* operator so they can be used later in the operators that calculate the validity measures. An iteration macro is set and this is used inside the loop. The approach of saving clusters makes the process more efficient because it avoids having to calculate clusters multiple times later on as the validity measures are calculated.

Generate Clusters (5)
Within the loop operator the process to create and store clusters is shown in Figure 11.4.

The *Materialize Data* operator is required. Without it, the process does not work properly. The reason is to do with copies of example sets inside loops being presented as input to subsequent iterations. This causes some confusion later on and simply materializing the

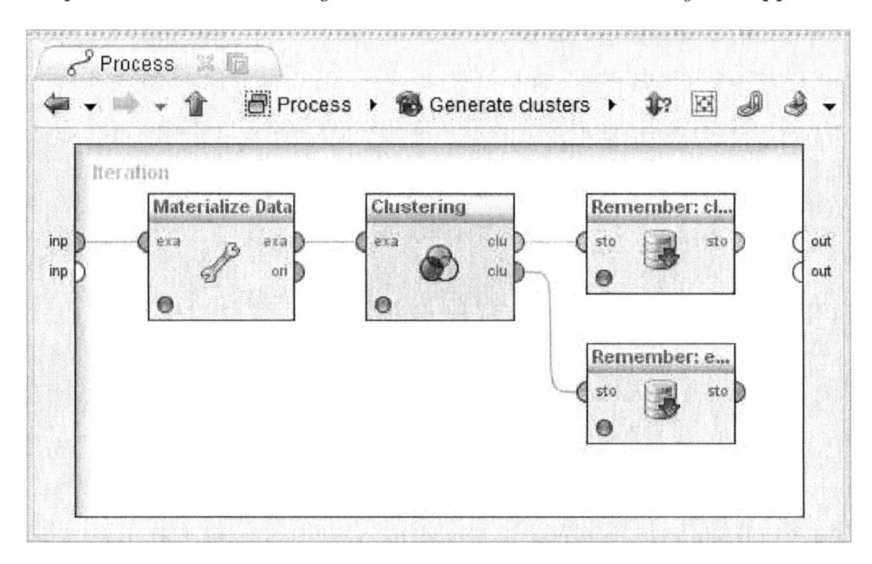

FIGURE 11.4: Detail of *Generate Clusters*. This is operator 5 from process section B in Figure 11.3.

data resolves it. As always, other resolution methods could be used such as storing the example set outside the loop and recalling it afresh for each iteration. The *Clustering* operator is the k-means operator with the value of k set from the loop iteration macro. The measure type is set to numerical Euclidean distance. Both the cluster model and the example set with the additional cluster attribute resulting from the cluster operation are stored for use later on.

11.6.3 Generate Ground Truth Validity Measures (C)

This process section generates various validity measures by comparing the *correct* or ground truth labels with each of the clusterings. This is done by making direct calls to R. The results are logged and converted to an example set with real values for the validity measures.

Generate Ground Truth Measures (6)
This process uses the *Optimize Parameters (Grid)* operator to make repeated calls to a chain of operators but with different parameters each time. In this case, the parameter settings are arranged so that the *Recall* operator inside the loop recalls the previously stored clustered example sets one by one that were stored in the *Generate Cluster* process section of Figure 11.4. The process in detail is shown in Figure 11.5.

The top row recalls one of the previously stored clustered example sets. The one recalled is determined by the iteration macro that is passed from the higher level containing operator. The operator chain selects the id and the cluster assigned by the k-means operator and renames the cluster attribute to cluster1. This name is needed later on by the R operator. The bottom row recalls one example and selects the id and cluster attribute provided by the original data-generation operator. This attribute is renamed to cluster2. The two example sets are joined and the result passed to the *Calculate groundTruthClusterValidityIndices* operator; an example of the resulting joined data is shown in Figure 11.6.

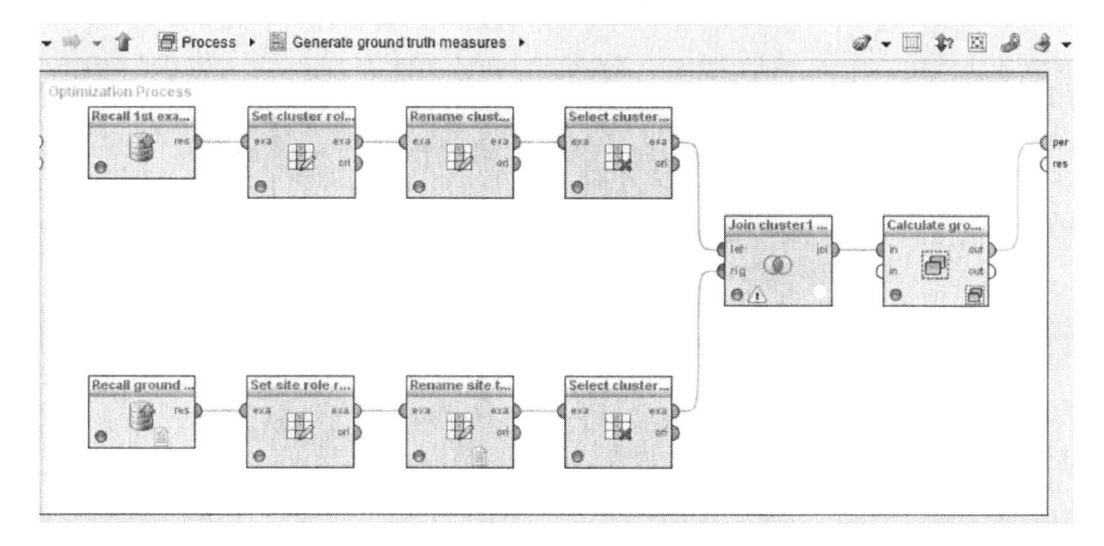

FIGURE 11.5: Generate ground truth measures detail. This is operator 6 within process section C from Figure 11.3.

The external validity measures are then calculated using R. The process that does this is shown in Figure 11.7.

The first operator executes the R script, which is described in more detail below; the second extracts the values returned in the R example set result as performance vectors, and the third logs these performance values and the value of k for each iteration. The log operator is a very important part of RapidMiner as it allows data to be recorded during the execution of a process and this can be related to the data being investigated or to the operation of RapidMiner itself.

The R script at the heart of the process uses two packages, mclust [15] and profdpm [16]. The first of these, mclust, provides the function to calculate the adjusted Rand index whilst the second provides the Rand, Fowlkes and Mallows, Wallace 10, Wallace 01, and Jaccard indexes. The details of R are beyond the scope of this chapter but the following notes are worthy of mention.

Firstly, the R process must download the two libraries. This is done using the install.packages command. The first time this is run, a dialog box will pop up asking the user to specify a mirror server from which to download the items. After the first run, the code uses the command find.package to locate the packages and this should detect that the packages are already installed. The install.packages command should therefore not run again.

Secondly, the process creates a data frame so that when returned to RapidMiner, the result is an example set with each attribute equal to the name of the calculated validity measure. For example, Figure 11.8 shows a typical example set. It contains a single row and each column corresponds to the calculated validity measure thereby making it straightforward to extract the data for use in result logging later. This is done in the Subprocess operator *Extract ground truth performance measures*, which uses repeated calls to *Extract Performance* on the example set returned from R to create macros which are then recorded by the log operator.

Ground Truth Measures (7)

ExampleSet (100 examples, 1 special attribute, 2 regular attributes)

Row No.	id	cluster1	cluster2
1	1	cluster_1	cluster1
2	2	cluster_0	cluster3
3	3	cluster_0	cluster7
4	4	cluster_1	cluster2
5	5	cluster_1	cluster6
6	6	cluster_0	cluster3
7	7	cluster_1	cluster6
8	8	cluster_1	cluster0
9	9	cluster_1	cluster1
10	10	cluster_0	cluster7
11	11	cluster_0	cluster3
12	12	cluster_0	cluster7

FIGURE 11.6: First few rows of k = 2 k-means clustering merged with original clusters for artificial data.

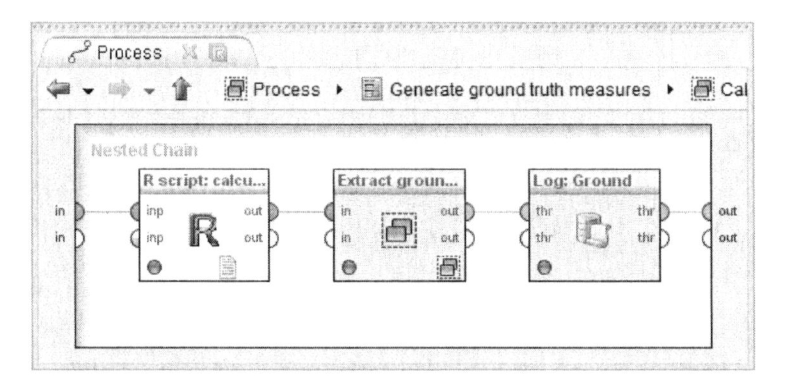

FIGURE 11.7: Operator Chain to calculate external validity measures using R. This is within operator 6 in process section C from Figure 11.3.

The validity measures are logged and to allow these values to be manipulated and stored, this operator converts the log entry to an example set.

Ground (8)
The values returned in the log are converted to real values, where necessary, to make analysis easier later on.

11.6.4 Generate External Validity Measures (D)

This process section generates various external validity measures by comparing each clustering with every other clustering. The idea behind this is to determine if there are any clusterings that retain a similarity despite the numbers of clusters changing. The processing is similar to that in Section 11.6.3 with some minor differences.

ExampleSet (1 example, 0 special attributes, 6 regular attributes)

Row No.	ARI	R	FM	W10	W01	J
1	0.346	0.675	0.586	1	0.344	0.344

FIGURE 11.8: Example set returned by R process.

Generate External Validity Measures (9)
This operator is similar to 11.6.3 but iterates over all possible pairs of clusterings. The intention is that a two-dimensional grid will be produced where each cell contains a validity measure. For example, a clustering with k = 2 and k = 3 would inevitably have similarities. By using one of the clusterings as the ground truth, it is possible to use the external validity measures to compare with the other clustering. This approach reveals clusters that are resilient to changing values of k and which may reveal interesting information about the data.

External Validity Measures (10)
The validity measures are logged and to allow these values to be manipulated and stored, this operator converts the log entry to an example set.

External (11)
The values returned in the log are converted to real values where necessary to make analysis easier later on.

11.6.5 Generate Internal Validity Measures (E)

This operator iterates over all the previously stored cluster models and clustered example sets to calculate various internal validity measures. As with the other validity measures, the log entries containing the values are converted to an example set and, where possible, numeric values are converted to real.

Generate internal validity measures (12)
The internal operators within this operator are shown in Figure 11.9.

Each iteration recalls the cluster model and clustered example set for each value of k and uses operators available within RapidMiner to calculate the following.

- Davies Bouldin cluster distance
- Average within centroid cluster distance
- Cluster density
- Sum of squares item distribution
- Gini Coefficient item distribution (this is added as a bonus feature but has not been described)

These values are logged using the *Log* operator.

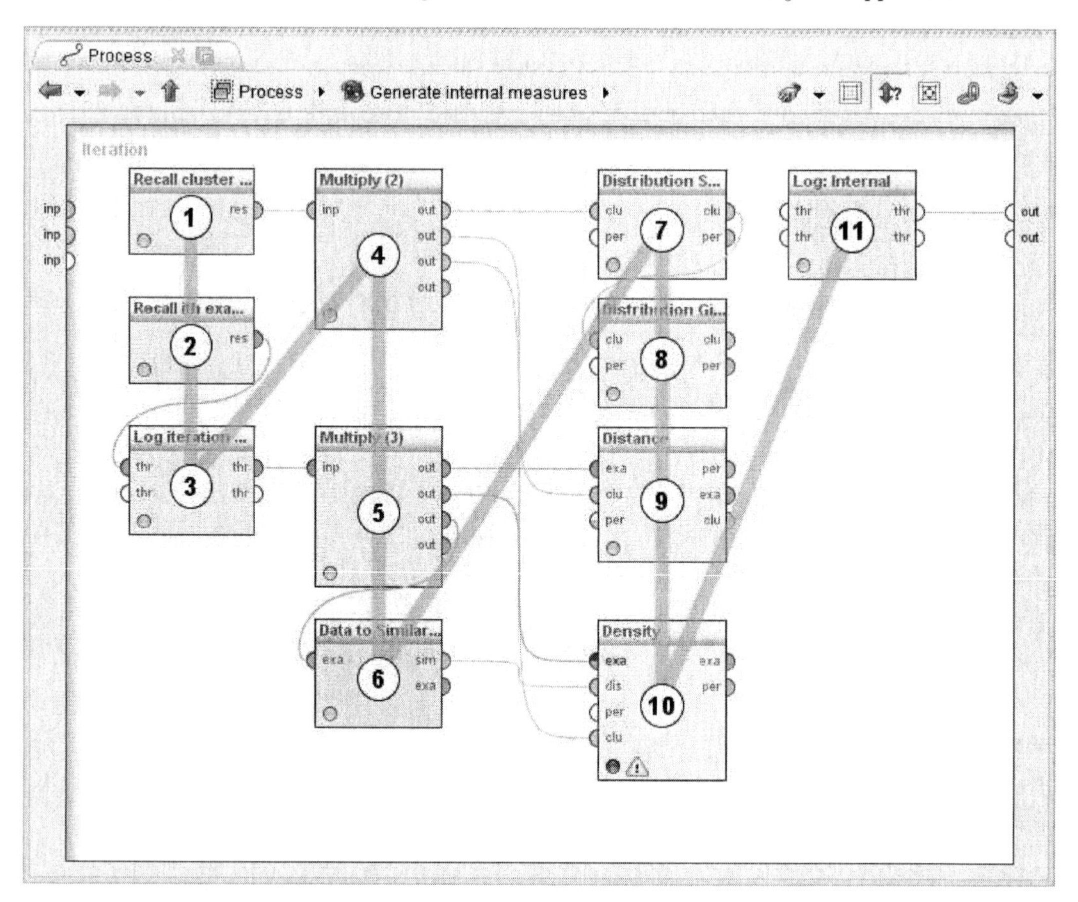

FIGURE 11.9: Internal validity measure calculation. This is operator 12 within process section E from Figure 11.3. The execution order is also shown.

Internal Validity Measures (13)
The validity measures are logged and to allow these values to be manipulated and stored, this operator converts the log entry to an example set.

Internal (14)
The values returned in the log are converted to real values, where necessary, to make analysis easier later on.

11.6.6 Output Results (F)

This process section recalls the cluster models and clustered example sets in order to allow them to be displayed.

Output Clusters and Partitioning (15)
This process is a loop operator that iterates over all the cluster models and clustered ex-

ample sets. Each is recalled and removed from memory. The output from this operator is a collection given that multiple models and example sets were previously calculated. Example set collections can be combined using the *Append* operator.

11.7 Running the Process and Displaying Results

With the clusterVisualisation process loaded, there are two investigations that can be performed. Firstly, investigation of the artificial data. This is done by connecting the output from the *Rename* operator to the *Generate Clusters* operator (see Figure 11.10).

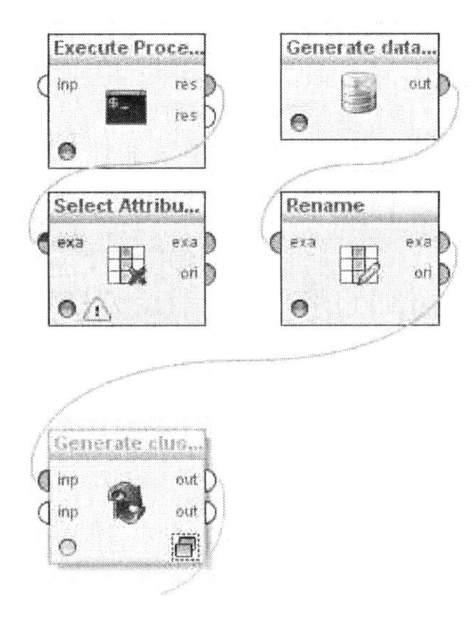

FIGURE 11.10: Process connection for artificial data. This is a close-up of process section A from Figure 11.3.

Pressing the run key will generate data and perform the various clusterings and validity measurements. Refer to Section 11.8.1 to see what happens. As an alternative, the output from the *E-coli* data processing can be used for the clustering investigation. In this case, the output from the *Select Attribute* operator is connected to the *Generate Clusters* operator (see Figure 11.11).

Results can be displayed using the built-in plotting functions of RapidMiner. These are accessed from the plot view in the results view where there are many plotter types to choose from. For example, 11.1 shows a 3-D scatter plot, 11.12, 11.13, 11.15, and 11.16 show multi-line series plots whilst 11.14 and 11.17 show block plots.

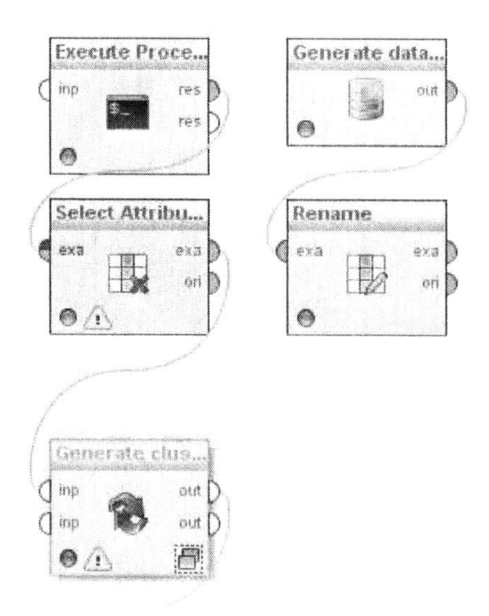

FIGURE 11.11: Process connection for *E-coli* data. This is a close-up of process section A from Figure 11.3.

11.8 Results and Interpretation

This section analyses the cluster validity measures starting with the artificial data in order to get a sense of how the measures should be interpreted given the domain expertise that inspection of Figure 11.1 gives.

This leads on to the *E-coli* dataset where domain expertise is not assumed but the interpretation of the validity measures would help guide an expert to further analysis and better understanding.

11.8.1 Artificial Data

Ground Truth

The ground truth validity measures are shown in Figure 11.12.

Each of these indexes tends to 1 as the match between two clusterings improves. As can be seen, all show that when k = 8, the clustering perfectly matches the ground truth answer. This is encouraging in that it shows that k-means is finding the true clusters, but is not realistic given that the answer was already known and is certain. If the ground truth answer was uncertain, this approach would help provide supporting evidence to confirm it but it would also highlight possible alternative candidate clusterings. In this case, there is a slight peak at k = 11. As it happens in this case, the k = 11 clustering corresponds to one cluster being split into 4 while the remaining 7 stay intact. This split is likely to be because there are only 100 data points and once the number of clusters increases, it gets more likely that the sparseness of the data would lead to clusters containing small numbers of examples.

Internal

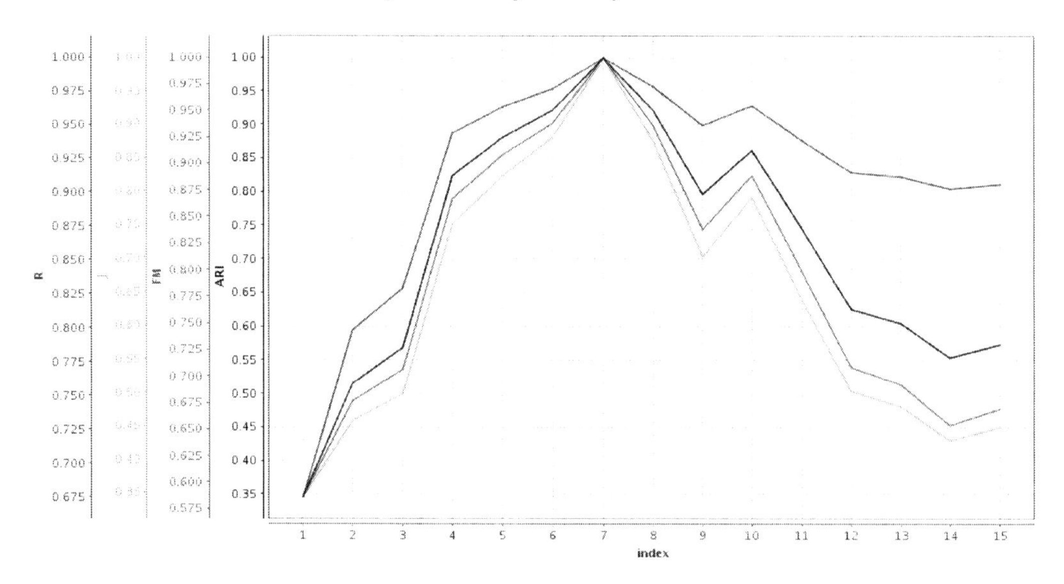

FIGURE 11.12 **(see color insert)**: Graph of Rand, Jaccard, Fowlkes-Mallow and adjusted Rand indexes as a function of cluster size compared with ground truth clustering of artificial data. The x axis is the value of k and the maximum is at k = 8. This graph was produced using the Series Multiple plotter, and consequently, the y axes are normalised to make the ranges of each series match.

The internal measures are shown in Figure 11.13. The Davies-Bouldin measure shows a clear minimum at k = 8 in agreement with the expected number.

The cluster density measure tends to zero as clusters get denser and examination of the graph demonstrates that somewhere around k = 8 the density stops increasing as sharply. This indicates that higher values of k are starting to split the real clusters up but this does not increase the density markedly, given that the clusters were already dense to start with.

The average within-distance measure tends to zero as clusters get smaller and more compact. At k = 8 the results shows that the decrease slows down indicating that already compact clusters are being split as k increases above 8.

The item distribution measure gives an indication about the relative sizes of clusters. The measure tends to 1 if there is one large cluster and tends to 1/k if the clusters are all equally sized. Between k = 5 and k = 8, this measure seems to show that the distribution is relatively constant and above these values, clusters seem to get smaller.

The Gini measure is shown for completeness. This is a more difficult measure to interpret and is not considered here.

All of these taken together indicate that k = 8 is a strong candidate for the best clustering. This is encouraging since in this case, the validity measure algorithm was not told the correct answer.

External

The final graph shows how the external validity measures vary as different clusterings are compared. The results are presented on a block plot and an example showing the adjusted Rand index is shown in Figure 11.14.

This measure tends to 1 when two clusterings are identical, as is the case when k1 = k2. It is interesting to note, however, that different clusterings can be quite similar. For

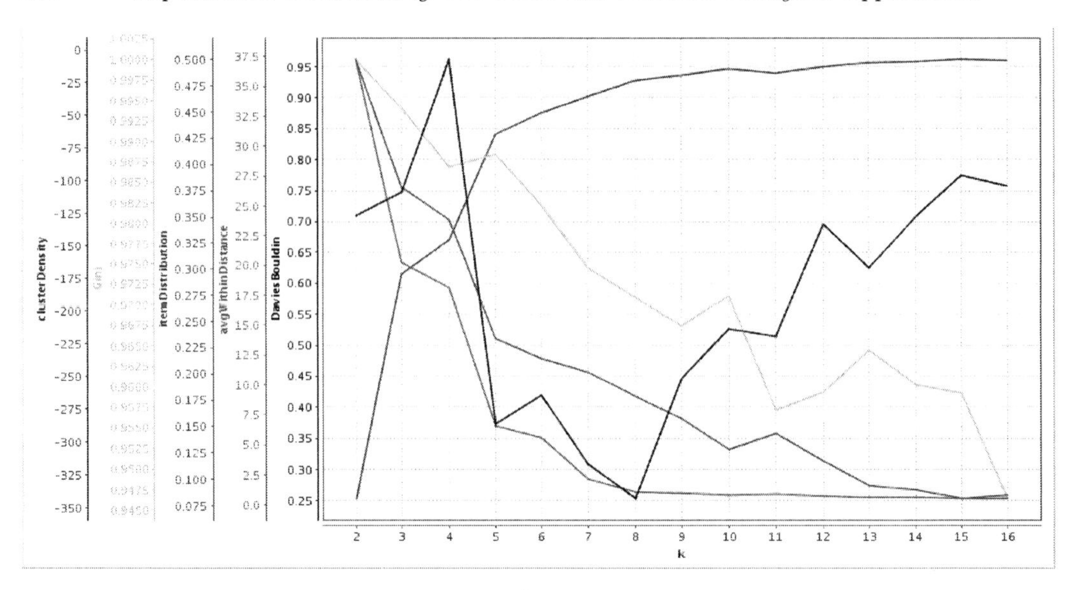

FIGURE 11.13 **(see color insert)**: Internal validity measures as a function of k for artificial data. The x axis is the value of k and the maximum is at k = 8. This graph was produced using the Series Multiple plotter and consequently, the y axes are normalised to make the ranges of each series match.

example, between k = 3 and k = 4, the clusterings are similar. In contrast, between k = 4 and k = 5, they are very different. This, perhaps, indicates that there is one cluster at k = 3 that remains largely intact at k = 4. For k = 5, however, the cluster is broken up. Between k = 5 and k = 9, there seem to be similarities in the clusters. For example, k = 6 and k = 7 appear similar as do k = 7, k = 8, and k = 9. This is not a perfect science of course, but examination of the results show that somewhere around k = 7 there seem to be clusters that remain intact and which disappear above k = 9. This is in agreement with the known clusters. This information allows focus to be given to more investigation.

11.8.2 *E-coli* Data

The *E-coli* data is not as well behaved as the artificial.

Ground truth

The adjusted Rand index for the *E-coli* data compared to the ground truth classification is shown in Figure 11.15. The best fit occurs when k = 4 with k = 3 not far behind. The expected result is 8 and this does correspond to a small peak, but significantly smaller than the k = 4 case. Examination of the real data shows that 4 clusters dominate and this may well explain the peak at 4.

Internal

The internal measures shown in Figure 11.16 are not as easy to interpret as the artificial data. The first thing to notice is the Davies Bouldin measure. This shows evidence that between k = 7 and k = 9, optimum clusters are found. It is interesting to note that there

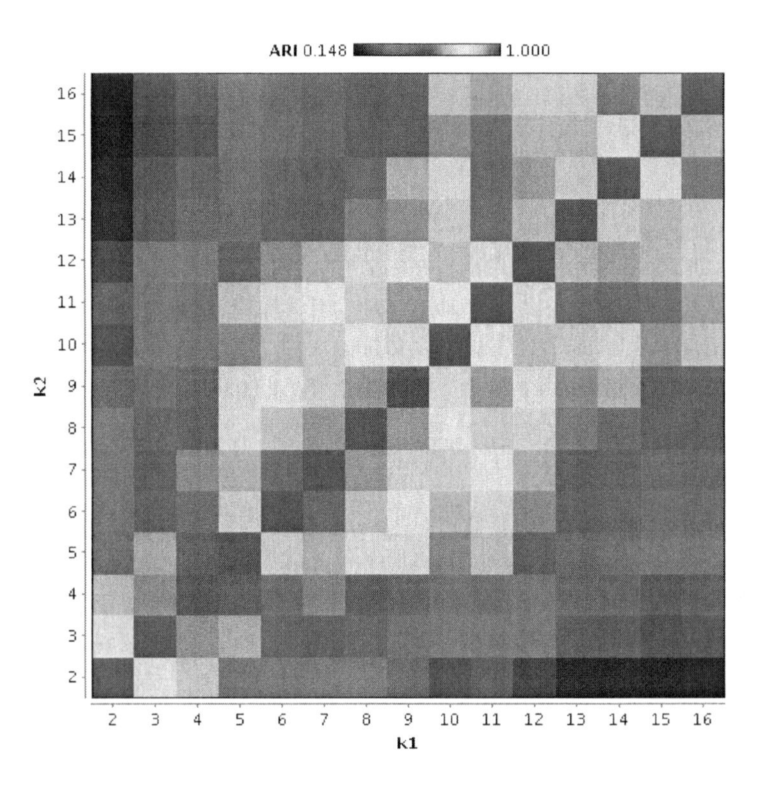

FIGURE 11.14 **(see color insert)**: Adjusted Rand index validity measure between different clusterings for artificial data.

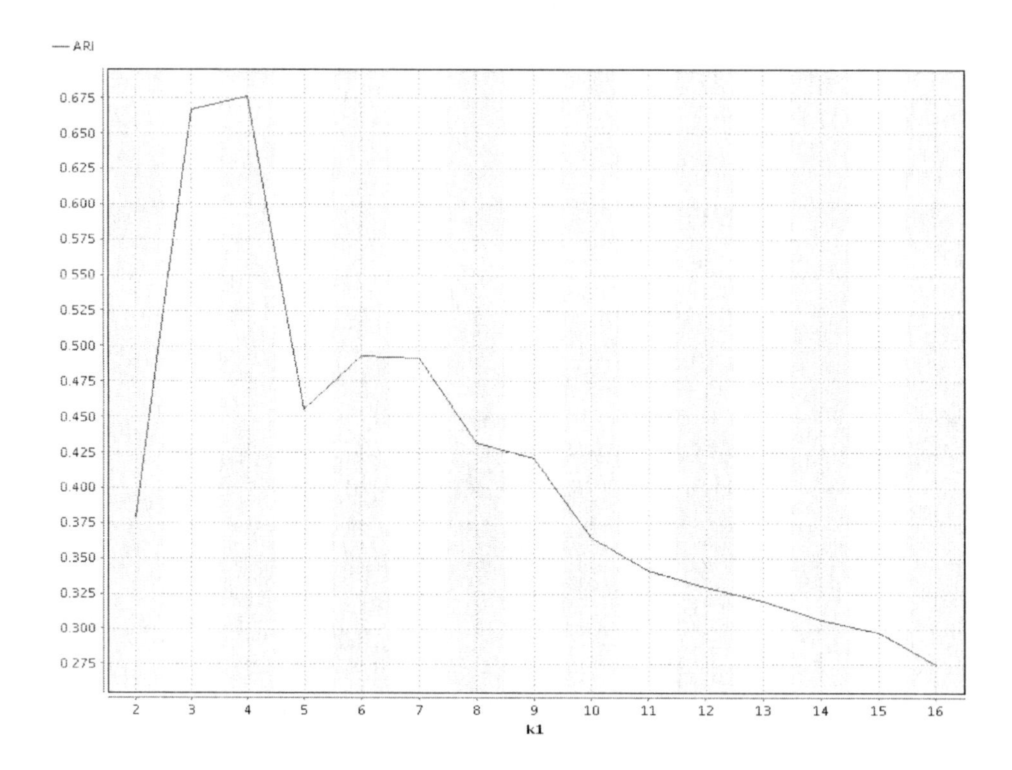

FIGURE 11.15: Adjusted Rand index for different values of k compared to ground truth for *E-coli* data.

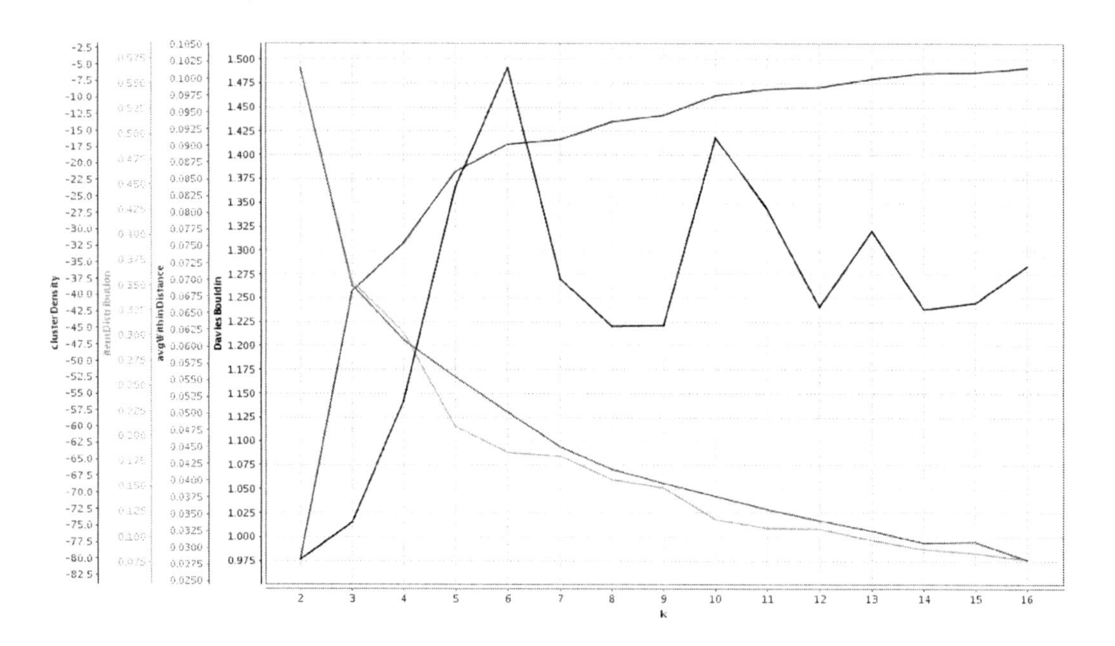

FIGURE 11.16 **(see color insert)**: Internal validity measures as k is varied for *E-coli* data.

is perhaps another candidate clustering at k = 12. This measure also shows that k = 6 and k = 10 are worse clusterings.

The cluster density and within-distance measures seem to flatten from about k = 6 although it is by no means conclusive. Similarly, the item distribution appears to have clusters of interest between k = 5 and k = 10.

External

The adjusted Rand index between the pairwise comparisons between different clusterings is shown in Figure 11.17, and this seems to show that between k = 6 and k = 7 as well as between k = 8 and k = 9, there are clusterings that appear to be preserved as k varies. The ground truth is 8 clusters and these results do not contradict this.

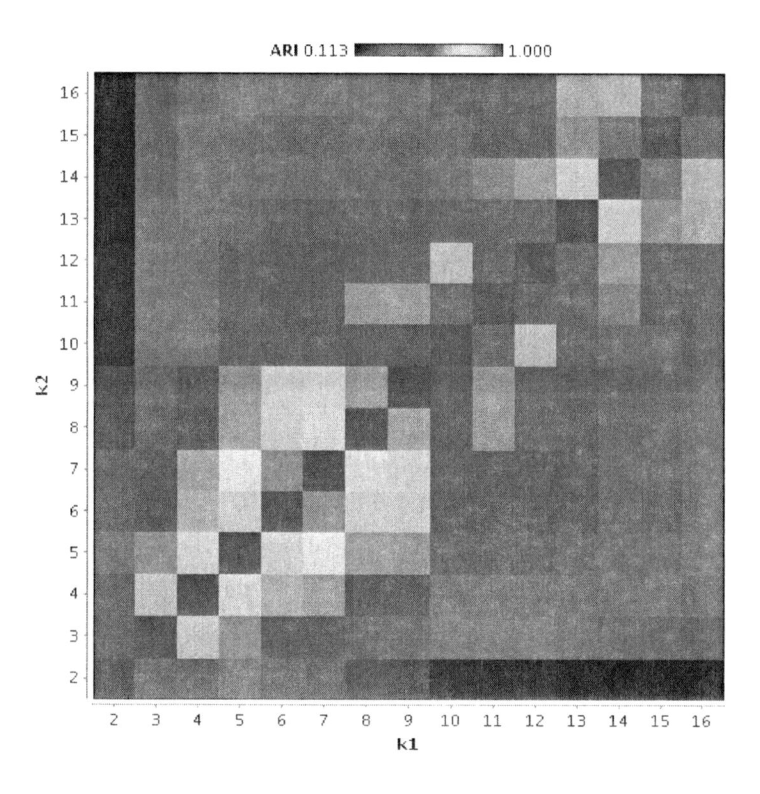

FIGURE 11.17 **(see color insert)**: Adjusted Rand index for different clusterings for *E-coli* data.

11.9 Conclusion

This chapter has given a detailed overview of a RapidMiner process that clusters data and generates cluster validity measures. The process has been designed in a relatively modular way so readers are encouraged to modify it for their own purposes. Owing to integration with R, many other powerful techniques can easily be realised without having to harm a software developer. By controlling everything from RapidMiner, the overall process is easy to understand and is easy to change.

In summary, the process helps a human to eliminate the meaningless and focus on the interesting.

Bibliography

[1] A.K. Jain, M.N. Murty, and P.J. Flynn. Data clustering: a review. *ACM Computing Surveys (CSUR)*, 31(3):264–323, 1999.

[2] P.N. Tan, M. Steinbach, and V. Kumar. Cluster analysis: Basic concepts and algorithms. *Introduction to Data Mining*, pages 487–568, 2006.

[3] R. Xu, D. Wunsch, et al. Survey of clustering algorithms. *Neural Networks, IEEE Transactions on*, 16(3):645–678, 2005.

[4] J. MacQueen et al. Some methods for classification and analysis of multivariate observations. In *Proceedings of the Fifth Berkeley Symposium on Mathematical Statistics and Probability*, volume 1, page 14. California, USA, 1967.

[5] M. Ester, H.P. Kriegel, J. Sander, and X. Xu. A density-based algorithm for discovering clusters in large spatial databases with noise. In *Proceedings of the 2nd International Conference on Knowledge Discovery and Data Mining*, volume 1996, pages 226–231. AAAI Press, 1996.

[6] E. Rendon, I. Abandez, A. Arizmendi, and E.M. Quiroz. Internal versus external cluster validation indexes. *International Journal of Computers and Communications*, pages 27–34, 2011.

[7] W.M. Rand. Objective criteria for the evaluation of clustering methods. *Journal of the American Statistical Association*, pages 846–850, 1971.

[8] K.Y. Yeung and W.L. Ruzzo. Details of the adjusted Rand index and clustering algorithms, supplement to the paper An empirical study on principal component analysis for clustering gene expression data. *Bioinformatics*, 17(9):763–774, 2001.

[9] E.B. Fowlkes and C.L. Mallows. A method for comparing two hierarchical clusterings. *Journal of the American Statistical Association*, pages 553–569, 1983.

[10] D.L. Wallace. A method for comparing two hierarchical clusterings: Comment. *Journal of the American Statistical Association*, 78(383):569–576, 1983.

[11] A. Frank and A. Asuncion. UCI machine learning repository, 2010.

[12] P. Horton and K. Nakai. A probabilistic classification system for predicting the cellular localization sites of proteins. In *Proceedings of the Fourth International Conference on Intelligent Systems for Molecular Biology*, pages 109–115. AAAI Press, 1996.

[13] R. Gelbard, O. Goldman, and I. Spiegler. Investigating diversity of clustering methods: An empirical comparison. *Data & Knowledge Engineering*, 63(1):155–166, 2007.

[14] H. Zhang. Exploring conditions for the optimality of naive bayes. *International Journal of Pattern Recognition and Artificial Intelligence*, 19(2):183–198, 2005.

[15] C. Fraley and A. E. Raftery. MCLUST Version 3 for R: Normal Mixture Modeling and Model-based Clustering. *Office*, (504):1–54, 2007.

[16] Matt Shotwell. *profdpm: Profile Dirichlet Process Mixtures*, 2010. R package version 2.0.

label ● cluster 0 ● cluster 1 ● cluster 2 ● cluster 3 ● cluster 4 ○ cluster 5 ● cluster 6 ● cluster 7

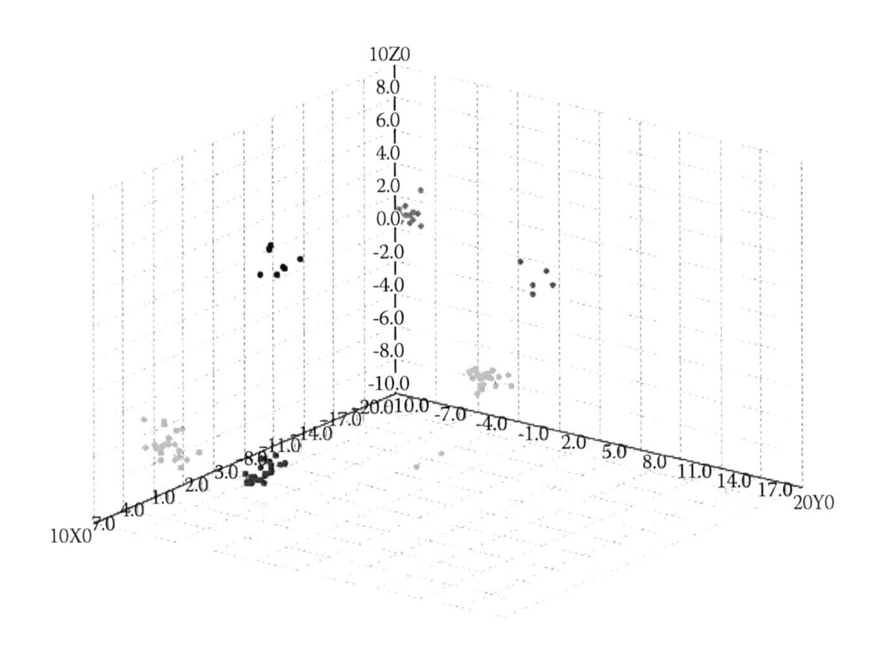

FIGURE 11.1: 3D scatter plot of articial data showing 8 clusters.

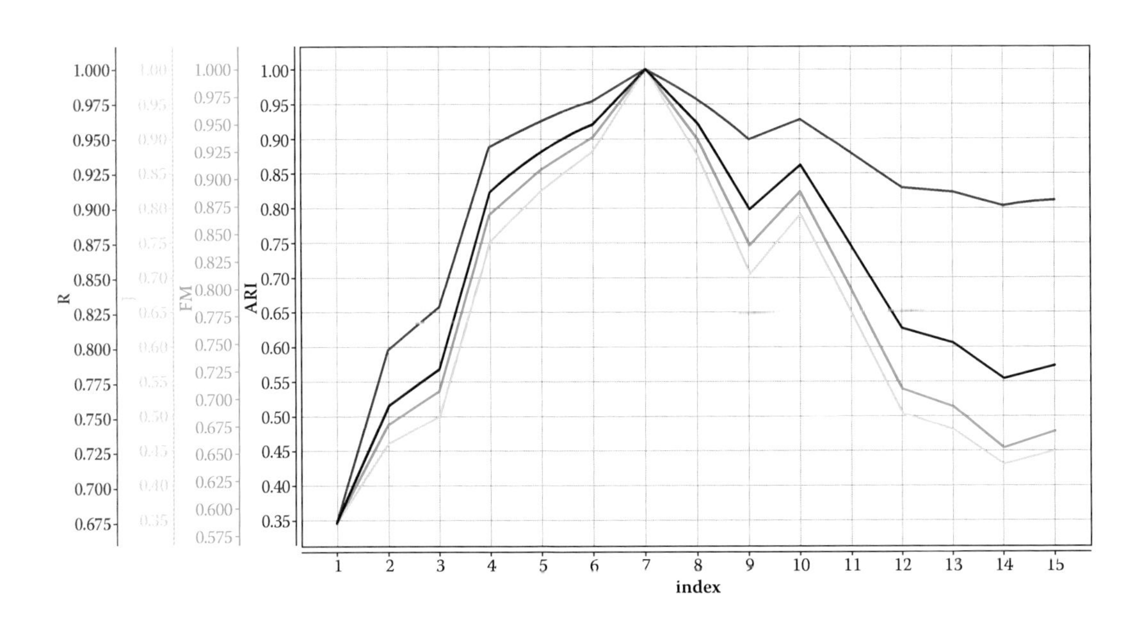

FIGURE 11.12: Graph of Rand, Jaccard, Fowlkes-Mallow and adjusted Rand indexes as a function of cluster size compared with ground truth clustering of articial data. The x axis is the value of k and the maximum is at k = 8. This graph was produced using the Series Multiple plotter and consequently, the y axes are normalised to make the ranges of each series match.

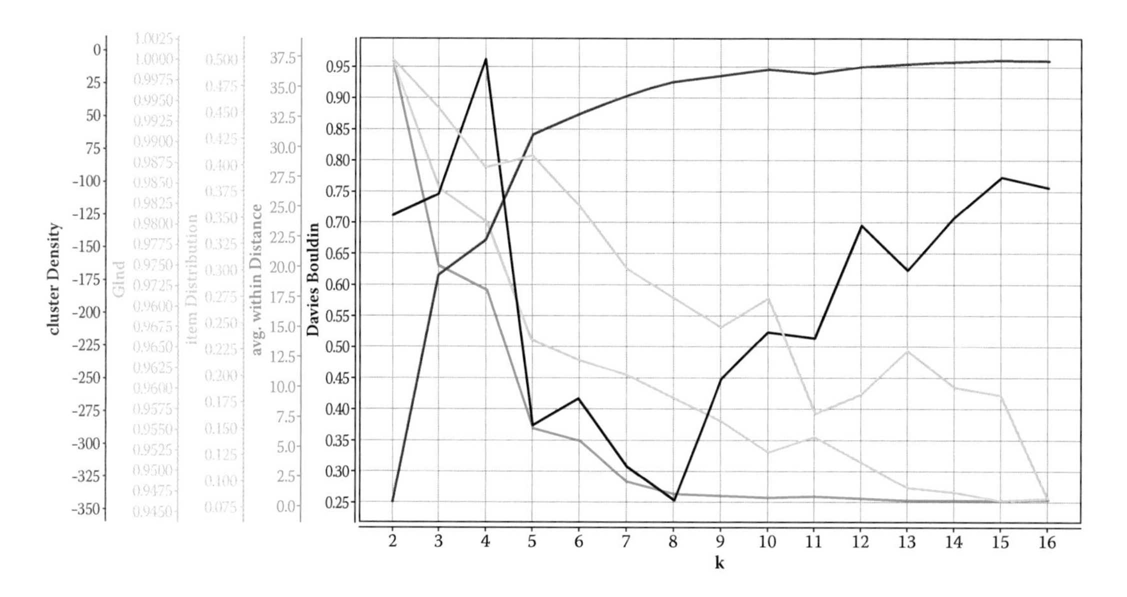

FIGURE 11.13: Internal validity measures as a function of k for artificial data. The x axis is the value of k and the maximum is at k = 8. This graph was produced using the Series Multiple plotter and consequently, the y axes are normalised to make the ranges of each series match.

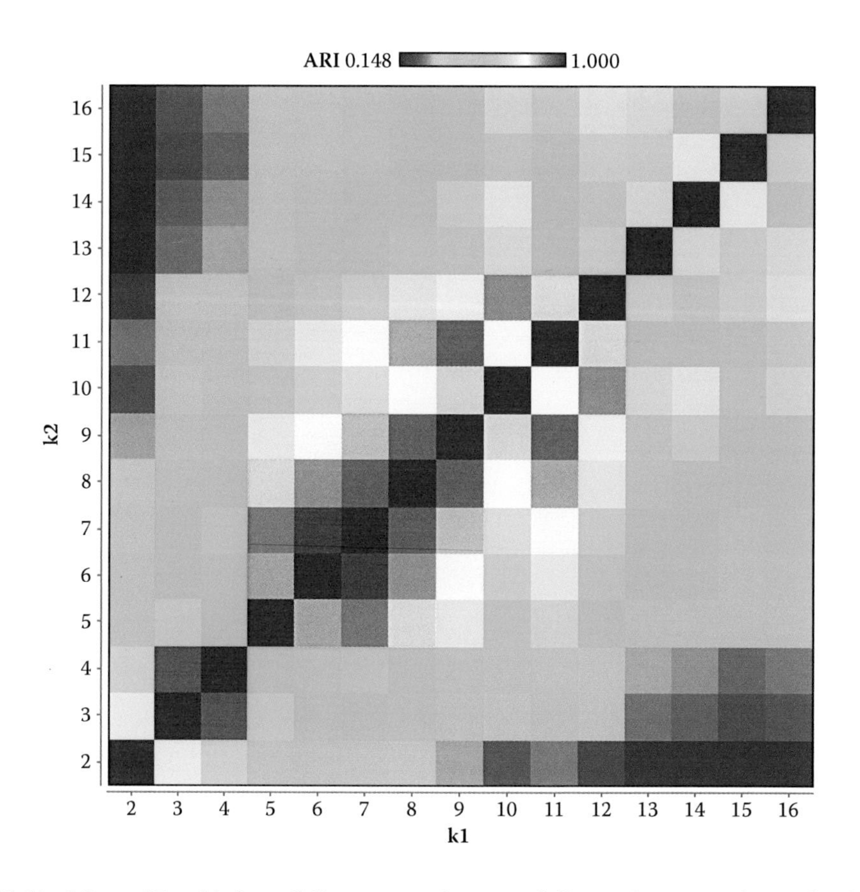

FIGURE 11.14: Adjusted Rand index validity measure between different clusterings for artificial data.

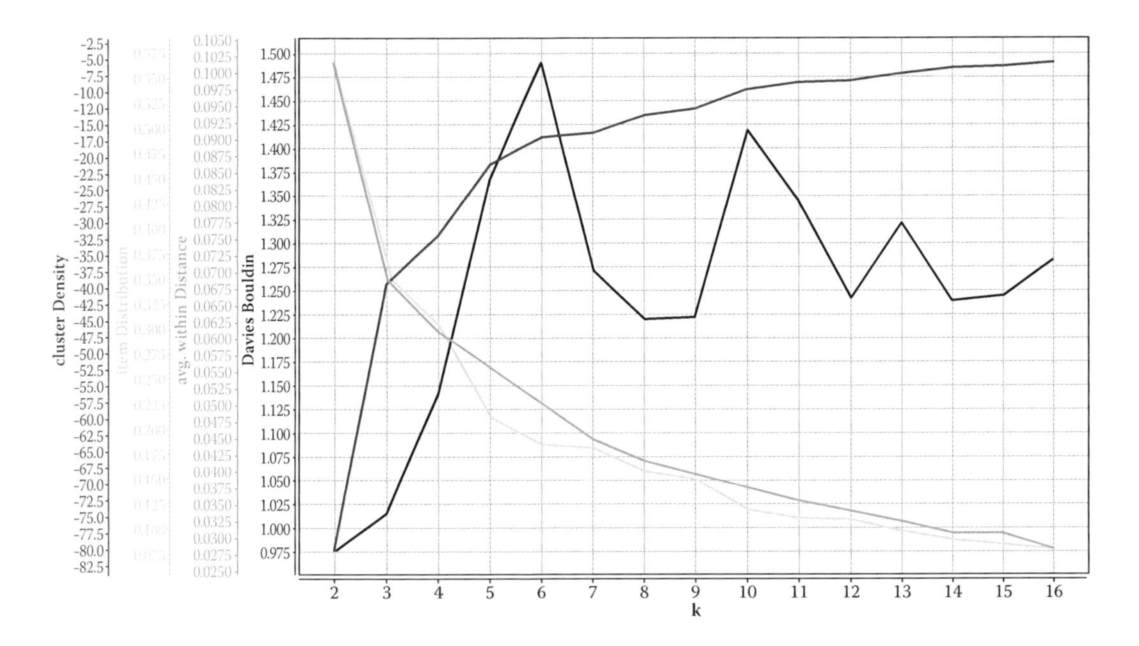

FIGURE 11.16: Internal validity measures as k is varied for *e-Coli* data.

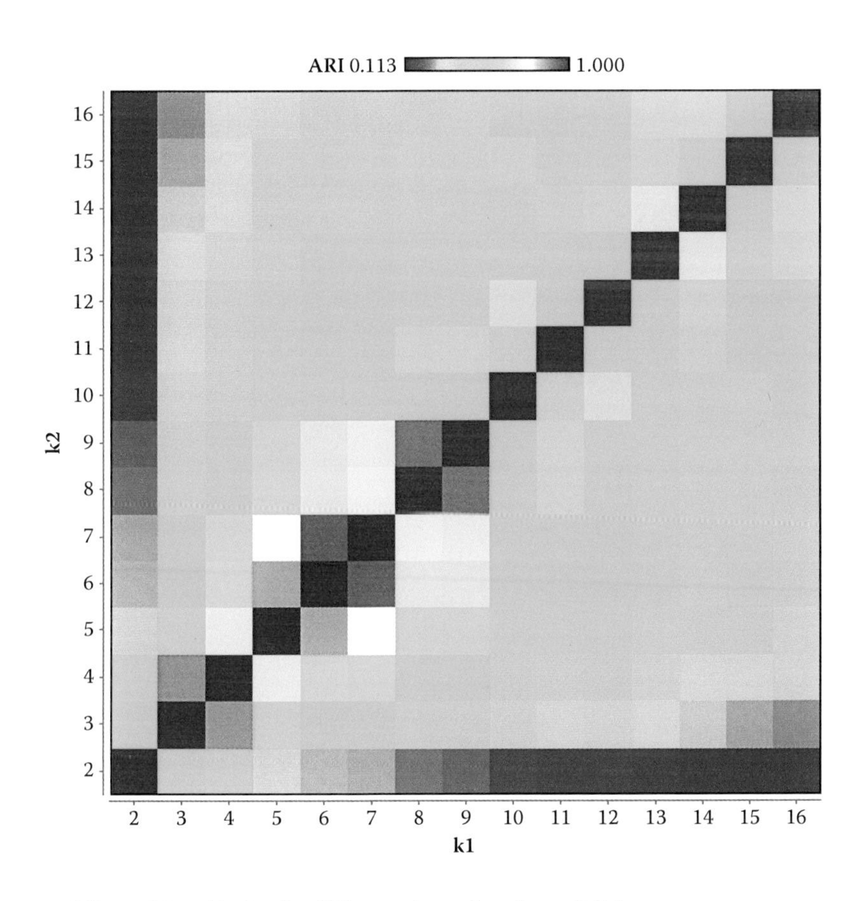

FIGURE 11.17: Adjusted Rand index for different clusterings for *e-Coli* data.

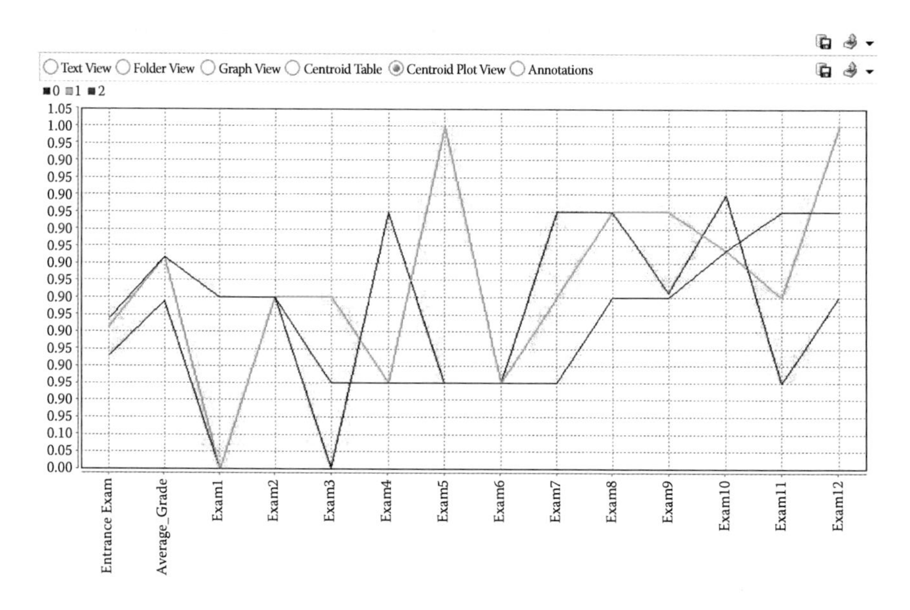

FIGURE 12.9: Centroid plot of K-medoids algorithm.

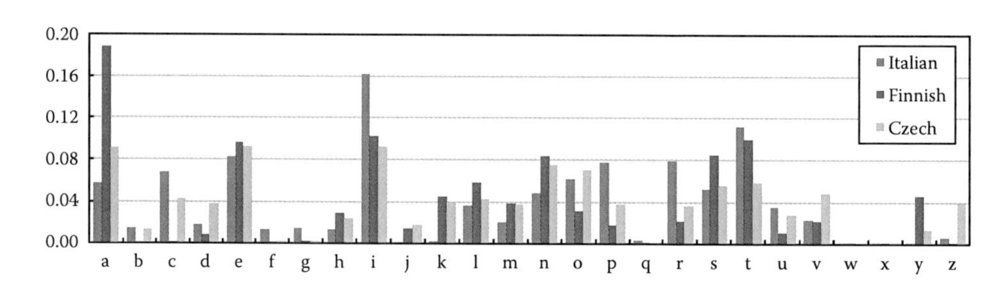

FIGURE 14.2: Distributions of unigrams in Italian, Czech and Finnish show clear differences in the use of distinct letters per language. In this example, accented letters are stripped of their accents, and non-alphabet symbols are ignored.

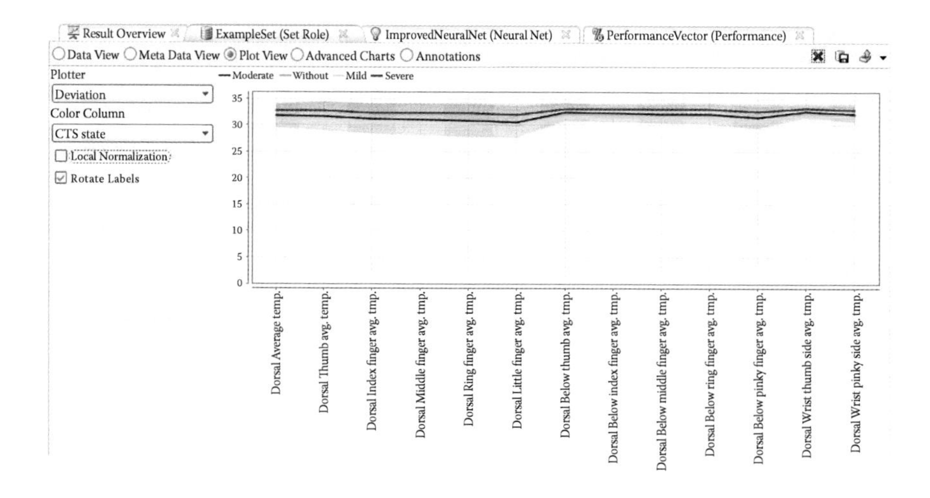

FIGURE 17.2: Simple data visualisation: Deviation Plot.

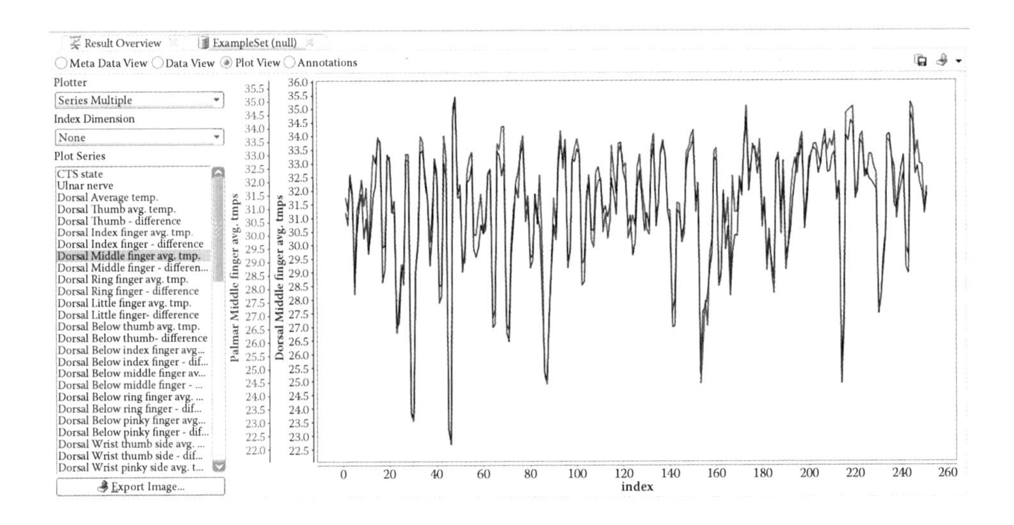

FIGURE 17.3: Simple data visualisation: Multiple Series Plot.

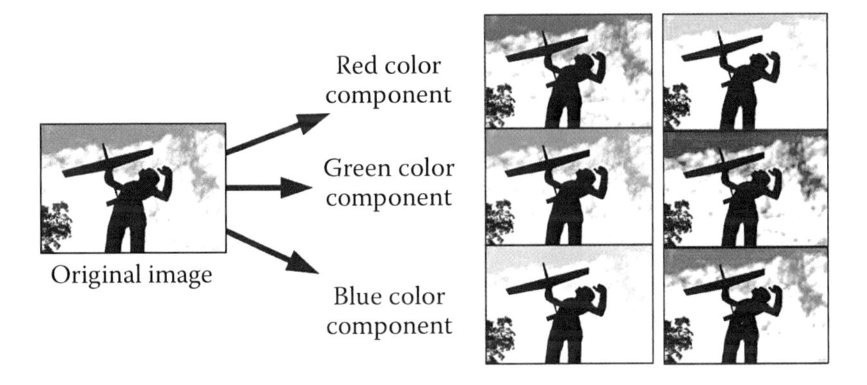

FIGURE 20.5: Example: Extraction of color component from image.

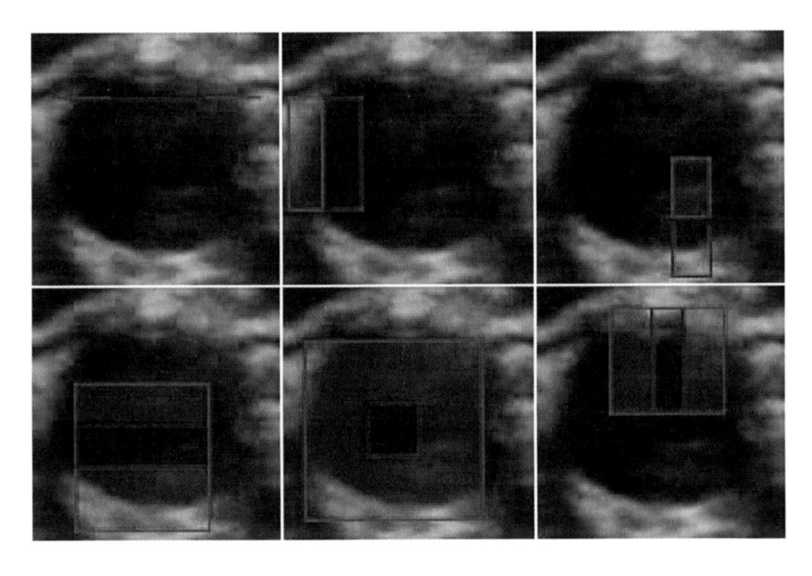

FIGURE 21.8: Haar-like features.

FIGURE 21.15: Example of segmentation.

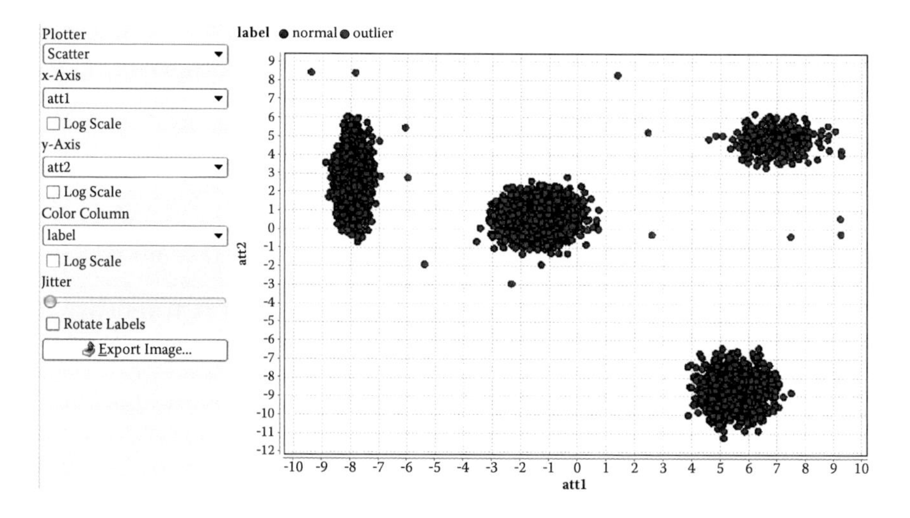

FIGURE 23.8: The scatter plot of the generated articial 2D data set for a unsupervised anomaly detection process. Blue color indicates normal data instances, red color the sampled outliers.

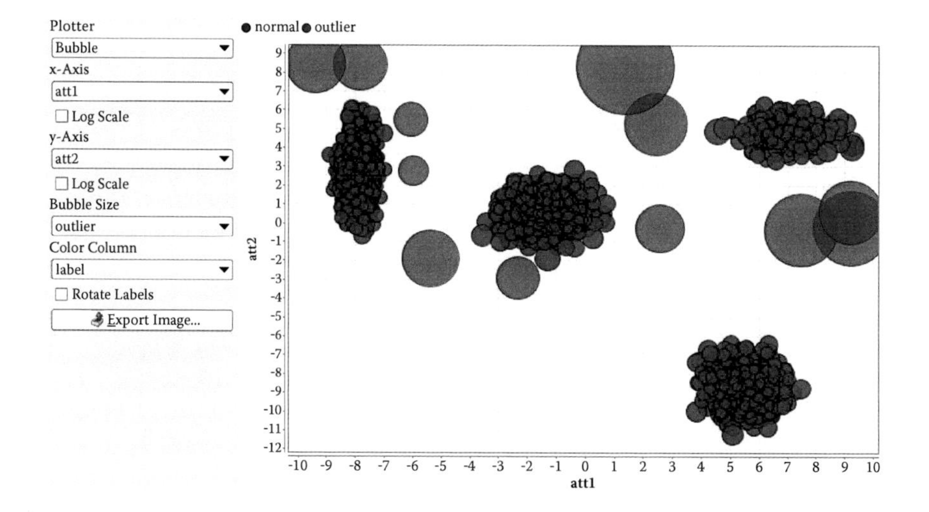

FIGURE 23.9: The result of the global k-NN anomaly detection.

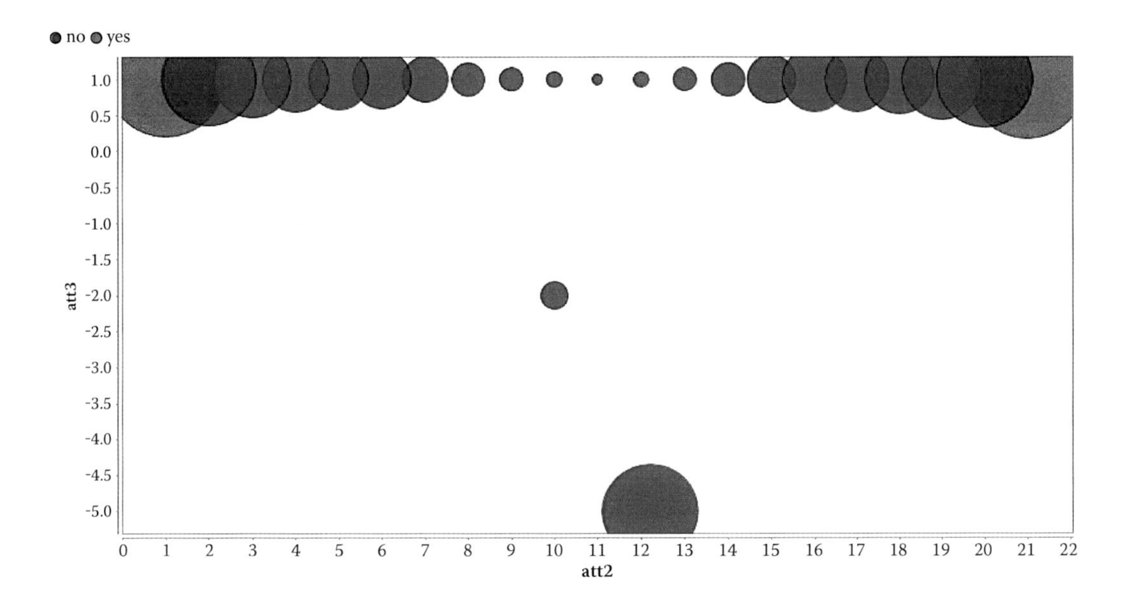

FIGURE 23.11: Results of a straight line data set with two outliers using LOF. The top two outlier scores are marked with red color.

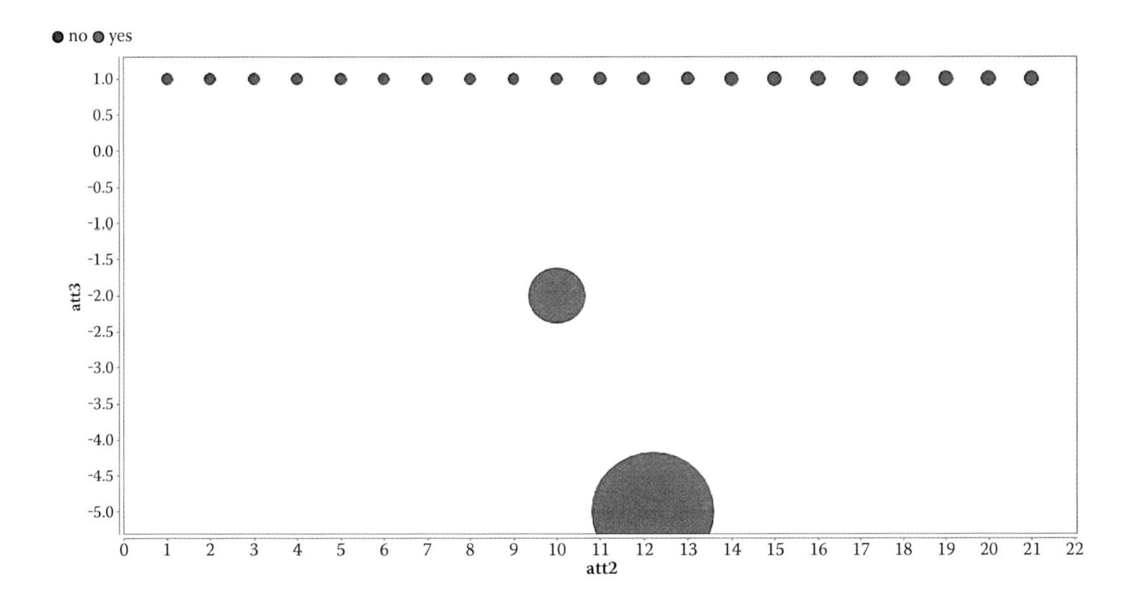

FIGURE 23.12: Results of a straight line data set with two outliers using COF. The top two outlier scores are marked with red color.

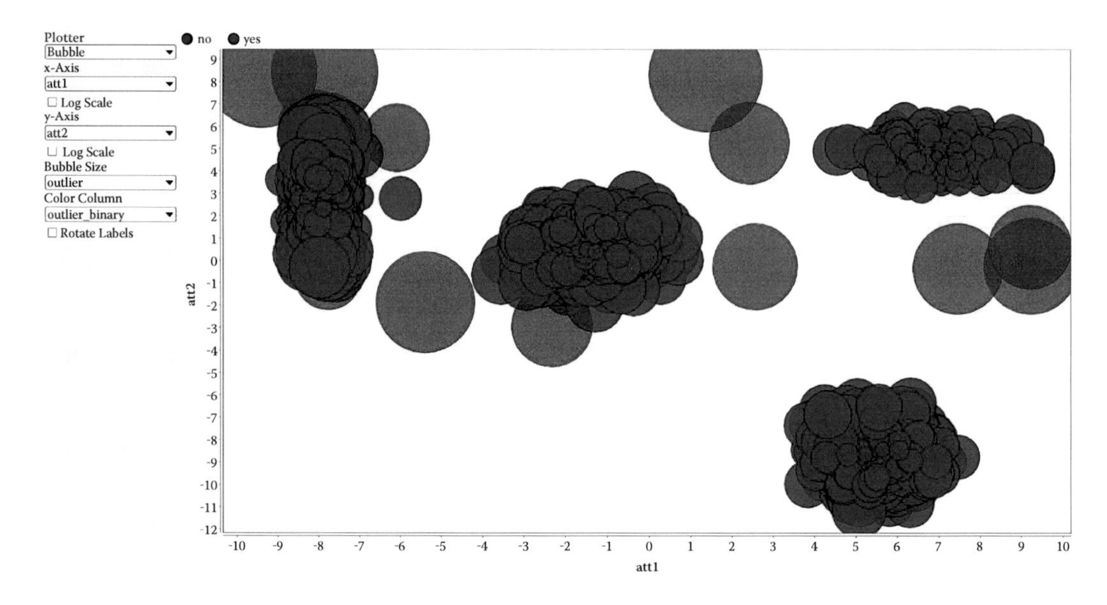

FIGURE 23.14: The result of the LDCOF anomaly detection algorithm using X-means clustering. The bubble size indicates the outlier score and the color indicates the binary outlier decision.

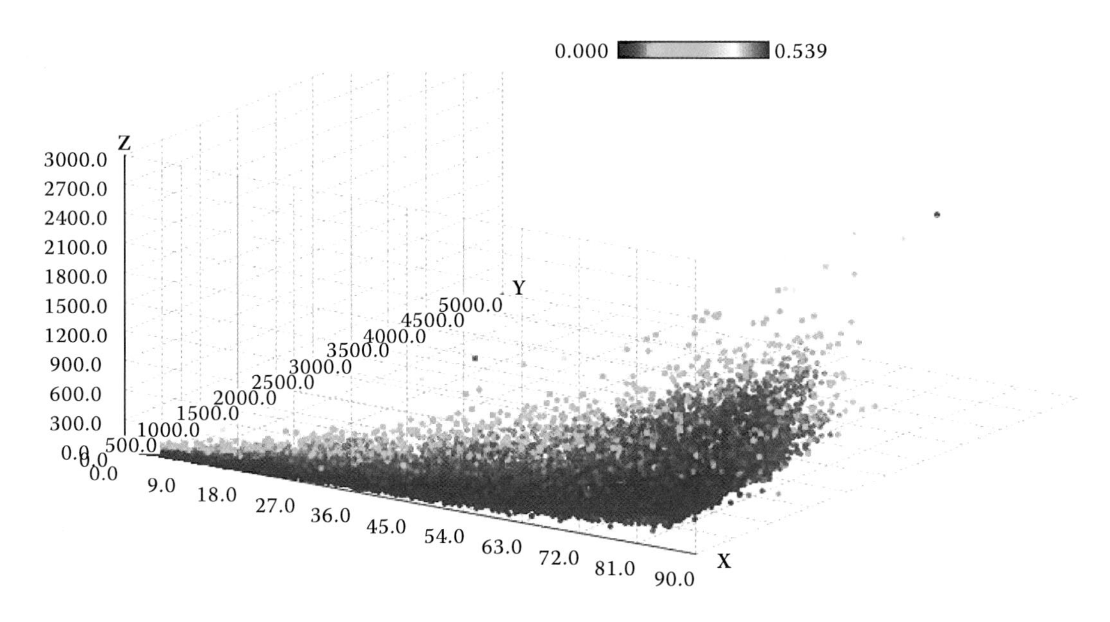

FIGURE 23.17: Visualizing the anomaly scores of the NBA data set. It can be seen that instances have higher scores which are outstanding in all dimensions - a property of a global anomaly detection algorithm.

Chapter 12

Grouping Higher Education Students with RapidMiner

Milan Vukićević

Faculty of Organizational Sciences, University of Belgrade, Belgrade, Serbia

Miloš Jovanović

Faculty of Organizational Sciences, University of Belgrade, Belgrade, Serbia

Boris Delibašić

Faculty of Organizational Sciences, University of Belgrade, Belgrade, Serbia

Milija Suknović

Faculty of Organizational Sciences, University of Belgrade, Belgrade, Serbia

Overview

Educational Data Mining (EDM) is a new emerging field that discovers knowledge from data originating from educational environments. In this research paper we used RapidMiner to cluster students according to their academic performance in order to recommend groups of students that would lead to general improvement of their performance.

12.1 Introduction

Grouping students is one of the frequent tasks in educational data mining. The objective is to create groups of students according to their customized features, personal characteristics, etc. Then, the clusters/groups of students obtained can be used by the instructor/developer to build a personalized learning system, to promote effective group learning, to provide adaptive contents, etc. [1]. In this chapter we present, RapidMiner [2] process for the analysis of higher education students' data based on clustering algorithms. The main idea of the proposed process is to find compact student groups in order to better adopt teaching materials and to provide direction for collaborative learning. RapidMiner provides a large clustering algorithm repository and preprocessing operators for clustering different types of data [3]. We used RapidMiner's possibility for design of complex processes based on inner operator structures, to define the process for automatic evaluation of several cluster algorithms from the RapidMiner algorithm repository. The quality of clusters is evaluated with internal evaluation measures (since the correct cluster structure is not known in advance). For the evaluation we used the "Cluster internal evaluation" operator, which is provided with the WhiBo plug-in [4] for RapidMiner. Additionally, we used RapidMiner's visualization techniques for interpretation and better understanding of characteristics of student groups.

Operators used: Read Aml, Set Role, Select Attributes, Replace Missing Values, Normalize, Loop Parameters, Select Sub-process, Cluster Internal Validation, K-means, Clustering, Support Vector Clustering, K-medoids, DBSCAN.

12.2 Related Work

Different clustering algorithms are used for improving the educational process from different aspects. Tang and McCalla [5] in their work suggested data clustering as a basic resource to promote group-based collaborative learning and to provide incremental student diagnosis. The clustering technique based on the implementation of the Bisection K-means algorithm and Kohonen's SOM algorithm [6] [7], [8] was used to group similar course materials with the aim of helping users to find and organize distributed course resources in the process of online learning. Also, the use of the K-means clustering algorithm for predicting students' learning activities was described in [9] work, where the information generated after the implementation of a data mining technique may be helpful for the instructor as well as for students. [10] used K-means and a model-based algorithm to group students with similar skill profiles on artificial data. Zakrzewska [11] used hierarchical clustering for grouping students based on their learning styles, to build individual models of learners and adjust teaching paths and materials to their needs. Perera et al. [12] used a technique of grouping both similar teams and similar individual members, and sequential pattern mining was used to extract sequences of frequent events. Tang and McCalla [13] proposed a clustering algorithm based on large generalized sequences to find groups of students with similar learning characteristics based on their traversal path patterns and the content of each page they visited. [14] used a K-means clustering algorithm for effectively grouping students who demonstrate similar behavior in an e-learning environment.

12.3 Using RapidMiner for Clustering Higher Education Students

In this section we will describe data for clustering students based on their academic performance, and the RapidMiner process for automatic evaluation algorithms for clustering of students' academic performance data. Finally, we discuss clustering results and provide some recommendations for improvement of the study process.

12.3.1 Data

The data is collected from the Faculty of Organizational Sciences, University of Belgrade, Serbia. The data involves 366 records about graduated students and their performance.

TABLE 12.1: Description of attributes for clustering students.

Attribute	Attribute Description
Sex	Student's sex Value: male/female type: binominal
Region	Region of the country from which student comes type: nominal
Score on entrance qualification exam	Points achieved in the entrance qualification exam for Faculty Values: 40–100 type: real
Grades at the first year of studies	Grades: marks on each of 11 exams in the first year of studies Values: 6–10 type: integer
Average grade	Average grade of the student after graduation Values: 6–10 type: continuous
Students' academic performance	Student's academic performance at the end of studies Values: "Bad", "Good", and "Excellent" type: polynomial

Attributes include the performance on the first-year exams (11 attributes), average grade of the student after graduation, number of points on entrance examination, sex, and the region of the country from which the student originates. Table 12.1 shows the description of attributes that are used in this study.

Students' academic performance: attribute is discretized from the "Average grade" where all students that had an average grade below 8 are classified as "Bad", between 8 and 9, "Good", and between 9 and 10 "Excellent". This attribute wasn't used in the clustering process, but it was used for the analyses and the discussion of the clustering results.

12.3.2 Process for Automatic Evaluation of Clustering Algorithms

The process for automatic evaluation of clustering algorithms is displayed in Figure 12.1. The "Read AML" operator is used for loading the data where all attributes had the "regular" role. The "Students academic performance" variable was used for analyses of the clustering results (not in a clustering process) and so we used "Set Role" to assign the "batch" role to this attribute. This way a clustering algorithms did not use this attribute for model creation.

We wanted to make cluster models based on students' behavior in the Faculty, and

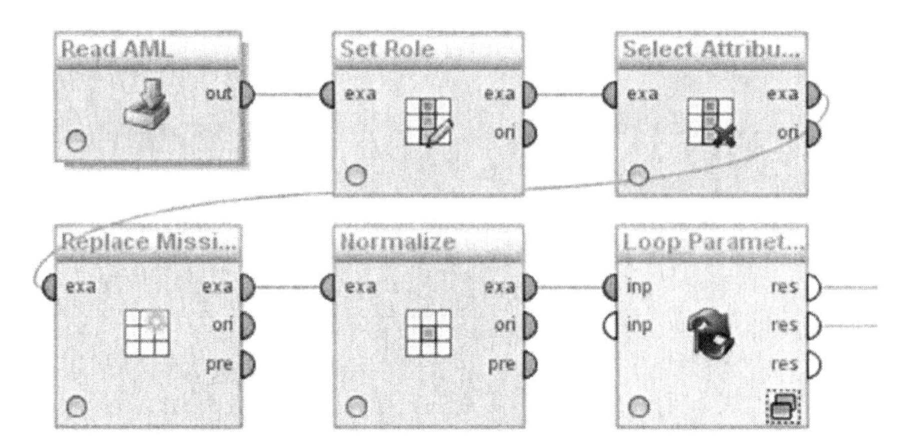

FIGURE 12.1: Main process for automatic evaluation of clustering algorithms.

not on their personal data, and so we used "Select Attributes" to remove "Sex", "Region" from the original dataset. The "Invert selection" option is activated and so all the selected attributes are removed from further process. Setting of the "Select Attributes" operator is shown in Figure 12.2.

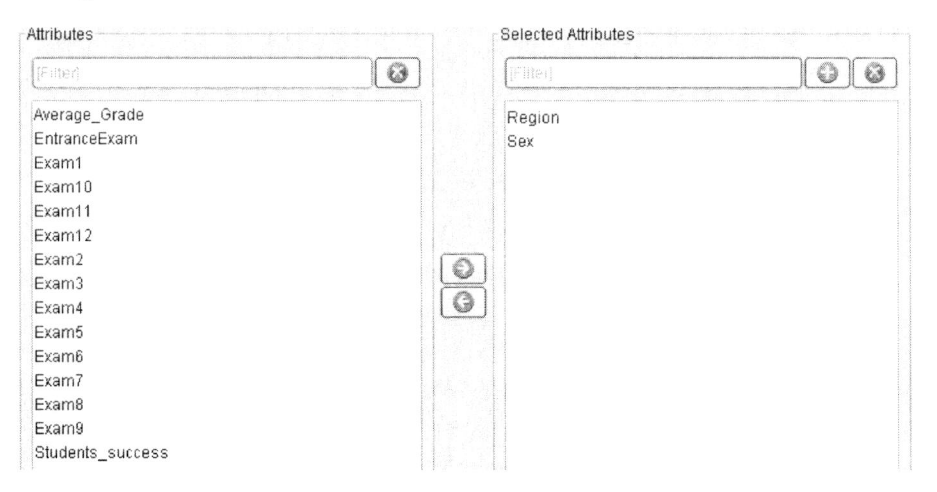

FIGURE 12.2: Attribute selection menu.

Many clustering algorithms do not support missing values. There are two ways of dealing with this problem. First, examples that contain missing values can be deleted from the dataset (this approach is effective only if there is small number of examples with missing values). Second, missing values can be replaced with some other value (imputation).

In this case, the "Replace Missing Values" operator is used. This operator replaces numerical values with a minimum, maximum, or average value. Nominal values are replaced with the value which occurs most often in the data. Since all of the attributes selected for the clustering process are numerical, all missing values are set to an average value of the attribute.

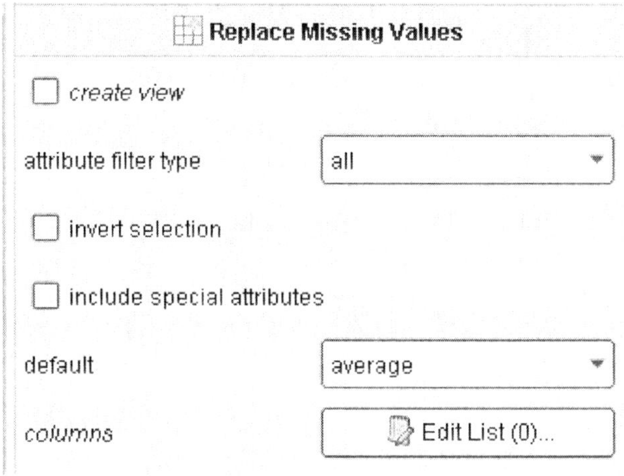

FIGURE 12.3: Replacing missing values with average values of the numerical attributes.

Normalization is used to avoid unwanted weighting of the attributes, just because they are measured in different scales (i.e., "Score on enter qualification exam" takes values from 40 to 100, while grades on the exams take values from 6–10).

The "Normalize" operator defines several options for normalization. "Z-transformation" where the mean value of the attribute is subtracted from the original value, and this difference is divided by the standard deviation. This way, the normalized attribute has standard Normal distribution with mean = 0 and variance = 1. "Range transformation" transforms an attribute into a user-defined range. "Proportional transformation" transforms an attribute as the proportion of the total sum of the respective attribute. Normalization is applied on all attributes, and all attribute values are transformed into 0–1 range (Figure 12.4).

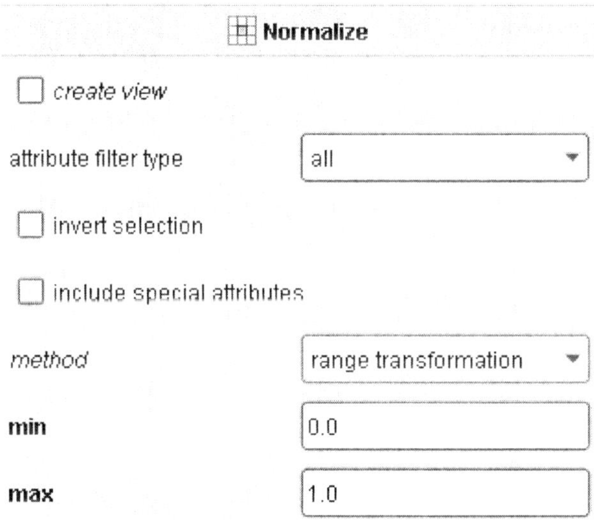

FIGURE 12.4: Normalization of the attributes.

Finally, the "Loop Parameters" operator is defined for iteration over the algorithms in the experiment. The operator "Loop Parameters" contains "Select Subprocess" as an inner operator. In every sub-process of "Select Subprocess" one clustering algorithm is defined.

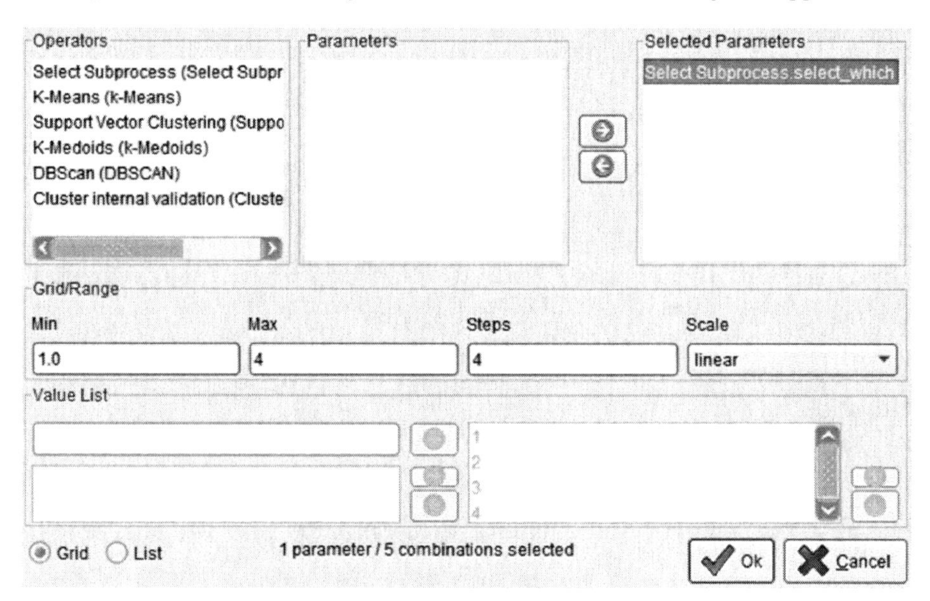

FIGURE 12.5: Loop parameters menu.

"Loop Parameters" loops over the "select_which" parameter of the "Select Subprocess", and this way a different algorithm is executed in every iteration. We used the range of the parameter from 1 to 4 with 4 steps and a linear scale because we evaluated exactly 4 algorithms.

Besides "Select Subprocess", "Cluster internal evaluation" operator is defined as an inner operator of "Loop Parameters".

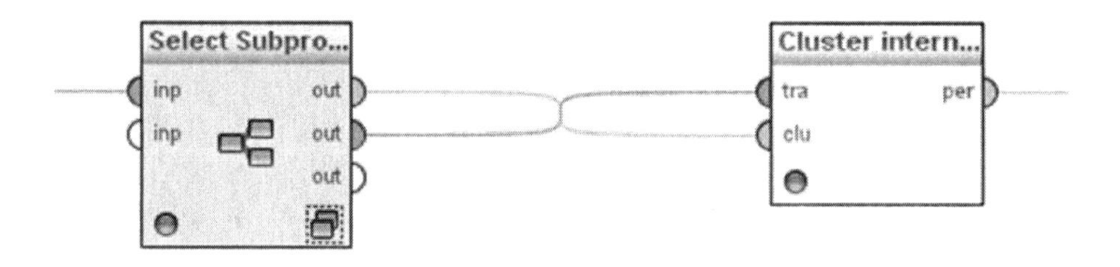

FIGURE 12.6: Sub-process of the "Loop Parameters" operator.

For inner operators of the "Select Subprocess", we used 4 clustering algorithms: K-means, Support Vector Clustering, K-medoids, and DBScan (Figure 12.7). In every iteration of "Loop Parameters", one of these algorithms is executed and its model is forwarded to the "Cluster internal evaluation" operator (Figure 12.6). In this case, all the algorithms are used with their default parameters. Note that optimization of the parameters (e.g., number of clusters, number of iterations, kernel type) can greatly influence the model performance. Additionally, most of the algorithms can use different distance measures, but there is no general recommendation for selection of the right distance measure [15]. Parameters and distance measure selection can be optimized with "Optimize parameters" operator in RapidMiner, but this is out of the scope of this chapter.

"Cluster internal evaluation" contains several internal evaluation measures and the dis-

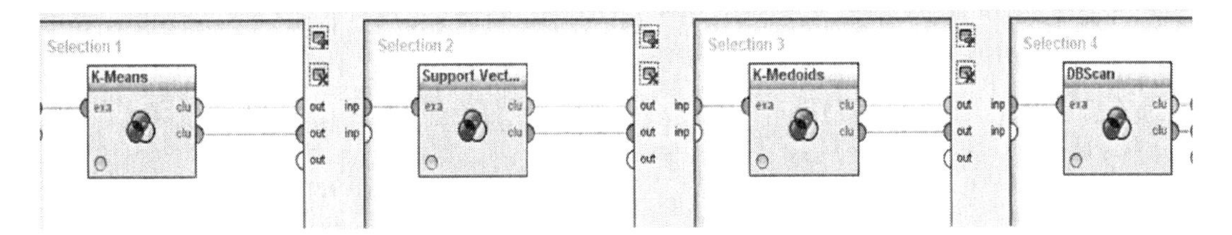

FIGURE 12.7: Sub-process of the "Loop Parameters" operator.

tance measure as the parameters (Figure 12.8). In one execution it is possible to evaluate one cluster model with several evaluation measures. All algorithms were evaluated with a silhouette index value [16] because it is widely used, fairly simple, and can be used for easy comparison of clustering quality between different datasets. The Silhouette index takes values between -1 and 1, where values closer to 1 indicate that objects are soundly clustered. By using only one evaluation measure, comparison of algorithms is adequate and consistent, as explained by Estivill-Castro [17]. For other evaluation measures the best algorithms could vary.

Distance measure is defined as a parameter in order to enable evaluation of the models with the same distance measure as was used for creation of models. In our case, all the algorithms used Euclidean distance.

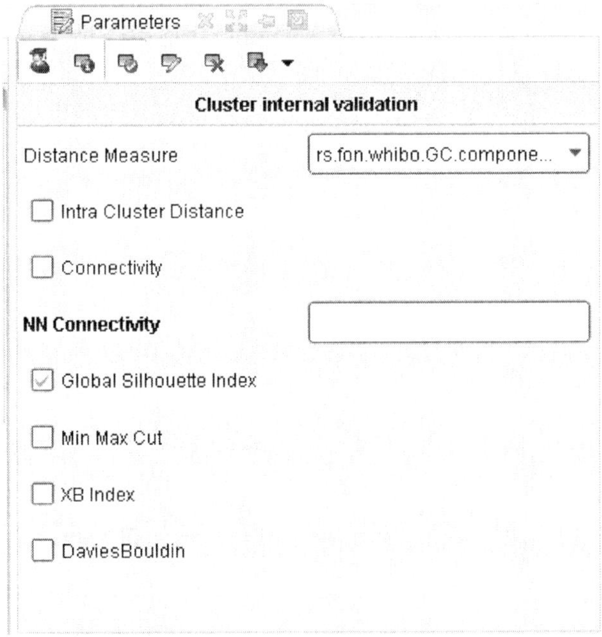

FIGURE 12.8: Parameters for selection of internal evaluation measures.

12.3.3 Results and Discussion

After execution of the process, we have the following values for the global silhouette index:

- K-means: -0.761

- Support Vector Clustering: −0.715
- K-medoids: −0.655
- DBScan: −0.856

Since the larger value of the global silhouette index indicates a better clustering model, we can conclude that K-medoids provided the best model. RapidMiner allows easy visualization of representative-based cluster models, we used "Centroid plot" to inspect clustering results The resulting model of the K-medoids algorithm is shown in Figure 12.9.

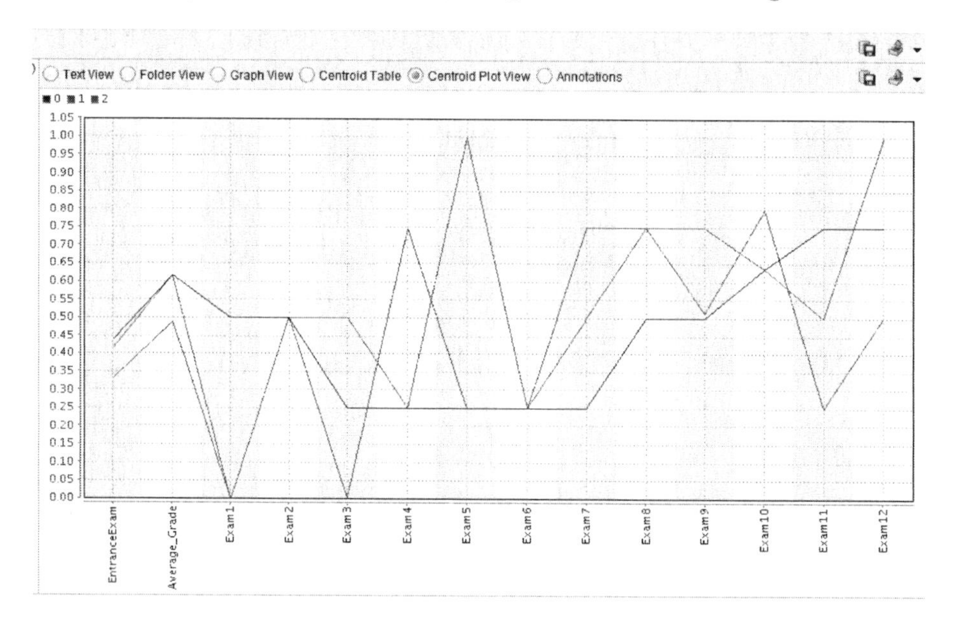

FIGURE 12.9 **(see color insert)**: Centroid plot of K-medoids algorithm.

It can be seen from Figure 12.9 that students that belong to Cluster_0 (blue line) had the best results on the entrance exam and also by average grade, Exam 1 and Exam 11. They had the worst results on: Exam 4, Exam 7, and Exam 8.

Students that belong to Cluster_1 (green line) had similar results to cluster_0, on the entrance exam and by the Average_Grade. These students achieved best results on Exam 3, Exam 5, Exam 8, Exam 9, and Exam 12.

Students that belong to Cluster 2 (red line) had the worst results on the entrance exam and by the average grade. They showed the best performance on Exam 4, Exam 7, and Exam 10, but the worst results one Exam 3, Exam 4, and Exam 11.

In order to make some recommendations for improving the studying process, we used "Histogram Color" to see how the students from different clusters are distributed by overall academic performance.

It can be seen from Figure 12.10 that most students that belong to the "Excellent" class also belong to cluster_1. Additionally students from cluster_1 very rarely belong to the "Bad" class. This indicates that "Excellent" performance depends on good grades on Exam 3, Exam 5, Exam 9, and Exam 12 (as shown in Figure 12.8). So, these exams should be pointed out as important at the beginning of studies.

Opposite conclusion could be made for cluster_2 where students are often classified as "Bad" or "Good" and very rarely as Excellent had the best results on Exam 4, Exam 7, and Exam 10, but the worst results one Exam 3, Exam 4, and Exam 11. Additionally, students

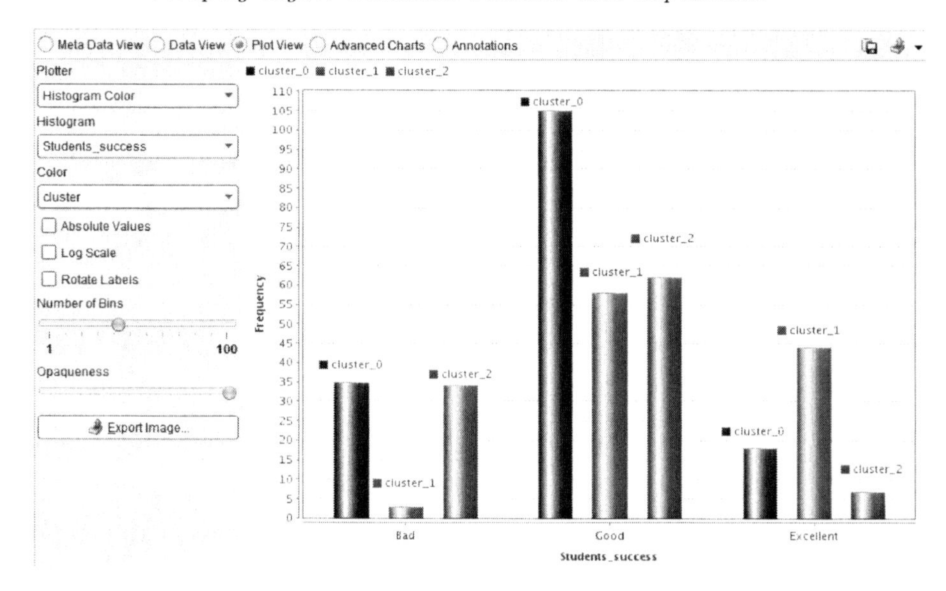

FIGURE 12.10: Clusters compared to students' academic performance at the end of studies.

from cluster_2 have the fewest points on the "Entrance exam" (compared to other two clusters). From this we can conclude that these students do not have enough background knowledge and that they should put in additional effort in order to achieve good academic performance at the end of studies.

Finally, majority of the students from cluster_0 belong to the "Good" class, and the number of "Bad" students is on the same level as "cluster_2". It can be seen from Figure 12.8, that these students have the worst results on the exams, where students from cluster_1 have the best results. Once again, this indicates that academic performance on Exam 3, Exam 5, Exam 9, and Exam 12 have influence on the students' overall academic performance.

From the above analysis we can assume that it would be good to adapt teaching materials (on the exams that are crucial for students' academic performance) for regular and advanced students separately. Students with less background knowledge should first learn from simpler materials.

12.4 Conclusion

In this chapter we proposed a RapidMiner stream for automatic evaluation of clustering algorithms. We applied this stream on data concerning higher education students' academic performance, but with minor modifications it could be used in any other application area.

Additionally, we used RapidMiner's visualization tools for inspecting the correlation between clusters and students' overall academic performance and to make some recommendations for general improvement of students' performance.

Bibliography

[1] C. Romero and S. Ventura. Educational data mining: A review of the state of the art. *Systems, Man, and Cybernetics, Part C: Applications and Reviews, IEEE Transactions on*, 40(6):601–618, 2010.

[2] I. Mierswa, M. Wurst, R. Klinkenberg, M. Scholz, and T. Euler. Yale: Rapid prototyping for complex data mining tasks. In *Proceedings of the 12th ACM SIGKDD International Conference on Knowledge Discovery and Data Mining*, pages 935–940. ACM, 2006.

[3] A. Chisholm and G. Gray. Cluster evaluation of highly dimensional data. In *Proceedings of the 2nd RapidMiner. Community Meeting and Conference. (RCOMM 2011).*, 2011.

[4] M. Vukićević, M. Jovanović, and B. Delibašić. WhiBo - RapidMiner plug-in for component based data mining algorithm design. In *Proceedings of the 1st RapidMiner. Community and Conference. (RCOMM 2010)*, 2010.

[5] T. Tang and G. McCalla. Smart recommendation for an evolving e-learning system: architecture and experiment. *International Journal on E-learning*, 4(1):105–129, 2005.

[6] A. Drigas and J. Vrettaros. An intelligent tool for building e-learning content-material using natural language in digital libraries. *WSEAS Transactions on Information Science and Applications*, 1(5):1197–1205, 2004.

[7] Hammouda, L. and M. Kamel. *Data Mining in E-learning.* 2006.

[8] J. Tane, C. Schmitz, and G. Stumme. Semantic resource management for the web: an e-learning application. In *Proceedings of the 13th International World Wide Web Conference on Alternate Track Papers & Posters*, pages 1–10. ACM, 2004.

[9] A. Shaeela, M. Tasleem, S. Ahsan-Raza, and K. Inayat. Data mining model for higher education system. *European Journal of Scientific Research*, 43(1):24–29, 2010.

[10] E. Ayers, R. Nugent, and N. Dean. A comparison of student skill knowledge estimates. In *2nd International Conference on Educational Data Mining, Cordoba, Spain*, pages 1–10, 2009.

[11] D. Zakrzewska. Cluster analysis for users modeling in intelligent e-learning systems. *New Frontiers in Applied Artificial Intelligence*, pages 209–214, 2008.

[12] D. Perera, J. Kay, I. Koprinska, K. Yacef, and O.R. Zaïane. Clustering and sequential pattern mining of online collaborative learning data. *Knowledge and Data Engineering, IEEE Transactions on*, 21(6):759–772, 2009.

[13] T. Tang and G. McCalla. Student modeling for a web-based learning environment: a data mining approach. In *Proceedings of the National Conference on Artificial Intelligence*, pages 967–968. Menlo Park, CA; Cambridge, MA; London; AAAI Press; MIT Press; 1999, 2002.

[14] C. Chen, M. Chen, and Y. Li. Mining key formative assessment rules based on learner portfolios for web-based learning systems. In *Proc. IEEE International Conference on Advanced Learning Technologies*, pages 1–5. ACM, 2007.

[15] M. Vukićević, K. Kirchner, B. Delibašić, M. Jovanović, J. Ruhland, and M. Suknović. Finding best algorithmic components for clustering microarray data. *Knowledge and Information Systems*, pages 1–20, 2012.

[16] P.J. Rousseeuw. Silhouettes: A graphical aid to the interpretation and validation of cluster analysis. *Journal of Computational and Applied Mathematics*, 20:53–65, 1987.

[17] V. Estivill-Castro. Why so many clustering algorithms: A position paper. *SIGKDD Explor. Newsl.*, 4(1):65–75, June 2002.

Part V

Text Mining: Spam Detection, Language Detection, and Customer Feedback Analysis

Chapter 13

Detecting Text Message Spam

Neil McGuigan
University of British Columbia, Sauder School of Business, Canada

Acronyms

CSV - Comma-separated values
SMS - Short Message Service
UTF - Universal Character Set Transformation Format

13.1 Overview

This chapter is about text classification. Text classification is an important topic in data mining, as most communications are stored in text format. We will build a RapidMiner process that learns the difference between spam messages, and messages that you actually want to read. We will then apply the learned model to new messages to decide whether or not they are spam. Spam is a topic familiar to many, so it is a natural medium to work in. The same techniques used to classify spam messages can be used in many other text mining domains.

Here are the main steps that we will follow:

1. Install the Text Processing extension.
2. Download the dataset, and load it into RapidMiner.
3. Examine the text.
4. Process the text.
5. Build a text classifier using the Naïve Bayes method.
6. Validate the model.
7. Apply the model to unseen data.

13.2 Applying This Technique in Other Domains

Text classification is used in a variety of industries, from human resource companies that use automated means to process incoming resumes and filter out good and bad candidates, to finance companies that process news stories to make stock price predictions. Other companies use text classification to monitor their reputations online, or to prioritize incoming customer complaints.

For example, as an HR company, you could take a sample of, say, 50 "good" resumes and 50 "bad" resumes, as rated by an HR manager. Run the resumes through a similar text mining process as below, to examine the word frequencies and create a model of the "goodness" or "badness" of the resume. In the future, use that model to screen incoming resumes, putting the "best" ones in front of a hiring manager first. Of course, one would want to ensure that the model does not break any anti-discrimination laws.

13.3 Installing the Text Processing Extension

To perform text classification in RapidMiner, you need to have the Text Processing extension installed. You can check whether you have it installed by following these steps:

1. Open RapidMiner. On the main menu, click Help >Manage Extensions.

2. Ensure that the Text Processing extension is listed and checked. Version 5.1.2 was used for this chapter.

If you do not have the Text Processing extension installed, you can install it by following these steps:

1. If you are using Windows, make sure that you are logged in with Administrator privileges
2. On the main menu, click Help >Update RapidMiner
3. Scroll down and double-click the Text Processing extension, and click the Install button
4. Accept the terms of use, and click the OK button. RapidMiner will restart

13.4 Getting the Data

This chapter uses a dataset of 5574 SMS (mobile phone text) messages hosted by the University of California, Irvine, Machine Learning Repository. You can read more about the dataset and download it from here:
`http://archive.ics.uci.edu/ml/datasets/SMS+Spam+Collection`

It contains 747 messages marked as "spam", and the remainder are non-spam messages marked as "ham". It is a tab-separated text file with one message per line, with UTF-8 encoding.

On the above site, click the Data Folder link and download and unzip the smsspamcollection.zip file.

13.5 Loading the Text

Start a new RapidMiner process, and save it in the Repository.

13.5.1 Data Import Wizard Step 1

1. On the main menu, click File >Import Data >Import CSV File.
2. Change the "Files of Type" drop-down to "All Files".
3. Select the unzipped SMSSpamCollection file that you downloaded in "Getting the Data".
4. Click Next.

13.5.2 Data Import Wizard Step 2

1. Change "File Encoding" to "UTF-8".
2. Set "Column Separation" to "Tab".
3. Uncheck "Use Quotes".

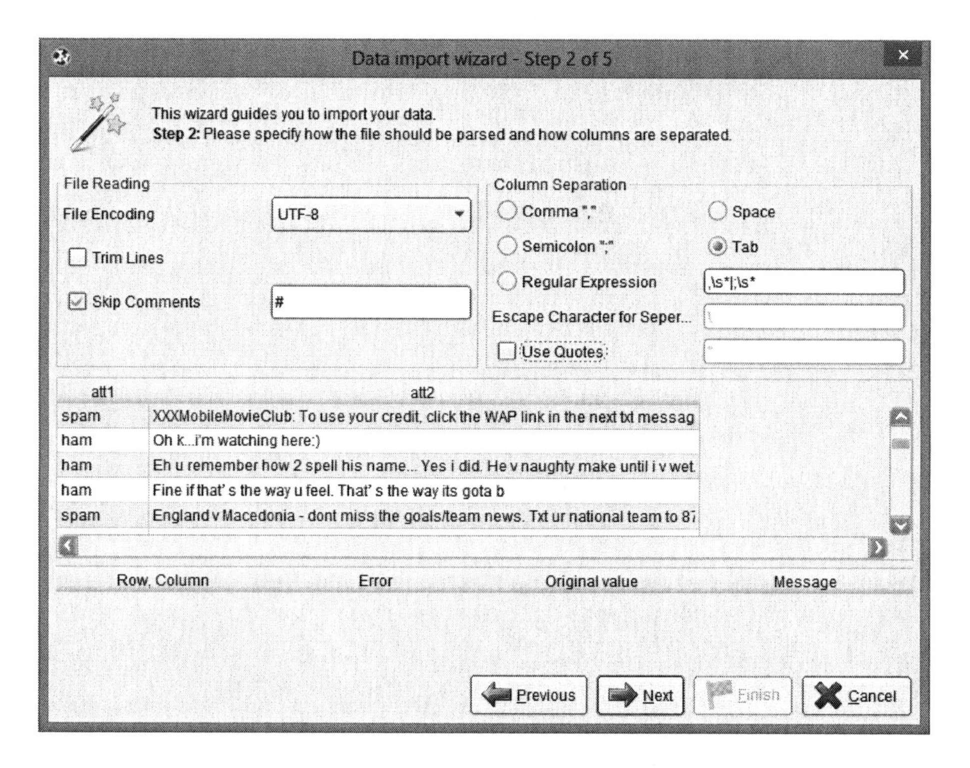

FIGURE 13.1: Data Import Wizard.

13.5.3 Data Import Wizard Step 3

1. You can preview your data here. Click Next.

13.5.4 Data Import Wizard Step 4

1. Change "att1" *role* from **attribute** to **label**. This tells RapidMiner that we would like to make predictions on this attribute.
2. Change "att2" *type* from **polynomial** to **text**. This tells RapidMiner that the attribute contains text that we would like to manipulate.

13.5.5 Step 5

1. Pick a repository location to save your imported data, and give it a name.
2. Click **Finish** to import the data.

3. Find the imported data file in the **Repository** tab, and drag it on to the Main Process window.

4. Click the **Play** button (on the main toolbar), and then the **Data View** button (in the Example Set tab) to view your data.

Note: You could use the **Read CSV** and **Store** operators in the process to read the data, but it can be faster to import the data into RapidMiner's repository once, and read it from there. Using the repository as the data source also makes it easier to select attributes in drop-down menus in certain operators. However, it can be easier to make changes "on the fly" if using the **Read CSV** operator.

13.6 Examining the Text

To get a better understanding of the text, it can be helpful to break the documents into individual words, and examine the frequency of the words. To do this, we will use the **Process Documents from Data** operator and examine its **Word List** (a list of word frequencies) and **Word Vector** (a cross-tabulation of word frequencies and documents). In this scenario, an individual line from the dataset is a single SMS message, which we will refer to as a document. RapidMiner can also use text from Excel, databases, the web, or folders on your computer.

1. If you have not already done so, drag the SMS data from the **Repository** tab onto the Main Process window. It will create a **Retrieve** operator (which retrieves data from the repository), and connect it to a **Results** node (labeled **res**) on the right of the process.

2. Click the **Operators** tab, and type "process documents from data" into the box labeled [Filter].

3. Drag the **Process Documents from Data** operator *onto the line* connecting the **Retrieve** operator and the **res** node in the process. This will automatically connect the new operator.

Tip: You can also enter the first letters of the words of operator names to find them in the Operators window. For example, to find the **Process Documents** operators, you could type "Pro Doc" or "PrDo". This is case-sensitive.

The **Process Documents from Data** operator is an *Inner Operator*, in that it is a process inside a process. To see "inside" an Inner Operator, double-click it. To exit an Inner Operator, click the blue "up" arrow.

13.6.1 Tokenizing the Document

The **Tokenize** operator takes a document and splits it into a set of words based on the selected mode. In this case, we will split the document on non-letters, meaning that every time the operator encounters a symbol, such as a space or hyphen character, it will split the document into a new token. Thus, "a stitch (in time), saves - nine", would turn into the tokens {a, stitch, in, time, saves, nine}.

1. Select the **Process Documents from Data** operator, and in the **Parameters** tab, change **vector creation** to **Term Occurrences**. This will let us see the number of times that each word appears in each document.

2. Check **keep text**. This will keep the original text in a column in the output table.

3. Using your mouse, drag a line from the **Process Documents from Data wor** port (for Word List) to a **res** node on the right of the process.

4. Double-click the **Process Documents from Data** operator to see its inner process.

5. Find the **Tokenize** operator in the **Operators** tab. Double-click the **Tokenize** operator to add it to the **Process Documents from Data** inner process. This will automatically connect the correct nodes.

6. Click the blue "up" arrow to exit the inner process.

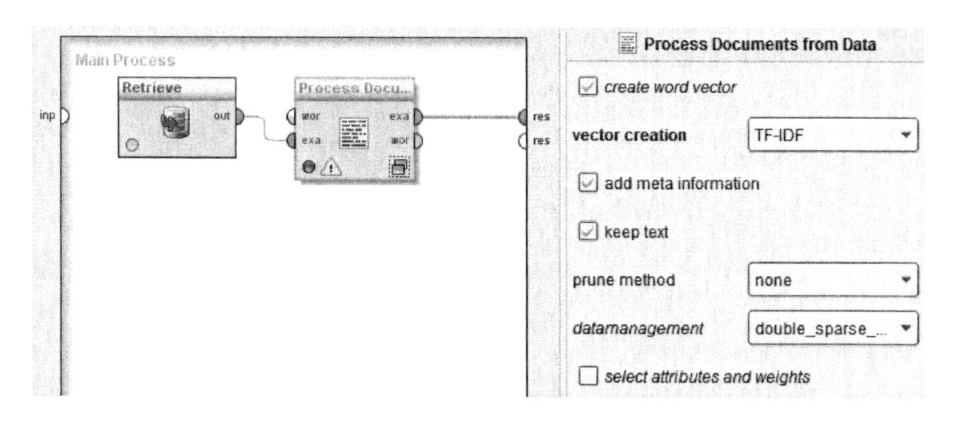

FIGURE 13.2: The basic Text Analysis Process.

13.6.2 Creating the Word List and Word Vector

1. Press the **Play** button on the main toolbar. This process took only a few seconds on an inexpensive computer.

2. Examine the **Word List**. This shows all of the unique words in all of the documents. You can click the column headers to sort the columns.

3. If you sort the **Total Occurrences** column, you will see that "I" is the second-most common word, with 1985 occurrences. You will also see that "I" occurs 1931 times in "ham" documents and 54 times in "spam", suggesting that it is an important discriminant between "ham" and "spam", due to its proportionally low occurrence in "spam" messages. This is intuitive because spammers want *you* to do something, whereas you are often talking about yourself when communicating with friends.

13.6.3 Examining the Word Vector

A word vector is just a fancy name for a table, where each row is a document (SMS message in this case), and each column is a unique word in the *corpus* (all of the words in all of your documents). The values inside the table depend on the type of word vector you are creating. In this case we are using **Term Occurrences**, meaning that a value in a cell represents the number of times that word appeared in that document. You could also use the **Binary Term Occurrences**, meaning the value in the cell will be zero if the word did not appear in that document, and one if the word appeared *one or more times* in that document. It is always a good idea to examine your data, in order to "get a feel" for it, and to look for strange anomalies.

1. Click the **ExampleSet** tab to view the word vector. You will see that the word vector has 9755 attributes, meaning that there are 9755 unique words in the corpus. Equivalently, there are 9755 columns in the word vector.
2. Look at the **Range** column for the "text" role attribute. You will note that:
   - "Sorry I'll call later" is the most common message.
   - Below that, you can see that there are 4827 "ham" messages, and 747 "spam" messages.
3. Click the **Data View** button. You will see that document 13 has a "1" under the word "A", meaning that the word "A" appears one time in that document. Actually, "a" and "A" appear in that document, but we are considering the *letter case* of the words in this process, so they are counted as distinct words.

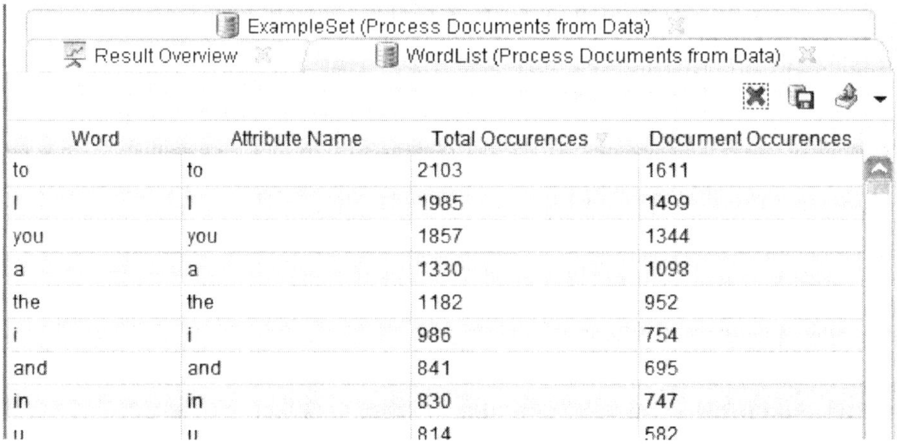

ExampleSet (Process Documents from Data)
Result Overview | WordList (Process Documents from Data)

Word	Attribute Name	Total Occurences	Document Occurences
to	to	2103	1611
I	I	1985	1499
you	you	1857	1344
a	a	1330	1098
the	the	1182	952
i	i	986	754
and	and	841	695
in	in	830	747
u	u	814	582

FIGURE 13.3: A table of word frequencies.

13.7 Processing the Text for Classification

If a dataset had one "spam" message and 999 "ham" messages, how accurate would you be by always guessing "ham"?

You would be 99.9% accurate (guessing correctly 999 out of 1000 times).

You now have a predictor with high accuracy, but it is useless in that it has no real predictive power (you know the answer is always "ham"). To solve this problem, we must *balance* the dataset, by using an equal number of "spam" messages and "ham" messages. Thus, we would expect an "always guess ham" predictor to have 50% accuracy, and a number greater than that is from real predictive power.

What is the disadvantage? We have to ignore a large portion of our "ham" messages. Such is life.

1. Add a **Sample** operator after the **Retrieve** operator.
2. Set its **sample** parameter to **absolute**. This lets us choose the sample size.
3. Check **balance data**. This lets us choose the sample size for each class of the label.
4. Click the **Edit List** button.
5. Add a "spam" **class** with **size** 747 (click the **Add Entry** button to add another row).
6. Add a "ham" **class** with **size** 747.

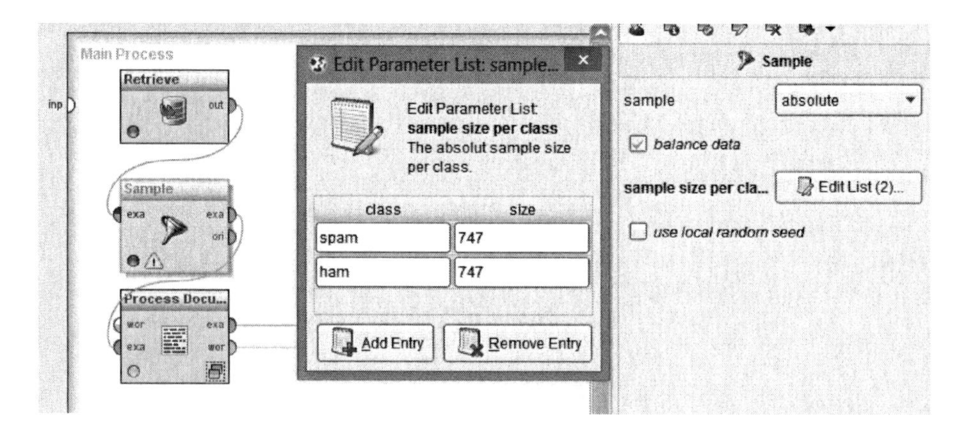

FIGURE 13.4: Balancing the Sample by the Label.

13.7.1 Text Processing Concepts

- Should "cat", "Cat", and "CAT" be counted as the same word? This is normally a good idea. In this case however, the letter case of words is a good predictor of a spam message, as spammers like to "shout" at their audience with upper-case words. In other processes, you can use the **Transform Cases** operator to force all words to lower-case.
- Should "organize", "organizes", and "organized" be counted as the same word? This is usually a good idea in most text mining processes. In this example though, it did not increase prediction accuracy, so was ignored for the sake of simplicity. You can use the **Stem (Porter)** operator to intelligently strip the suffixes of words. The stem operators put all words into lower case.
- Should short sentence fragments be counted as distinct items? For example, along with individually using "quick", "brown", and "fox", we could include the fragment "quick brown fox", and count that as its own attribute. Perhaps the fragment is a better predictor than the individual words. You can use the **Generate N-Grams (Terms)** operator to do this. In this case, it only increased accuracy by a couple of decimal points, so was neglected as it increased processing time.

- Most times, you will want to remove **Stop Words** (such as "and" and "the"), as they have little predictive power, and add to processing time. In this case, however, stop words actually helped the accuracy of the process. You can use the **Filter Stopwords** operator to remove common words from your data.

13.8 The Naïve Bayes Algorithm

The Naïve Bayes algorithm is an old and simple classification method, but can be very useful in text mining. The main reason that it is useful is that text mining tends to create a very large number of attributes (9755 in our case), and this algorithm works well with a large number of attributes. The reason that it can handle so many attributes quickly is that it treats the attributes as independent, and thus does not need to worry about calculating the relationship between them. Assuming that the attributes are independent may be "naïve", but the algorithm works well regardless.

13.8.1 How It Works

The model works by examining the relative frequency ("probability") of the attributes with respect to the relative frequency of the label. Consider the following fictional historical data on car theft:

TABLE 13.1: Fictional historical data on car theft.

Stolen	**Red**	**Yellow**	**Sports**	**SUV**	**Domestic**	**Imported**
No	40%	60%	33%	75%	40%	60%
Yes	60%	40%	67%	25%	60%	40%

Also: 50% of cars are stolen.

If you bought a yellow, domestic, SUV, is it more likely to be stolen or not stolen? To figure this out, you multiply the probabilities of the class with the matching attributes:

The probability of the car being stolen:

Pr(Stolen |Yellow, Domestic, SUV) = Pr(Stolen) * Pr(Yellow |Stolen) * Pr(SUV |Stolen) * Pr(Domestic |Stolen)

$$= \quad 0.50 \quad * \quad 0.40 \quad * \quad 0.25 \quad * 0.40 \quad = \quad 2\%$$

The probability of the car being not stolen:

Pr(Not Stolen |Yellow, Domestic, SUV) = Pr(Not Stolen) * Pr(Yellow |Not Stolen) * Pr(SUV |Not Stolen) * Pr(Domestic |Not Stolen)

$$= \quad 0.50 \quad * \quad 0.60 \quad * \quad 0.75 \quad * 0.60 \quad = \quad 13.5\%$$

Therefore, it is more likely that the car will not be stolen—whew!

Note: In the real world, an adjustment is made to account for the case when a probability is **zero**, so that their product is not zero. Also, the method is somewhat different for continuous attributes versus categorical ones, however the concept remains the same.

This method works the same way in text mining. One calculates the most probable class ("spam" or "ham") based on the multiplication of the probabilities of the values of the attributes.

13.9 Classifying the Data as Spam or Ham

To build the Naïve Bayes model, add the **Naive Bayes** operator after the **Process Documents from Data** operator.

To find the model's predictive accuracy, we must apply the model to data, and then count how often its predictions are correct. The **accuracy** of a model is the number of correct predictions out of the total number of predictions.

Add the **Apply Model** operator after the **Naïve Bayes** operator and connect their two nodes together. Add a **Performance** operator after the **Apply Model** operator and connect it to a **res** node.

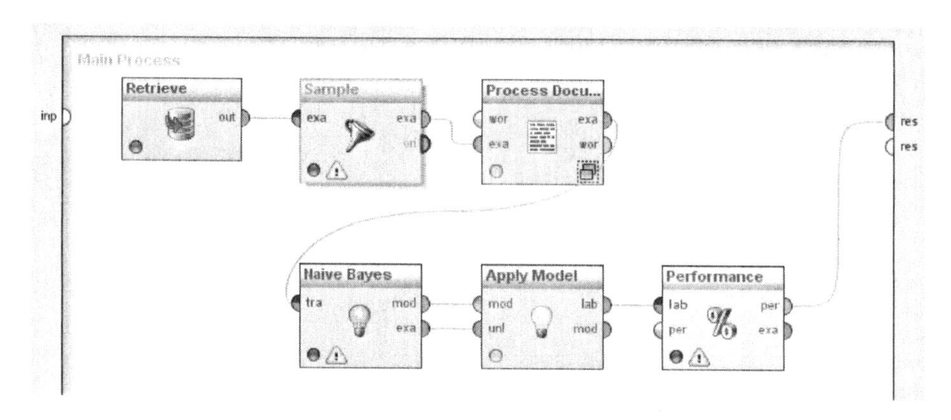

FIGURE 13.5: The Text Classification Process.

Click the **Play** button and wait for the process to complete, and examine the **Performance Vector** tab in the **Results Perspective**.

You will see that the accuracy is 99.6%, which is pleasantly high.

Unfortunately, when you try this model on unseen data, you are very likely to get that lower accuracy. The problem is that the model was learned on all of the data and was never tested on unseen data.

13.10 Validating the Model

To be able to predict a model's accuracy on unseen data, we must hide some of the data from the model, and then test the model on that unseen data. One way to do this is to use K-fold Cross-Validation.

When using, say, 10-fold Cross-Validation, we would hide $1/10^{th}$ of the data from the model, build the model on the remaining $9/10^{ths}$ of the data, and then test the model on the whole dataset, calculating its accuracy. We would do this again, hiding a different $1/10^{th}$

of the data from the model, and test again. We would do this 10 times in total, and take the average of the accuracies. This provided a better idea of how the model will perform on data that it has not seen before.

1. Remove the **Naïve Bayes**, **Apply Model**, and **Performance** operators from the Main Process window.
2. Connect an **X-Validation** operator to the **Process Documents from Data** operator, and connect its **ave** (for average performance) node to a **res** node.
3. Double-click the **X-Validation** operator. Put a **Naïve Bayes** operator in the left side of this inner process, and an **Apply Model** operator and a **Performance** operator in the right side of the process. Connect all required nodes.

Run the process again. You will see that the accuracy has dropped, but not by much. It is now 95.7% +/− 1.5%. Thus, we should expect an accuracy of about 96% on data that the model has never seen before. The "95.7" figure is the average accuracy across the 10 validations, and the "1.5" figure is the standard deviation.

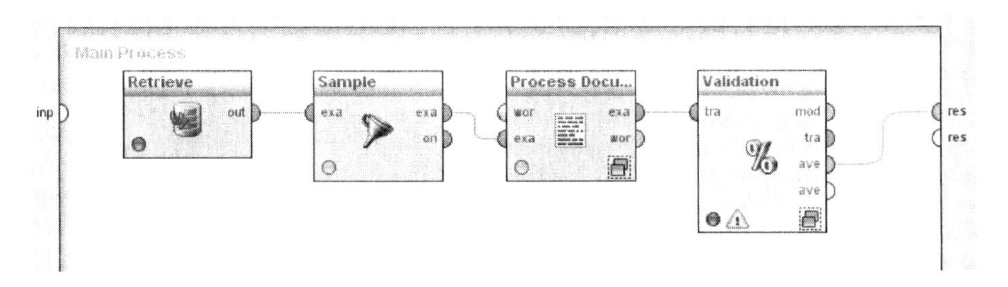

FIGURE 13.6: Cross-Validation.

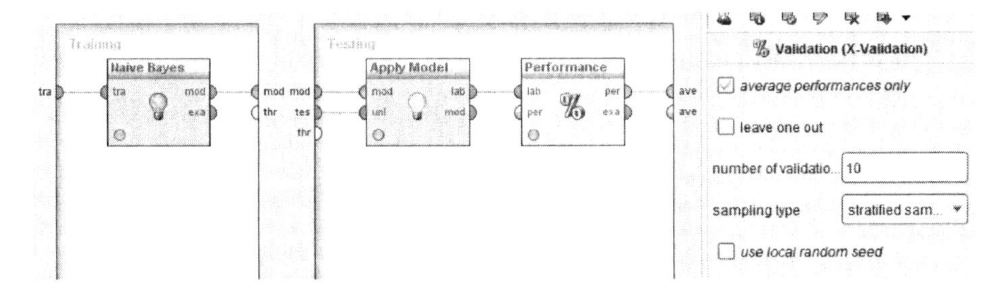

FIGURE 13.7: Inside the X-Validation operator.

13.11 Applying the Model to New Data

To apply the learned model to new data, we must first save the model, so that we can use it again on new data. We also have to save the word list. The reason that we need to save the word list is that we need to compare apples to apples. When we are estimating the probability that a new message is "spam" or "ham", we have to use the same attributes (words) that we used in the original process.

You need the same word list, same model, and need to process the new data in exactly the same way as you processed the learning data. The only thing different is the new data.

1. Connect a **Store** operator to the **wor** port on the **Process Documents from Data** operator. Set the **repository entry** parameter to something memorable. This will save the word list for later.

2. Connect a **Store** operator to the **mod** (for model) port of the **X-Validation** operator. Set the **repository entry** parameter to something memorable. This will save the model for later.

3. Run the process again.

13.11.1 Running the Model on New Data

1. Create and save a new process. We will use this process to apply the model on new data to predict whether a new message is spam or not.

2. Import new, unseen data into your repository, as you did at the beginning of this chapter. It should be in the same format as the other data. Add a **Retrieve** operator to retrieve the data.

3. Copy and paste the **Process Documents From Data** operator from the previous process into this process.

4. Connect an **Apply Model** operator to the **Process Documents** operator.

5. Connect a **Retrieve** operator to the left-side **wor** port of the **Process Documents** operator, and set its **repository entry** parameter to the name of the wordlist that you saved previously. This will load the previous wordlist.

6. Connect a **Retrieve** operator to the left-side **mod** port of the **Apply Model** operator, and set its **repository entry** parameter to the name of the model that you saved previously. This will load the previously learned model.

7. Click the **Play** button to run the process.

8. You will see the predictions of the model in the output.

13.12 Improvements

We only looked at one feature of the words in the documents—their frequency. But there are many other features in words that we could also examine. For example, we could examine their length, upper case to lower case ratio, number of symbols, and similar features in the previous and next words. This might lead to a more accurate classifier.

13.13 Summary

In this chapter we built a RapidMiner process to examine and learn to classify spam messages. Several thousand messages were analyzed, and a Naïve Bayes learner was able to classify messages with about 96% accuracy in a very simple process. We discussed how to examine the frequency of words in documents. The basics of the Naïve Bayes method were explained, as well as cross-validation, and dataset balancing.

Chapter 14

Robust Language Identification with RapidMiner: A Text Mining Use Case

Matko Bošnjak

University of Porto, Porto, Portugal; Rudjer Boskovic Institute, Zagreb, Croatia

Eduarda Mendes Rodrigues

University of Porto, Porto, Portugal

Luis Sarmento

Sapo.pt - Portugal Telecom, Lisbon, Portugal

Acronyms

API - Application Programming Interface

ETL - Extract, Transform and Load

HTTP - HyperText Transfer Protocol

k-NN - k Nearest Neighbours

NLP - Natural Language Processing

SVM - Support Vector Machines

TF-IDF - Term Frequency - Inverse Document Frequency

UTF-8 - Unicode Transformation Format – 8-bit

XML - eXtensible Markup Language

14.1 Introduction

Language identification, the process of determining the language of machine-readable text, is an important pre-processing step in many information retrieval and web mining tasks. For example, the application of natural language processing (NLP) methods may require prior language identification, if the language of the text at hand is unknown. In order to properly execute stemming, sentence tokenization or named entity recognition, we need to identify the language of the text to successfully apply appropriate language technologies. In addition, language identification is essential in machine translation tasks. Some text classification tasks, such as sentiment analysis in social media, may also require language identification for filtering content written in a specific language.

There are numerous proprietary and open-source solutions for language identification, including stand-alone solutions and APIs. However, proprietary solutions are usually costly, and APIs may cease to exist, while open-source solutions may require deeper understanding of the code for further modification and implementation. Moreover, most solutions are usually fixed on a pre-defined set of languages.

This chapter aims to provide a walk-through of language identification fundamentals by first introducing the theoretical background, followed by a step-by-step guide on model building, according to the standard practice of data mining. We will put emphasis on understanding the problem, and solving it in an effective and costless manner within the open source environment RapidMiner. We will become familiar with RapidMiner's "Text Mining Extension", and learn how to create several simple and fast workflows for language identification. The methods we will use are generic, and applicable to any dataset, with the appropriate pre-processing. In addition, we will learn how to use the implemented models in custom projects, by exporting them as web services using RapidAnalytics, and deploying them in custom applications. We will go through a case study of a web application for language identification of web pages, using RapidMiner's "Web Mining Extension". Special emphasis will be given to the optimization of the exported workflow, to enable faster execution, and harmonious integration in other applications.

14.2 The Problem of Language Identification

The task of language identification can be simply described as discerning the language of a given text segment. For example, given a set of six sentences in different languages:

How many languages do you speak well?

Wie viele Sprachen können Sie gut sprechen?

Combien de langues parles-tu bien?

Πόσες γλώσσες μιλάς καλά;

¿Cuántas lenguas hablas bien?

Koliko jezika govorite dobro?

Quantas línguas você fala bem?[1]

the goal is to identify the language of each sentence. This task may be somewhat trivial for humans. We might be able to easily discern the language and possibly identify some of them, even though we might not speak any of those languages. The first sentence is easy, since we are reading this book written in English. We may say that the second sentence "feels" German, or that the third one "sounds" French. The alphabet of the fourth sentence looks completely different from the others, so we might guess it "looks like" Greek, while the fifth sentence might "lead to" Spanish. We might find the sixth sentence somewhat tricky since it looks different from the previous ones. If we are vaguely familiar with the Slavic family of languages we might say it "relates to" that family, since the sentence is written in Croatian. Finally, although we might find the last sentence "similar to" Spanish, it is, in fact, written in Portuguese.

When identifying languages, we use our knowledge of languages, acquired either actively or passively. To design algorithms for achieving the same task, we first need to systematize the knowledge needed for language identification. There are several indicators we can rely on when identifying languages, without knowing those languages at all. These indicators are the following:

Alphabet Differences in symbols a language is written in are the most obvious feature for language identification. Even a person unfamiliar with languages at hand can discern different alphabets, and can easily be taught to identify different scripts due to their specificity. This, of course, does not hold for languages written in the same alphabet. Examples of several different alphabets are given in Figure 14.1.

Typical words Words can be a strong language indicator, whether those words are unique for the language, or the most frequent ones. Unique words, such as *fika*[2] in Swedish, or *saudade*[3] in Portuguese, are an excellent indicator of their corresponding languages. Nevertheless, they may occur infrequently in text. The most frequent words, like *niet*

[1] Source: Tatoeba Project: Open, collaborative, multilingual dictionary of sentences: `http://tatoeba.org`, Sentence $n°$682402.

[2] "Coffee-break"

[3] "A deep emotional state of nostalgic longing for an absent something or someone that one loves", source: Wikipedia

Latin	data mining
Greek	εξόρυξη δεδομένων
Cyrillic	добыча данных
Hebrew	כריית נתונים
Arabic	استخراج البيانات
Japanese	データマイニング
Chinese	数据挖掘

FIGURE 14.1: Examples of different alphabets.

in Dutch, *ne* in Slovene, or *ikke* in Norwegian[4] are the next obvious choice, and we will discuss them further in the chapter.

Accented letters Languages in Latin alphabets often contain various accented letters, which can be present in only several languages, or can even be unique to a language. These letters can be used for narrowing down the choice of languages in question. The downside of this approach is that accented letters may not at all be present in the text, depending on the encoding used or the style of communication. For example, in informal communication, people frequently omit accents. Examples of accented letters are *Ě* in Czech, *Ö* in German, Swedish, Icelandic, Turkish, etc., and *Š* in Slavic and Baltic languages.

Special symbols If a language uses special symbols, rarely present or not present at all in other languages, they can be used as a strong language indicator. A weakness of this indicator is that, similar to accented letters, these symbols can be used rarely, or not used at all. Examples of such symbols are the inverted question mark *¿*, and exclamation mark *¡*, used only in Spanish, the "Scharfes S", *ß* used in German, *ħ* used in Maltese, or semicolon *;* used as a question mark in Greek, among others.

Letter combinations Combinations of n consecutive letters found in substrings of words are called character n-grams, and are powerful language indicators. By finding these combinations we find digraphs, word beginnings and endings, prefixes, suffixes, or any other frequent letter combinations typical for a language. Examples of these combinations are the digraphs *Lj*, *Nj*, *Dž* in Croatian; *Zs*, *Gy*, *Ny* in Hungarian; typical word beginnings like *be-*, *re-*, *sh-* in English, and *des-*, *pr-*, *da-* in Portuguese; and typical word endings like *-ns*, *-ur*, *-ont* in French, and *-ung*, *-eit* in German.

Word length Some languages tend to form long words by concatenating distinct words, suffixes, and prefixes into single words with a compound meaning. Such words can frequently be found in Finnish, German, and Hungarian, among others. An example of such a composite word is *Geschwindigkeitsbegrenzung*[5] in German.

N-gram distribution Distribution of character n-grams per language is one of the most powerful indicators used for language identification. Different languages exhibit different character n-gram distributions, starting with single letters, also called 1-grams or

[4] All of them mean "no".
[5] "Speed limit".

unigrams. An example of the distribution of unigrams is given in Figure 14.2. In this example we observe a much bigger usage of letter *a* in Finnish than in the other two languages, an almost non-existent usage of letter *j* in Italian, and far greater usage of letter *z* in Czech than in both Finnish and Italian.

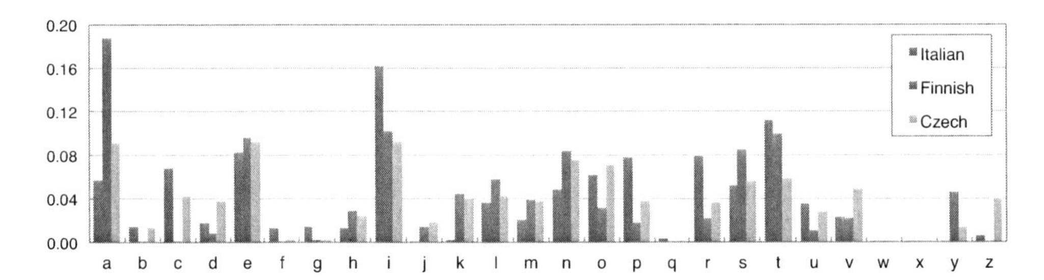

FIGURE 14.2 **(see color insert)**: Distributions of unigrams in Italian, Czech, and Finnish show clear differences in the use of distinct letters per language. In this example, accented letters are stripped of their accents, and non-alphabet symbols are ignored.

In the remainder of this chapter, we show how to implement several language identification methods using the most popular techniques and features, both in literature and in practice. Namely, we implement the frequent words, the character n-gram, and the similarity-based techniques. Although some researchers [1] argue that the problem of language identification has been resolved, issues are still ongoing [2, 3], from the language identification of web pages [4, 5], short queries [6, 7], and micro-blog messages [8], to differentiation of similar languages [8, 9]. However, the extent of these issues go beyond the scope and the intention of this chapter. We continue by presenting the fundamentals of text representation and classification models.

14.3 Text Representation

In order to start the text analysis process, we first need to represent the text as a vector of numbers in a consistent way. Prior to explaining the representation format, we need a word of notice regarding digital alphabet representations — encodings.

14.3.1 Encoding

Simply put, character encoding is a convention that maps a number to a symbol, which enables numerical storage of characters on a digital media. Due to historical reasons, there are many different ways to represent different characters. There are two main ways of encoding characters: single- and multi-byte character encodings.

Single-byte character encodings represent symbols in a single-byte of information. The most known such encoding is ASCII, which uses only 7 bits to encode up to 128 different symbols. The inclusion of the 8th bit already brings many troubles since there are many different character encoding tables which will map numbers higher than 127 to different symbols. The most widespread single-byte character encodings are the ISO 8859, and Microsoft Windows sets of encodings. Due to single-byte encodings' limited capacity of only 256 symbols, multi-byte encodings are more frequent in use today. Multi-byte character encodings utilize several bytes of data to represent a symbol, and are therefore able to encode much larger spectra of symbols. The most widespread multi-byte encoding is UTF-8.[6]

The choice of encoding is very important since a badly chosen encoding might make the text illegible. For example, the Greek word for data, *δεδομένα* encoded in UTF-8, if mistakenly shown with the ISO 8859-1 encoding, will appear as *Î"ÎµÎÎ?'Î¼ÎÎ½Î±*.

Throughout the rest of this chapter, we use the UTF-8 encoding, due to its widespread usage and the availability of major language scripts in it. While some language identification methods operate on bytes of data rather than characters, they cannot cope with code overlap in different encodings, and need to employ encoding detection. More information about encoding detection for language identification can be found in [10].

In addition, we limit our language identification to languages written in the Latin alphabet. Other alphabets can be detected by checking the availability of letters in certain code ranges, under the assumption of single encoding like the UTF-8. For example, we can detect the Greek alphabet by executing the regular expression `[\u0370-\u03FF]*`. This regular expression will include all the symbols ranging from the code point 0370 to the code point 03FF, expressed in hexadecimal, which in UTF-8 represents all the new and old Greek alphabet symbols. In RapidMiner, this regular expression can be applied to text with the *Keep Document Parts* operator. Note that this regular expression will not include Latin letters nor numerals. To include them and the standard punctuation, `[\u0370-\u03FF\u0020-\u007E]*` should be used.

With the encoding explained, and set to UTF-8, we continue on the level of text representation for data mining tasks.

14.3.2 Token-based Representation

The token-based representation is built on top of basic meaningful elements of text, called tokens. Tokens can be words separated by delimiters; logograms—signs and characters representing words or phrases, such as Chinese letters; idioms, or fixed expressions, such as named entities—names of places, people, companies, organisms, etc.

The process of token extraction is called tokenization. For most of the languages and uses, tokenization is done by splitting sentences over whitespaces and punctuation characters, sometimes ignoring numbers, or allowing specific punctuation in a word, e.g., *wasn't*, *off-line*. For example, the following sentence:

```
El mundo de hoy no tiene sentido, así que ¿por qué debería pintar
cuadros que lo tuvieran?
```
[7]

separated by non-alphabet characters, results in the following set of tokens:

```
El mundo de hoy no tiene sentido así que por qué debería pintar
cuadros que lo tuvieran.
```

In this chapter we use this simple tokenization over non-letter characters. However, in languages in which words are not delimited by space, such as Chinese, Japanese, Thai, Khmer, etc., tokenization is a more complicated process, and it requires complex machine learning models outside of the scope of this chapter.

Note that when tokenizing over non-letter characters, apostrophes are used as points of separation, which results in *wasn't* tokenized as *wasn* and *t*. We can argue that this is not

[6]Usage Statistics of Character Encodings for Websites `http://w3techs.com/technologies/overview/character_encoding/all/`.

[7]"Today's world doesn't make any sense, so why should I paint pictures that do?" ---Pablo Picasso

a desired behaviour in our process, since *wasn't* should be one token. However, due to the simplicity of setting parameters in RapidMiner, we opted for this solution. Easy tweaking of RapidMiner options allows inclusion of any other symbol, including the apostrophe. Later in the chapter, we explain how to set these parameters.

When dealing with specialized forms of text, like blog or micro-blog messages, for specific purposes like sentiment analysis, tokenization becomes a more complicated process in which non-letter tokens, such as smileys and interjections,[8] are retained. For more information on tokenization, consult further literature [11, 12, 13].

Other than splitting the text in tokens, and continuing to build a data structure for text mining, we can also split the text on the character basis. This results in a character-based representation.

14.3.3 Character-Based Representation

Character-based representation is a text representation based on top-of-character n-grams, previously defined as word substrings consisting of n consecutive characters. In general NLP usage, n-grams can also denote an ordered set of words, however is this chapter when mentioning the term n-gram, we specifically refer to character n-grams. For example, the following sentence:

```
Ce n'est pas assez d'avoir l'esprit bon, mais le principal
est de l'appliquer bien.
```
[9]

cut into 2-grams, also called bigrams, after tokenization results in the following set:

```
_a(3) _b(2) _c  _d(2) _e(3) _l(3) _m  _n  _p(2) ai al  ap as(2) av
bi  bo  ce  ci  d_  de e_(3) en  er  es(3) ez ie  in  ip  iq
ir is  it  l_(3) le  li  ma  n_(3) nc  oi on pa(2) pl  pp  pr(2)
 qu  r_(2) ri(2) s_(2) se  sp ss st(2) t_(3) ue  vo  z_
```
[10]

Extraction of n-grams can be executed after tokenized, or even non-tokenized text. The difference is that when extracting n-grams from non-tokenized texts, depending on the number n, n-grams can catch both the ending of one, and the beginning of another word. For example, when extracting 3-grams from the term *data mining*, 3-gram "*a m*" will also be extracted. This does not happen if we execute tokenization prior to n-gram extraction.

N-grams are normally used to catch word beginnings and endings, as well as typical combinations of consecutive letters. Whereas the number of words is practically unlimited, the number of n-grams is limited to the power of number of letters n. In real scenarios, the number of observed n-grams is actually much lower than the maximum possible number of n-grams, since not all letter combinations are present in natural languages. This enables compressed format of the text and creates helpful features in language identification.

Whether we choose to represent the text with tokens or n-grams, prior to continuing the text mining process, we convert them to a bag-of-words vector.

14.3.4 Bag-of-Words Representation

The bag-of-words is a representation used in NLP and text mining, in which a text, such as a sentence or a full document, is represented by a set of words—an unordered collection of

[8] Words used to express emotions and sentiments.

[9] "It's not enough to have a good mind; the most important thing is to use it well." ---René Descartes.

[10] Apostrophes replaced by empty space, added underscores to empty spaces to emphasize word beginnings and ends shorter than n, bigram cardinality shown in parenthesis, where greater than one.

words.[11] Note that although this representation completely disregards grammar and word order, it suffices for our needs. For example, the following sentences:

```
The following sentence is true.
The preceding sentence is false.
```

consists of the following words, without repetition:

```
1. the, 2. following, 3. sentence, 4. is, 5. true, 6. preceding, 7. false
```

By using this set of words as an index or dictionary, we create a fixed representation of a sentence, consisting of an occurrence of a specific word, at its appropriate index. This results in creating a word vector. For example, the previous two sentences, using the extracted dictionary, result in these word vectors:

```
(1 1 1 1 1 0 0)
(1 0 1 1 0 1 1)
```
[12]

The second vector in our example denotes a sentence containing one element of each of indexes $1, 3, 4, 6, 7$ in the dictionary, corresponding to words *the*, *sentence*, *is*, *preceding*, and *false*. With this approach we lose grammatical structures and word order, but we simplify the resulting model, making it tractable from the analytic point of view, and as we will demonstrate, sufficient for our case studies. Upon obtaining word vectors of the bag-of-words representation, we are ready to use classifiers for language identification.

14.4 Classification Models

Our main goal, developing a system for language identification, requires an applicable classification model. An example of a model might be a simple string matching—if a language contains one or more words, or n-grams of a certain language, it belongs to that language. This type of model leads to rule-based systems. Another approach is to represent the text segment in an n-dimensional space, where each dimension corresponds to a single word metric like the occurrence, TF-IDF score, etc. This leads to vector-space models, that are frequently used with Support Vector Machines (SVMs). In practice, statistical language models, like Naïve Bayes or Markov models, are frequently used since they yield high performance, and in the case of Naïve Bayes, low computational cost. Though the Naïve Bayes classifier is often outperformed by other classifiers, like SVM, in terms of prediction performance, we use it since it is fast, and does not need parameter optimization. Feel free to experiment with other classifiers to find the best performing one.

The first classification model we employ is the Naïve Bayes classifier. The Naïve Bayes classifier is a generative probabilistic model that assigns a probability to a segment of a text, composed of either single or multiple tokens or parts of tokens [14]. This is done through statistical analysis of the text, by translating frequencies of word occurrences in a document into probabilities, and employing a statistical framework for classification [15]. This is one of the models we use in our case studies.

The second model we employ is the similarity-based model. The similarity-based model

---

[11] The precise term would be multiset, a generalization of set, which allows repeating members.

[12] The example is considered case-insensitive; words are ordered in the sentence order.

relies on creating a distribution of elements, such as tokens or n-grams per language, and using a similarity measure to determine the language with the most similar distribution to the queried example. While the classification approach requires a large number of examples per language, the similarity approach relies on a single large example per language, for example, all the texts of a single language concatenated together. This model, in its essence, is the k Nearest Neighbours (k-NN) model. Having a large text per language enables us to extract a distribution of tokens, or n-grams, that closely approximates the true distribution of the language.

One of the most widely used approaches is the ranking approach proposed by Cavnar and Trenkle [16] in which they implement the ranking similarity between the query text profile and language profiles. Since the ranking similarity is not yet implemented in RapidMiner, we can use other similarity measures frequently utilized, such as the Euclidean distance, dot product, and cosine similarity [10, 17, 18].

Having token and character-based text representation as the ground representation, out-of-bag word vectors, and the Naïve Bayes and k-NN classifiers at hand, we are in the position to start discussing the implementation of possible language identification workflows in RapidMiner.

14.5 Implementation in RapidMiner

In this section, we move on to building RapidMiner workflows by presenting three different models for language identification, built on top of the previously introduced theoretical background. The models we present are the frequent words, character n-gram, and the similarity-based models [19]. Of course, additional operators in RapidMiner can be used to build more complex models, and improve the ones we will present; however, we do not cover all the advanced details in this chapter.

The first step in the implementation process is to obtain a suitable dataset for experimenting with language identification techniques. If you already have a dataset available, and you wish to skip on to model construction, you can proceed to Section 14.5.3. However, if you do not have a dataset prepared, and cannot easily get a hold of one, there are several freely available datasets you can use. In the next subsection we introduce a few such datasets, which you can use to re-create and experiment with RapidMiner workflows.

14.5.1 Datasets

When creating a dataset for developing language identification models, one has to consider several important factors:

Time Language is dynamic, and changes over time. Commonly used words and phrases differ between centuries, decades, and even years. Obtaining datasets from comparable eras ensures similar language properties and stability of used words.

Source Different sources of text exhibit different properties. For example, news articles, web pages, dictionaries, social media like Twitter, and classical writing greatly differ in the vocabulary, punctuation, and even word case. Thus, the choice of dataset should favor texts from languages coming from the same or comparable sources. This ensures that the language has a fairly stable style.

Size Datasets may differ in size. Small datasets might not exhibit a very representative distribution of words or n-grams, while a large dataset might. However, computations with large datasets are more costly, in terms of time, complexity, and memory requirements. Obtaining comparable sizes of datasets per language is also important to represent all the languages with approximately the same number of sentences, words, or tokens.

Length Length of texts used for training or for querying has an impact on performance. Building classifiers on small sentences for training might decrease performance if the number of sentences is also small. Likewise, identifying language of a short sentence is a difficult task since only a small number of features are observed.

Therefore, when creating a dataset, we should collect texts of comparable sizes, coming from comparable sources, created in near time, and of adequate text length compared to query texts, in order to ensure higher performance of the language identifier.

There are multiple different datasets which can be used for this purpose, that abide by most of the previous recommendations.

Wikipedia Wikipedia[13] is a free Internet encyclopaedia with more than 21 million articles written in 284 languages. The articles can be downloaded and used for language identification. Wikipedia database dumps can be found at `http://dumps.wikimedia.org/`. in various formats and levels of information.[14] For language identification tasks, we recommend abstracts of web pages in XML format.

Leipzig Corpora Collection The Leipzig Corpora Collection[15] is a collection of textual corpora in 155 different languages using the same format and comparable sources [20]. The corpora consist of randomly selected sentences in the corpus language, and are available in sizes varying from tens of thousands up to tens of millions of sentences. The sources of sentences are news articles and texts, randomly collected from the web.

Project Gutenberg Project Gutenberg[16] is the oldest digital library of written cultural work. This web page offers full text of public domain books in various formats, out of which UTF-8 plain-text is the most straightforward for our use. In May 2012, Project Gutenberg claimed to have in excess of 39,000 free ebooks in more than 50 languages, with English being the most predominant language. Note that these books are older belletristic works.

European legislature datasets There are several datasets originating from the official documents of the European Union (EU), published in the official languages of the EU. Download

EuroGOV[17] is the document collection of web documents crawled from European governmental sites, with restricted access.

European Parliament Proceedings Parallel Corpus[18] [21] is a parallel corpus extracted from the proceedings of the European Parliament 1996–2010. It consists of 21 languages and 20 language pairs, English versus the rest. Though conceived as a standard dataset for statistical machine translation systems, it can also be used for language identification.

---

[13] `http://www.wikipedia.org/`.

[14] Details on database download are available at `http://en.wikipedia.org/wiki/Wikipedia:Database_download`.

[15] `http://corpora.informatik.uni-leipzig.de/download.html`.

[16] `http://www.gutenberg.org/`.

[17] `http://ilps.science.uva.nl/WebCLEF/EuroGOV/`.

[18] `http://www.statmt.org/europarl/`.

The JRC-Acquis Multilingual Parallel Corpus[19] is the corpus representing the total body of European Union law applicable in the EU member states [22]. These texts are available in 22 languages.

In this chapter, we use the Leipzig Corpora Collection dataset of 30.000 sentences in English, German, French, Spanish, and Portuguese. The dataset is easy to download, we just need go to the Leipzig Corpora Collection web page at `http://corpora.informatik.uni-leipzig.de/download.html`, scroll down to "Download Corpora" type in the captcha text, and click on "check". When a table with various corpus sizes and formats appears, we select the corpora we find suitable and download them by clicking on them. The next step after downloading the data is to prepare it and import into RapidMiner.

14.5.2 Importing Data

Upon downloading the corpora for English, German, French, Spanish, and Portuguese, we extract the compressed file contents and isolate files with a suffix `-sentences.txt`. These sentence files are tab delimited files containing a line number and a sentence in each line of the file. After downloading a dataset per language, and extracting its content, we rename the downloaded text file according to the language of the data, in our case *eng*, *deu*, *fra*, *spa* and *por* respectively, and save the files in a single directory. In our repository, these files are located in the `data/corpus` directory.

In order to import the Leipzig Corpora Collection, we create a data importing workflow, presented in the upper part of Figure 14.3. We loop over the files we previously saved using the *Loop Files* operator (1), by setting the *directory* property to the path to data, in our case `data/corpus`. The *Loop Files* operator will iterate over all files in the directory we set it to, and execute its nested process for each iterated file.

The content of the *Loop Files* operator is shown in the lower part of Figure 14.3. Essentially, in every iteration of the *Loop Files* operator, we read the text file, and fit it to specifications for further processing, creating the appropriate label, and finally splitting the data into two datasets: training and test datasets. The training dataset is further split into two forms: i) example per sentence and ii) all the text concatenated per language.

In more detail, the *Read CSV* operator (1) receives the file to open from the nested process of the *Loop Files* operator. We set the *column separator* property to `\t` since the text is tab delimited, uncheck the *use quotes* property since sentences are not delimited by quotes, uncheck the *first row as names* property since there are no row names in the first row of the file, and set the *encoding* property to UTF-8. The *Rename* operator (2) renames the default `att2` attribute to `text`, since the *Read CSV* operator returns default values for row names, when the first row of the file does not convey names of attributes. With the *Select Attributes* operator (3), we select only the `text` attribute, thus ignoring the line number. The *Generate Attributes* operator (4) generates the `language` attribute, and sets its value to the `replace("%{file_name}",".txt","")` expression. This expression essentially uses the name of the file to populate the `language` attribute. This is why we renamed the file names to short names of languages, to automate the whole process. The `replace("%{file_name}",".txt","")` expression relies on the macro *%{file_name}* constructed by the *Loop Files* operator per each iteration, and contains the name of the file in the current iteration. The next step is to set the label role to the newly created `'language'` attribute with the *Set Role* operator (5). The output of the *Set Role* operator is now a dataset containing two attributes: the *text* attribute containing each sentence text, and the *language* attribute containing the label of each of those sentences, fixed to the name of

[19] `http://langtech.jrc.it/JRC-Acquis.html`.

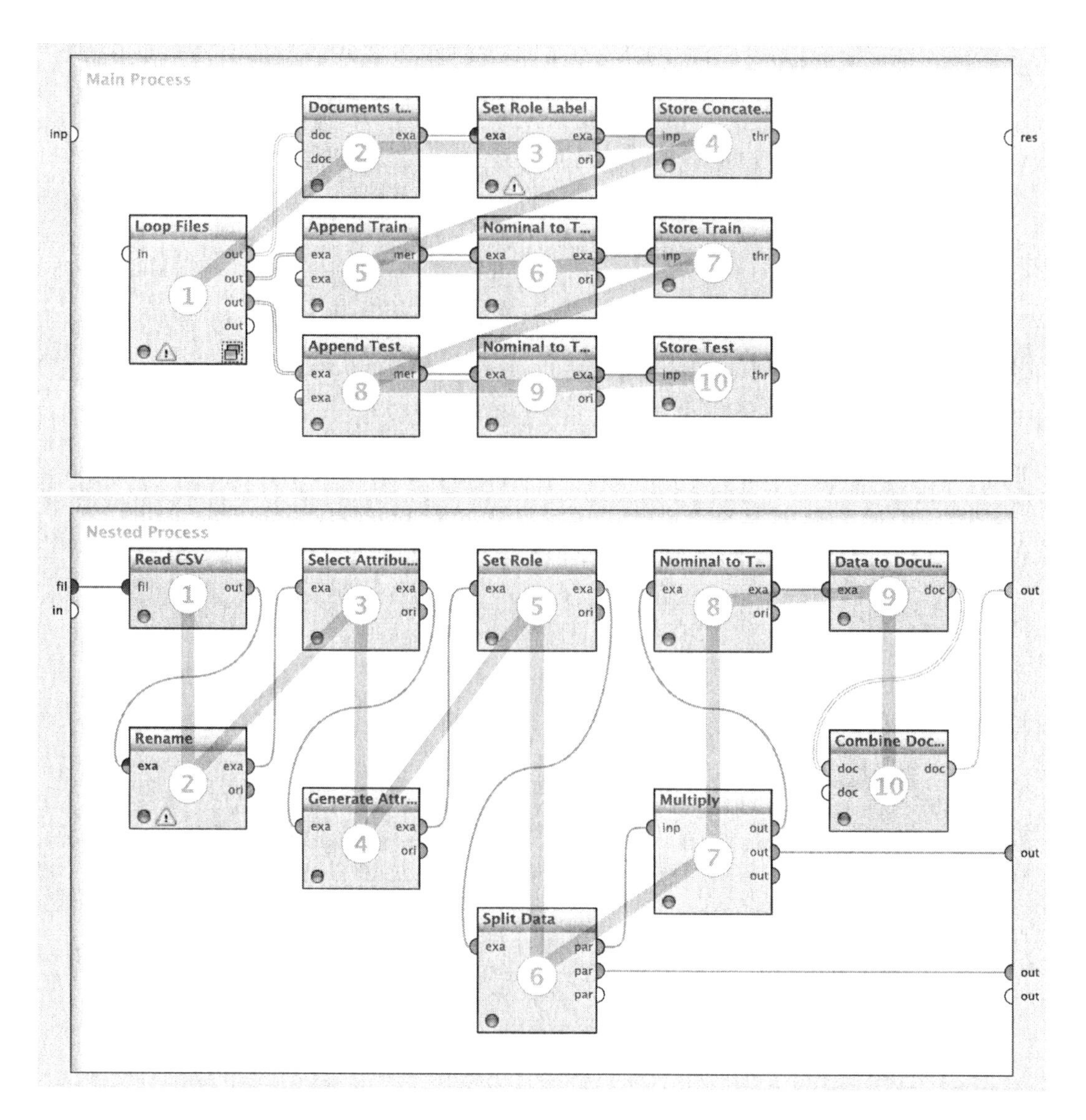

FIGURE 14.3: The workflow for loading the Leipzig Corpora dataset. The upper image depicts the main workflow process, whereas the lower image depicts the content of the *Loop Files* operator (1) in the upper image.

the file which is a short code for the language. With the *Split Data* operator (6) we split the dataset randomly with the ratios of 0.3 and 0.7 into the train and the test datasets, respectively. This kind of split is unusual in real-world setting—we would usually create a bigger training set than the test set. However in this case, we opted for a smaller training set to increase the speed of training. We output the test dataset without any change, while we process the train set into two modes. The first mode is simply the standard sentence-per-example dataset, outputted for processing and saving, just as the test set. The second mode is the concatenated dataset, created by first converting the `text` attribute from nominal to text type, using the *Nominal to Text* operator (8), since the nominal type cannot be processed as a textual value. In the end, the *Data to Documents* (9) and the *Combine Documents* (10) operators convert the dataset first to a collection of documents, and then combine that collection in a single resulting document.

Back to the main process. The datasets are outputted from the *Loop Files* operator (1) in the following order: the collection of documents containing one document of concatenated train data, the collection of train data, and the collection of test data, all three per iterated file. The collection of documents containing documents of concatenated data per language is converted to data with the *Documents to Data* operator (2), the role of the *language* attribute is set to label via the *Set Role* operator (3). The resulting dataset is the concatenated dataset, a dataset containing two columns: a label column denoting the language, and the text column, containing concatenated texts of all the sentences corresponding to each language. This dataset is stored to the repository as the `language_concatenated` dataset, using the *Store* operator (4). The collection of train and test data, on the other hand, both go through the identical process. They are first flattened in a single dataset by the *Append* operator (5), and the `text` attribute is converted to the text type via the *Nominal to Text* operator (6). The resulting datasets contain two columns: a label column denoting the language, and the text column, containing one sentence per line of the dataset. These datasets are stored in the repository as `language_train` and `language_test` with the *Store* operators (7) and (10), respectively.

Thus far, we obtained the data for language modeling, and we are ready to construct the first language identification system. The simplest and the most naïve approach to language identification is the dictionary-based identification. By enumerating all the possible words in every single language, we can try to match the word with its language and identify the language of the text. However, this method is intractable, since the language is a dynamic complex system with no bound on the number of words. In addition to the complexity, the creation and maintenance of a large dictionary set would create another set of problems, from computational load, to the fact that different languages share same words. Alternatively, we can select a fixed number of words and use them for language identification. Choosing the most frequent words for each language is a reasonable option. This is why the next model we introduce is the frequent words model.

14.5.3 Frequent Words Model

The frequent words approach identifies the most frequent words in the text, and uses them for the language identification. The most used words across different languages have similar meanings, and regarding the type of words, they are usually articles, prepositions, pronouns, and some frequently used verbs, and adjectives. Due to their high frequency of occurrence, they are suitable even for shorter texts. For example, it might be easy to see that the following sentences are written in three distinct languages:

```
Det gäller att smida medan järnet är varmt.[20]
Het is niet alles goud wat er blinkt.[21]
Nu masura pe altii cu palma ta.[22]
```

Nevertheless, it is difficult to identify specifically the languages of these sentences without being acquainted with them, especially if those languages are not widespread. Still, given a list of the most frequent words in Dutch,[23] Romanian, [24] and, Swedish[25] as in Table 14.1, it is straightforward to identify the language of each of these sentences as Swedish, Dutch, and Romanian, respectively.

TABLE 14.1: Lists of the 20 most frequent words in Dutch, Romanian, and Swedish.

Dutch	ik, je, het, de, dat, is, een, niet, en, wat, van, we, in, ze, hij, op, te, zijn, er, maar
Romanian	sa, nu, de, o, ca, şi, a, ce, în, e, am, pe, te, la, mai, cu, un, asta, ai, ma
Swedish	i, och, en, av, som, är, att, den, på, till, med, för, det, han, de, ett, var, har, från, under

Building a statistical model for language identification using the frequent words method in RapidMiner is straightforward. First, we must analyze the training set text, extract all the words from it, and select the most frequent ones. Second, we represent the analyzed text by the most frequent words, converting the text to the word-vector representation in which a value of an attribute denotes the occurrence or a frequency measure of a given word in the text. Finally, we feed the word-vector representation of the text to the Naïve Bayes classifier, and build a model that can classify languages by the most frequent words exhibited in the training set.

We create a single RapidMiner workflow for the purpose of building the frequent words language identification, evaluating it, and applying it on a test dataset. This workflow is presented in Figure 14.4.

The first three operators (1), (2), and (3) execute data loading, textual processing, and model training and performance estimation via 10-fold cross-validation. First we load the `language_train` dataset using the *Retrieve* operator (1). The dataset we load contains two attributes, the *text* attribute containing sentences, and the *language* attribute containing the language of the text as the label. After loading the data, we analyze it with the *Process Documents from Data* operator (2). This operator created word vectors based on term frequencies with the property *Vector creation* set to `Term Frequency`. The inner Vector Creation Subprocess of the Process Documents from Data operator, as can be observed in the middle workflow of the Figure 14.4, contains the *Transform Cases* and the *Tokenize* operators. These two operators are set to the default properties; therefore each sentence of our dataset is first transformed to lower-case, and then tokenized over non-letter characters. We want to limit the resulting number of words in the analysis, to speed up the model learning, as well as simplifying the process. We choose to extract the top 0.1% of all the words in our example set by setting the *prune method* parameter to `ranking`, with the

[20] "You should forge while the iron is hot."
[21] "All that glitters is not gold."
[22] "Don't measure others with your hand."
[23] `http://en.wiktionary.org/wiki/Wiktionary:Frequency_lists/Dutch_wordlist.`
[24] `http://en.wiktionary.org/wiki/Wiktionary:Frequency_lists/Romanian.`
[25] `http://www.lexiteria.com/word_frequency/swedish_word_frequency_list.html`

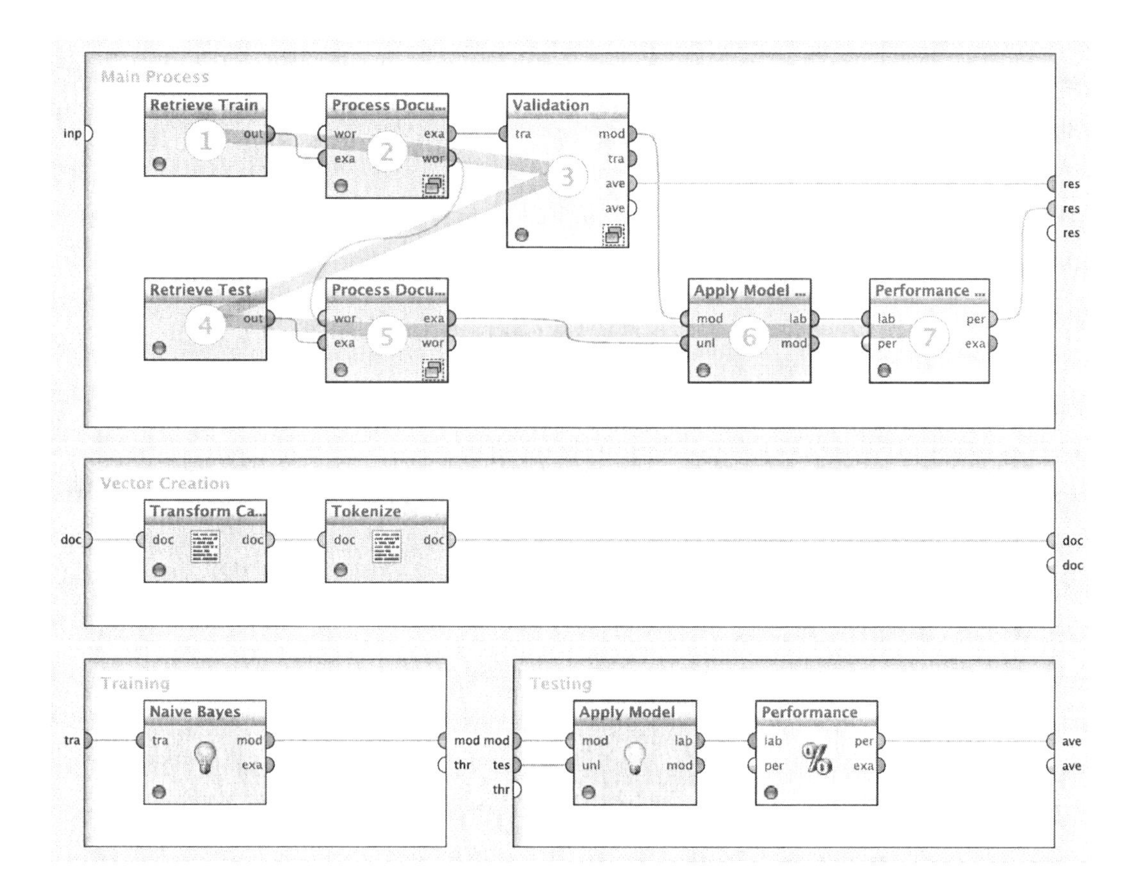

FIGURE 14.4: Workflow for the language identification system based on the most frequent words in languages. The main process is depicted in the top image. The middle image depicts the contents of both *Process Documents from Data* operators (2), (5). The bottom image depicts the content of the *Validation* operator (3).

prune below ranking parameter set to `0.001`, and the *prune above ranking* set to `0.0`. The final stage of the modeling phase is the *X-Validation* operator (3), which executes the Naïve Bayes classifier, and estimates its performance through 10 folds.

The next four operators (4), (5), (6), and (7), constitute the application phase of the workflow, meaning that they are responsible for loading the test data, transforming it in the way in which the previously built model was trained, applying the model, and calculating its performance. First, we load the **`language_test`** dataset using the *Retrieve* operator (4). The test dataset is of the same format as the train dataset. After loading the train dataset, we analyze it in the *Process Documents from Data* operator (5) by passing the WordList from the *Process Documents from Data* operator (2). This ensures that the same words are used for analyzing the test set, and thus ensuring that the dataset is of the same format as the dataset used for model building. The operators inside the *Process Documents from Data* operator (5) are the same as in the model building phase, namely the *Transform Cases* and the *Tokenize* operators. The result of the *Process Documents from Data* operator is a word-vector dataset. The *Apply model* operator (6) then receives the Naïve Bayes model, outputted by the *Validation* operator (3), and applies it on the newly created word-vector dataset. The process ends with the display of the performance.

Note that if we want to separate the model building and the model application into distinct workflows, we should save the resulting WordList, and the model, both built during the model building phase, and load them in the application phase of the workflow of interest.

The cross-validation estimation of the accuracy of the frequent words workflow is equal to 96.19% ±0.35%, with the 96.23% accuracy achieved on the test set, as shown in Figure 14.5. It is possible to further improve these results by increasing the percentage of the words included in the model, however this would increase the model complexity. For example, including the top 1% of words instead of the top 0.1%, results in the 99.05% accuracy on the test set.

Table View Plot View

accuracy: 96.19% +/- 0.35% (mikro: 96.19%)

	true deu	true eng	true fra	true por	true spa	class precision
pred. deu	8575	11	16	6	0	99.62%
pred. eng	62	8736	8	3	10	99.06%
pred. fra	13	22	8749	5	35	99.15%
pred. por	10	17	18	8324	53	98.84%
pred. spa	340	214	209	662	8902	86.20%
class recall	95.28%	97.07%	97.21%	92.49%	98.91%	

Table View Plot View

accuracy: 96.23%

	true deu	true eng	true fra	true por	true spa	class precision
pred. deu	20013	28	27	18	3	99.62%
pred. eng	152	20436	11	12	22	99.05%
pred. fra	25	38	20402	30	77	99.17%
pred. por	35	50	39	19403	113	98.79%
pred. spa	775	448	521	1537	20785	86.37%
class recall	95.30%	97.31%	97.15%	92.40%	98.98%	

FIGURE 14.5: Estimated performance of the frequent words method with 10-fold cross-validation (upper image) and performance of the method on the test set (lower image).

Taking a look at the top 10 words extracted from the training set gives us a glimpse of the dataset, and can provide a quick check if everything is in order. This is done by taking a look at the WordList output of the *Process Documents from Data* operator (2), and sorting it over *Total Occurrences*, as presented in Figure 14.6. Most of the top words are, by themselves, bad identifiers of a single language, since they are shared among different languages. In our example, *a* is shared among all languages analyzed, except German, and *en* is shared among French and Spanish, though *the* is specific only to the English language. We also note that most of the top 10 words do appear in all languages, since it is possible for a sentence in, for example, French to contain foreign named entities such as "El País" or "The New York Times". Though most of the top words are not reliable indicators of a language by themselves, when taken together they achieve a higher accuracy. Note here that the most frequent words are usually discarded in various text mining applications on a single language since those words appear often, and as such do not have the needed discriminating value for various other purposes, though they do in the case of language identification.

Word	Attribute Name	Total Occurences	Document Occurences	deu	eng	fra	por	spa
de	de	27488	15832	30	17	10193	6809	10439
a	a	13553	10592	31	4116	1066	4935	3405
la	la	12868	8542	4	11	5891	39	6923
the	the	9629	5606	13	9525	77	6	8
que	que	9565	7429	0	0	1048	4066	4451
en	en	7372	5738	2	6	2897	1	4466
in	in	5502	4644	2335	3138	18	6	5
el	el	5462	3892	5	6	19	6	5426
o	o	4995	3899	5	19	23	4672	276
le	le	4884	3706	5	1	4605	3	270

FIGURE 14.6: List of the top 10 most frequent words in the language identification dataset, obtained with the frequent words workflow.

The advantages of the frequent words method are its speed, due to relatively small set of words, and high performance, greater than 95%. Higher accuracy can be achieved by including more words, thus increasing the model complexity and its processing time, or by using the two methods we present next. The disadvantages of the frequent words method are problems with similar languages due to similar top words, and bad results on sentences which do not include the top frequency words—usually short words and phrases.

A variation of this technique, short frequent words, uses only words of short length, usually four or five letters long. The rationale of the short frequent words technique is the same, with added shortness criteria which additionally filters out longer words, thus limiting itself to a smaller subset of the most frequent words. This filters out longer words which might be frequent due to linguistic properties of the source of the dataset. Creating a workflow for this technique is simple; it is only necessary to add a *Filter Tokens (by Length)* operator in the *Process Documents from Data* operator in Figure 14.4 (upper image), just after the *Tokenization* operator. The usual length of the token, set for this method, is 5 to 6 characters.

In the following subsection, we present the next method, which differs from the frequent words model on the level of the unit of text used for modeling; it uses n-grams rather than words. This is the character n-gram method.

14.5.4 Character n-Grams Model

We observed that the frequent words method achieves high accuracy, although it performs poorly in cases when the text to be classified does not contain any of the most frequent words. In cases of short sentences, or single words, we want a method capable of achieving high accuracy, without using words as the basis for classification. This is why we turn to character n-grams. As mentioned before, character n-grams are sequences of n consecutive characters, and as such are a powerful feature for language identification. They dissect the text in its substrings and can catch beginnings and endings of words, as well as some typical character combinations, characteristic to a language. For example, the following sentences:

```
I cannot make out the meaning of this sentence.
Ich kann die Bedeutung dieses Satzes nicht erkennen.
```
[26]

are easily classified into English and German, respectively, knowing that the character n-grams *I*, *can*, *the*, *-ing*, and *of* appear more often in English, and *ich*, *die*, *be-*, *-ung*, *-cht* in German. Notice that n-grams catch not only typical word beginnings and endings, or the combinations of letters, but also the short frequent words of length less than or equal to n.

Building a character n-gram language identification model is essentially an upgrade of the frequent words model, with the resulting workflow differing only slightly from the workflow in Figure 14.4. The conceptual difference is only in employing character n-gram creation after the tokenization, when analyzing the text. The rest of the process is the same: we convert the text to the word-vector representation, and build a classifier to classify it.

The workflow of the character n-gram method, depicted in Figure 14.7, is divided into two parts, i) the model creation consisting of operators (1), (2), and (3) and ii) the model application consisting of operators (4), (5), (6), and (7). As stated, the difference between the character n-gram and the frequent words workflows is minor. In the model building phase, the difference is in the two new operators in the *Process Documents from Data* operator (2), and a modification in properties of two other operators in it. The *Replace Tokens* operator, as seen in the lower image of Figure 14.7, adds underscores around a word, which enables differing cases when an n-gram is found at the beginning or at the end of a word. This is done by replacing the expression `([\w]+)` with the expression `___$1___`. This expression will replace any string of word characters longer than one, with the same string surrounded by three underscores from both the left and the right. The *Generate n-Grams (Characters)* operator creates character n-grams, in our case, n-grams of length 4. Note here that should we tick the *keep terms* option, we would keep the words, in addition to character n-grams, and thus effectively create a hybrid between the frequent words and the character n-gram methods. Two important modifications are apparent in the *Process Documents from Data* operator (2). First, the *prune method* property is set to `by ranking`, pruning all n-grams which occur in more than 10%, and less than 1% of sentences. Second, the *mode* of the tokenization is set to `regular expression` with the *expression* equal to `[^a-zA-Z_]+`. This regular expression describes all the splitting points. In this particular case it defines that the splitting points are everything NOT (`^`) containing strings of all the lower- and upper-case letters, and the underscore (`a-zA-Z_`), of string length equal to at least one character (`[]+`). By adding additional characters or character ranges in this expression, we influence tokens used for the tokenization. For example, by adding `0-9` to it, we enable numbers and words with numbers and tokens, by adding `'` we include apostrophes in words, and by adding various other punctuation characters we can catch smileys and interjections. The rest of the workflow stays the same as the workflow for the most frequent words model.

When executed, the character n-gram workflow, depicted in Figure 14.7, achieves 98.26%±0.14% accuracy of the estimated performance with 98.31% accuracy on the test set, as observed in Figure 14.8. By increasing the number of n-grams included in the analysis, we improve the accuracy of the method.

When compared to the frequent words method, character n-grams exhibit an important property—they create a compact representation of text into a finite number of n-grams. This representation is a priori limited by the number of n-grams, whereas the word count in a language is not limited at all, due to the fact that new words are being produced constantly, i.e., language is dynamic.

[26] Both of these sentences have the same meaning. Source: Tatoeba Project.

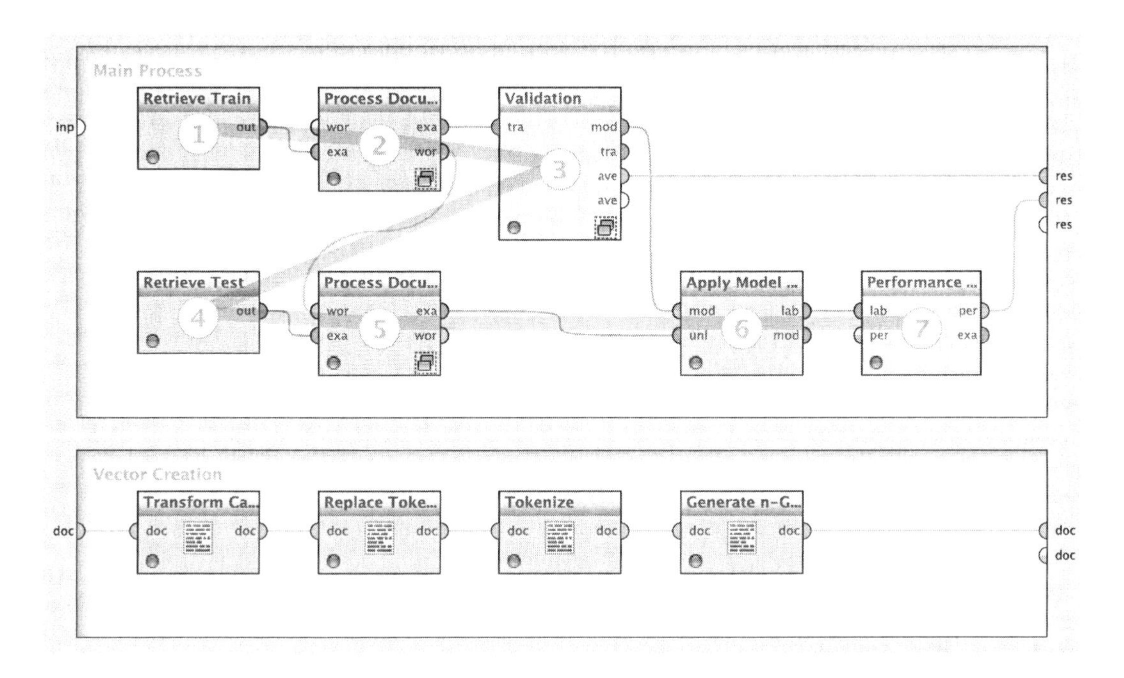

FIGURE 14.7: The workflow for the language identification system based on character n-grams depicted in the upper image. The lower image depicts the content of both *Process Documents from Data* operators (2), (5). The content of the *Validation* operator (3) is the same as in Figure 14.4.

Table View Plot View

accuracy: 98.26% +/- 0.14% (mikro: 98.26%)

	true deu	true eng	true fra	true por	true spa	class precision
pred. deu	8965	14	11	7	32	99.29%
pred. eng	5	8906	18	10	32	99.28%
pred. fra	24	64	8929	32	87	97.73%
pred. por	4	6	15	8764	196	97.54%
pred. spa	2	10	27	187	8653	97.45%
class recall	99.61%	98.96%	99.21%	97.38%	96.14%	

Table View Plot View

accuracy: 98.31%

	true deu	true eng	true fra	true por	true spa	class precision
pred. deu	20923	20	28	14	75	99.35%
pred. eng	18	20823	56	17	88	99.15%
pred. fra	42	118	20832	62	205	97.99%
pred. por	7	19	19	20475	458	97.60%
pred. spa	10	20	65	432	20174	97.45%
class recall	99.63%	99.16%	99.20%	97.50%	96.07%	

FIGURE 14.8: Estimated performance of the n-grams method with 10-fold cross-validation (upper image) and performance of the method on the test set (lower image).

By inspecting the top 10 n-grams in Figure 14.9, we confirm our assumptions of the characteristics and the use of n-grams. We can observe typical word beginnings and endings characteristic for each language. For example, *_die* in German; `ed__`, `ing_`, and `and_` in English; `no__`, and `na__` in Portuguese and Spanish; `__d_`, and `_le_` in French; and `_el_` in Spanish. We also observe that the character n-gram method catches some short frequent words.

Word	Attribute Name	Total Occurences	Document Occurences	deu	eng	fra	por	spa
_die	_die	5815	4267	5616	44	77	13	65
ed__	ed__	5634	4150	88	5319	76	41	110
ing_	ing_	5571	4065	179	5174	115	43	60
and_	and_	5513	4487	600	4528	333	25	27
_el_	_el_	5500	3922	5	6	35	14	5440
no__	no__	5387	4483	45	324	93	2443	2482
__d_	__d_	5373	4089	113	62	4663	198	337
_d__	_d__	5373	4089	113	62	4663	198	337
_le_	_le_	5175	3847	11	1	4864	14	285
na__	na__	5109	4254	128	194	220	2118	2449

FIGURE 14.9: List of the top 10 most frequent n-grams in the language identification dataset, obtained with the character n-gram workflow.

The advantages of the character n-gram method are the previously stated compact representation of words, ability to classify single words, as opposed to the frequent words method unless the word in question is frequent, and high accuracy. The disadvantage lies in the fact that the number of n-grams grows exponentially with n, thus resulting in a larger number of attributes, and longer processing, and learning times.

In order to lower the processing, and learning times, and reduce the size of the resulting model, we use the similarity-based approach.

14.5.5 Similarity-based Approach

Up to this moment, we treated every single sentence or a piece of text as a single example, and extracted features, namely words and character n-grams, for model-building. When we discussed possible language indicators, we mentioned the distribution of character n-grams as one of them.

The similarity-based approach is based on concatenating all of the sentences of a single language into a single body of text, calculating the distribution of character n-grams as a "profile" of a language, and using that profile for comparison to the profiles of query texts. In this way, we obtain the most compact representation of a model—a single language profile is the model of that language, and the classification process is only the similarity calculation.

The similarity-based approach results in a faster analysis, and a persistent profile which stays for later comparison. In addition, building the profile by using the n-gram count enables incremental updating with new data.

The similarity-based model workflow, depicted in Figure 14.10, is similar to the character n-gram, and the frequent words workflows are divided into two parts, i) the model creation, consisting of operators (1), (2), (3), (4), and (5), and ii) the model application part, consisting of operators (6), (7), (8), and (9). After loading the concatenated training data `language_concatenated` with the *Retrieve* operator (1), we process it with the *Process Documents from Data* operator (2). This operator contains exactly the same textual analysis operators as the same operator in the character n-gram workflow, depicted in the lower part of Figure 14.7: the *Transform Cases*, the *Replace Tokens*, the *Tokenize*, and the *Generate n-grams (Characters)* operators. The *Process Documents from Data* operator is

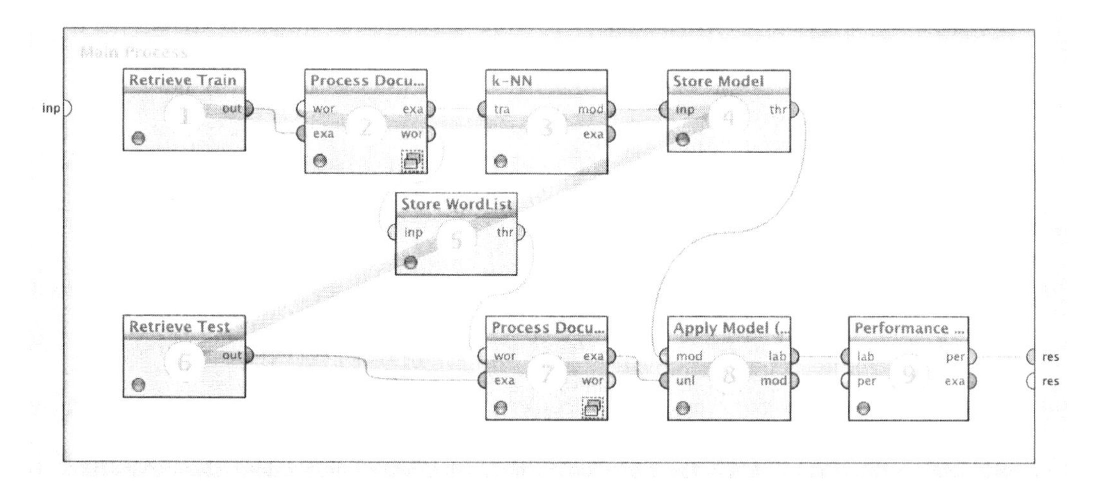

FIGURE 14.10: Workflow for the language identification system based on n-grams profile similarity. The content of both *Process Documents from Data* operators (2), (7) is the same as in Figure 14.7.

set to use a *prune method* `by ranking`, with the *prune below rank* property set to `0.01` and the *prune above rank* property set to `0.0`. The next step is the model-building using the *k-NN* operator (3), with k set to `1`, *measure types* set to `NumericalMeasures`, and the *numerical measure* set to `CosineSimilarity`. The 1-NN model essentially stores language profiles built from the word vectors, and uses cosine similarity to find the profile most similar to the query profile. Note that we cannot use cross-validation with the 1-NN model since the whole example set contains only one example per language. At this moment, we store the word list `similarity_wordlist` and the model `similarity_model` via the *Store* operators (4) and (5), respectively. We do this for later use of the word list and the similarity model in Section 14.6. This finishes the model-building phase, which executes faster since there is a low number of examples to analyze and learn; k-NN is fast at learning since it only needs to save the instance and the similarity measure used for the model application phase.

The model application part of the workflow starts with loading the test data `language_test` with the *Retrieve* operator (6). The *Process Documents from Data* operator (7) converts the test data into word vectors on which the 1-NN model was trained, by executing the analysis with the same word list and the same text analysis operators as the operator (2). After the analysis, the model application is executed using the *Apply Model* operator (8) and the previously built 1-NN classifier. The final step of the workflow is measuring the performance with the *Performance* operator (9), and outputting the results.

Table View Plot View

accuracy: 94.69%

	true deu	true eng	true fra	true por	true spa	class precision
pred. deu	20273	276	218	14	174	96.75%
pred. eng	146	20297	89	26	122	98.15%
pred. fra	545	322	20482	311	624	91.91%
pred. por	20	61	21	19977	1687	91.78%
pred. spa	16	44	190	672	18393	95.23%
class recall	96.54%	96.65%	97.53%	95.13%	87.59%	

FIGURE 14.11: Performance of the n-gram similarity method on the test set.

The similarity-based workflow achieves 94.69% on the test set, as reported in Figure 14.11. The observed accuracy is lower due to several reasons. There is no optimization performed, like the optimization of the number of features used, and more concretely, the similarity-based approach uses a different number of features due to differences in both the pruning methods and the starting dataset used. Other similarity measures [16] yield higher accuracy than the used cosine similarity, however they are not currently implemented in RapidMiner. Still, there are two big advantages of the similarity-based method. First, the method is fast since it only requires N computations (where N is the number of languages trained) per example. Second, the resulting model has a small memory footprint—it only requires storage of a single vector per language.

In this section we showed how to implement three different language identification techniques in RapidMiner. In the following section, we use the similarity-based approach as a basis for a language identification web service.

14.6 Application

Up to this point, we showed how to construct several different RapidMiner operating workflows for successful language identification of several European languages. In this section we concentrate on how to utilize the previously built models in custom projects, by demonstrating a concrete case study of a web page language identification with RapidAnalytics. The result of this demonstration is a working web service which can be used in custom projects.

14.6.1 RapidAnalytics

RapidAnalytics is an open source business and data mining analytics server, built around RapidMiner. It enables full RapidMiner functionality in a server environment, including Extract, Transform and Load (ETL), data mining, and reporting. In addition, it includes remote and scheduled execution of processes, and exporting them as web services, all in a web-based environment, open to collaborative work. The RapidAnalytics' Fact Sheet, screenshots, and download locations can be found at its web page `http://rapid-i.com/content/view/182/192/`. In addition, the User and Installation Manual, as well as video tutorials can be found under *Documentation* > *RapidAnalytics* in the main menu on the Rapid-I web page.

In this section, we use RapidAnalytics for exporting RapidMiner workflows as web services. This enables the usage of workflows we implemented in the previous section, in other applications outside of RapidMiner, via a simple HTTP call, either from a web browser, or from a custom application. Building on the knowledge from the previous section, we create a similarity-based language identification method for web pages, and invoke it via a common web browser.

14.6.2 Web Page Language Identification

When web crawling forums for opinion mining or language-dependent link analysis, language identification is necessary to apply further, language-dependent NLP tools. We learned how to create a language identification model, therefore we can utilize one of the workflows presented in the previous section to execute language identification on arbitrary texts. In this case we use the n-gram similarity-based model. Since we are doing web page identification, we are using the operators from the *Web Mining Extension* for fetching and processing web pages. In order to optimize the workflow runtime performance, we make sure to minimize all the necessary processing, and therefore use the saved word list and the model to carry out the task. This is a good practice when exporting workflows as services—always minimize the processing time by saving all the data, and all the models which do not change through time. If they do change, we can create a scheduled process in RapidAnalytics to periodically update the model and the data used, but we will not cover that in this chapter. For more details on scheduling, please consult the *RapidAnalytics User and Installation Manual.*

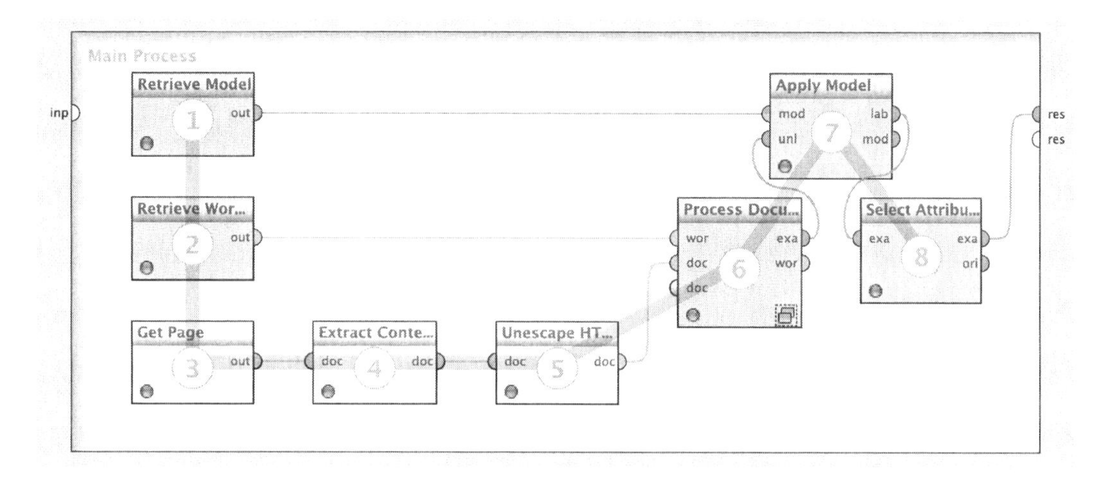

FIGURE 14.12: The workflow of the web page classification application.

The workflow in Figure 14.12 begins by first loading the similarity model `similarity_model` (1) and the WordList `similarity_wordlist` (2). The next step is downloading the web page at the URL defined by the context macro `%{URL}`, by using the *Get Page* operator (3). In case the context tab is hidden, we can easily turn it on by clicking on *View → Show View → Context*, and clicking on the *Context* tab. The next two operators extract the content of the web page (4), and unescape HTML sequences in the resulting document (5). This results in a textual document analyzed by the *Process Documents* operator (6) in the same manner as the n-gram similarity model in Figure 14.7: the operator receives the dataset and the loaded WordList, and processes the data to fit it in a form on which the similarity model was trained. By applying the `similarity_model` model via the *Apply Model* operator (7) on the processed text data of the requested page, we obtain the classification result—the language identification of the web page. The final step is to use the *Select Attribute* operator (8) to select only the `prediction(language)` attribute, while including the special attribute types, and redirect the result to the output of the workflow.

Prior to exporting the web page classification workflow, we first make sure to copy all the required files—the model, the word list, and the process itself—to the RapidAnalytics server by simple means of drag-and-drop to a remote repository set up for RapidAnalytics. The following step is to log into the administration console of RapidAnalytics, and navigate to *Repository > Browse Repository*. Browsing through the repository, we navigate until we get to our `application_webpage` workflow, and select it. On the next web page, by selecting `Export as service` under `Execution`, we invoke the exporting web page, as seen in Figure 14.13. The next step is setting up the needed parameters of the service. First, we set the *Output format* to `XML_PLAIN`. Then by setting the *URL query parameter* and *Target (macro/operator parameter)* to URL, we name the query parameter of the web service and the corresponding macro, which will obtain the value passed by that query parameter, and check it as a *Mandatory* parameter. The final step is to submit the newly created service settings.

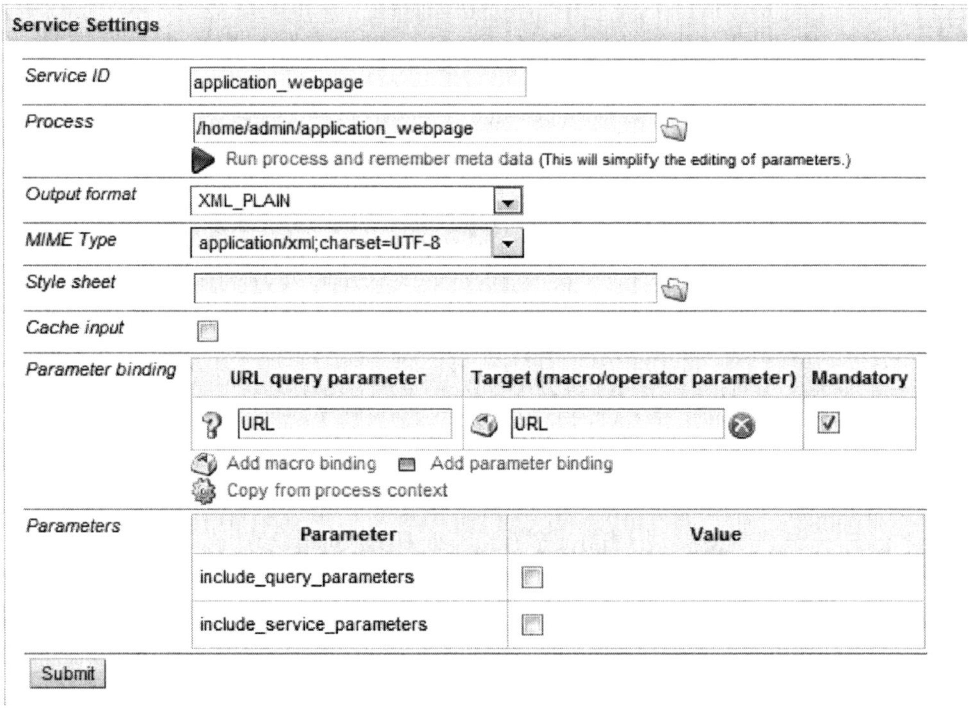

FIGURE 14.13: Exporting the web page classification workflow as a service.

The following step is to test the service by invoking a query to it via a regular web browser. The standard URL where the service should be found by default is `http://localhost:8080/RA/process/application_webpage`. Adding

`?URL=http://www.zeitung.de`

to it equals setting the query string of the web service URL to contain the previously defined *URL* query parameter, and setting that parameter to `http://www.zeitung.de`. The resulting XML the invoked service returned can be easily parsed and processed in an application for further use. We observe the result of invoking the web service from an Internet browser in Figure 14.14. The language of the web page of our choice is correctly classified as German.

The web service we prepared in this section can be used in standalone applications for identification of web pages, by simply calling the service URL with the *URL* parameter provided. Similarly, other language identification models can be employed in this workflow, and easily exported as a web service.

localhost:8080/RA/process/application_webpage?URL=http://www.zeitung.de

This XML file does not appear to have any style information associated with it. The document tree is shown below.


```
▼<results serviceId="application_webpage">
  ▼<example-set>
    ▼<example>
       <prediction-language- role="prediction">deu</prediction-language->
     </example>
   </example-set>
 </results>
```


FIGURE 14.14: Using the web page classification service.

14.7 Summary

Language identification is an important pre-processing step in many text mining applications, and is in its basic form easy to implement in RapidMiner. In this chapter, we demonstrated three different basic approaches to the problem. The frequent words method identified languages over the most frequent words in the corpus, per language. We saw that this method yields relatively high accuracy, but is unable to correctly classify sentences which do not contain frequent words, since the number of words in a language is practically unlimited. The character n-gram method, on the other hand, uses character n-grams instead of whole words in order to do the language identification. This method yields higher accuracy than the frequent words method, and possibly results in a smaller and faster model. The similarity-based method presented in the end, simplifies the resulting model and the computation, though at a cost of a lower accuracy. We implemented and tested these methods using RapidMiner, on an easily obtainable dataset. In the end, we showed how to use the similarity-based workflow in order to create a workflow for web page language identification, and export it as a web service using RapidAnalytics. Having a language identification workflow as a web service enabled us to apply it in various custom applications by simply issuing an HTTP call.

By getting familiar with the problem, its possible solutions, and using the solution in a real-world problem, we hope you mastered language identification and are ready for further challenges. Further challenges might include experimentation with method improvement, optimization of accuracy, and computational performance. Though it might seem that language identification is a solved problem, there is still room for further research and improvement, especially in the application areas like tweet, multilingual document, or language-variant language differentiation. We hope that we successfully gave you the ground knowledge, and inspired you to tackle such challenges and venture deeper into the new horizons of language identification.

Acknowledgment

The authors would like to thank Marin Matijaš, Anabela Barreiro, and Nives Škunca for the invaluable discussions, and comments that added to the quality of this chapter.

Glossary

Token - Token is a continuous string of characters, representing a meaningful element of text such as a word, phrase, symbol, and logogram.

Tokenization - Tokenization is the process of token extraction, i.e. the process of splitting up text into meaningful elements.

Character n-grams - Character n-gram is a substring of a token, containing n consecutive characters.

Bibliography

[1] Paul McNamee. Language identification: A solved problem suitable for undergraduate instruction. *Journal of Computing Sciences in Colleges*, 20(3):94–101, 2005.

[2] Joaquim Ferreira da Silva and Gabriel Pereira Lopes. Identification of document language is not yet a completely solved problem. In *Proceedings of the International Conference on Computational Intelligence for Modelling Control and Automation and International Conference on Intelligent Agents Web Technologies and International Commerce*, CIMCA '06, pages 212–219, 2006.

[3] Baden Hughes, Timothy Baldwin, Steven Bird, Jeremy Nicholson, and Andrew Mackinlay. Reconsidering language identification for written language resources. In *Proceedings of LREC2006*, pages 485–488, 2006.

[4] Eda Baykan, Monika Henzinger, and Ingmar Weber. Web page language identification based on urls. *Proc. VLDB Endow.*, 1(1):176–187, 2008.

[5] Bruno Martins and Mário J. Silva. Language identification in web pages. In *Proceedings of the 2005 ACM Symposium on Applied Computing*, SAC '05, pages 764–768, 2005.

[6] Hakan Ceylan and Yookyung Kim. Language identification of search engine queries. In *Proceedings of the Joint Conference of the 47th Annual Meeting of the ACL and the 4th International Joint Conference on Natural Language Processing of the AFNLP: Volume 2*, ACL '09, pages 1066–1074, 2009.

[7] Thomas Gottron and Nedim Lipka. A comparison of language identification approaches on short, query-style texts. In *Proceedings of the 32nd European Conference on Advances in Information Retrieval*, ECIR'2010, pages 611–614, 2010.

[8] Gustavo Laboreiro, Matko Bošnjak, Eduarda Mendes Rodrigues, Luís Sarmento, and Eugénio Oliveira. Determining language variant in microblog messages, 2013. To appear in the Proceedings of The 28th ACM Symposium on Applied Computing, SAC 2013.

[9] Nikola Ljubešić, Nives Mikelić, and Damir Boras. Language indentification: How to distinguish similar languages? In *Information Technology Interfaces, 2007. ITI 2007. 29th International Conference on*, pages 541–546, 2007.

[10] Anil Kumar Singh. Study of some distance measures for language and encoding identification. In *Proceedings of the Workshop on Linguistic Distances*, LD '06, pages 63–72, 2006.

[11] Gregory Grefenstette and Pasi Tapanainen. What is a word? What is a sentence? Problems of Tokenization. In *Third International Conference on Computational Lexicography (cOMPLEX'94)*, pages 79–87, 1994.

[12] Hans van Halteren. *Syntactic Wordclass Tagging*. Kluwer Academic Publishers, 1999.

[13] Jonathan J. Webster and Chunyu Kit. Tokenization as the initial phase in NLP. In *Proceedings of the 14th Conference on Computational Linguistics. Volume 4*, COLING '92, pages 1106–1110, 1992.

[14] Christopher D. Manning, Prabhakar Raghavan, and Hinrich Schtze. *Introduction to Information Retrieval*. Cambridge University Press, 2008.

[15] Ted Dunning. Statistical Identification of Language. Technical report, Computing Research Lab, New Mexico State University, 1994.

[16] William B. Cavnar and John M. Trenkle. N-gram-based text categorization. In *Proceedings of SDAIR-94, 3rd Annual Symposium on Document Analysis and Information Retrieval*, pages 161–175, 1994.

[17] Timothy Baldwin and Marco Lui. Language identification: The long and the short of the matter. In *Human Language Technologies: The 2010 Annual Conference of the North American Chapter of the Association for Computational Linguistics*, HLT '10, pages 229–237, 2010.

[18] John C Schmitt. Trigram-based method of language identification, October 29 1991. US Patent 5,062,143.

[19] E. DeLuca L. Grothe and A. Nürnberger. A comparative study on language identification methods. In *Proceedings of the Sixth International Conference on Language Resources and Evaluation (LREC'08)*, 2008.

[20] Uwe Quasthoff, Matthias Richter, and Christian Biemann. Corpus portal for search in monolingual corpora. In *Proceedings of the Fifth International Conference on Language Resources and Evaluation, LREC*, pages 1799–1802, 2006.

[21] Philipp Koehn. Europarl: A parallel corpus for statistical machine translation. In *Proceedings of the Tenth Machine Translation Summit*, pages 79–86. AAMT, 2005.

[22] Ralf Steinberger, Bruno Pouliquen, Anna Widiger, Camelia Ignat, Tomaž Erjavec, and Dan Tufis. The JRC-Acquis: A multilingual aligned parallel corpus with 20+ languages. In *Proceedings of the 5th International Conference on Language Resources and Evaluation (LREC'2006)*, pages 2142–2147, 2006.

Chapter 15

Text Mining with RapidMiner

Gurdal Ertek
Sabancı University, Istanbul, Turkey

Dilek Tapucu
Sabancı University, Istanbul, Turkey

Inanc Arin
Sabancı University, Istanbul, Turkey

15.1 Introduction

The goal of this chapter is to introduce the text mining capabilities of RapidMiner through a use case. The use case involves mining reviews for hotels at TripAdvisor.com, a popular web portal. We will be demonstrating basic text mining in RapidMiner using the text mining extension. We will present two different RapidMiner processes, namely Process01 and Process02, which respectively describe how text mining can be combined with association mining and cluster modeling. While it is possible to construct each of these processes from scratch by inserting the appropriate operators into the process view, we will instead import these two processes readily from existing model files.

Throughout the chapter, we will at times deliberately instruct the reader to take erroneous steps that result in undesired outcomes. We believe that this is a very realistic way of learning to use RapidMiner, since in practice, the modeling process frequently involves such steps that are later corrected.

15.1.1 Text Mining

Text mining (also referred to as *text data mining* or *knowledge discovery from textual databases*), refers to the process of discovering interesting and non-trivial knowledge from text documents. The common practice in text mining is the analysis of the information extracted through text processing to form new facts and new hypotheses, that can be explored further with other data mining algorithms. Text mining applications typically deal with large and complex datasets of textual documents that contain significant amount of irrelevant and noisy information. Feature selection aims to remove this irrelevant and noisy information by focusing only on relevant and informative data for use in text mining. Some of the topics within text mining include feature extraction, text categorization, clustering, trends analysis, association mining, and visualization.

15.1.2 Data Description

The files required for this chapter, including the data and pre-built processes reside within a folder titled `LocalRepository`. The data used in this chapter comes from TripAdvisor.com, a popular web portal in the hospitality industry. This publicly available dataset contains the reviews and ratings (1 through 5) of clients or customers for 1850 hotels. The original data was extracted by The Database and Systems Information Laboratory at the University of Illinois at Urbana-Champaign, and is available under `http://sifaka.cs.uiuc.edu/~wang296/Data/`. There are 1850 text documents in the original dataset, corresponding to the reviews of 1850 hotels. Each document contains all the reviews for that hotel. While it is possible to run text mining processes with the original data, we will be using a subset of the data containing only the first 100 text documents. The dataset used in this chapter may be downloaded from `http://people.sabanciuniv.edu/ertekg/papers/supp/09.zip` or the book's website (`http://www.rapidminerbook.com`).

15.1.3 Running RapidMiner

When running RapidMiner, it is strongly recommended to right-click the mouse button on the start menu and Run as administrator, rather than simply double-clicking the executable or shortcut icon. By running as administrator, we are granted the permissions to install extension packages. Running the software without the administrator rights may cause errors when trying to install updates, including the extension packages.

15.1.4 RapidMiner Text Processing Extension Package

RapidMiner is the most popular open source software in the world for data mining, and strongly supports text mining and other data mining techniques that are applied in combination with text mining. The power and flexibility of RapidMiner is due to the GUI-based IDE (integrated development environment) it provides for rapid prototyping and development of data mining models, as well as its strong support for scripting based on XML (extensible mark-up language). The visual modeling in the RapidMiner IDE is based on the defining of the data mining process in terms of operators and the flow of process through these operators. Users specify the expected inputs, the delivered outputs, the mandatory and optional parameters, and the core functionalities of the operators, and the complete process is automatically executed by RapidMiner. Many packages are available for RapidMiner, such as text processing, Weka extension, parallel processing, web mining, reporting extension, series processing, Predictive Model Markup Language (PMML), community, and R extension packages. The package that is needed and used for text mining is the *Text Processing* package, which can be installed and updated through the Update RapidMiner menu item under the Help menu.

15.1.5 Installing Text Mining Extensions

We will initiate our text mining analysis by importing the two previously built processes. However, even before that we have to check and make sure that the extensions required for text mining are installed within RapidMiner. To manage the extensions, select the Help menu and then the Manage Extensions menu item.

Select Help menu and then the Updates and Extensions (Marketplace) menu item. RapidMiner will display a window that mainly lets you select updates and search for extensions. The extension packages (use the *Search* or *Top Downloads* tabs) needed for text mining are Text Processing, Web Mining, and Wordnet Extension. After you find an extension, click on the icon and tick *Select for Installation* before clicking on the install button. Accept the Terms and Conditions.

15.2 Association Mining of Text Document Collection (Process01)

15.2.1 Importing Process01

We will now initiate the text mining analysis by importing the processes supplied with this chapter. For this, select Files and then the Import Process menu item. Navigate to the folder location that contains the files associated to this chapter. Inside the `Processes` folder, click on `Process01.rmp`. The RapidMiner process `Process01` is now displayed in the *design perspective*, as shown in Figure 15.1.

15.2.2 Operators in Process01

The parameters for the operators in this process are given on the right-hand side of the process, listed under the Parameters tab. For the first process, the parameter text directories

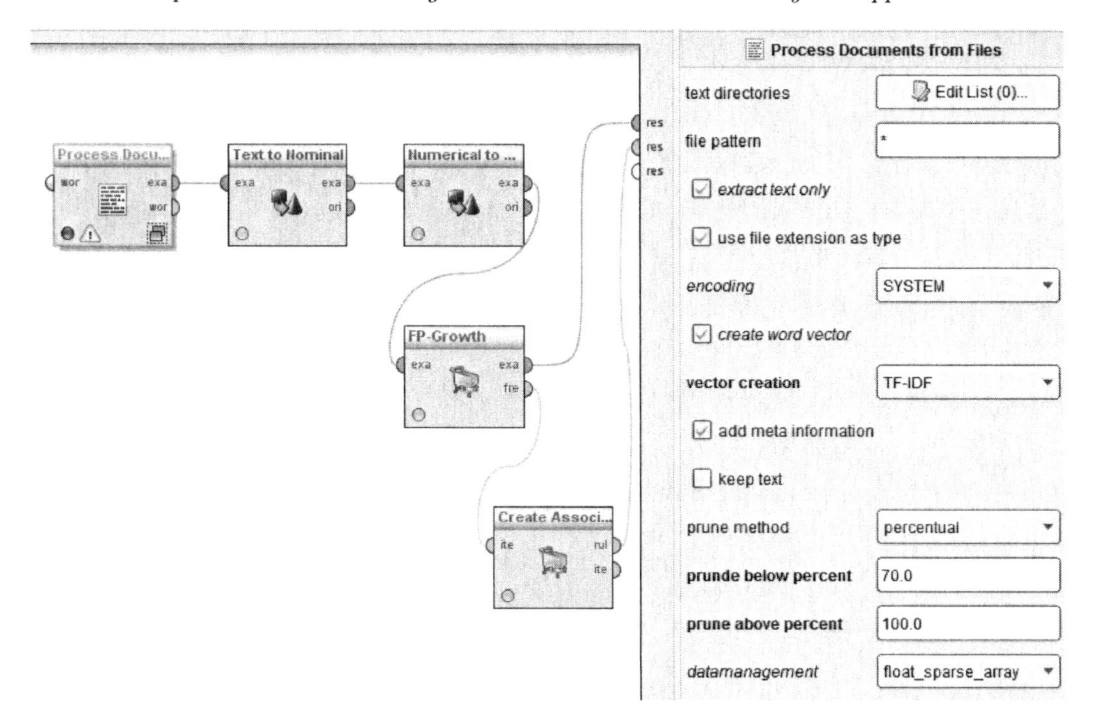

FIGURE 15.1: Operators for Process01 and the Parameters for Process Documents from Files operator.

specifies where to read the text data from. A very important parameter is the vector creation used. In Process01, the selected vector creation method is *TF-IDF* (*Term Frequency-Inverse Document Frequency*) and is a term-weighting method. It gives higher weight to terms that appear frequently in the document, but not many times in other documents. However, this may yield too many words, even up to tens of thousands of them. Too many words would make it prohibitive to carry out the successive data mining steps due to lengthy running times for the data mining algorithms. Hence, it is a very good idea to prune the resulting word set using a prune method, by selecting the method and its parameters within the Parameters view.

Click on the Process Documents from Files operator, as in Figure 15.1. In Process01, words that appear in less than 70.0 % of the documents are pruned, as can be seen from the value of 70.0 for the prune below percent parameter. It is also possible to prune the words that appear in too many documents, but this was not done in this example, as can be seen from the value of 100.0 for the prune above percent parameter. In association mining, we are interested in the items (words in text mining context) that appear in 100.0% of the transactions (documents in our text mining context), since they can form interesting frequent itemsets (word lists) and association rules, that provide actionable insights. Thus, it is appropriate to set prune above percent to 100.0 in Process01 (Figure 15.1), including the items (words) that appear in every document.

Process01 consists of five operators (Figure 15.1). The Process Documents from Files operator performs text processing, which involves preparing the text data for the application of conventional data mining techniques. The Process Documents from Files operator reads data from a collection of text files and manipulates this data using text processing algorithms. This is a nested operator, meaning that it can contain a sub-process consisting of a multitude of operators. Indeed, in Process01, this nested operator contains other operators inside. Double-click on this operator, and you will see the sub-process inside it, as outlined in Figure 15.2. This sub-process consists of six operators that are serially linked (Figure 15.2):

- Tokenize Non-letters (Tokenize)
- Tokenize Linguistic (Tokenize)
- Filter Stopwords (English)
- Filter Tokens (by Length)
- Stem (Porter)
- Transform Cases (2) (Transform Cases)

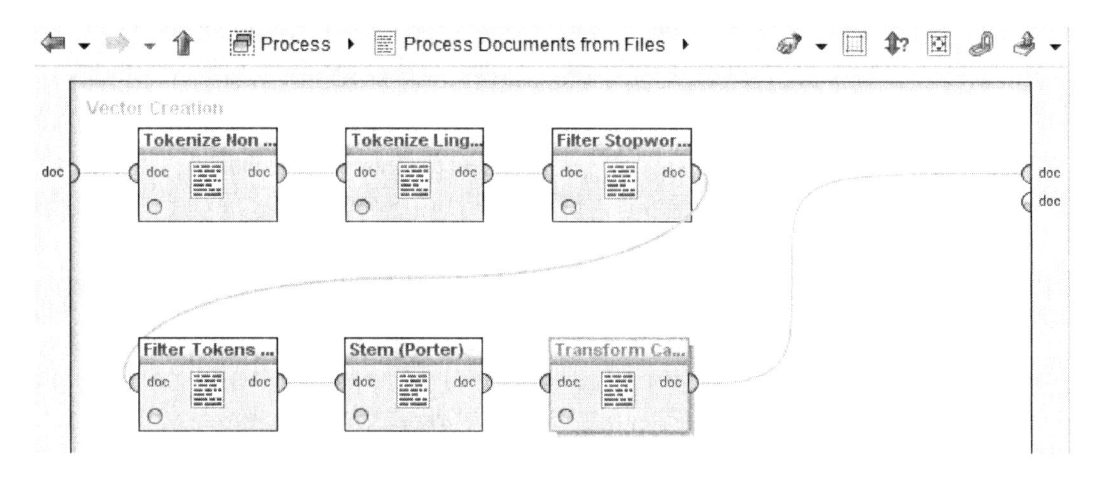

FIGURE 15.2: Operators within the Process Documents from Files nested operator.

The sub-process basically transforms the text data into a format that can be easily analyzed using conventional data mining techniques such as association mining and cluster modeling. The parameters for each of the operators in this sub-process (within the Process Documents from Files operator) are displayed in Figure 15.3. Notice that Figure 15.3 was created manually by combining multiple snapshots.

In this sub-process, the Tokenize Non-letters (Tokenize) and Tokenize Linguistic (Tokenize) operators are both created by selecting the Tokenize operator, but with different parameter selections. The former operator tokenizes based on non letters whereas the latter operate tokenizes based on the linguistic sentences within the English language. The Filter Stopwords (English) operator removes the stop words in the English language from the text dataset. The Filter Tokens (by Length) operator removes all the words composed of less than min chars characters and more than max chars characters. In this example, words that have less than 2 characters or more than 25 characters are removed from the dataset. The Stem (Porter) operator performs stemming, and the Transform Cases(2) (Transform Cases) operator transforms all the characters in the text into lower case. It should be noted that the name of this last operator is not a good name, and we, as the modelers, have forgotten to rename the operator after its name was automatically assigned by RapidMiner. This mistake should be avoided in constructing the processes.

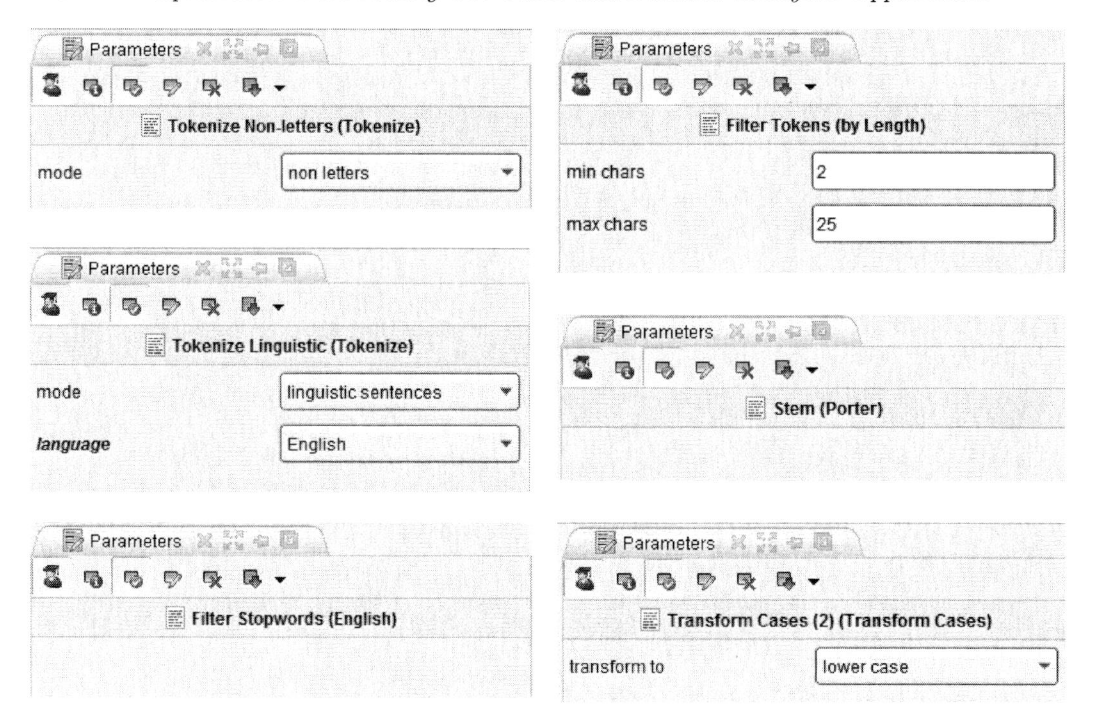

FIGURE 15.3: Parameters for the operators within the Process Documents from Files operator.

FIGURE 15.4: Parameters for the operators in Process01.

The parameters for each of the operators in Process01 (Figure 15.1) are displayed in Figure 15.4. Notice that Figure 15.4 was created manually by combining multiple snapshots. The Text to Nominal operator transforms the text data into nominal (categorical) data. The Numerical to Binomial operator then transforms the data into binominal form. This means that each row represents a document, a few columns provide meta-data about that document, and the remaining columns represent the words appearing in all the documents, with the cell contents telling (true or false) whether that word exist in that document or not. The FP-Growth algorithm for identifying frequent item sets. In this example, the min support parameter is 0.7 (Figure 15.4), meaning that the operator generates a list of the frequent sets of words (itemsets) that appear in at least 70 % of the documents. Notice that it will be computationally efficient to select the min support value in the FP-Growth operator to be equal to the prune below percent value for the Process Documents from Files operator (Figure 15.1) divided by 100. Also, the max items parameter is 2, meaning that the generated list is limited to pairs of words (2-itemsets), and the list will not contain frequent word sets (itemsets) with 3 or more words in them. The final operator in Process01, namely Create Association Rules, receives the list of frequent word sets from the FP-Growth operator, and computes the rules that satisfy the specified constraints on selected association mining criteria. In this example, the association rules are computed according to the the criterion of confidence, as well as gain theta and laplace k. The specified minimal values for these 3 criteria are 0.8, 1.0, and 1.0, respectively.

15.2.3 Saving Process01

So far Process01 has been imported into the workspace of RapidMiner. Now it is a good time to keep it in the LocalRepository, so that we will not have to import it again next time we work with RapidMiner. Click on the Repositories view, right-click on the LocalRepository and select Configure Repository as shown in Figure 15.5. This is actually an initialization step before using RapidMiner for the first time, but we will still go through this step to ensure that we are saving everything to a preferred folder directory in our computer. In this chapter, we will be saving everything under the Root directory of `C:\RapidMiner`. Next, we can save Process01 in the LocalRepository. Right-click on the LocalRepository text and select Store Process Here. When the Store Process dialog window appears, click OK. Our importing and saving of Process01 is now completed.

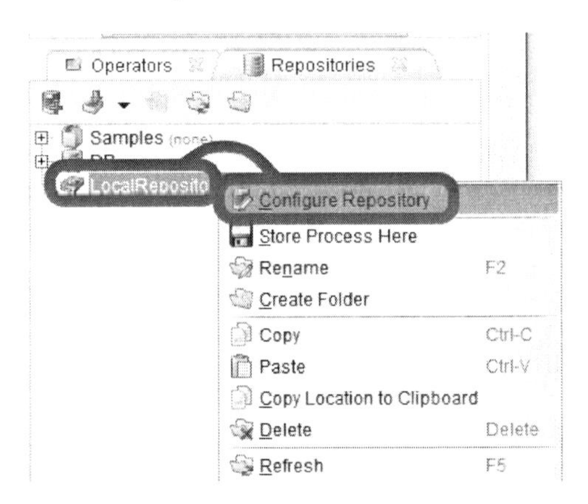

FIGURE 15.5: Configuring LocalRepository.

15.3 Clustering Text Documents (Process02)

The second text mining process that we will introduce in this chapter is Process02, and involves the clustering of the 100 documents in the text collection.

15.3.1 Importing Process02

To import this process, select File menu, Import Process menu item, select `Process02.rmp` within the Import Process window, and click Open.

15.3.2 Operators in Process02

Process02 is given in Figure 15.6. Similar to Process01, Process02 begins with the Process Documents from Files operator, whose parameters are given in Figure 15.6. In Process02, the selected vector creation method is different; it is Term Frequency. The impact of this new selection will be illustrated later. For now, it should be noted that this selection results in the computation of the relative frequencies of each of the words in each of the documents in the dataset. For example, if a word appears 5 times within a document that consists of 200 words, then the relative frequency of that word will be $5/200 = 0.025$. This value of 0.025 will appear under the column for that word, at the row for that document. In Process02, the prune method is again percentual, just as in Process01 (Figure 15.1). However, the value for the prune below percent parameter is now different.

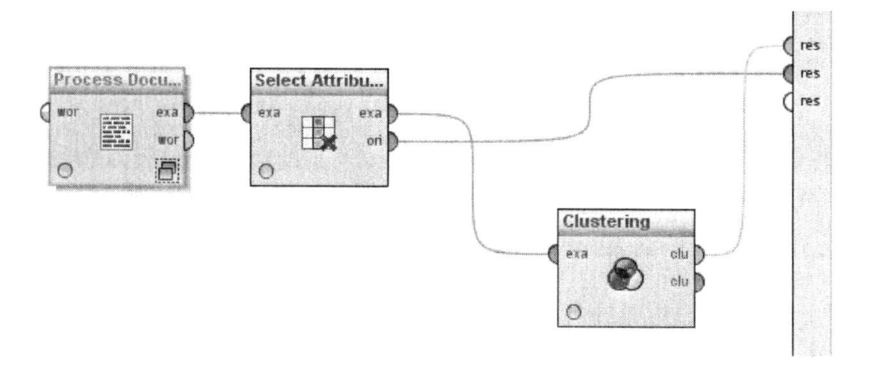

FIGURE 15.6: Operators in Process02.

The prune below percent parameter in Process01 was 70.0, whereas it is now 20.0 in Process02. The reason for this change is due to the fact that the applied data mining technique in Process01 is association mining, whereas in Process02 it is clustering. The former technique is computationally much more expensive than the latter, meaning that running association mining algorithms on a dataset takes much longer running time compared to the k-means clustering algorithm on the same dataset. Therefore, with the same amount of time available for our data mining process, we have to work with much smaller datasets if we are carrying out association mining, as in Process01. The values displayed in Figures 15.1 and 15.6 have been determined through trial and error, such that the running time for the processes do not exceed 30 seconds on a laptop with an Intel i5 processor and 4GB RAM.

Figure 15.7 shows the contents of this operator. While there were six operators within

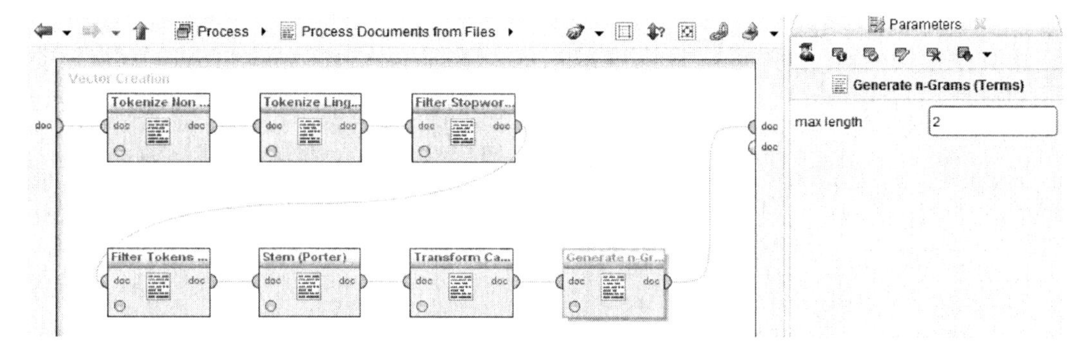

FIGURE 15.7: Operators within the Process Documents from Files nested operator and the Parameters for the Generate n-Grams (Terms) operator.

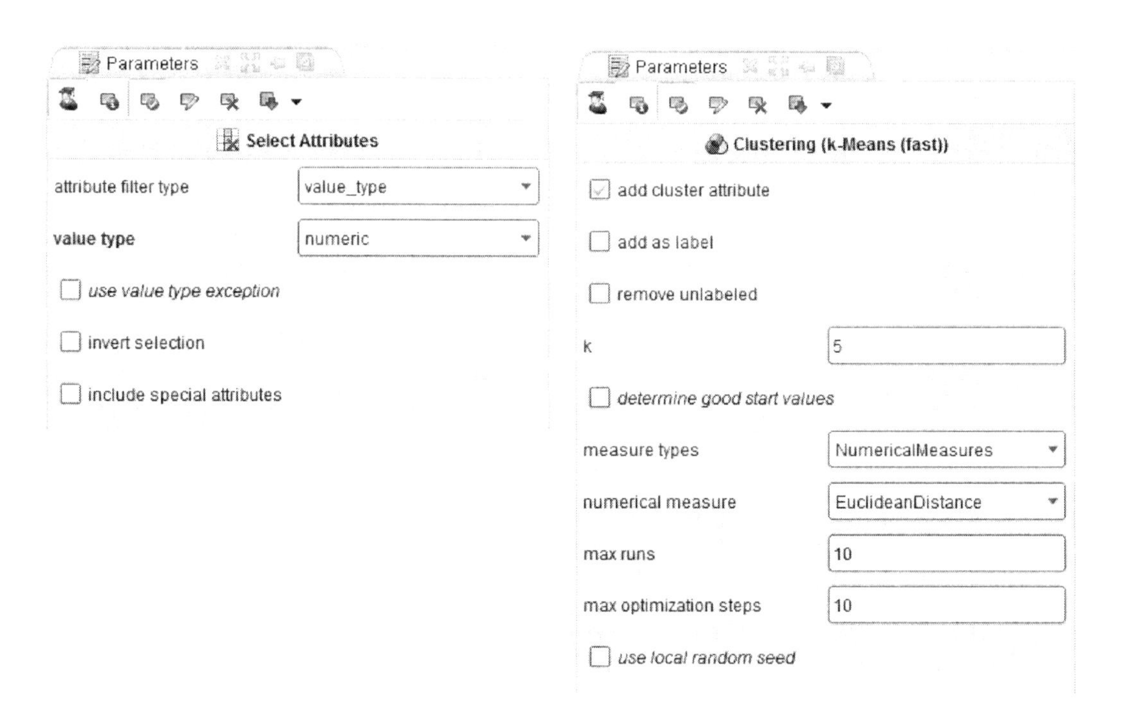

FIGURE 15.8: Parameters for the operators in Process02.

this nested operator in Process01, there are now seven operators in Process02. The newly added operator is Generate n-Grams (Terms). The only parameter for this operator is max length, which is set equal to 2 in our example (Figure 15.7).

The parameters for the Select Attributes and Clustering (k-means(fast)) operators within Process02 are displayed in Figure 15.8. The Select Attributes operator takes the complete dataset and transforms it into a new one by selecting only the columns with numeric values, i.e., columns corresponding to the words. This transformation is required for the next operator, which performs clustering based on numerical values. Clustering (k-means (fast)) operator carries out the k-means clustering algorithm on the numerical dataset. Since each row of the data (each document in the text collection) is characterized in terms of the occurrence frequency of words in it, this operator will place together the documents that have a similar distribution of the word frequencies. The k value, which denotes the number of clusters to be constructed, is set to 5 in our example.

15.3.3 Saving Process02

Now, let us save Process02 into LocalRepository: Click on Repositories view, right-click on LocalRepository view and select Store Process Here. When the store process dialog box appears, click OK. Both processes are now saved under local repository. Since we had earlier defined the directory root for local repository as `C:\RapidMiner`, the saved processes will appear as `.rpm` (RapidMiner process) files under that directory.

Unfortunately, there is a problem with the naming of the files: the file names also contain the extension .rpm, in addition to the `.rpm` extension itself. For example, the correct name for `Process01.rmp.rmp` under `C:\RapidMiner` should be `Process01.rmp`. Right-click on Process01.rmp text within RapidMiner, and select rename, then remove the .rmp extension from the file name and click OK. Do the same for renaming `Process02`.

15.4 Running Process01 and Analyzing the Results

Having imported and saved the processes into the local repository, now we will run these processes.

15.4.1 Running Process01

For running Process01, double click on Process01.rmp under the Repositories view. Process01 will be loaded into the Process view. Click the run button in the tool bar, as in Figure 15.9. You may be prompted and asked whether you would like to close all results before starting process (Figure 15.9). Click Yes. When the process run is completed, you will be prompted again and asked whether you would like to switch to the result perspective (Figure 15.10). Click Yes and you will have the screen in Figure 15.11. This is indeed the *result perspective*, displaying the result obtained through your process. Until now, you worked on your process in the *design perspective*, where you could change your process by adding/removing/editing operators. Now, in the result perspective, you are not able to modify our process. In the tool bar, you can switch to the design perspective or the result perspective by clicking their respective buttons.

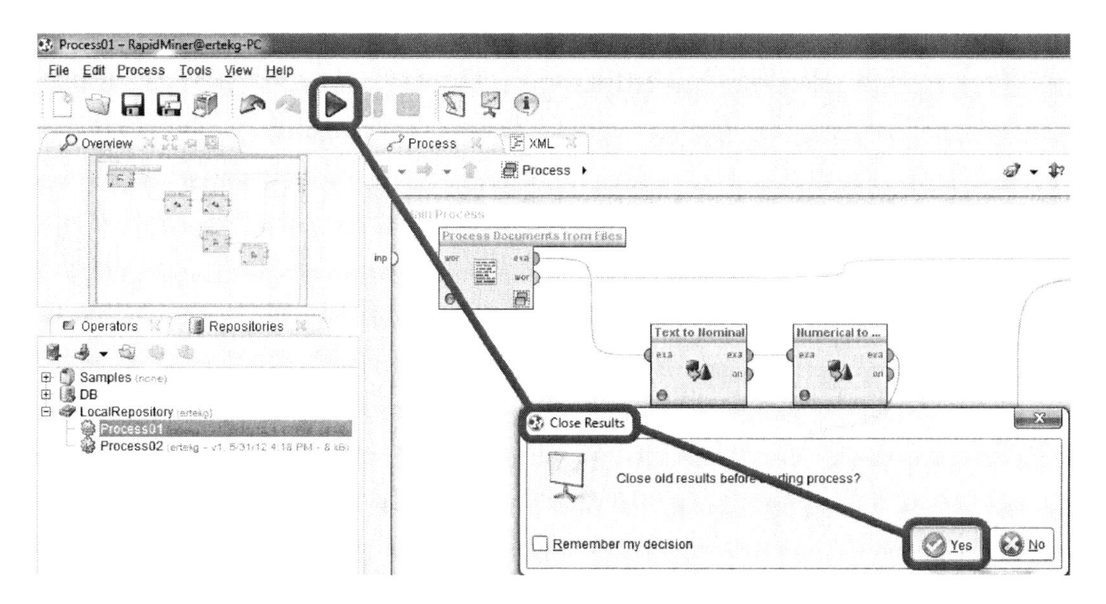

FIGURE 15.9: Opening and running Process01.

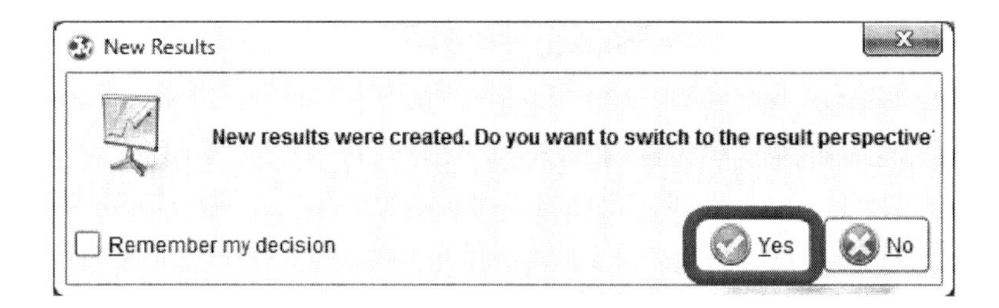

FIGURE 15.10: Dialog box alerting the switch to the result perspective.

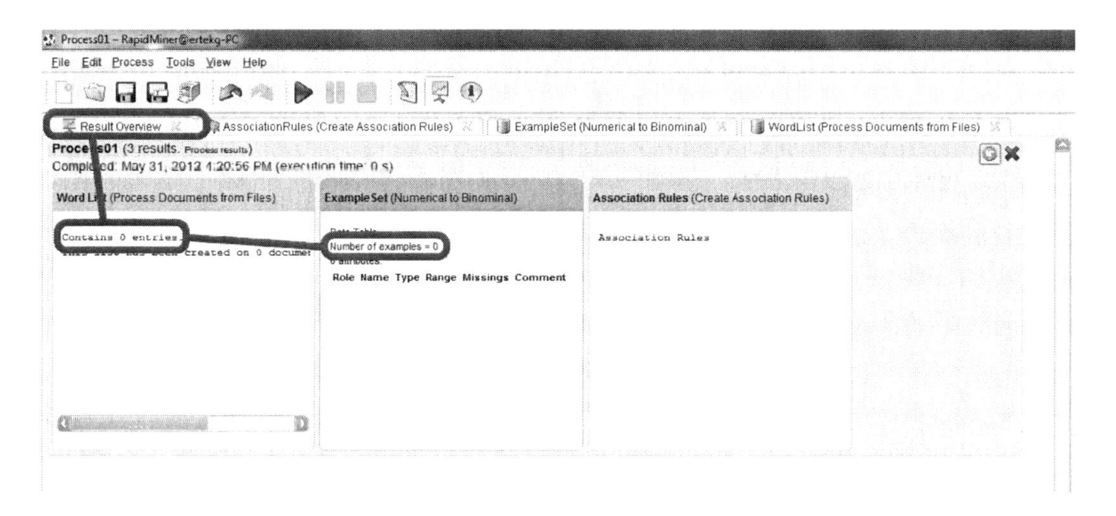

FIGURE 15.11: Result Overview for Process01 results.

15.4.2 Empty Results for Process01

In the result perspective, we can see the latest results at the bottom of the result overview. To see the details, click inside the bar with the most recent date and time, and you will see an expanded view as in Figure 15.11. In the extended view there are three types of results obtained as a result of running Proces01. These three results correspond precisely to the three result ports for Process01, which are represented by the three half-circles on the right of the process view in Figure 15.1 and are labeled with res. The first result is the wordlist (wor) generated from the Process Documents From Files operator (Figure 15.1). The second result is the example set exa, generated from the FP-Growth operator. The third and final result is the list of association rules (rul), generated from the Create Association Rules.

In Figure 15.11 the word list contains 0 entries and the number of examples is 0. But why? Was there something wrong with our processes? We will answer this question next.

15.4.3 Specifying the Source Data for Process01

Running Process01 resulted in no tangible results and the reason is very simple. We have not yet specified the text dataset that will be processed and mined in Process01. To specify the data source, as in Figure 15.12, switch to the design perspective by clicking the notepad button, click on the Process Documents from Files operator, and click on the box to the right of the parameter text directories.

In the dialog box, click on the directory button, select the source data folder, double click TripAdvisor_First_100_Hotels folder, and click the open button. Now the directory for the source text data appears in the dialog box. Give it a class name such as *TripAdvisor_First_100_Hotels* and click OK.

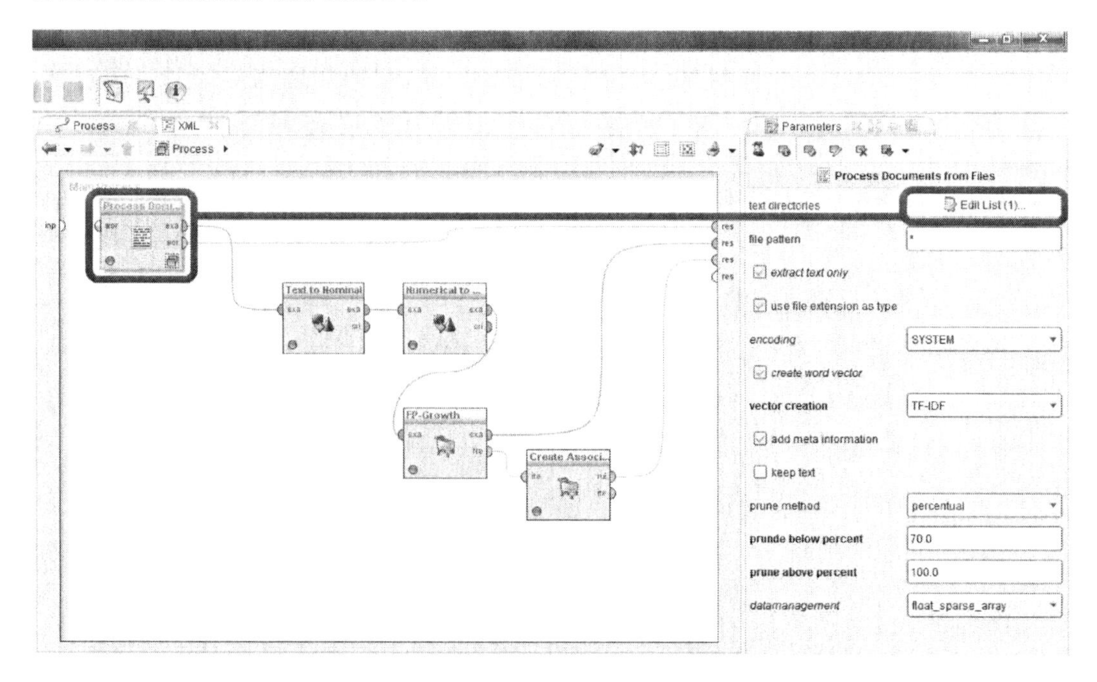

FIGURE 15.12: Specifying the data source text directories for Process01.

15.4.4 Re-Running Process01

Having specified the data source, we can run Process01 again. Click on the run button (or F11 in Windows). When the process is running you can observe which operator is currently being used. A small green triangle appears on the operator's lower left corner when it is being processed. Further information, such as processing times or iteration numbers, can be seen in the status bar located in the bottom left segment of the screen.

15.4.5 Process01 Results

When the new results are created, click Yes to switch to the result perspective. In the result perspective you will now see a see a blue bar. Click inside the bottommost blue bar and you will see an overview of the results, as in Figure 15.13. This time the word list is not empty, it Contains 631 entries. The Number of examples in the example set is equal to 100, and there are 250 attributes (Figure 15.13).

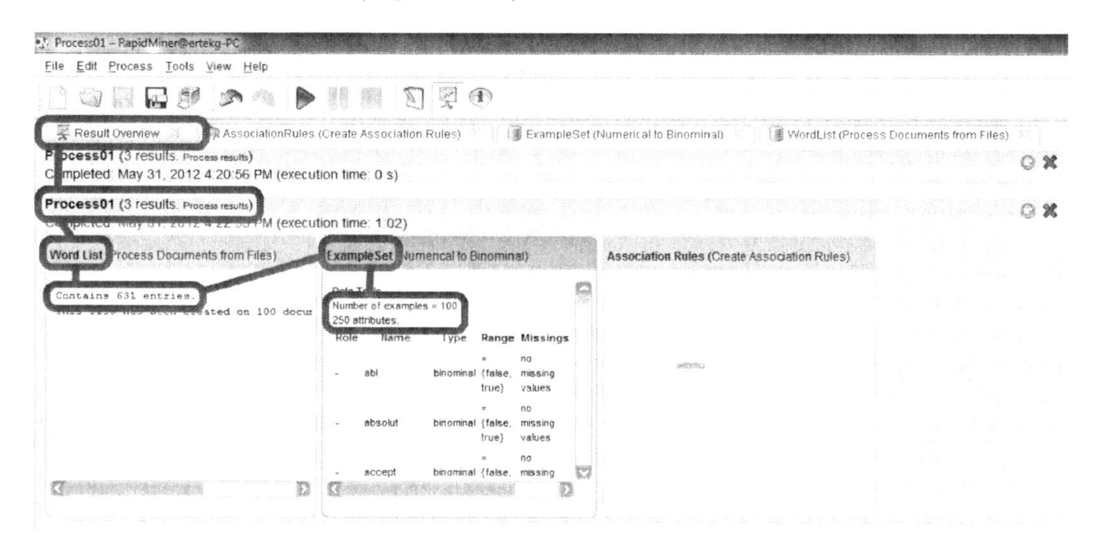

FIGURE 15.13: Result Overview for Process01 results.

Process01 – RapidMiner@ertekg-PC

File Edit Process Tools View Help

Result Overview | AssociationRules (Create Association Rules) | ExampleSet (Numerical to Binominal) | WordList (Process Documents from Files)

Word	Attribute Name	Total Occurences	Document Occurences	TripAdvisor_First_100_Hotels
abl	abl	749	91	749
absolut	absolut	546	92	546
accept	accept	214	78	214
access	access	1118	100	1118
accommod	accommod	464	90	464
accomod	accomod	499	88	499
actual	actual	726	94	726
ad	ad	214	73	214
addit	addit	331	84	331
adequ	adequ	407	86	407

FIGURE 15.14: WordList generated by Process01.

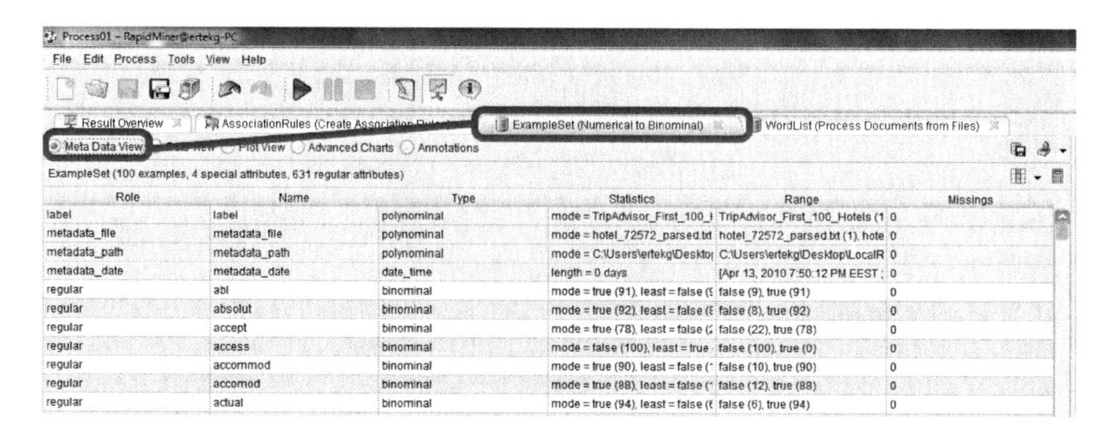

FIGURE 15.15: Meta Data View for the ExampleSet generated by Process01.

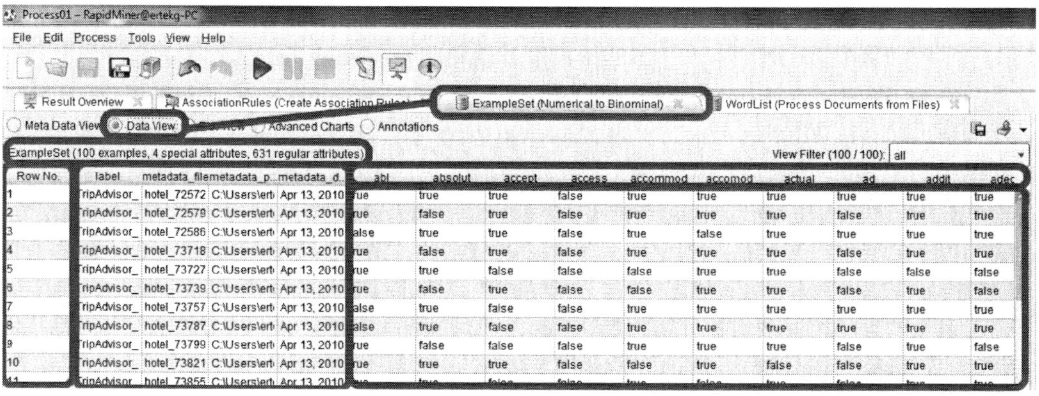

FIGURE 15.16: Data View for the ExampleSet generated by Process01.

Click on the word list view to display the words found by the Process Documents from Files operator, as in Figure 15.14.

Click on the ExampleSet (Numerical to Binomial) view to display the example set. Figure 15.15 presents the Meta Data View, which gives the metadata (information on the data attributes) of the ExampleSet. Now click on the Data View button to display the data itself (Figure 15.16). Each row corresponds to a text document and the first column is an automatically generated key attribute. The extra four columns contain the label, file name, file location, and the file creation data. The remaining columns correspond to the word list. The cells under the word attributes take the binominal value of true or false, denoting whether the word exists in that document.

The final result is the set of association rules. Click on the AssociationRules (Create Association Rules) tab to display the association rules (Figure 15.17). In the Table View, the table grid presents the generated association rules, with one rule in each row. For example, the second row states "IF worth THEN holidai" with a Support level of 0.760 and Confidence level of 0.800. This rule means that in 76 of the 100 documents, words with the stem worth and holidai appear together. Furthermore, in 80% of the documents where a word derived from the stem worth appears, at least one word derived from the stem holidai is observed. On the left-hand side of the grid we can select and Show rules matching a particular set of words (Figure 15.17).

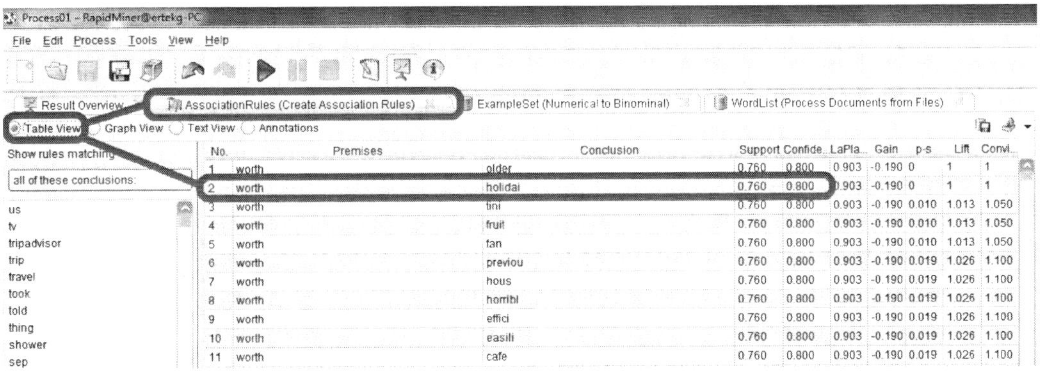

FIGURE 15.17: Table View for the AssociationRules generated by Process01.

These association rules can also be displayed in the form of a graph. For this, click on the Graph View radio button. The initial graph visualization may be cluttered due to the Node Labels. Click on the check box Node Labels to uncheck that option and eliminate or at least reduce the clutter. In this window, filtering can again be carried out by selecting words from the list box on the left-hand side. Once a filtered word is selected you will want to again display the Node Labels as in Figure 15.18. At this point, select a word from the list box and check the Node Label check box (Figure 15.18).

The final result that we would like to share is displayed in Figure 15.19. Click on the WordList (Process Documents from Files) tab to return to the word list. Click on the attribute label Document Occurrences to obtain a descending order. Then move the vertical scroll bar to see the border line where the value changes from 100 to 99. From Figure 15.19 we can observe that the word work appears in all 100 documents, whereas the word big appears in 99 documents.

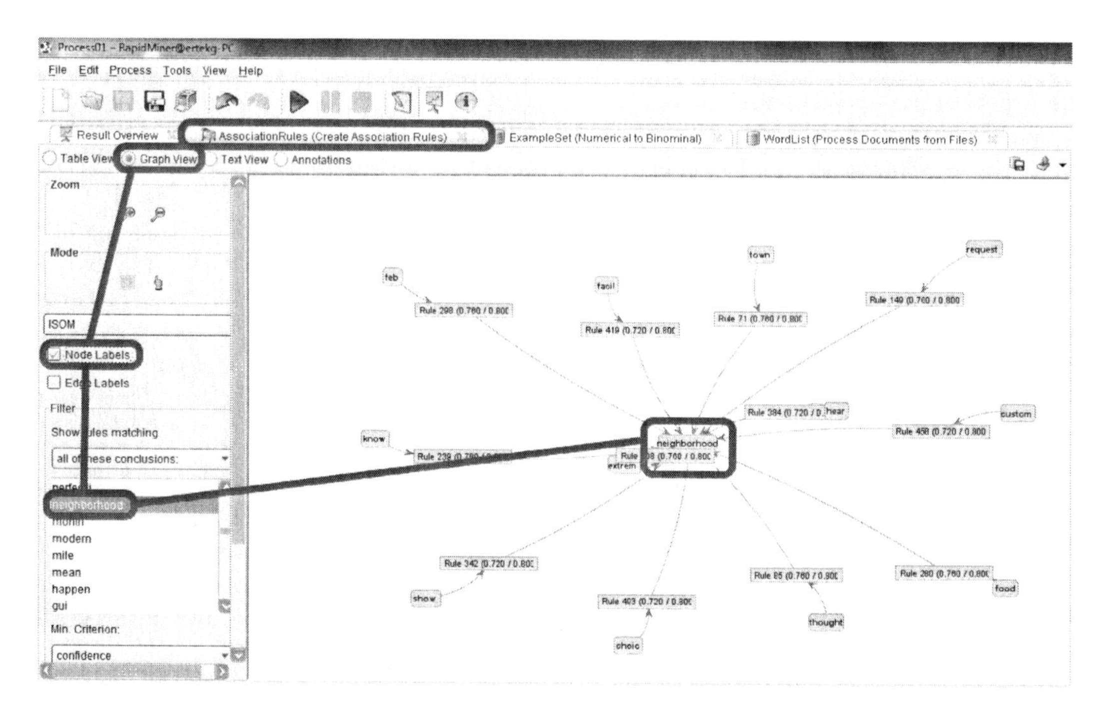

FIGURE 15.18: Filtering rules in the Graph View for the AssociationRules.

Word	Attribute Name	Total Occurences	Document Occurences	TripAdvisor_First_100_Hotels
servic	servic	5078	100	5078
sleep	sleep	1098	100	1098
staff	staff	7401	100	7401
stai	stai	17341	100	17341
take	take	2038	100	2038
time	time	4837	100	4837
walk	walk	5273	100	5273
want	want	2726	100	2726
went	went	1435	100	1435
work	work	1740	100	1740
big	big	1355	99	1355
book	book	3074	99	3074
car	car	3730	99	3730
comfort	comfort	3576	99	3576
conveni	conveni	1324	99	1324
definit	definit	1553	99	1553

FIGURE 15.19: Document Occurrences of the words in the WordList.

15.4.6 Saving Process01 Results

Having generated results of Process01, we will save these results into LocalRepository, so that we can save time when we want to analyze the results next time. The three types of results, namely the Word List, ExampleSet, AssociationRules, have to be saved individually. First, click on the save button in the upper right corner of the results, click on LocalRepository in the appearing window, and type in an appropriate name inside the Name text box, then click OK.

Next, click on the ExampleSet (Numerical to Binominal) tab and repeat the same steps.

Finally, click on the AssociationRules. Using this last result we would like to show you a common possible mistake. After clicking the save button, omit clicking on LocalRepository and click OK. Please note that in this case we did not specify a location for the generated results. This will cause an error . When you see the error dialog box, click Close. Then, follow the correct steps; Click on the save button, click on LocalRepository, give an appropriate unique name and click OK.

Now that all the data for the generated results are saved, let us also see how we can save (export) the data visualizations. Click on the export button to the right of the save button. Then specify the directory where the visualization will be saved (exported) and select the image file type. In this example, select the .png image format, and then click OK. You can check that the visualization is now saved as an image file, as specified.

15.5 Running Process02 and Analyzing the Results

In this section we will run Process02 and analyze the generated results.

15.5.1 Running Process02

To switch to the design view, click on the design view button (Notepad icon) or press F11. Then double-click on Process02 inside the Repositories view, and click the run button. You will be prompted with an error dialog box outlining that the example set does not contain enough examples.

15.5.2 Specifying the Source Data for Process02

The reason for this error is that we have not specified the dataset for Process02. To resolve this, click on the Process Documents from Files operator and click on the button next to the text directories parameter inside the Parameters view. Then, select the same dataset as you have selected for Process01, as shown in Figure 15.12. Now the data source has been specified and we can run Process02 to obtain tangible results.

15.5.3 Process02 Results

The results of Process02 are shown in Figure 15.20. There are three types of results generated after running Process02:

Firstly, the Word List contains the filtered words from the documents in our sample; the Word List generated Contains 5175 entries.

The second result is the Example Set, which contains data regarding the word composition of each of the documents in the database.

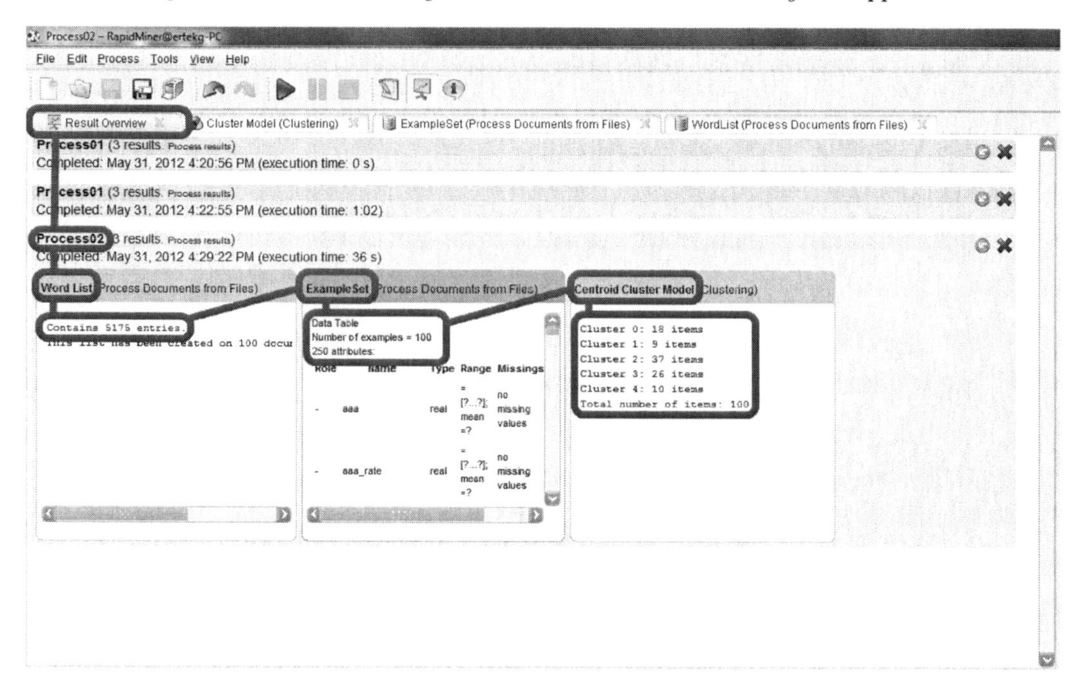

FIGURE 15.20: Result Overview for Process02 results.

The third result of Process02 is the Centroid Cluster Model. In our example five clusters (Cluster 0, Cluster 1, Cluster2, Cluster 3, Cluster 4) have been generated, where each cluster contains a subset of the 100 documents.

The Word List is very similar to the word list of Process01. However, there is a difference due to the additional Generate n-Grams (Terms) operator (Figure 15.7). This additional operator extracts n-grams (sequences of words) as well as single words. To see these generated n-grams as a part of the word list, you can click on the WordList view or the Example view. Click on the Example view, as in Figure 15.21, and you will see several of these n-grams. For example, besides the word absolut, the n-gram words absolut_love, absolut_perfect, and absolut_stai are part of the WordList, and appear as attributes in the ExampleSet (Figure 15.21). Now, to see the data itself, click on Data View radio button as in Figure 15.22. In this grid, each number tells the relative frequency of a word in a document. For example, the word absolut (the words that have absolut as their stem) constitutes 0.030 proportion (3%) of all the words in document number 13, which contains the reviews for hotel_73943.

The main contribution of Process02 is the cluster model, which clusters documents, and thus the hotels, according to the similarity of the frequency distribution of the words contained in their TripAdvisor reviews. Now click on Cluster Model (clustering) view to view the cluster model results. Then, to view the content of each of the clusters, click the Folder View radio button. In our example the first cluster of hotels is cluster_0, which contains 28 hotels, including hotels with row numbers 47, 51, 55, 95, 100.

How do these clusters differ from each other? To get an answer to this question, click the Centroid Table radio button (Figure 15.23). In this Centroid table we can observe the average frequency of each word in the documents of each cluster. For example, the word absolut appears much more frequently in cluster_3, compared to the other clusters: In cluster_3, the average frequency for absolut is 0.013, whereas it is at most 0.008 in the others.

Finally, save the results of Process02, just as you did for Process01.

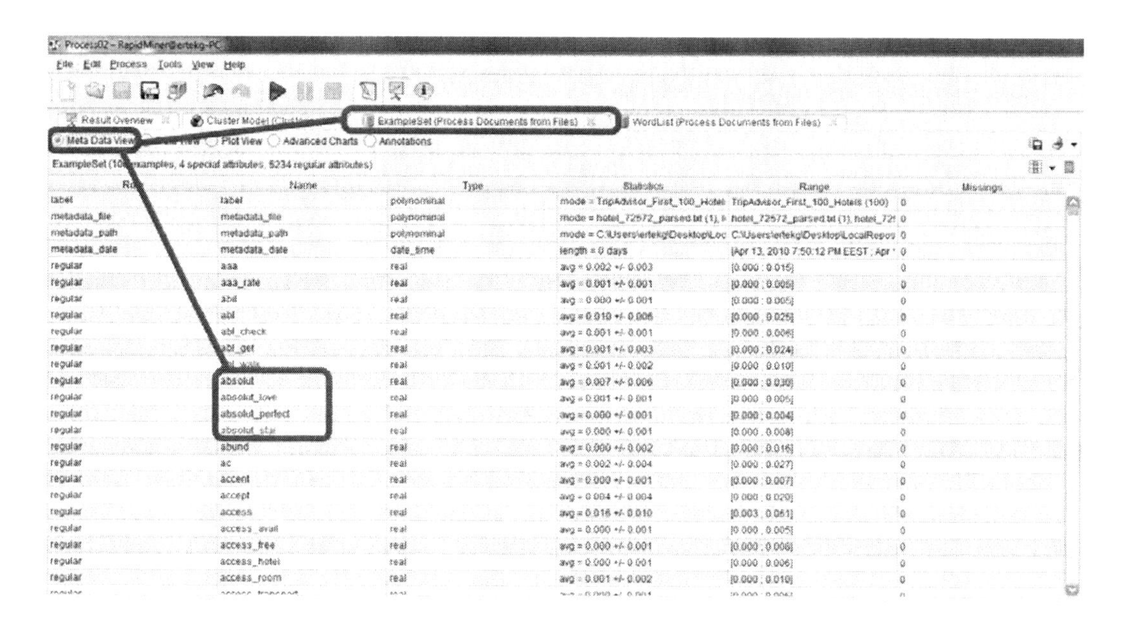

FIGURE 15.21: Meta Data View for the ExampleSet generated by Process02, including the n-Grams.

File Edit Process Tools View Help

Result Overview | Cluster Model (Clustering) | ExampleSet (Process Documents from Files) | WordList (Process Documents from Files)

Meta Data View | Data View | Plot View | Advanced Charts | Annotations

ExampleSet (100 examples, 4 special attributes, 5234 regular attributes) View Filter (100 / 100): all

Row No.	label	metadata_file	metadata_p	metadata_d	aaa	aaa_rate	abil	abl	abl_check	abl_get	abl_walk	absolut	absolut_love	absolut_per	absolut_star	abund
1	TripAdvisor_	hotel_72572	C:\Users\ert	Apr 13, 2010	0.003	0	0	0.008	0	0	0	0.00	0	0	0	0
2	TripAdvisor_	hotel_72579	C:\Users\ert	Apr 13, 2010	0	0	0	0.025	0	0	0.010	0	0	0	0	0
3	TripAdvisor_	hotel_72586	C:\Users\ert	Apr 13, 2010	0.004	0.004	0	0	0	0	0	0.00	0.004	0	0	0
4	TripAdvisor_	hotel_73718	C:\Users\ert	Apr 13, 2010	0	0	0	0.007	0	0	0	0	0	0	0	0
5	TripAdvisor_	hotel_73727	C:\Users\ert	Apr 13, 2010	0	0	0	0.017	0	0.008	0	0.00	0	0	0	0
6	TripAdvisor_	hotel_73739	C:\Users\ert	Apr 13, 2010	0.008	0	0	0.008	0	0	0	0	0	0	0	0.016
7	TripAdvisor_	hotel_73757	C:\Users\ert	Apr 13, 2010	0	0	0	0	0	0	0	0.00	0	0	0	0
8	TripAdvisor_	hotel_73787	C:\Users\ert	Apr 13, 2010	0	0	0	0	0	0	0	0.00	0	0	0	0
9	TripAdvisor_	hotel_73799	C:\Users\ert	Apr 13, 2010	0	0	0	0.005	0	0	0	0	0	0	0	0
10	TripAdvisor_	hotel_73821	C:\Users\ert	Apr 13, 2010	0	0	0	0.005	0	0	0	0.00	0	0	0.005	0
11	TripAdvisor_	hotel_73855	C:\Users\ert	Apr 13, 2010	0.003	0	0	0.029	0.006	0	0.006	0.00	0	0	0	0
12	TripAdvisor_	hotel_73923	C:\Users\ert	Apr 13, 2010	0	0	0.005	0.005	0	0	0	0	0	0	0	0
13	[illegible]	hotel_7394	[illegible]	[illegible]	[illegible]	[illegible]	[illegible]	[illegible]	[illegible]	[illegible]	[illegible]	0.030	0.004	0.001	0	0
14	TripAdvisor_	hotel_73947	C:\Users\ert	Apr 13, 2010	0	0	0	0	0	0	0	0	0	0	0	0
15	TripAdvisor_	hotel_73950	C:\Users\ert	Apr 13, 2010	0	0	0	0.016	0	0	0	0.008	0	0	0	0
16	TripAdvisor_	hotel_73957	C:\Users\ert	Apr 13, 2010	0.002	0.001	0	0.018	0	0.002	0	0.009	0.001	0	0.001	0.001
17	TripAdvisor_	hotel_73985	C:\Users\ert	Apr 13, 2010	0	0	0	0.024	0	0.024	0	0.024	0	0	0	0
18	TripAdvisor_	hotel_74051	C:\Users\ert	Apr 13, 2010	0	0	0	0	0	0	0	0.013	0	0	0	0
19	TripAdvisor_	hotel_74583	C:\Users\ert	Apr 13, 2010	0.014	0	0	0	0	0	0	0.014	0	0	0	0
20	TripAdvisor_	hotel_74626	C:\Users\ert	Apr 13, 2010	0	0	0.002	0.016	0.002	0	0	0.016	0	0.002	0	0
21	TripAdvisor_	hotel_75662	C:\Users\ert	Apr 13, 2010	0	0	0	0.012	0	0	0	0.006	0	0	0	0
22	TripAdvisor_	hotel_76061	C:\Users\ert	Apr 13, 2010	0.001	0.001	0	0.006	0	0	0	0.003	0	0	0	0
23	TripAdvisor_	hotel_76083	C:\Users\ert	Apr 13, 2010	0	0	0	0.004	0	0	0	0.006	0.004	0	0	0
24	TripAdvisor_	hotel_76442	C:\Users\ert	Apr 13, 2010	0.003	0.002	0.001	0.015	0.006	0.003	0.001	0.007	0	0	0.001	0

FIGURE 15.22: Data View for the ExampleSet generated by Process02, and the relative concurrence frequency of the word absolut in `hotel_73943.txt`.

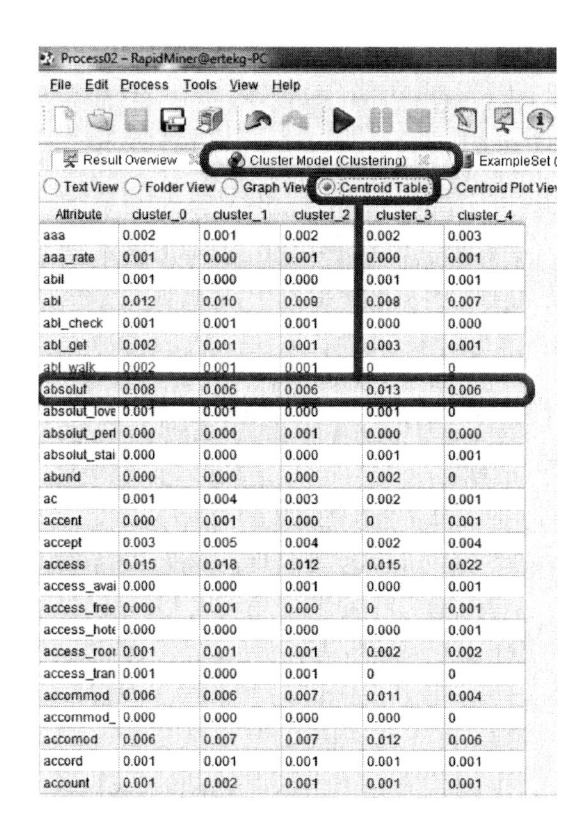

Attribute	cluster_0	cluster_1	cluster_2	cluster_3	cluster_4
aaa	0.002	0.001	0.002	0.002	0.003
aaa_rate	0.001	0.000	0.001	0.000	0.001
abil	0.001	0.000	0.000	0.001	0.001
abl	0.012	0.010	0.009	0.008	0.007
abl_check	0.001	0.001	0.001	0.000	0.000
abl_get	0.002	0.001	0.001	0.003	0.001
abl_walk	0.002	0.001	0.001	0	0
absolut	0.008	0.006	0.006	0.013	0.006
absolut_love	0.001	0.001	0.000	0.001	0
absolut_perf	0.000	0.000	0.001	0.000	0.000
absolut_stai	0.000	0.000	0.000	0.001	0.001
abund	0.000	0.000	0.000	0.002	0
ac	0.001	0.004	0.003	0.002	0.001
accent	0.000	0.001	0.000	0	0.001
accept	0.003	0.005	0.004	0.002	0.004
access	0.015	0.018	0.012	0.015	0.022
access_avai	0.000	0.000	0.001	0.000	0.001
access_free	0.000	0.001	0.000	0	0.001
access_hote	0.000	0.000	0.000	0.000	0.001
access_roor	0.001	0.001	0.001	0.002	0.002
access_tran	0.001	0.000	0.001	0	0
accommod	0.006	0.006	0.007	0.011	0.004
accommod_	0.000	0.000	0.000	0.000	0
accomod	0.006	0.007	0.007	0.012	0.006
accord	0.001	0.001	0.001	0.001	0.001
account	0.001	0.002	0.001	0.001	0.001

FIGURE 15.23: Centroid Table for the Cluster Model generated by Process02, displaying the average frequency of each word in each cluster.

15.6 Conclusions

In this chapter, we have shown the basic RapidMiner operators for text mining, as well as how they can be used in conjunction with other operators that implement popular conventional data mining techniques. Throughout the chapter, we have discussed two processes, Process01 and Process02, which complement text processing with association mining and cluster modeling, respectively. We have shown how these text mining processes, which are formed as combinations of text processing and data mining techniques, are modeled and run in RapidMiner, and we discussed how their results are analyzed. RapidMiner has an extensive set of operators available for text processing and text mining, which can also be used for extracting data from the web, and performing a multitude of other types of analysis. While most of these operators were not discussed in this chapter due to space limitations, the reader is strongly encouraged to explore and experiment other text processing and mining operators and capabilities within RapidMiner, its official extensions, and the processes posted by the RapidMiner user community.

Acknowledgment

This work is financially supported by the UbiPOL Project within the EU Seventh Framework Programme, under grant FP7-ICT 248010. The authors thank Assoc. Prof. Yücel Saygın for supporting and helping in the writing of this chapter. The authors also acknowledge the earlier help of Sabancı University alumni Yücel Balıklılı and Emir Balıkçı with the design of the first process in this chapter.

Part VI

Feature Selection and Classification in Astroparticle Physics and in Medical Domains

Chapter 16

Application of RapidMiner in Neutrino Astronomy

Tim Ruhe
TU Dortmund, Dortmund, Germany

Katharina Morik
TU Dortmund, Dortmund, Germany

Wolfgang Rhode
TU Dortmund, Dortmund, Germany

16.1 Protons, Photons, and Neutrinos

Ever since the groundbreaking discovery of cosmic radiation by Victor Hess in 1912 [1], one question has been driving the field of astroparticle physics: How does nature manage to accelerate particles up to energies, several orders of magnitude larger, than can be achieved in man-made accelerators (e.g., the Large Hadron Collider at CERN)? Even today, more than 100 years after Hess's world-famous balloon flights, this question remains unanswered. Nowadays, however, several experiments around the globe are dedicated to answering this question by studying the different components of cosmic radiation [2].

Cosmic rays impinging on the Earth's atmosphere consist mainly of protons (90%) and helium nuclei (9%). The rest are heavier nuclei, electrons, and photons. Although protons are very abundant, they cannot be observed directly for energies $\gg 1$ TeV. When entering

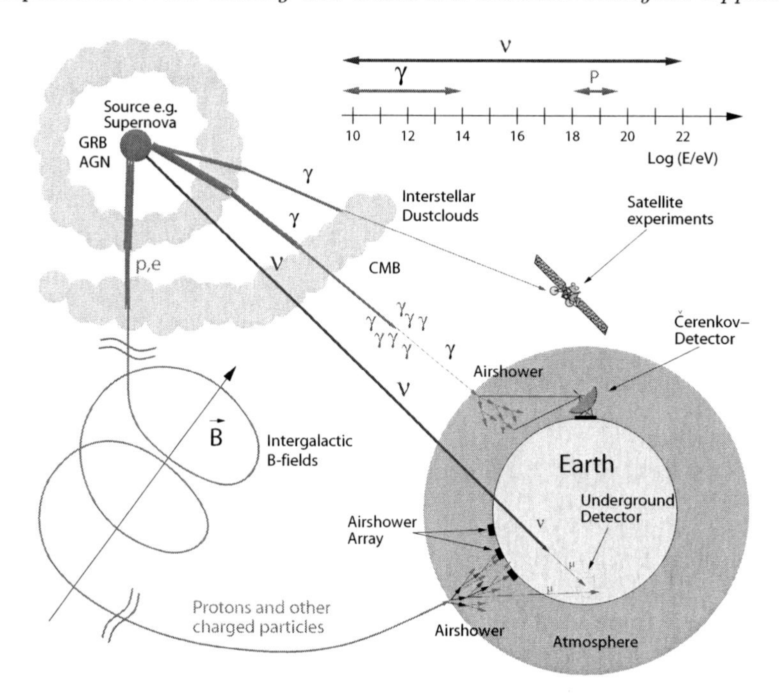

FIGURE 16.1: Overview over the wide field of astroparticle physics: Several types of messenger particles are observed using different experimental techniques. Protons are observed indirectly by studying air showers induced by their interaction with nuclei in the Earth's atmosphere. Due to their electrical charge, however, their trajectories are bent by interstellar and intergalactic magnetic fields. High-energy photons can be observed in satellite experiments and by so-called Cherenkov telescopes. They offer the advantage of unbent trajectories, but their observation is limited in energy, due to interaction with the Cosmic Microwave Background (CMB). Neutrinos, like high-energy photons, directly point back to their sources. Their observation is also not limited in energy. Due, to their small interaction probability, however, large underground detectors are required for their observation.

the atmosphere they interact with nitrogen and oxygen nuclei in air, producing a whole cascade of interactions. Such a cascade can contain thousands of particles and is, in general, observed by an array of surface detectors or in balloon experiments [3]. Moreover, protons get deflected by interstellar and intergalactic magnetic fields due to their electrical charge. Thus, upon the point of observation, the proton does not contain any directional information (see Figure 16.1). This means that despite the fact that the proton itself and even its energy can be measured on Earth, it cannot be assigned to come from a particular cosmic source. The same argument can, of course, be made for all charged particles.

For photons, the situation is somewhat different. As they do not carry any electrical charge, they can traverse cosmic distances without being deflected by magnetic fields. Thus, upon detection they can be clearly allocated to certain sources in the sky. The detection of photons, however, suffers from the fact that the atmosphere is only transparent for photons of certain wavelengths, e.g., optical light.

One option avoiding the influence of the atmosphere is the use of satellite-bound experiments. Several satellite experiments have been carried out with great success and are still under operation. The downside to this kind of experiment is the large cost and the fact that the detectors are limited in size, due to weight restrictions.

The second option for observing high-energy photons is Earth bound and similar to the detection of protons. Like protons, photons interact with nuclei in the Earth's atmosphere producing a particle shower. The shape of the shower, however, is different. Images of the showers can be taken by so-called Cherenkov telescopes. Nowadays, several Cherenkov telescopes are in operation, including MAGIC [4], HESS [5], and FACT [6].

Although today high-energy gamma particles are routinely detected, there is one challenge associated with propagation. When high-energy photons exceed a certain energy threshold they begin to interact with the Cosmic Microwave Background (CMB). The CMB consists of relic photons from the big bang, having a particular spectral shape and temperature (details on the discovery of the CMB can be found in Chapter 3 of [2]). Thus, the observation of photons is limited in energy. Moreover, the detection of photons from cosmic sources does not enable the observer to distinguish between certain models of acceleration.

Probably the most interesting component of cosmic radiation, which has not been discovered so far, are neutrinos. Atmospheric neutrinos are produced in extended air showers where cosmic rayinteract with nuclei of the Earth's atmosphere. Within these interactions, mainly pions and kaons are produced, which then subsequently decay into muons and neutrinos. Like photons, neutrinos do not carry any electrical charge, which means they are undeflected by magnetic fields and, upon detection, directly point back to their sources. Moreover, the interaction probability of a neutrino with matter is very small, compared to other particles, e.g., photons. Thus, neutrinos can travel cosmic distances without being scattered, deflected, or absorbed. In addition, the observation of photons and neutrinos from the same source could help to increase our understanding of cosmic acceleration processes.

Neutrinos can only be detected when interacting in or close to a detector. So, their small interaction probability, which helps to get them from a source to an observer, is a huge disadvantage, at the same time. This disadvantage has to be compensated by using larger detectors. Large detectors require large volumes of detection media. Again, the detection is based on the bluish light called Cherenkov light. This means that large-scale neutrino telescopes require detection media that are both cheap and transparent. In principal, this leaves two choices: deep seawater and South Pole ice.

16.2 Neutrino Astronomy

Today, the world's largest neutrino telescope, IceCube, is located at the geographic South Pole utilizing the Antarctic ice cap as a detection medium. The detector was completed in December 2010 and consists of 86 vertical cables called strings. Each string holds as many as 60 digital optical light sensors called DOMs. The light sensors are buried at depths between 1450 m and 2450 m, thus instrumenting a total volume of 1 km^3 [7, 8].

One of the main scientific goals of neutrino telescopes is the detection of neutrinos originating from astrophysical sources, as well as a precise measurement of the energy spectrum of neutrinos produced in cosmic ray air showers in the Earth's atmosphere. These so-called atmospheric neutrinos, however, are hidden in a noisy background of atmospheric muons produced in air showers as well. The first task in rejecting this background is the selection of upward-going tracks since the Earth is opaque to muons but can be traversed by neutrinos up to very high energies.

This procedure reduces the background by roughly three orders of magnitude. For a detailed analysis of atmospheric neutrinos, however, a very clean sample with a purity larger than 95% is required. The main source of remaining background at this stage are muon tracks, falsely reconstructed as upward going. These falsely reconstructed muon tracks

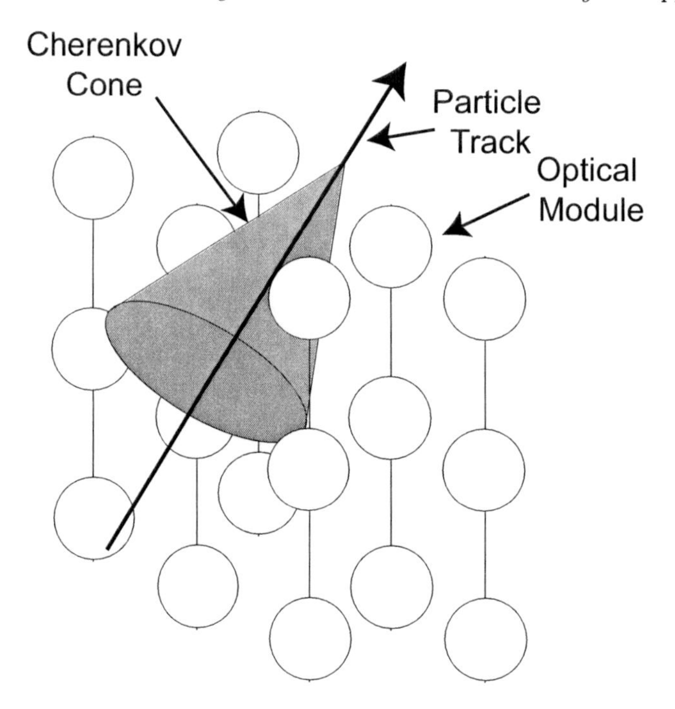

FIGURE 16.2: Schematic view of the detection principle. A cone of Cherenkov light is emitted along the track of the particle. Track parameters, e.g., the angle of the incident neutrino, can be reconstructed from the light pattern recorded by the optical modules.

still dominate the signal by three orders of magnitude and have to be rejected by the use of straight cuts or multivariate methods.

Due to the ratio of noise (muons) and signal (neutrinos), about 10,000 particles need to be recorded in order to catch about 10 neutrinos. Hence, the amount of data delivered by these experiments is enormous and it must be processed and analyzed within a proper amount of time. Moreover, data in these experiments are delivered in a format that contains more than 2000 attributes originating from various reconstruction algorithms. Most of these attributes have been reconstructed from only a few physical quantities. The direction of a neutrino event penetrating the detector at a certain angle can, for example, be reconstructed from a pattern of light that is initiated by particles produced by an interaction of the neutrino close to or even in the detector.

Due to the fact that all of the 2000 reconstructed attributes are not equally well suited for classification, the first task in applying data mining techniques in neutrino astronomy lies in finding a good and reliable representation of the dataset in fewer dimensions. This is a task which very often—possibly almost always—determines the quality of the overall data analysis. The second task is the training and evaluation of a stable learning algorithm with a very high performance in order to separate signal and background events. Here, the challenge lies in the biased distribution of many more background (negative) examples than there are signals (positive) examples. Handling such skewed distributions is necessary in many real-world problems.

The application of RapidMiner in neutrino astronomy models the separation of neutrinos from background as a two-step process, accordingly. In this chapter, the feature or attribute selection is explained in the first part and the training of selecting relevant events from the mass of incoming data is explained in the second part.

16.3 Feature Selection

One of the main challenges for the application of data mining techniques in neutrino astronomy lies in the large number of reconstructed attributes available at the analysis level. Reconstruction algorithms range from a simple line fit to more sophisticated methods such as likelihood reconstructions of the track in the detector. At this stage, one could, of course, utilize knowledge about the detector itself and the physics involved to find a good representation of the data in lower dimensions. This would, without doubt, result in a very good representation and has already been used in several analyses (see for example [9]). However, it will not necessarily result in the *best* representation of data. In order to find better representations at lower dimensions, feature selection algorithms can be applied.

Due to the many reconstruction algorithms, many parameters of the example set are either redundant or highly correlated. Even if all of these features strongly indicate the separation of signal and background, they all carry either the same or not much additional information. A method taking into account not only the relevance of an attribute but also its redundancy with respect to all features already selected in previous iterations is Maximum Relevance Minimum Redundancy, or MRMR for short [10].

MRMR is not part of the standard RapidMiner installation, but it comes with a neat extension package for feature selection. The installation, together with the use of this extension, will be described in the following sections.

16.3.1 Installation of the Feature Selection Extension for RapidMiner

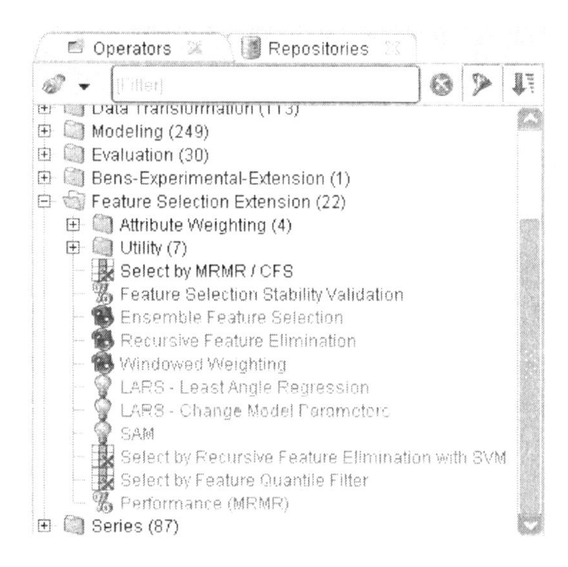

FIGURE 16.3: View of the Feature Selection Extension after a successful installation.

The feature selection extension offers a large variety of methods that validate feature sets independent of a learning task or with respect to a learning task [11].

Installation on Linux and Mac

- Download the RapidMiner-Feature-Selection extension via: `http://sourceforge.net/projects/rm-featselext`

- Copy the RapidMiner-Feature-Selection-extension to the plug-in folder of your copy of RapidMiner by opening a shell and typing:

  `cp rapidminer-Feature-Selection-extension-version.jayour-path-to-rapidminerlibplug-ins`

- The start or restart of RapidMiner, respectively, will then load the plug-in and make all its operators available from the list of operators (see Figure 16.3).

Installation on Windows

The installation of a plug-in on Windows is a little more difficult. The challenge lies in finding the location of the RapidMiner folder. All remaining individual steps are similar to an installation on Mac and Linux.

- Download the RapidMiner-Feature-Selection extension via: `http://sourceforge.net/projects/rm-featselext`.

- Locate the RapidMiner installation. Most likely it can be found in: `/programs/Rapid-I/RapidMiner/`.

  Utilizing the Windows search might be a helpful option if the folder is located elsewhere.

- Inside the RapidMiner folder, one must open the *lib* folder and from there the *plug-ins* folder. The jar file can now be copied to the *plug-ins* folder by a simple drag-and-drop operation.

- The start or re-start of RapidMiner will then load the plug-in and make all its operators available from the list of operators (see Figure 16.3).

16.3.2 Feature Selection Setup

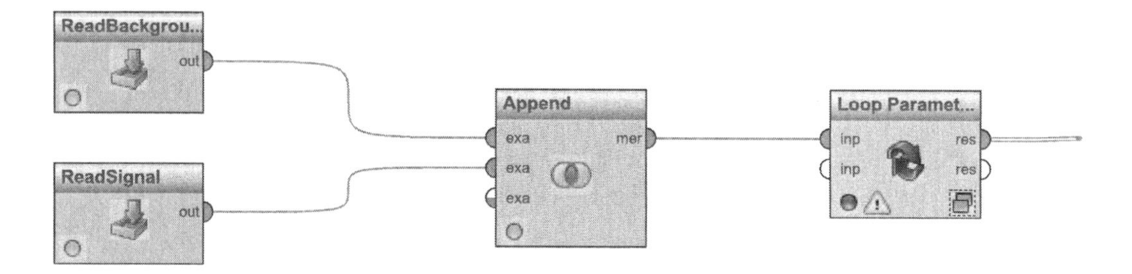

FIGURE 16.4: View of the Feature Selection Setup.

The Feature Selection setup starts with two simple READAML operators; one retrieving the signal simulation and the other retrieving the background simulation. Both operators have been re-named to READSIGNAL and READBACKGROUND, respectively, in order to organize the process more clearly. Operators, in general, may be re-named by selecting *Rename* after right-clicking a certain operator.

Using the READ AML operators, one encounters a variety of possible parameter selections with the most important ones being *attributes*, *sample ratio*, and *sample size*. The parameter *attributes* must contain the proper path to the location of the aml file to be used. The aml file solely contains meta-information on all the attributes such as attribute name and attribute type. The examples are stored in a dat file associated with the aml file. The dat file may be placed at a totally different location than the aml file, once its path is encoded. Note, that an incorrect path to the dat file is one of the most common reasons for processes to fail when transferring them from one workstation to another.

The *sample ratio* specifies which ratio of examples will be read from file. A sample ratio of 0.25, for example, will read in one quarter of the examples within the file. Using the parameter *sample size*, allows reading a specified number of examples. The default setting for this parameter is −1, which means that all examples are read. When both *sample size* and *sample ratio* are set, *sample ratio* will be over-ruled and exactly the number of examples as specified in *sample size* will be read in.

If the aml file has not yet been created, one can use the *Data Loading Wizard*, which will provide detailed guidance in creating an aml file from a simple ascii or csv file, including more advanced options like settings for attribute types.

The outputs of the READSIGNAL and READBACKGROUND operators are connected to an APPEND operator that merges both datasets. Using this particular operator, one has to ensure that both example sets contain the exact same attributes in the exact same order. Otherwise, the operator might not be able to append them. Although the APPEND operator only has two input ports, one can easily append more than two datasets. A new input port will appear automatically once all existing ports have been used.

Now, the data of background and neutrino signals have been retrieved together with their meta-data. The next step is to work on the attributes in order to obtain the appropriate set of features for learning. The feature selection takes place within a loop. We have connected the output of the APPEND operator to the input of the LOOP PARAMETERS operator. This operator will loop over selected parameters of its inner operators. The setup of the inner operators as well as the settings for the LOOP operator, are described in the following.

16.3.3 Inner Process of the Loop Parameters Operator

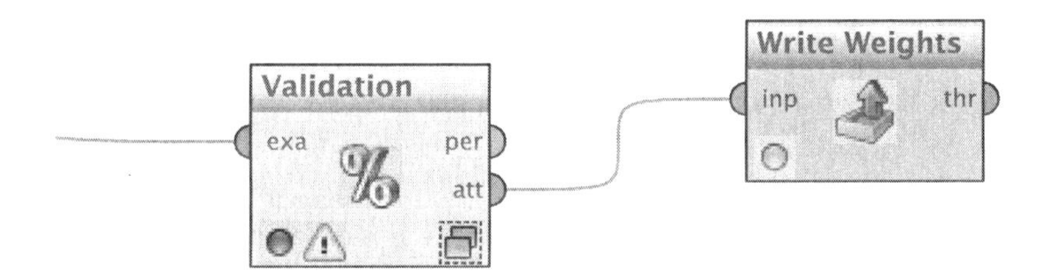

FIGURE 16.5: Inner process of the LOOP PARAMETERS operator.

Entering the inner process of the LOOP PARAMETERS operator, or any other operator for that matter, is simply done by double-clicking on the two overlapping squares in the lower right corner of the operator icon. The inner process for LOOP PARAMETERS in our case is rather simple, since it consists of only two operators, the first one being a WRAPPER X-VALIDATION and the second one being WRITE WEIGHTS.

The WRITE WEIGHTS operator stores the weights assigned to attributes in any preceding weighting process in a pre-defined file ending in wgt. Selecting an attribute means to assign the weight 1; dropping an attribute is expressed by its weight being zero. The location of the file can be selected using the option *attribute weights file*. For further simplification, this operator only has one input port that is connected to the lower output port of the WRAPPER X-VALIDATION (see Figure 16.5) as this is the one delivering the attributes and their weights. The output port of WRITE WEIGHTS outputs the attributes and weights without any further alteration.

The WRAPPER X-VALIDATION contains several inner operators. They will select an adequate and stable subset of features. Their usage and user options for the WRAPPER X-VALIDATION are explained in the next section.

16.3.4 Inner Operators of the Wrapper X-Validation

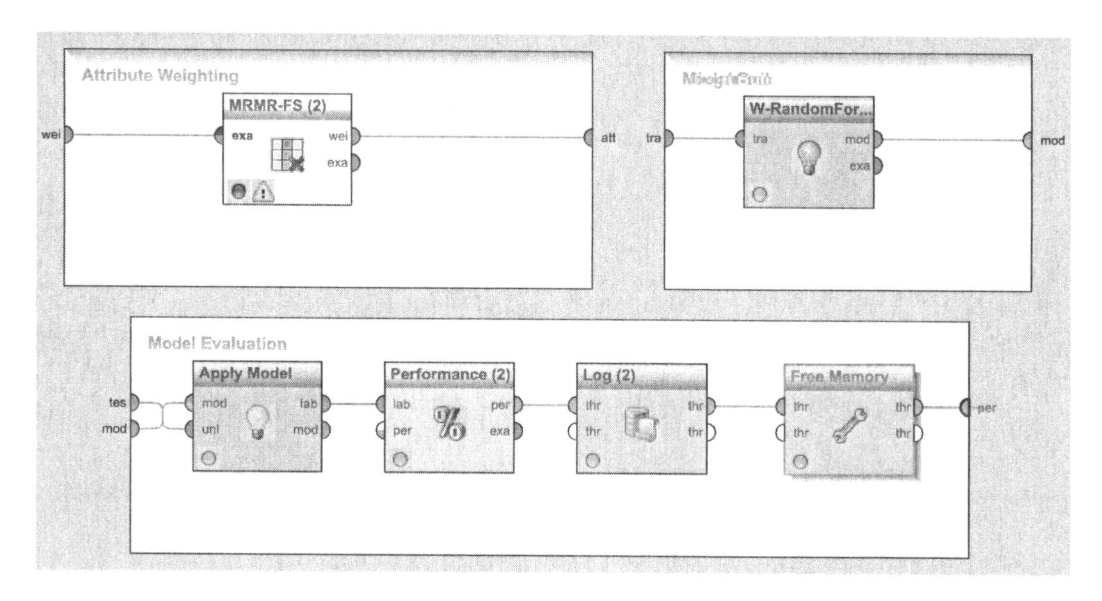

FIGURE 16.6: Inner process of the WRAPPER X-VALIDATION. This validation operator consists of three panels: Attributes Weighting, Training, and Testing.

In contrast to a regular cross validation which only contains a training and a testing process, the WRAPPER X-VALIDATION consists of an additional inner process dedicated to attribute weighting. Using a WRAPPER X-VALIDATION, one is capable of directly studying the influence of a certain attribute weighting algorithm on the stability and performance of any learner. The WRAPPER X-VALIDATION offers the advantage of optimizing attribute weighting and learning in one single process. It is a wrapper approach where the attribute selection is wrapped around a learner. A wrapper assesses subsets of features against the quality of a learned model. Since the attribute weighting, the model-building using the attributes with weights equal 1, and the model testing is done in a cross validation style, the feature selection gives a statistically valid and possibly stable result.

The *attribute weighting* in our case is performed using a Maximum Relevance Minimum Redundancy (MRMR) Feature Selection [10]. Within an MRMR feature selection, attributes are added iteratively (*forward selection*) according to a quality criterion Q that is given either as:

$$Q(x) = R(x,y) - \frac{1}{j}\sum_{x \in F_j} D(x,\hat{x}) \tag{16.1}$$

or

$$Q(x) = \frac{R(x,y)}{\frac{1}{j}\sum_{\hat{x} \in F_j} D(x,\hat{x})}. \tag{16.2}$$

In both cases, y represents the target variable or label, and $\hat{x} \in F_j$ is an attribute selected in one of the previous iterations. F_j depicts the set of all selected attributes. $R(x, y)$ indicates the relevance of x for the class y. By introducing the redundancy of two variables $D(x, \hat{x})$ as a penalty term $\frac{1}{j} \sum_{\hat{x} \in F_j} D(x, \hat{x})$, MRMR aims at excluding attributes that demonstrate high separation power, but do not contain any additional information.

In neutrino astronomy, for example, the zenith angle of the reconstructed track is a parameter which shows high separation power. However, as several advanced reconstruction algorithms are run, several zenith angles with little or no difference are being reconstructed. Using MRMR will then prevent the construction of a feature set that consists of 10 different zenith angles. This example may be somewhat constructed but does show the advantages of an MRMR feature selection for applications where diverse features have been created independently.

The MRMR feature selection in RapidMiner is part of the FEATURE SELECTION EXTENSION FOR RAPIDMINER [12]. The operator itself goes by the name SELECT BY MRMR/CFS and is simply placed inside the attribute weighting panel of the WRAPPER X-VALIDATION by a drag-and-drop operation. The input port expecting an example set with one label attribute must then be connected to the one input port in the attribute weighting panel. The upper output port of the MRMR selection, which is the one delivering a set of weighted attributes, has to be connected to the output port of the overall weighting process (see Figure 16.6).

The *model-building* environment determines the relevance of attributes for classifying an example. It can be filled with any learning algorithm which handles binary target variables. If one simply wishes to test the setup or the performance of the selected feature set, something simple like Naive Bayes is a good choice. For the purpose of instruction, however, one should use the W-RANDOM FOREST as a learner. At present, it is not necessary to be concerned about the detailed options and selection possibilities of this particular operator as these things will be described further down in this section. In order to correctly include the W-RANDOM FOREST into the process, its training set input needs to be connected to the training set knob of the model-building environment. From the available outputs, the model output has to be connected to the output of the model-building environment (see Figure 16.6).

The *model evaluation* environment tests in cross validation manner the performance of the learned model. For the current problem, the process consists of three basic operators: APPLY MODEL, PERFORMANCE, and LOG (see Figure 16.6). Using the FREE MEMORY operator is optional.

Both input ports of the APPLY MODEL must be connected to the color code corresponding input ports of the evaluation panel. The APPLY MODEL output holding the labeled example set must then be connected to the input port of the PERFORMANCE operator, for further evaluation. As the case at hand is a two-class problem, one might be interested in true and false positives or true and false negatives, respectively. To ensure that these measures can be evaluated, one should use the operator PERFORMANCE (BINOMINAL CLASSIFICATION).

A LOG operator may be used to save the results of any RapidMiner process. Using a LOG operator, a path including the name of the file must be inserted in the *filename* panel. Clicking the *Edit list* button, one can edit the list of variables and parameters that are of interest for the process. After inserting a column name under which a certain value will be logged, two drop-down menus will allow the selection of an operator and the corresponding value or parameter of interest. The option *persistent* is another interesting and helpful feature that, if set to *true*, will output every new logfile entry to the RapidMiner result screen. Using this option enables the user to very closely follow the performance of the process.

Note: When using loop operators on very large example sets, one may, at times, run the risk of using too much memory, because too much unnecessary information is retained. To avoid this, the FREE MEMORY operator can be utilized. It will simply free unused memory and can prevent the process from crashing when large amounts of data are handled, for example, in a loop.

16.3.5 Settings of the Loop Parameters Operator

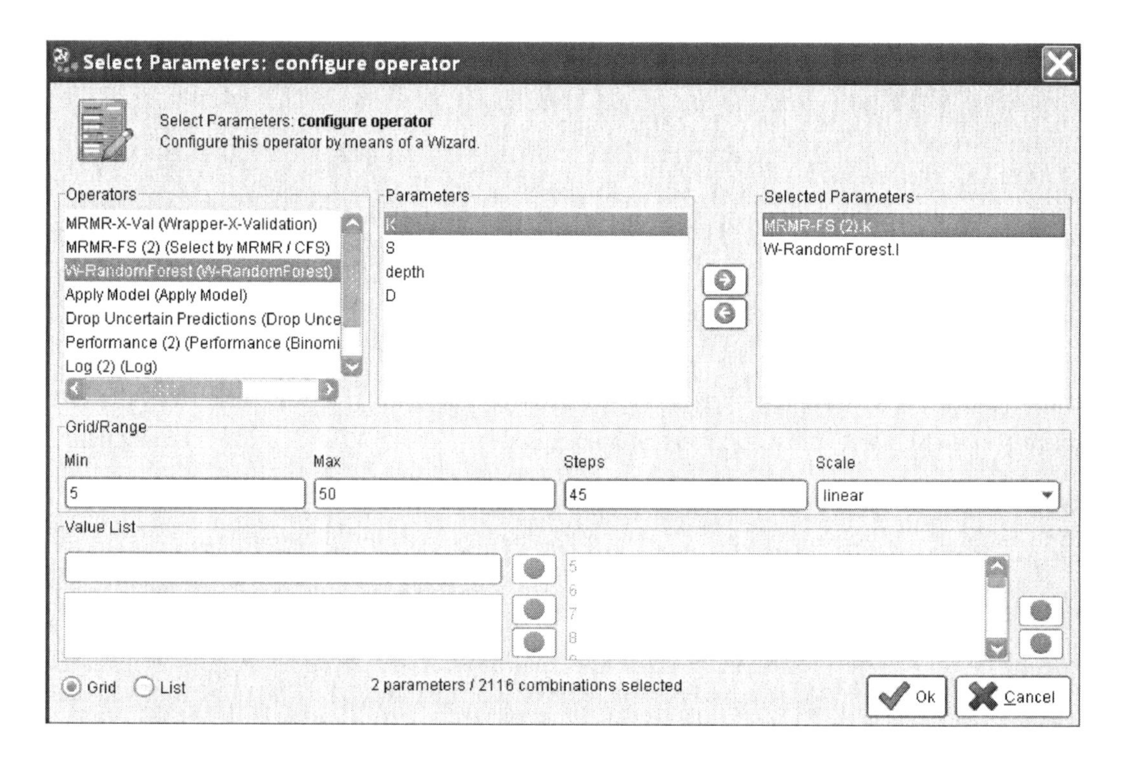

FIGURE 16.7: Editing the settings of the LOOP PARAMETERS operator.

The loop operator is general and can be tailored to diverse needs. Everything is a parameter in the general case: an operator as well as its parameters. Hence, the specifications of the general loop are parameter settings. To select the parameters for the loop, one must simply click the *Edit Parameter Settings* button. A window that will allow the configuration of the parameters will pop up (see Figure 16.7).

The upper part of this pop-up contains three panels: *Operators*, *Parameters*, and *Selected Parameters* from left to right. As a first step, one must choose the operator containing the parameter to loop. In Figure 16.7, this is the W-RANDOM FOREST. Once the operator has been selected, all its parameters (here: the parameters of W-RANDOM FOREST) are displayed in the middle panel. The specific parameters can now be selected or deselected using the two arrow buttons. Selected parameters (here, the feature selection and the learning operators) are displayed in the far right field.

Any selected parameter can now be configured individually by using the lower part of the window. The minimum value, the maximum value, and the number of steps can be selected. If one is interested in using nonlinear step sizes, *quadratic*, *logarithmic*, and *logarithmic(legacy)* are available from the *scale* drop-down menu.

All individual parameter settings that are run in successive order are now displayed in the lower right panel. One should check this field at least briefly to make sure one did not miscount the number of steps.

Occasionally, one might want to utilize user-defined steps that cannot be covered by a linear or a logarithmic scale, or one wants to loop over parameters encoded in string format. In this case, the *Value List* panel in the lower left part of the pop-up window has to be used. To enter a value in the corresponding field, *list* must be checked instead of *grid*. Any parameter value—string or numeric—must be entered manually. Then this parameter can be added to the list of steps using the corresponding arrow button (see Figure 16.8). If a certain parameter value needs to be deselected, this can be done using the arrow buttons (see Figure 16.8).

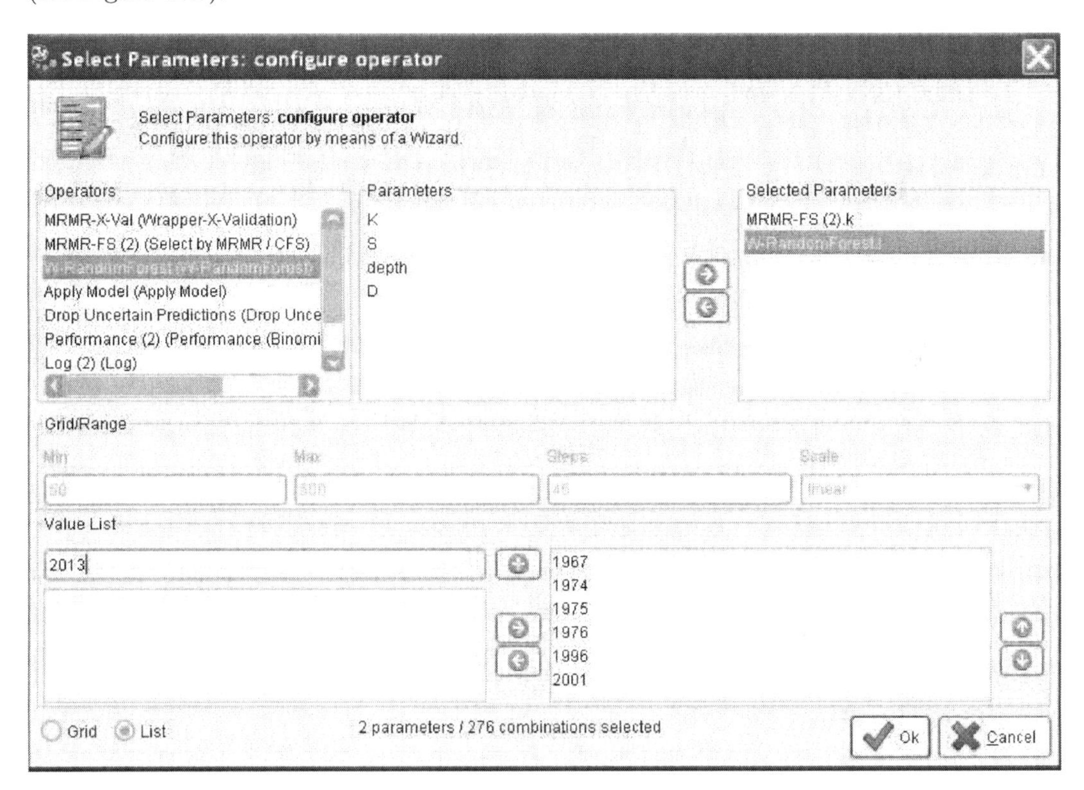

FIGURE 16.8: Looping over a list of values instead of a grid.

Finally, if two or more operators have the same number of steps in the LOOP operator, the looping can be synchronized by checking the corresponding option of the operator. This is useful, if one is only interested in certain pairs of parameters. For the problem at hand, this is the case, because the number of trees in the forest needs to be matched to the number of parameters selected by the MRMR Feature Selection. Synchronizing might thus save a valuable amount of time when running the process as no useless settings are iterated.

The settings for the parameter k of the *Select by MRMR/CFS* operator have to be chosen as follows: *Min=5*, *Max=50*, and *Steps=45*.

16.3.6 Feature Selection Stability

When running a feature selection algorithm, not only the selection of attributes itself is important, but also the stability of the selection has to be taken into account. The stability indicates how much the choice of a good attribute set is independent of the particular

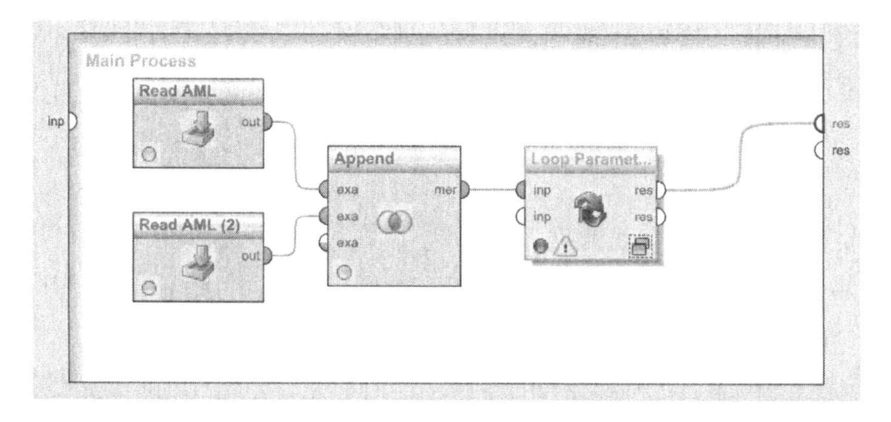

FIGURE 16.9: Setup for the evaluation of the Feature Selection Stability.

sample of examples. If the subsets of features chosen on the basis of different samples are very different, the choice is not stable. The difference of feature sets can be expressed by statistical indices.

Fortunately, an operator for the evaluation of the feature selection is also included in the FEATURE SELECTION EXTENSION FOR RAPIDMINER. The operator itself is named FEATURE SELECTION STABILITY VALIDATION.

This operator is somewhat similar to a usual cross validation. It performs an attribute weighting on a predefined number of subsets and outputs two stability measures. Detailed options as well as the stability measures will be explained later in this section.

In order to reliably estimate the stability of a feature selection, one should loop over the number of attributes selected in a specific algorithm. For the problem at hand, the process again commences with two READ AML operators that are appended to form a single set of examples. This single example set is then connected to the input port of a LOOP PARAMETERS operator. The settings of this operator are rather simple, and are depicted in Figure 16.10.

The FEATURE SELECTION STABILITY VALIDATION (FSSV) is placed inside the LOOP PARAMETERS operator accompanied by a simple LOG operator (see Figure 16.11). The two output ports of the FSSV are connected to the input ports of the LOG operators. A LOG operator stores any selected quantity. For the problem at hand, these are the Jaccard index [13] and Kuncheva's index [14]. The Jaccard index $S(F_a, F_b)$ computes the ratio of the intersection and the union of two feature subsets, F_a and F_b:

$$S(F_a, F_b) = \frac{|F_a \cap F_b|}{|F_a \cup F_b|}.$$

The settings for the LOG operator are depicted in Figure 16.12. It consists of two fields, the first one being *column name*. Entries can be added and removed using the *Add Entry* and *Remove Entry* buttons, respectively. The entry for *column name* can be basically anything. It is helpful to document the meaning of the logged values by a mnemonic name.

The second field offers a drop-down menu from which any operator of the process can be selected. Whether a certain value that is computed during the process or a process parameter shall be logged, is selected from the drop-down menu in the third panel. The fourth field offers the selection of output values or process parameters, respectively, for the selected operator.

An operator for attribute weighting is placed inside the FSSV. For the problem at hand,

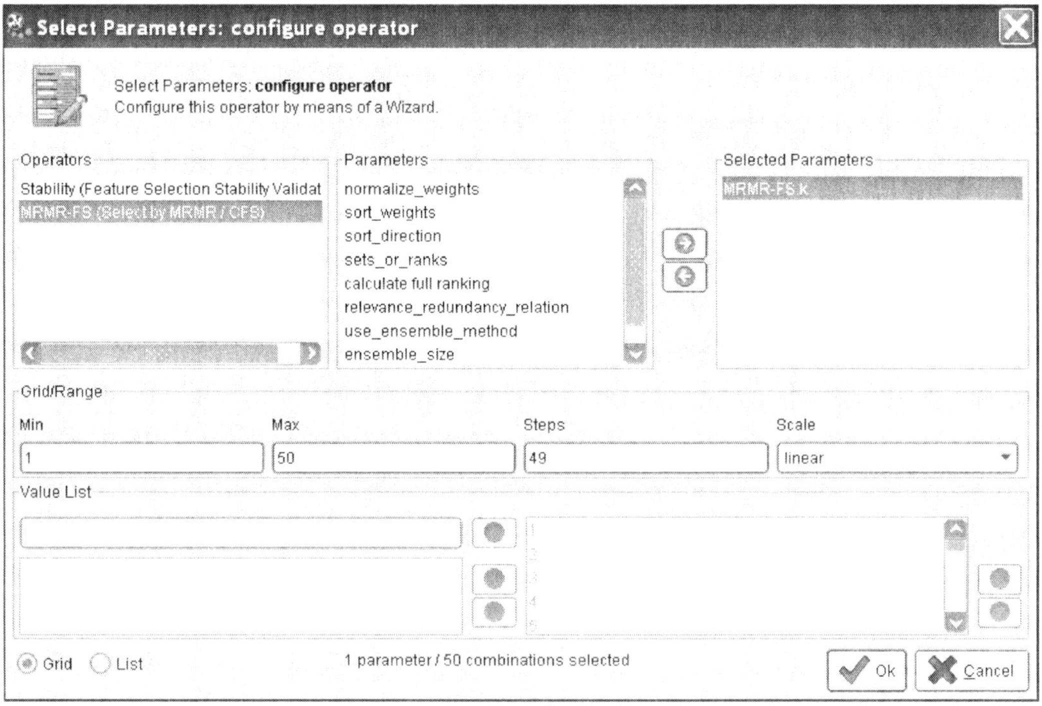

FIGURE 16.10: Settings of the LoopParameters operator used for the stability evaluation.

Select by MRMR/Cfs is used. However, any other feature selection algorithm can be used as well.

As can be seen, the process for selecting features in a statistically valid and stable manner, is quite complex. However, it is also very effective. Here, for a number of attributes between 30 and 40, both stability measures Jaccard and Kuncheva's index lie well above 0.9. Both indices reach the maximum of 1.0, if only one attribute is selected. This indicates that there is one single attribute for the separation of signal and background that is selected under all circumstances. Since other attributes also enhance the learning performance, about 30 more attributes are selected. This substantially decreases the original number of dimensions.

In the next section, we'll see how the selected features are used for modeling the event extractions from the observations.

16.4 Event Selection Using a Random Forest

The overall learning process is shown in Figure 16.14. The model-building is performed on the basis of data from Monte Carlo simulations. Hence, we can simulate diverse distributions of signals and background. The actual learning setup again starts out with two Read AML operators, which again have been renamed to ReadSignal and ReadBackground, respectively, and an Append operator. The resulting example set is then passed to the Generate Attributes operator. This operator can be used to create new attributes from the existing ones by using several mathematical operations. For example, one might think of:

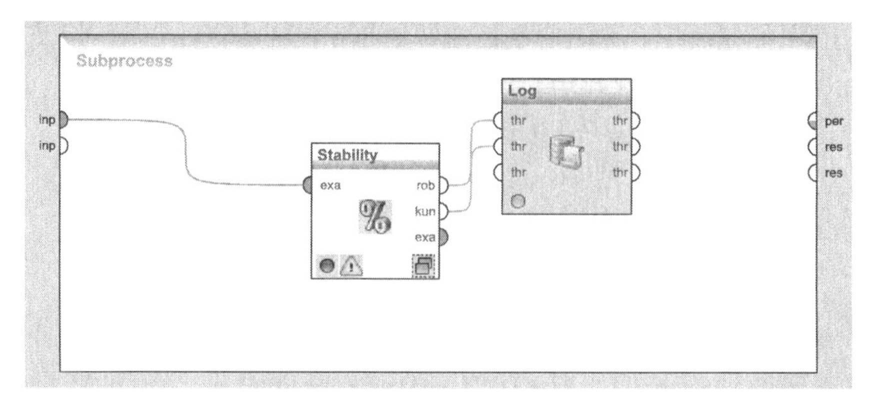

FIGURE 16.11: Inner process, FSSV, of the LOOP PARAMETERS operator.

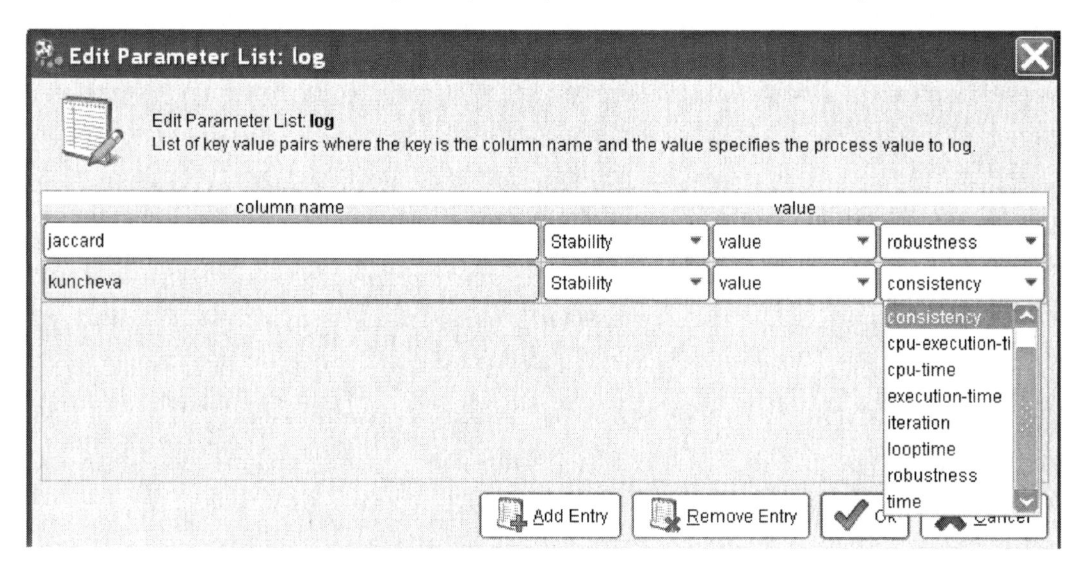

FIGURE 16.12: Settings of the LOG operator.

$$\texttt{attribute4} = \frac{\texttt{attribute1} + \texttt{attribute2}}{\texttt{attribute3}}. \tag{16.3}$$

The settings for the feature constructions at hand are depicted in Figure 16.15. From the GENERATE ATTRIBUTES operator, the example set is passed on to the REMAP BINOMINALS operator. This operator is used to define the positive and negative class of the example set. By not using this operator, RapidMiner will simply commence with an assumption about the positive and negative class, which in some cases might lead to unpleasant surprises when evaluating learning processes.

The settings of the REMAPBINOMINALS operator are depicted in Figure 16.16.

16.4.1 The Training Setup

The training setup starts out with a small work-around required in order to vary the ratio of signal and background events that enter the learner. This can, of course, be done by varying the number of events in the READ AML operators, but that is only the second-best

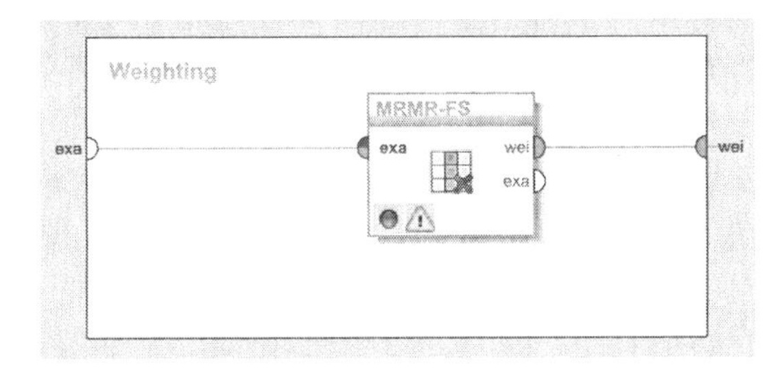

FIGURE 16.13: Weighting operator placed inside the FEATURE SELECTION STABILITY VALIDATION.

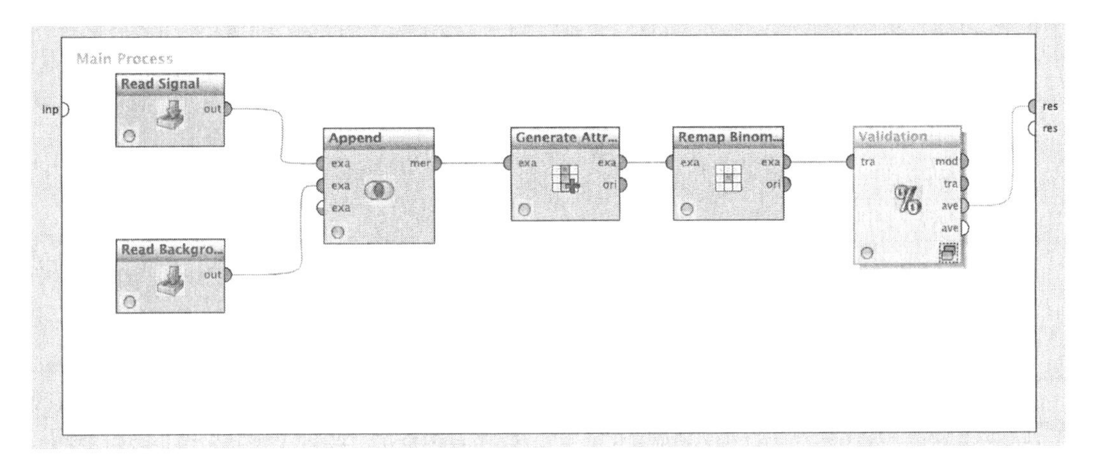

FIGURE 16.14: Setup of the complete learning process using a 5-fold cross validation.

solution, since as many examples of both classes as possible need to be kept for the testing process.

Since the X-VALIDATION operator contains only one input port, the signal and background sets must be appended before entering the cross validation and then be split up again. The splitting is done as follows: First, the example set is divided using the MULTIPLY operator. This operator simply takes the input and delivers it to its output ports, without alteration. Once an output port has been connected, a new one will pop up.

Two output ports of the MULTIPLY operator need to be connected to FILTER EXAMPLES operators, one for each class in the sample. In Figure 16.17 the operators were renamed to GET SIGNAL and GET BACKGROUND for better distinguishing between the two example inputs. The settings for the FILTER EXAMPLES operators are *condition class: attribute value filter* and *parameter string: label=1* or *parameter string: label=0*, respectively.

Both FILTER EXAMPLES operators then have to be connected to SAMPLE operators.

There are five different sampling operators available in RapidMiner: SAMPLE, SAMPLE (STRATIFIED), SAMPLE (BOOTSTRAPPING), SAMPLE (MODEL BASED), and SAMPLE (KENNARD-STONE). Here, the different sampling methods are not detailed, and we refer to [13] for further reading.

One can simply use the plain SAMPLE operator. However, should one wish to test very different numbers of signal and background examples, one might end up in a case where a certain over-sampling cannot be avoided. In this case, the SAMPLE (BOOTSTRAPPING) operator should be a good choice.

Since the number of examples is somewhat uneven in the example sets, *absolute sampling*

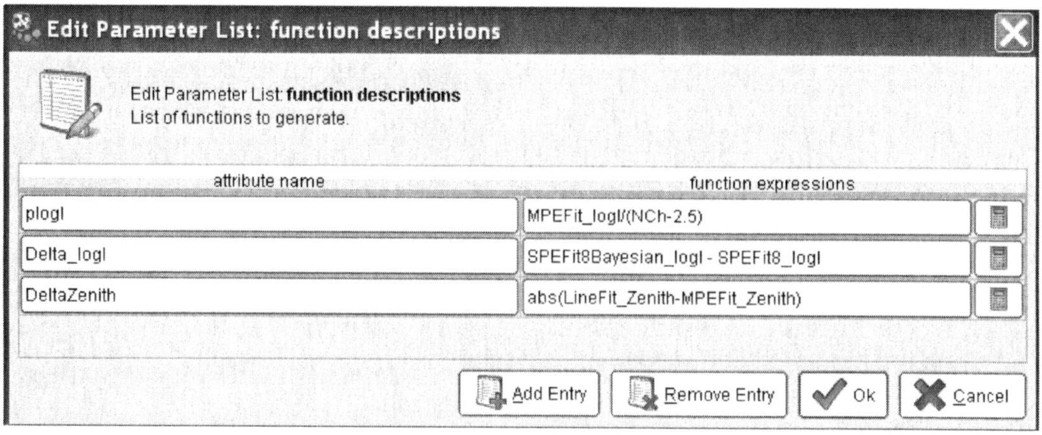

FIGURE 16.15: Settings for the GENERATE ATTRIBUTES operator.

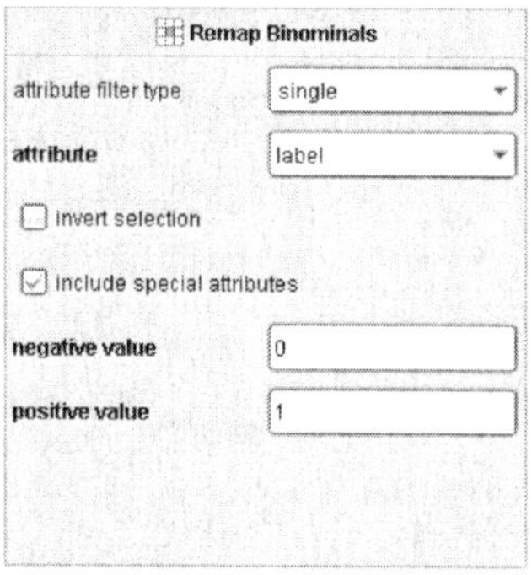

FIGURE 16.16: Settings for the REMAP BINOMINALS operator.

should be used. For relative sampling, a relative number needs to be entered. Entering 0.1, for example, would sample 10% of the examples from a specific example set. Using two SAMPLE operators allows us to fine-tune the ratio of signal and background examples manually.

At this stage, both example sets are again appended using the APPEND operator. From there on, the created example set is passed to the actual learner, which in our case is a Random Forest, which will be explained in greater detail in the following section.

16.4.2 The Random Forest in Greater Detail

A Random Forest is a multivariate method for classification and regression developed by Leo Breiman in 2001 [15]. It utilizes an ensemble of weak classifiers (decision trees) and the final classification is achieved by averaging over all individual classifications of the trees in the forest.

When a decision tree is grown, it starts with all training examples in one node. Then, a number of attributes are picked at random from the set of available attributes. To further grow the decision tree, a split has to be applied. Both the split attribute and the actual split value are determined in an optimization procedure. The criterion for optimization might differ between individual implementations. Once the split has been carried out, one gets new nodes, one for each value, that already show a better separation of classes.

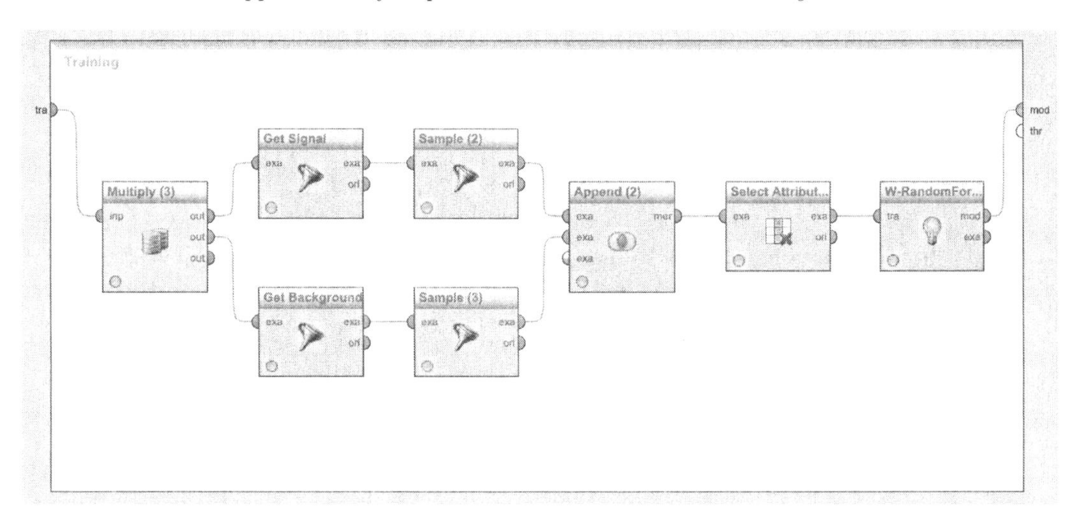

FIGURE 16.17: Setup of the training process inside the CROSS VALIDATION operator.

Further nodes are created due to further splits by repeating the procedure described above. The algorithm will stop when either a certain depth is reached or when a node is pure, i.e., it contains only observations of one class. Nodes where no further splitting takes place are called terminal nodes.

Labels are assigned to terminal nodes. Two variations are possible: the first is a simple majority vote that assigns the label of the majority class to all events in the node, and the second is an average that assigns the average label to all events in the terminal node. Using the average, however, might not be a good choice when using polynomial instead of binomial target classes.

In neutrino astronomy, only pure nodes are used and the label assigned to a terminal node is then the label of the class present in that very node.

For the actual classification, previously unseen examples are processed down each tree until they end up in a terminal node of the tree. The label previously assigned to the node is now assigned to the example. As one example might get assigned different labels in different trees, the final classification is achieved by averaging over all labels assigned to an example by all individual trees in the forest. The signalness s of an observation is determined by its label in all the trees normalized by the number of trees:

$$s(x) = \frac{1}{n_{trees}} \sum_{i=1}^{n_{trees}} s_i(x). \tag{16.4}$$

In neutrino astronomy, a simple classification by the Random Forest is not sufficient. In order to achieve the purity required for a detailed and reliable analysis, one must take into account the confidence of the positive class (confidence(1)). The actual value of this cut again depends strongly on the goals of the analysis and the analysis level one started out from.

16.4.3 The Random Forest Settings

There are two RANDOM FOREST operators in RapidMiner, the first one being simply RANDOM FOREST and the second one being part of the Weka extension implemented as W-RANDOM FOREST. Studies showed that for our problem, the W-RANDOM FOREST showed

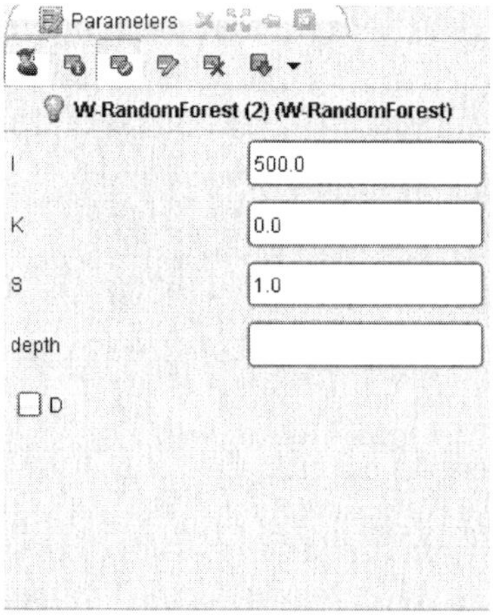

FIGURE 16.18: Settings for the W-RANDOM FOREST operator.

a somewhat more stable performance. As stability is crucial for classification problems in neutrino astronomy, this is the main motivation for using the W-RANDOM FOREST.

The W-RANDOM FOREST has four basic parameters, depicted in Figure 16.18. Probably the most important one is the number of trees encoded in the simple letter I. The number of trees has to be expressed by a real number between 1.0 and $+\infty$, although only natural numbers make sense. As performance and runtime vary strongly with the number of trees, this parameter should be carefully selected. In most cases, performance measures such as *precision* and *recall* go into saturation once the number of trees exceeds a certain threshold. In those cases, the improvement achieved by adding additional trees to the forest is very small. Therefore, one should think twice if the extra amount of required resources is justified by the small improvements. For the separation of signal and background in neutrino astronomy, 500 trees were found to be sufficient.

K represents the number of attributes that are considered for splitting at each split level. This again is a real number that should be smaller than the number of attributes present in the example set. In principle, it can also be equal to the number of attributes, but in this case, one ends up building more or less the same tree over and over again and all advantages of forests over simple decision trees vanish. The default value is $K = 0.0$, which means $log_2 M + 1$ attributes are considered for splitting with M being the total number of attributes available in the example set.

The third parameter, S, sets a local random seed for the forest. The default setting is $S = 1.0$, which means the Random Seed of the entire process is used. At times, it can be useful to set S to a different value if one wishes to deliberately grow exactly the same forest several times while changing the surrounding process.

The parameter *depth* represents the depth to which the trees in the forest are outgrown. Setting $depth = 10$, for example, would grow trees with 10 layers. The default setting is $depth = 0$, which grows trees of unlimited depth, meaning splitting will only stop if no further splits are possible.

Checking the parameter D will run the W-RANDOM FOREST in debug mode.

16.4.4 The Testing Setup

The testing setup starts out with an APPLY MODEL operator. Its input ports are connected to the corresponding knobs of the testing environment. The APPLY MODEL operator

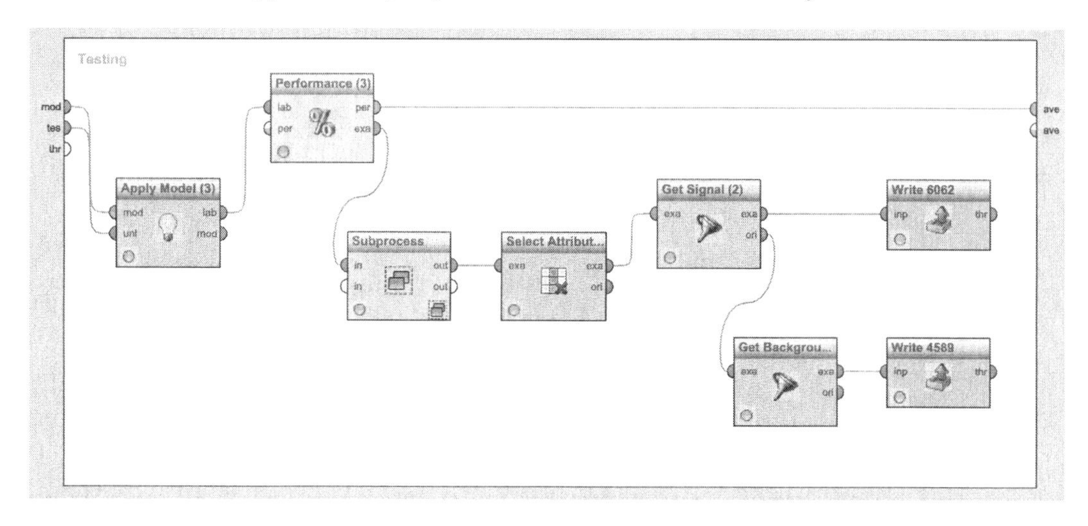

FIGURE 16.19: Overview over the complete testing setup.

now applies the model created within the training environment on an unseen set of examples. The size of the testing set will be the nth part of all examples that entered the cross validation, with n being the total number of iterations. A prediction is assigned to every example in the set, and the labeled set can be accessed from the *lab* port.

The labeled example set is passed on to an operator that will evaluate the performance of the learner. There are several operators available, the majority of which are very suited to specific purposes. For the problem at hand, the PERFORMANCE (BINOMINAL CLASSIFICATION) should be used, as this operator can deliver a couple of measures specific for a binary classification problem such as *true positives* and *false positives*. The variety of performance measures that can be accessed using this operator is rather large, and one may pick favorite ones. The ones that should be checked in an example process are: *accuracy*, *true positives*, and *false positives*. The *per* output port of the performance operator can now be connected to one of the output ports of the testing environment. At this point, one could actually stop. The performance of a specific learning setup can be examined using the result view. Doing so is probably fine, but the challenges in neutrino astronomy are somewhat higher so that the labeled example set should be stored for further evaluation using other analysis tools for high-energy physics such as ROOT [16]. As root files cannot (yet) be directly read and written from RapidMiner processes, one has to think of a work-around. One such work-around is presented in the following.

All valuable information that can be extracted from a RapidMiner process, such as *prediction* or *confidence* can be written to an ascii file. The iteration in which a certain example was tested inside the cross validation is also accessed and written to file. This ascii file can then be used to append all those values to the original root files. Prediction and confidences are already delivered by the APPLY MODEL operator. The iteration, however, has to be accessed differently. Ideally, this information would be available, but since it is not, one simply assigns the same number to all examples in the same iterations. The actual value of the number is not of great importance since one only has to distinguish between different iterations.

This is nicely capsuled in a small operator called SUBPROCESS which, in general, makes the whole process a little cleaner.

Assigning *iteration* as a new attribute can be broken down into two steps. Firstly, the value to be assigned has to be extracted using the EXTRACT MACRO operator. As can be

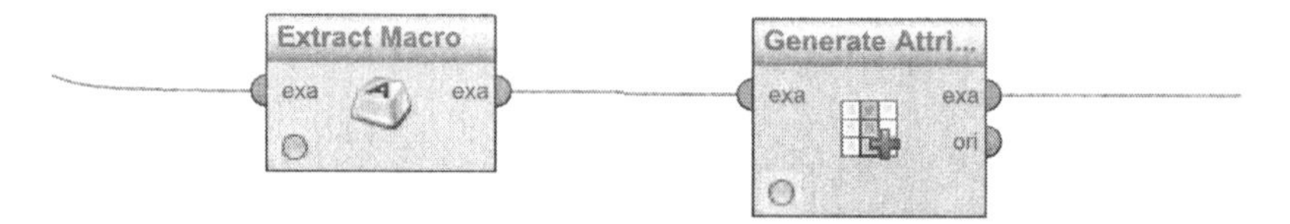

FIGURE 16.20: Overview over both operators included in the subprocess.

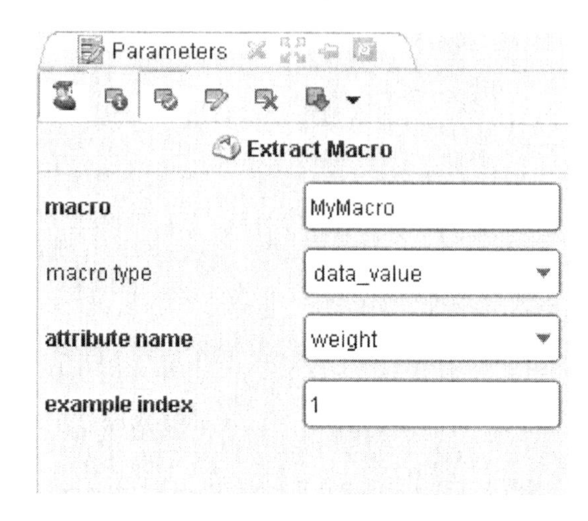

FIGURE 16.21: Settings of the EXTRACT MACRO operator.

easily inferred from the name, EXTRACT MACRO is used to extract a macro that can be used further on in the process. The operator has four settings, the first one being *macro*, which is simply the name of the macro. The *macro type* can be accessed from a drop-down menu. In this case, *data value* must be selected. As a third option, the name of the attribute to extract a value from must be set. This can easily be achieved by setting *attribute name* to *weight*. The last option is *example index*. With this option, one specifies which example is to be used to extract a certain value. For the problem at hand, the first example is used, so *example index* is set to 1, but in principle, any index can be used. When selecting the attribute, one should make sure to choose one that is different for every example. If this is not possible, one should at least pick one which is different for most of the examples.

After the macro has been extracted, the GENERATE ATTRIBUTES operator is used to create a new attribute using the macro. This new attribute, in this case called *iteration*, is now part of the example set and can be used to distinguish whether two examples were tested in the same cross validation or not. The operator settings, and especially the use of the macro, are depicted in Figure 16.22.

One should use a SELECT ATTRIBUTES operator after the SUBPROCESS, as not all attributes need to be written to file. For the problem at hand, only the confidences (*confidence(0)* and *confidence(1)*), the event number and the newly created *iteration* are of importance. These attributes will be written to file to append them to existing ROOT files that carry the full information on any event.

Before writing the attributes to file using WRITE AML operators, the example set has to be divided into signal and background events, as this makes the further use in ROOT scripts a lot easier. This is done using two FILTER EXAMPLES operators. Again, the operators have been renamed to GET SIGNAL and GET BACKGROUND to better distinguish between both streams of data. The first one filters out all the signal events by setting *condition class* to *attribute value filter* and *parameter string* to *label* = 1. The lower output port of the

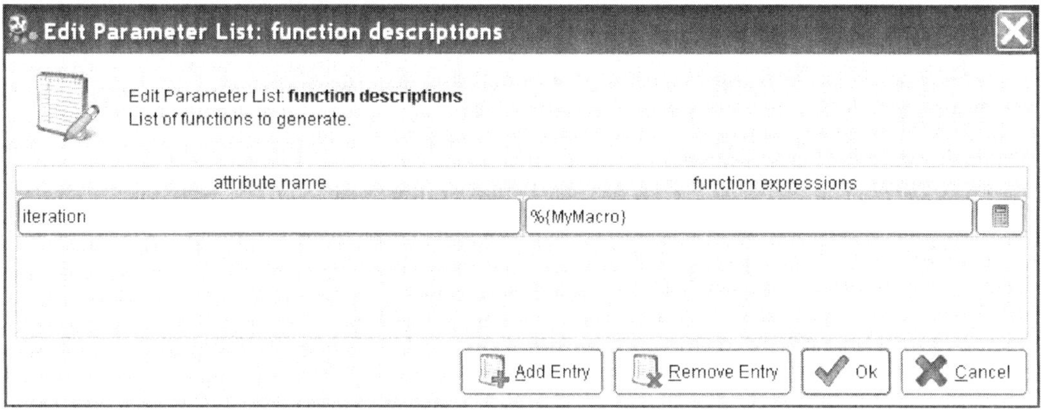

FIGURE 16.22: Setting to create a new attribute using a macro.

FILTER EXAMPLES will return the full and unchanged example set. This output port can now be plugged into the input of another FILTER EXAMPLES operator. The settings for this operator are basically the same as for the first one. Only *parameter string* has to be set to *label* = 0.

The output ports of the FILTER EXAMPLES now have to be connected to WRITE AML operators. The WRITE AML operators have several options that need to be set. The first one is *example set file*, which has to contain the complete path to the .dat file to contain all of the examples. One can use the browser option to browse through the file system in order to find the optimal location for the file.

The second option is *attribute description file*. This option can contain the path to the .aml file. This setting, however, is optional, and one might leave it blank as well. In that case, of course, no .aml file is written.

Several options for the file format are available from a drop-down menu, but *dense* is the recommended default setting. One might check *zipped* in case zipped files are preferred.

Running the WRITE AML operators in a cross validation *overwrite mode* should be set to *overwrite first append then*. In this case, an already existing file is overwritten in the first iteration. In all other iterations, however, new examples will be successively appended to the file.

16.5 Summary and Outlook

In summary, the application of RapidMiner in neutrino astronomy has been successful. The complex process using two subprocesses which each show a sophisticated control flow is well documented by the RapidMiner operators together with their parameter settings. Zooming into and out of processes allows the developer to focus attention on the particular subprocess at hand. The goal of transparency and reproducibility of scientific analyses is well supported by the system.

The combination of an MRMR feature selection and a Random Forest yielded very positive results. In terms of background rejection as well as in terms of signal recall, our results were found to improve the results of straight cuts and Boosted Decision Trees [17].

Further details on the performance of the event selection for the problem at hand can be found in [18, 19].

For the ongoing application of RapidMiner in an IceCube analysis, several improvements of the process have been discussed. Probably the most promising one is the use of a multi ensemble approach which subdivides the detector according to physical aspects. Then feature selection, training, and evaluation of the learning process are carried out for each sub-detector, individually. From studies on Monte Carlo simulation, an increase in recall between 10% and 20% can be expected from the application of this method [20].

Bibliography

[1] V. F. Hess. *Beobachtungen der durchdringenden Strahlung bei sieben Freiballonfahrten.* Kaiserlich-Königlichen Hof-und Staatsdruckerei, in Kommission bei Alfred Hölder, 1912.

[2] B. Falkenburg and W. Rhode. *From Ultra Rays to Astroparticles: A Historical Introduction to Astroparticle Physics.* Springer, 2012.

[3] T. K. Gaisser. *Cosmic Rays and Particle Physics.* Cambridge University Press, 1991.

[4] J. Aleksić et al. Performance of the MAGIC stereo system obtained with Crab Nebula data. *Astroparticle Physics*, 35:435–448, February 2012.

[5] J. A. Hinton. The status of the HESS project. *New Astronomy Reviews*, 48:331–337, April 2004.

[6] H. Anderhub et al. FACT - The first Cherenkov telescope using a G-APD camera for TeV gamma-ray astronomy. *Nuclear Instruments and Methods in Physics Research A*, 639:58–61, May 2011.

[7] T. DeYoung. Neutrino Astronomy with IceCube. *Modern Physics Letters A*, 24:1543–1557, 2009.

[8] J. Ahrens et al. Sensitivity of the IceCube detector to astrophysical sources of high energy muon neutrinos. *Astroparticle Physics*, 20:507–532, February 2004.

[9] W. Huelsnitz. *Search for Quantum Gravity with IceCube and High Energy Atmospheric Neutrinos.* PhD thesis, University of Maryland, College Park, MD, USA, 2010.

[10] C.H.Q. Ding and H. Peng. Minimum redundancy feature selection from microarray gene expression data. *J. of Bioinformatics and Computational Biology*, 3(2), 2005.

[11] S. Lee, B. Schowe, V. Sivakumar, and K. Morik. Feature selection for high-dimensional data with rapidminer. Technical Report 1, TU Dortmund University, 2011.

[12] B. Schowe. Feature Selection for High-Dimensional Data in RapidMiner. In S. Fischer and I. Mierswa, editors, *Proceedings of the 2nd RapidMiner Community Meeting And Conference (RCOMM 2011)*, Aachen, 2011. Shaker Verlag.

[13] T. Hastie, R. Tibshirani, and J. H. Friedman. *The Elements of Statistical Learning: Data Mining, Inference, and Prediction.* New York: Springer-Verlag, 2001.

[14] L.I. Kuncheva. A stability index for feature selection. In *Proceedings of the 25th IASTED International Multi-Conference*, 2007.

[15] L. Breiman. Random forests. *Machine Learning*, 45, 2001.

[16] http://root.cern.ch/drupal/.

[17] The IceCube Collaboration. Time dependent searches for astrophysical neutrino sources with the combined data of 40 and 59 strings of IceCube. In *Proceedings of the 32nd International Cosmic Ray Conference*, 2011.

[18] T. Ruhe, K. Morik, and B. Schowe. Data mining on ice. In *Proceedings of the GREAT Conference on Astrostatistics and Data Mining*, 2011.

[19] T. Ruhe and K. Morik. Data Mining IceCube. In *Proceedings of the ADASS XXI Meeting*, 2011.

[20] F. Scheriau. Data mining für den IceCube Detektor, 2011.

Chapter 17

Medical Data Mining

Mertik Matej

Faculty of information study Novo mesto, Slovenia

Palfy Miroslav

University Medical Centre Maribor, Slovenia

17.1 Background

The successful application of data mining in highly visible fields like e-business, marketing and retail have led to the popularity of its use in knowledge discovery in databases (KDD) in other industries and sectors. Among sectors that are just discovering data mining are the fields of medicine and public health. Applying data mining in the medical field is a very challenging process due to the idiosyncrasies of the medical profession. [1] cites several inherent conflicts between the traditional methodologies of data mining approaches and medicine. Those inherent conflicts refer to specifics in medical research. In this sense data mining in medicine differs from standard data mining practise in the following aspects:

(a) medical data mining usually starts with a pre-hypothesis and then the results are adjusted to fit the hypothesis, and

(b) whereas traditional data mining is concerned about patterns and trends in datasets, data mining in medicine is more interested in the minority that do not conform to the patterns and trends.

This difference in approach highlights the fact that most standard data mining methods are concerned mostly with describing but not explaining the patterns and trends, where in contrast, medicine needs those explanations because a slight difference could change the balance between life and death. Therefore, standardised data mining techniques are often not applicable [2] in medicine; however, there are four types of data mining techniques that are frequently documented:

- Classification
- Production of association rules
- Clustering
- Trend temporal pattern analysis

These techniques are applied to provide generalised functions of:

- Prediction
- Expert decision support
- Diagnosis

In this chapter we will describe use cases and data mining processes in the medical field with RapidMiner. We will introduce frequently used techniques of data mining in medicine and RapidMiner operators focused on those techniques. We will present two different medical examples where the data mining process was successfully applied in medical research, a) the knowledge discovery in dataset from Carpal Tunnel Syndrome and b) survey of knowledge on diabetes that was produced during clinical studies at University Medical Centre Maribor. The chapter will introduce KDD use cases from accompanied datasets based on practical research work at University Medical Centre Maribor. While reading the chapter, the reader will be able to follow and implement those use cases in RapidMiner 5.3.

17.2 Description of Problem Domain: Two Medical Examples

Two pilot studies from the University Medical Centre Maribor, concerning Carpal Tunnel Syndrome and Diabetes, are explained in this section for a better understanding the construction of the data mining process with RapidMiner.

17.2.1 Carpal Tunnel Syndrome

Carpal tunnel syndrome (CTS) is one of the most common compressive neuropathies in the upper extremities [3] and a frequent cause of pain, paraesthesia, and impaired hand function. A syndrome is, by definition, a collection of signs and symptoms. Its clinical symptoms depend on duration and degree of the compression of the median nerve (MN). At first the sensory nerve fibres are affected, but as the compression persists, large-calibre myelinated nerve fibres (sensory and motor) undergo damage as well. Clinical symptoms and signs alone are not sufficient to confirm the diagnosis and surgical release; electrodiagnostic methods are needed for this purpose [4]. Within the last decades, CTS reached epidemic proportions in many occupations and industries. It represents a large expense because of absence from work and wage compensation due to temporary incapacity for work. For this reason, numerous researchers, not only in the field of medicine, are looking for a non-invasive diagnostic method for determining those loads in workplaces that facilitate the development of CTS [5], [6], [7], [8], [9]. Thermography received attention [10] in the area of determining entrapment neuropathies. It is a procedure for remote determination of temperature of objects, based on the detection of infrared radiation that the observed objects emit. A thermographic camera can determine an object's temperature on the basis of the object's energy flux density. In this manner, it can create a thermal image of observed objects, where warmer objects usually appear in a warmer and lighter colour (light red to white) and colder objects appear in a colder and darker colour (blue to black). Its advantages are that it is a completely safe method, passive investigation, without contact, painless, can be easily repeated with low costs of use [11]. In the study of carpal tunnel syndrome (CTS), tomographic images of hands for constructing a predictive model for CTS were collected. Based on the images, dataset with average values of temperatures of different areas of patients' hands was constructed. The data mining process was applied for constructing the predictive model that could identify the typical CTS pattern from the thermogram of a hand. In this chapter the construction of the data model in RapidMiner capable of diagnosing CTS on the basis of very discrete temperature differences invisible to the eye on a thermographic image is explained in detail.

17.2.2 Diabetes

The second example of medical data mining is a study of diabetes, implemented at University Medical Centre Maribor. The aim of this study was to evaluate the level of knowledge and overall perceptions of diabetes mellitus Type 2 (DM) within the older population in North-East Slovenia. As a chronic disease, diabetes represents a substantial burden for the patient. In order to accomplish good self-care, patients need to be qualified and able to accept decisions about managing the disease on a daily basis. Therefore a high level of knowledge about the disease is necessary in order for the patient to act as an equal partner in management of the disease. Various research questions were posed for the purpose of the study, for example: What is the general knowledge about diabetes among diabetic patients 65 years and older? and What is the difference in knowledge about diabetes with regard to education and place of living on (1) diet, (2) HbA1c, (3) hypoglycaemia management, (4) activity, (5) effect of illness and infection on blood sugar levels, and (6) foot care? Moreover, the hypothesis about the level of general knowledge of diabetes in the older population living in urban and rural areas was predicted and verified through the study. A cross-sectional study of older (age $>$ 65 years), non-insulin-dependent patients with diabetes mellitus Type 2 who visited a family physician, DM outpatient clinic, a private specialist practice, or were living in a nursing home was implemented. The Slovenian version of the Michigan Diabetes Knowledge test was then used for data collection [12]. In this example some descriptive methods of data mining were applied for constructing a hypothesis, and a prediction regression model of the data was constructed for the aim of the study. The construction of experiments in RapidMiner is explained in detail.

17.3 Data Mining Algorithms in Medicine

This section explains some of the most common and well-known data mining techniques used in a different setting—medicine. As we stress, the medical data mining usually starts with a pre-hypothesis and then the results are used to confirm or reject this hypothesis. Data mining in medicine is also interested in the minority cases that do not conform to the patterns and trends, whereas traditional data mining is mostly concerned with patterns and trends in datasets. Moreover, most standard data mining methods usually describe but not explain the patterns and trends, where in contrast, medicine often needs those explanations for better understanding and diagnosis. In the following subsections, we explain data mining techniques and associated algorithms that are often implemented in medicine. We concisely describe the aim of classification and regression techniques, clustering, and association rules.

17.3.1 Predictive Data Mining

The predictive models are supervised learning functions that predict a target value. RapidMiner supports many predictive functions and associated algorithms, and under operator modelling classification and regression we can find many different techniques. For data mining in medicine the following predictive functions and associated algorithms available in RapidMiner are frequently used. Table 17.1 include the algorithms (RapidMiner operators) that will be presented later in this chapter.

TABLE 17.1: Predictive functions and associated algorithms for medical data mining.

Functions	**Associated Algorithms**
Classification	*Decision Trees*, *Neural Networks*, Naive Bayes, Support Vector Machines
Regression	Linear Regression, Support Vector Machines

Classification and Regression

In a classification problem, there are typically historical data (labelled examples) and unlabelled examples. Each labelled example consists of multiple predictor attributes and one target attribute, dependent variable. The value of the target attribute is a class label. The unlabelled examples consist of the predictor attributes only and the goal of classification is to construct a model using the historical data that accurately predicts the label (class) of the unlabelled examples. A classification task begins with training data for which the target values (class assignments) are known. Different classification algorithms use different techniques for finding relations between the predictor attribute's values and the target attribute's values. These relations are summarized in a model, which can then be applied to new cases with unknown target values to predict target values. A classification model can also be used on data with known target values, to compare the predictions to the known answers. Such data is known as test data or evaluation data. This technique is called testing a model and it measures the model's predictive accuracy. The application of a classification model to new data is called applying the model. The classification task in medicine is used for prediction, expert decision support, and diagnosis. For example, a physician may wish to have additional support for decision making based on the patients that were treated previously. In this case each patient corresponds to a case. Data for each case might consist of a number of attributes that describe the patient's health, treatment parameters, demographic attributes, etc. These are the predictor attributes. The target attribute typically indicates the patient's state, disease, and treatment. The data is used to build a model that physicians then use to predict the target value for new cases. In many cases such a model can improve decision making, help in making a diagnosis or, as in one of our examples, it could help indeveloping non-invasive methods for treatments. Regression, similarly to classification, creates predictive models. The difference between regression and classification is that regression deals with numerical (continuous) target attributes, whereas classification deals with discrete (categorical) target attributes, that is, if the target attribute contains continuous (floating-point) values, a regression technique is required. If the target attribute contains categorical (string or discrete integer) values, a classification technique is called for. The most common form of regression is linear regression, in which a line that best fits the data is calculated. This line minimizes the average distance of all the points from the line. This line becomes a predictive model when the value of the dependent variable is not known.

17.3.2 Descriptive Data Mining

Descriptive data mining models are unsupervised learning functions. These functions do not predict a target value, but focus more on the intrinsic structure, relations, interconnectedness, etc. of the data. The RapidMiner interface supports many descriptive models and associated algorithms. Under Clustering and segmentation and Association and item set mining, we can find many different models. For data mining in medicine the following descriptive functions and associated algorithms available in RapidMiner are frequently used and documented:

TABLE 17.2: Descriptive functions and associated algorithms for medical data mining

Functions	Associated Algorithms
Clustering	*K-means*
Association	Association rules - FPGrowth algorithm

Clustering

Clustering is a technique that is useful for exploring data. It is particularly useful where there are many cases and no obvious natural groupings. Here, clustering data mining algorithms can be used to find whatever natural groupings may exist. Clustering analysis identifies clusters embedded in the data. A cluster is a collection of data objects that are similar in some sense to one another. A good clustering method produces high-quality clusters to ensure that the inter-cluster similarity is low and the intra-cluster similarity is high; members of a cluster are more like each other than they are like members of a different cluster. Clustering can also serve as a useful data-preprocessing step to identify homogeneous groups on which to build predictive models. Clustering models are different from predictive models in that the outcome of the process is not guided by a known result, as there is no target attribute. Predictive models predict values for a target attribute, and an error rate between the target and predicted values can be calculated to guide model building. Clustering models, on the other hand, uncover natural groupings (clusters) in the data. The model can then be used later to assign grouping labels (cluster IDs) to data points.

Association Rules

The Association model is often associated with a "market basket analysis", which is used to discover relationships or correlations in a set of items. It is widely used in data analysis for direct marketing, catalogue design, and other business decision-making processes. A typical association rule of this kind asserts the likelihood that, for example, "70% of the people who buy spaghetti, wine, and sauce also buy olive bread." Association models capture the co-occurrence of items or events in large volumes of customer transaction data. It is now possible for retail organizations to collect and store massive amounts of sales data, referred as "basket data" Association models were initially defined on basket data, even though they are applicable in several other applications. Finding all such rules is valuable for cross-marketing and mail-order promotions, but there are other applications as well. In medicine, they can be used for discovering relationships between different parameters in various segments of data. There are several measures available for generating association models. Most used are:

- Support: Support of a rule is a measure of how frequently the items involved in it occur together. Using probability notation, support (A implies B) = P(A, B).
- Confidence: Confidence of a rule is the conditional probability of B given A; confidence (A implies B) = P (B given A), which is equal to P(A, B) or P(A).

RapidMiner provides more measures (e.g., conviction, gain). These statistical measures can be used to rank the rules and hence the predictions.

17.3.3 Data Mining and Statistics: Hypothesis Testing

Hypothesis testing is one of the most important tasks in clinical medicine. Statistics, therefore, plays an important role in the field. Generally the fields of data mining and statistics have many common points as data mining developed from statistics and machine learning. However in medicine, these two fields have two different aspects. Statistical analysis starts with a hypothesis and then later the aim of research is to accept or reject this hypothesis, whereas data mining is a process in which we primarily search for various relationships among the data. In medicine these two aspects are interrelated. Data mining serves as a valuable process on which a hypothesis may be constructed, and on the other hand, it is useful for constructing predictive models on the data that may provide decision support. In the following sections we use two medical examples to present and explain in detail data mining processes with RapidMiner.

17.4 Knowledge Discovery Process in RapidMiner: Carpal Tunnel Syndrome

The knowledge discovery (data mining) process with RapidMiner on datasets containing information on patients suffering from carpal tunnel syndrome and diabetes is presented in these sections. Explained are RapidMiner operators with associated algorithms and the iterative process of constructing different models. The reader can use RapidMiner and generate the models on described datasets.

17.4.1 Defining the Problem, Setting the Goals

In the study of carpal tunnel syndrome (CTS), the thermal images of patients' hands were collected. The purpose of the study was to try to use the data derived from these thermal images for constructing the predictive model for CTS. This could be helpful when looking for a non-invasive diagnostic method. The temperatures of different areas of a patient's hand were extracted from the image and saved in the dataset. The aim of data mining was to construct a data model in RapidMiner, capable of diagnosing CTS on the basis of very discrete temperature differences that are invisible to the eye on a thermographic image. Different appropriate well-known machine learning and data mining techniques were used.

17.4.2 Dataset Representation

Figure 17.1 presentsthe Meta Data View in RapidMiner of the provided dataset. There are 70 attributes, two polynomial (categorical) CTS state and ulnar nerve, and 68 real attributes that represent temperatures from different parts of the hand (palmar and dorsal side) and their differences. The dataset consists of 251 clinical examples.

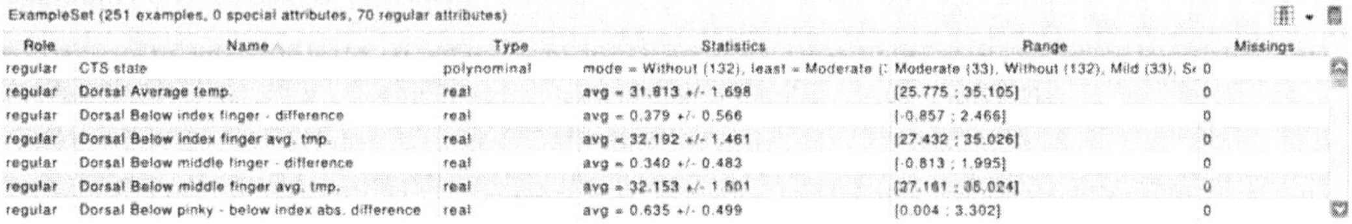

ExampleSet (251 examples, 0 special attributes, 70 regular attributes)

Role	Name	Type	Statistics	Range	Missings
regular	CTS state	polynominal	mode = Without (132), least = Moderate (	Moderate (33), Without (132), Mild (33), S	0
regular	Dorsal Average temp.	real	avg = 31.813 +/- 1.698	[25.775 ; 35.105]	0
regular	Dorsal Below index finger - difference	real	avg = 0.379 +/- 0.566	[-0.857 ; 2.466]	0
regular	Dorsal Below index finger avg. tmp.	real	avg = 32.192 +/- 1.461	[27.432 ; 35.026]	0
regular	Dorsal Below middle finger - difference	real	avg = 0.340 +/- 0.483	[-0.813 ; 1.995]	0
regular	Dorsal Below middle finger avg. tmp.	real	avg = 32.153 +/- 1.501	[27.161 ; 35.024]	0
regular	Dorsal Below pinky - below index abs. difference	real	avg = 0.635 +/- 0.499	[0.004 ; 3.302]	0

FIGURE 17.1: Meta Data View on CTS dataset.

To import the data in RapidMiner we used the Wizard Import Excel sheet from Import

Data in the File menu and created the appropriate Repository. Some very useful tools in RapidMiner that can be used before we start experimenting with construction of knowledge discovery processes (different data models) are visualization tools on ExampleSet. They are available immediately after importing the data: we can create an empty process on which we only connect input and output ports on both sides.

After running the process, we can investigate the landscape of the data. With the plot engine we can see different statistical properties of data such as dependencies among attributes (variables), their dissemination, correlations, etc. Two examples are presented in figure 17.2. We can see deviation of attribute (label) CTS state considering a) temperatures and b) series plot, where palmar and dorsal middle finger avg. temp. attributes were selected.

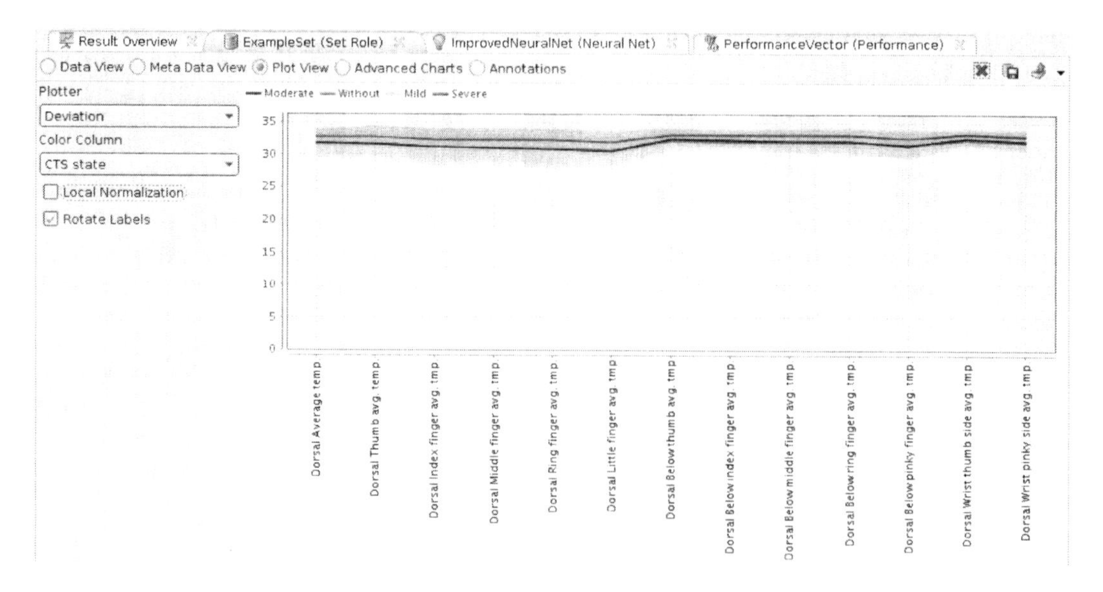

FIGURE 17.2 (**see color insert**): Simple data visualisation: Deviation Plot.

17.4.3 Data Preparation

Before building the model, it is necessary to prepare the data for appropriate methods. This is the first phase of the knowledge discovery process, the so-called data preprocessing phase, which consists of data filtering, data cleansing, and data transformation. RapidMiner supports many different operators for this phase (more than one hundred); some are more general and some more specific. Use of these operators depends on the dataset that you have and the model that you want to apply. If the dataset consists of different types of attributes, all possible types of conversion are possible. You can reduce and transform the attributes, change their values, generate new attributes, aggregate, sort, filter, and more. Usually different combinations of operators are necessary for various models. Here we describe some operators that we have used in our final data model. Considering the CTS dataset example, these were FilterExample, RemoveUnusedValues, SelectAttributes, and SetRole. Note that during the iterative process of constructing the model, we used some other operators like operators for discretization, type conversion, and map operator, depending on data mining techniques that we tried to apply. Figure 17.4 shows operators that were used in the pre-processing phase for CTS.

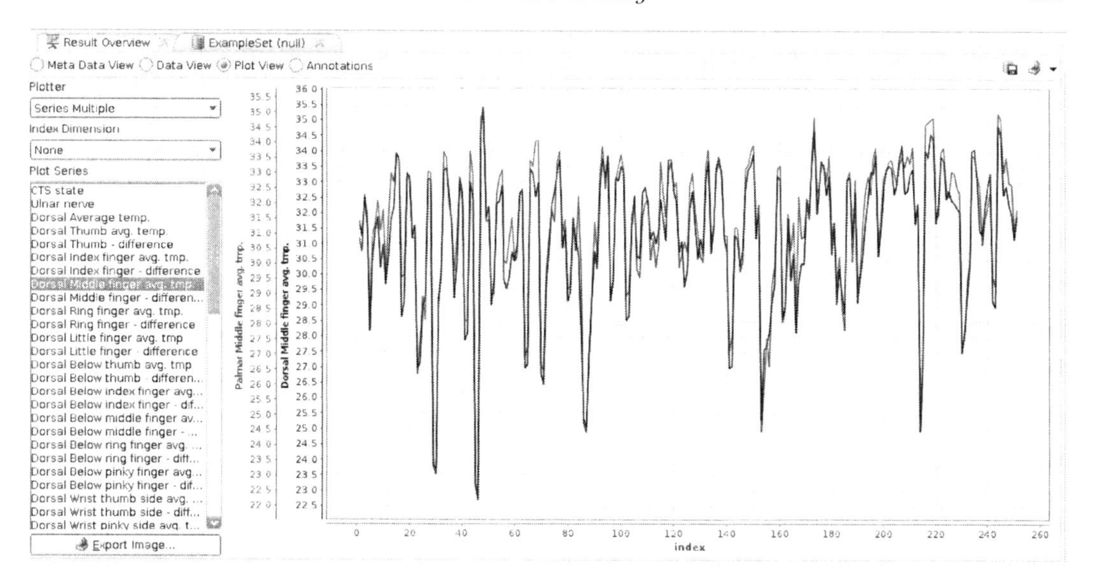

FIGURE 17.3 (**see color insert**): Simple data visualisation: Multiple Series Plot.

FilterExample Operator

This operator was used to filter examples from the dataset. During the experiments it was shown that it is not possible to construct the prediction model with more than 55% accuracy for four different classes of CTS state. Therefore the consulting expert decided to construct the experiment with only two groups of examples, belonging to classes with most significant differences: those that have CTS state equal to "Without" and those with CTS state equal to "Severe". The FilterExample operator with Attribute Value filter was selected from the available classes of operators. Then examples with defined CTS states were selected by setting the appropriate condition for the filter. This was achieved by defining the parameter string as follows: *parameter string* CTS state = Severe || CTS state = Without. We can see that we define the attribute and its values that we want to filter. We can combine different conditions for attribute values. For this purpose we used the logical operators AND (&&) and OR (||).

RemoveUnusedValues Operator

This operator was used to remove those values in attributes that were not needed anymore after filtering the examples. In our case those values were Mild and Moderate for CTS state. As we filtered out the examples with the values Without and Severe, the operator RemoveUnusedValues removed Mild and Moderate from the range of values of the CTS state attribute. Attribute filter type single was selected as we wanted to operate on a specific attribute. The Include Special Attributes parameter was set to true, since it could happen that this could be the target attribute. As we will see later this was not the case here as we defined this attribute as a target with the Set Role operator.

SelectAttributes Operator

This operator provides us with the various possibilities for attribute selection from the dataset. It allows us to select different combinations of attributes for building the model. Usually these combinations can be defined by an expert, familiar with the aims of research. On the other hand, when searching for knowledge, different statistical techniques and visu-

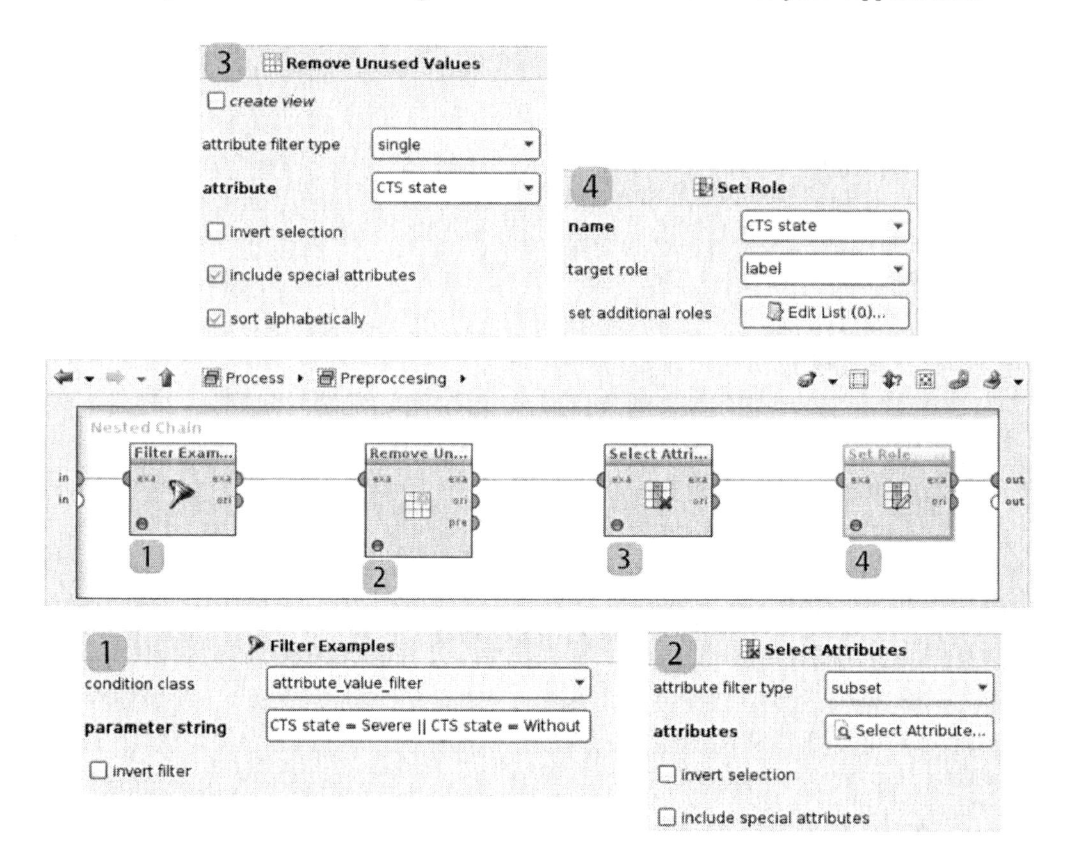

FIGURE 17.4: Operators used in pre-proccessing.

alisations are useful for defining the subset of attributes. In our case attributes were selected based on expert's expectations and visualisations of data that were made in the previous step. The Select Attributes operator allows us to define different filters for selection of attributes like single, subset, regular expression, value type, and no missing values. In our case the subset was selected and an appropriate subset of attributes was chosen for the experiment. Figure 17.5 shows a subset of selected attributes on the CTS dataset.

SetRole Operator

This operator is necessary to define a target attribute when building a prediction model. This is possible by the Set Role operator, where *CTS state* was defined for a target, as the aim of the study was to construct the data model capable of diagnosing CTS.

17.4.4 Modeling

17.4.5 Selecting Appropriate Methods for Classification

After the pre-processing step, various techniques of data mining are available for constructing the model. As we mentioned in the beginning, in medicine most documented techniques and associated algorithms belong to classification, production of association rules, clustering, and trend temporal pattern analysis. Considering the aim of the study—constructing the data model capable of diagnosing CTS—classification and regression tech-

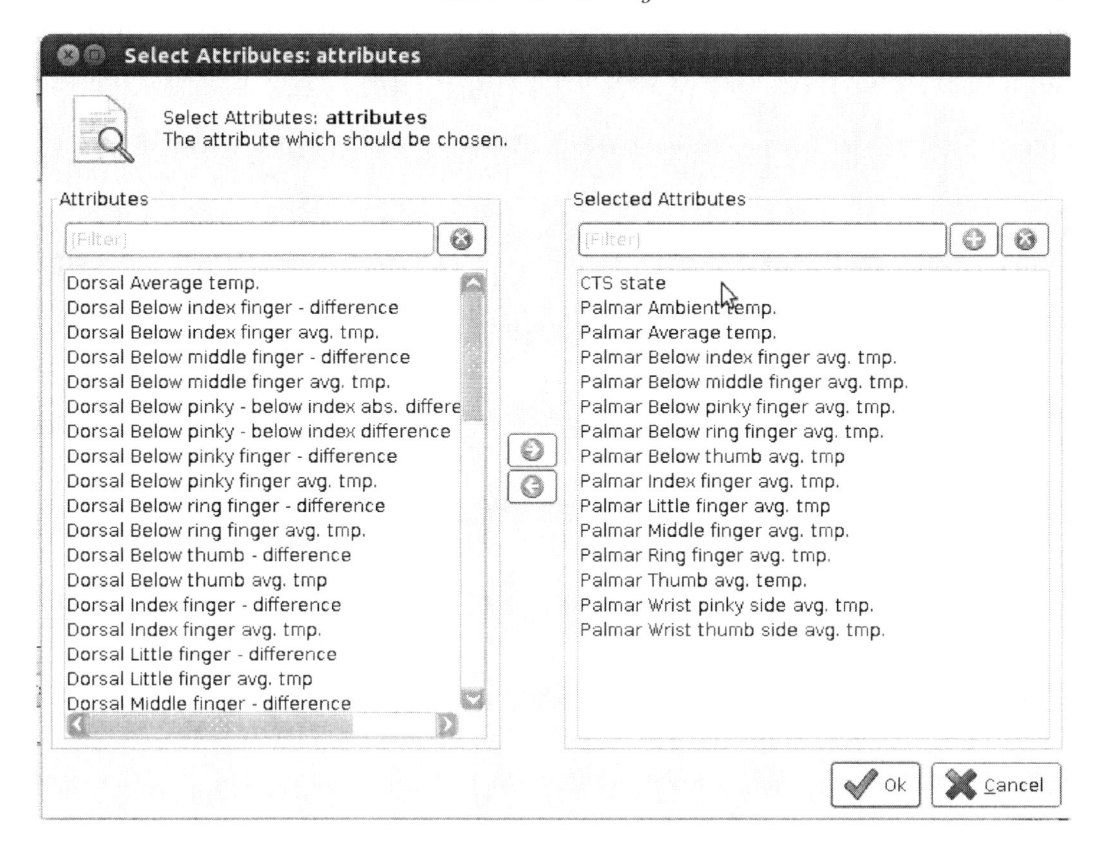

FIGURE 17.5: Select attributes dialog.

niques are the classes of data mining algorithms that are suitable for this task. In RapidMiner these techniques are offered by various associated algorithms like k-NN nearest neighbour, Naive Bayes, Decision Trees, Neural Networks, Support Vector Machines, Linear Regression, and their varieties. Most common algorithms used in medicine are presented in Table 17.4. Each algorithm has particular properties and capabilities for handling the data. The selection of appropriate techniques therefore depends on the

- nature of the data (types of attributes), and
- problem definition (type of problem, function).

In our case we chose the appropriate methods based on mentioned findings. Firstly, most of the attributes represent the temperatures of different areas of a hand, so we were working mostly with numerical data (continuous values). Secondly, we have defined a classification problem for predicting the CTS state. Put differently, it was not the aim of the research to study possible dependencies between those temperatures, but rather to use them to construct a black box which would be capable of predicting the state of the medical condition. Based on our conclusions, the following methods were appropriate: Artificial Neural Network and Support Vector Machine. Tables 17.3 and 17.4 present associated algorithms and their selection considering the nature of the data and the type of problem.

Note that the pre-processing step described in Section 17.4.3 depends on the selection of an appropriate algorithm in the data mining step and that process of constructing the final model is an iterative process where we often change operators in pre-processing considering selected techniques by systematically testing and experimenting. Usually we try different

TABLE 17.3: Selection of DM algorithm considering type of (medical) data.

Function	Algorithm	Data Properties		
		Continuous Data	Discrete Data	Size of Dataset
Prediction	Regression Trees	x		medium
	Neural Networks	x		large
	Support Vector Machines	x		large
	kNN	x		medium
	Naive Bayes	x		large
Knowledge	Decision Trees		x	medium
searching	Association rules		x	large

TABLE 17.4: Selection of DM algorithm considering type of problem.

Function	Algorithm	Type of Algorithm	
		Black box approach	Symbolical representation approach
Prediction	Regression Trees	x	
	Neural Networks	x	
	Support Vector Machines	x	
	kNN	x	
	Naive Bayes	x	
Knowledge	Decision Trees		x
searching	Association rules		x

combinations of pre-processing operators with a combination of the first appropriate single data mining operator (see Table 17.4). In the second step we try other data mining operators that might be suitable for the problem. Note also that the pre-processing step is of great importance as with the right data representation we can significantly improve the performance of the model, in most cases significantly better than with selection of suitable data mining techniques in Table 17.4 (pre-processing plays a crucial role in data mining). This is also the reason why the co-operation of a physician and a data mining expert is crucial for constructing successful prediction models.

In Figure 17.6 we can see the whole RapidMiner process where, based on previous conclusions, Artificial Neural Network (ANN) was selected as the learning method for the defined problem. In this figure we can see the pre-processing subprocess and data mining step for constructing an ANN model which is implemented through the X-Validation operator.

In Figure 17.7 the data mining step is presented in the nested XValidation operator.

We can see that for construction and estimation of the prediction model, we used the XValidation operator, Apply Model operator, and Performance operator. In the following paragraphs, the description and settings of each operator are described in detail.

XValidation Operator

When the classification model is used on data with known target values, to compare the predictions to the known answers, we are talking about test data or evaluation data. The XValidation operator is a Validation operator in RapidMiner that performs a cross-validation process. The input ExampleSet S is split up into a chosen number of subsets S_i.

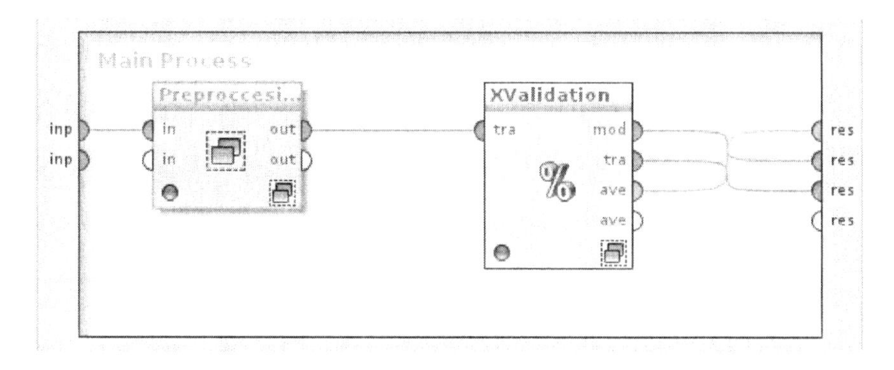

FIGURE 17.6: Whole KDD process.

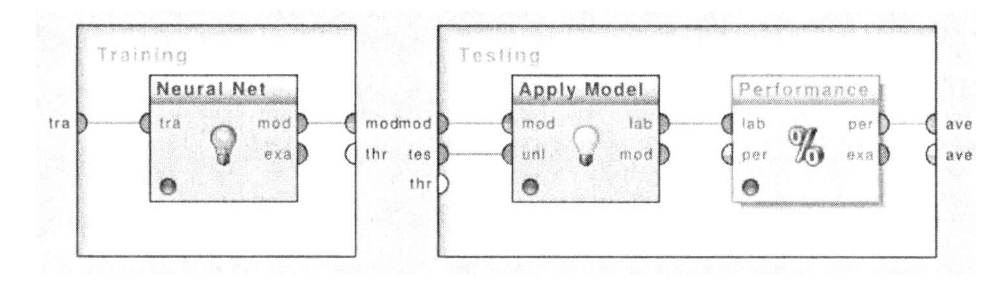

FIGURE 17.7: Learning in X-Validation operator.

The inner subprocesses are applied for each of those subsets using S_i as the test set and $S \setminus S_i$ as training set. In our case, Number of validations was set to 10 and sampling type was set to shuffled sampling.

Apply Model Operator

The application of a classification model to new data is called applying the model. Apply Model is an operator that takes our model generated on the training set and applies this model to the testing set. As we are using the XValidation operator, this is done for the chosen number of validation times and the achieved performance metrics are obtained by averaging the individual performance results. Note that we cannot perform evaluation of data without this operator as we are not applying the model to the testing data.

Neural Network Operator

Artificial Neural Networks (ANN) are one of the data mining techniques that are suitable for data analysis [13]. They are increasingly used in problem domains involving classification and are capable of finding shared features in a set of seemingly unrelated data. An ANN is an abstract computational model of the human brain. Similar to the brain, an ANN is composed of artificial neurons and interconnections. When we view such a network as a graph, neurons can be represented as nodes and interconnections as edges. There are many known variations of ANNs, differing in their topology, orientation of connections, and approaches to learning [14]. Probably the most established are "feed-forward backpropagation" neural networks (multilayer perceptrons). A Neural network is a very efficient algorithm and has a relatively simple training process. However following parameters of ANN are crucial:

- Training cycles: The number of training cycles used for the neural network training.

- Learning rate: The learning rate determines by how much we change the weights at each step. May not be 0.
- Momentum: The momentum simply adds a fraction of the previous weight update to the current one (prevent local maxima and smoothes optimization directions).
- Decay: Indicates if the learning rate should be decreased during learning.
- Shuffle: Indicates if the input data should be shuffled before learning (increases memory usage but is recommended if data is sorted before).
- Normalize: Indicates if the input data should be normalized between -1 and $+1$ before learning (increases runtime, but is in most cases necessary).
- Error epsilon: The optimization is stopped if the training error gets below this value.

Figure 17.8 presents parameters that were optimal for our example of building the model for CTS.

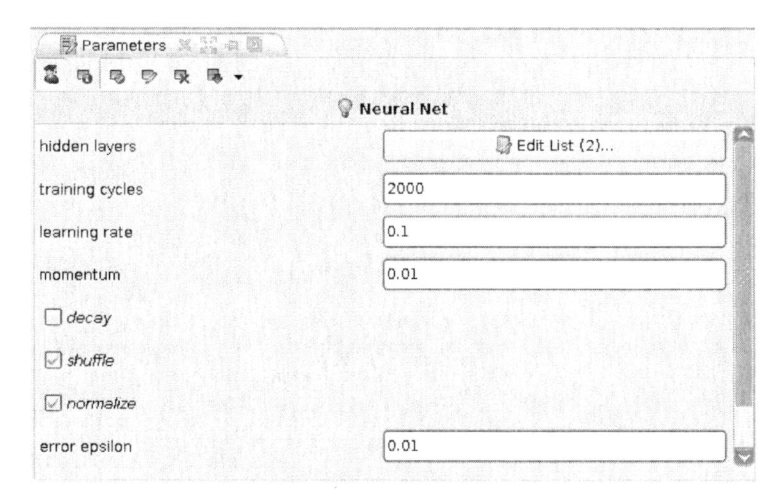

FIGURE 17.8: Parameters for ANN.

Note: The learning rate and momentum are very important parameters that are crucial for achieving the optimal performance of the model. If weights change too much or too little with each training example, the ANN will not learn appropriately. Since the amount by which weights change is determined by learning rate and momentum, estimating the appropriate values of those parameters is of great importance.

Performance Operator

This operator evaluates the results on a selected set of data. In our case we used the classification performance operator that is used for classification tasks, i.e. in cases where the label (target) attribute has a polynominal (categorical) value type (CTS state). The operator expects a test set that is provided by the Validation operator as an input, containing an attribute with the role label and attribute with the role prediction. Usually we define this when we first apply the learning method. On the basis of these attributes, the results are calculated in a PerformanceVector operator, which contains the values of the various available performance criteria. All of the performance criteria can be switched on using logical operators. The main criterion is used for comparisons and needs to be specified only for processes where performance vectors are compared (meta optimization process). When no main criterion is selected, as in our case, the first criterion in the resulting performance vector is assumed to be the main criterion.

17.4.6 Results and Data Visualisation

RapidMiner provides a useful tool for presentation of the results. The result view is a view in the RapidMiner environment which presents all output values of the process that was designed in the environment. In our case the outputs of the following operators were linked from the Validation operator to the output (see Figure 17.7):

- Model
- Training
- Average

Based on these parameters, different views are presented within the environment. Figures 17.9, 17.10, and 17.11 show the generated model, results in the performance vector, and the example set. We can see that our model is capable of predicting the new cases with 74,59% accuracy, measured on the cross-validation operator on the test set (unseen examples for the learner). These results are promising as the prediction model might be used as a supporting tool for CTS diagnosis.

ExampleSet (185 examples, 1 special attribute, 13 regular attributes)

Role	Name	Type	Statistics	Range	Missings
label	CTS state	polynominal	mode = Without (132), least = Severe (53)	Severe (53), Without (132)	0
regular	Dorsal Average temp.	real	avg = 31.943 +/- 1.559	[26.315 ; 35.105]	0
regular	Dorsal Thumb avg. temp.	real	avg = 32.023 +/- 1.956	[23.899 ; 35.302]	0
regular	Dorsal Index finger avg. tmp.	real	avg = 31.507 +/- 2.020	[24.090 ; 35.268]	0
regular	Dorsal Middle finger avg. tmp.	real	avg = 31.415 +/- 2.120	[23.724 ; 35.425]	0
regular	Dorsal Ring finger avg. tmp.	real	avg = 31.324 +/- 2.204	[23.636 ; 35.195]	0
regular	Dorsal Little finger avg. tmp	real	avg = 31.017 +/- 2.289	[23.581 ; 35.032]	0
regular	Dorsal Below thumb avg. tmp	real	avg = 32.430 +/- 1.306	[28.455 ; 35.075]	0
regular	Dorsal Below index finger avg. tmp.	real	avg = 32.299 +/- 1.364	[27.566 ; 35.026]	0
regular	Dorsal Below middle finger avg. tmp.	real	avg = 32.270 +/- 1.393	[27.161 ; 35.024]	0
regular	Dorsal Below ring finger avg. tmp.	real	avg = 32.251 +/- 1.470	[26.710 ; 35.196]	0
regular	Dorsal Below pinky finger avg. tmp.	real	avg = 31.783 +/- 1.569	[26.108 ; 34.844]	0
regular	Dorsal Wrist thumb side avg. tmp.	real	avg = 32.563 +/- 1.200	[29.726 ; 35.011]	0
regular	Dorsal Wrist pinky side avg. tmp.	real	avg = 32.224 +/- 1.289	[28.296 ; 34.875]	0

FIGURE 17.9: Preprocessed example set.

17.4.7 Interpretation of the Results

The findings of the medical pilot study at University Medical Centre Maribor speak in favour of the applicability of the prediction model generated in RapidMiner for diagnosing CTS from thermal images. In a smaller number of study and test cases (n = 112) the efficiency of classification exceeded expectations of physicians. Therefore realistic evaluation of efficiency of classification was implemented on a significantly larger data base (n = 364), as explained in this chapter. To use a model in a medical practice as a support tool for CTS diagnosis from thermal images, additional experiments should be performed. Besides a larger dataset than provided in these experiments, the misclassification costs could be estimated and then optimized, when using a model. An expert could for example estimate the cost of error when the model would misclassify examples as false positive or false negative. For this purpose the expert could set the weights for the false positive (CTS = Severe) or false negative predictions (CTS = Without). In RapidMiner these can be implemented using the

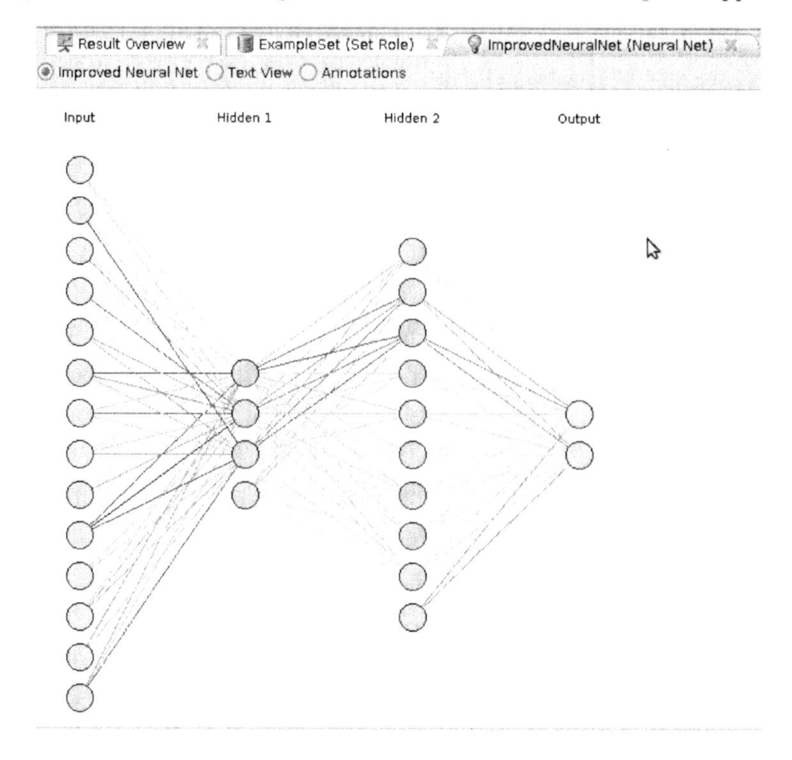

FIGURE 17.10: Neural Network Model.

accuracy: 76.14% +/- 7.48% (mikro: 76.22%)

	true Severe	true Without	class precision
pred. Severe	26	17	60.47%
pred. Without	27	115	80.99%
class recall	49.06%	87.12%	

FIGURE 17.11: Performance vector.

CostPerformance operator. Furthermore, the MetaCost operator can be used later during the training process with the goal of improving the prediction model.

17.4.8 Hypothesis Testing and Statistical Analysis

In this section we show the process of testing a hypothesis with RapidMiner. We wanted to compare the differences in average temperatures of segments on the palmar side of the hand with corresponding segments on the dorsal side. The null hypothesis "there are no significant differences between hand sides" was chosen at the beginning. For hypothesis testing we used the ANOVA operator, with which we compared average temperatures of chosen hand segments on the dorsal and palmar side. To do the ANOVA we first need to prepare data for analysis. Figure 17.12 presents a process for hypothesis testing.

We can see that our process contains a main subprocess where we combined RapidMiner operators. In the pre-processing subprocess we prepare data for ANOVA and reconstruct the dataset from two subsets by adding a new attribute, Hand. The following operators were used: FilterExample (1), RemoveUnusedValues [2], Multiply (3), SelectAttributes (4.1, 5.1,

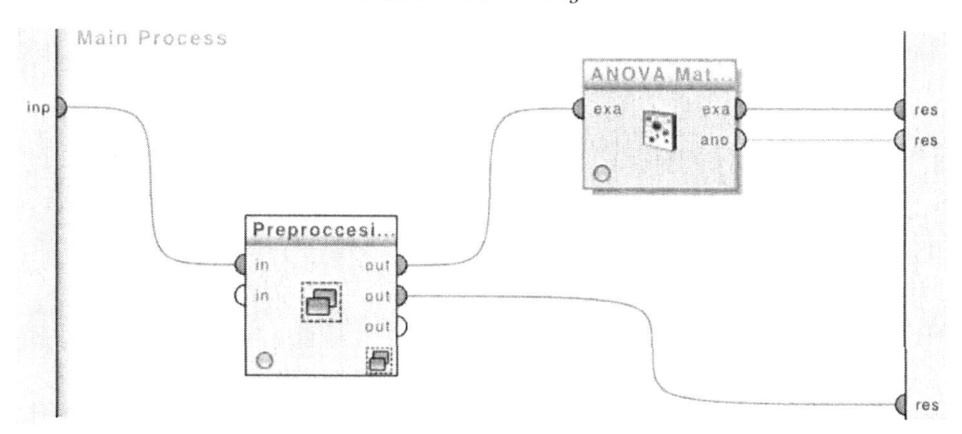

FIGURE 17.12: Statistical process for hypothesis testing.

7), Rename (4.1, 5.2), Generate Attributes (4.3, 5.3), Numerical2Nominal (8), Append (6), and Aggregate (9). Figure 17.13 shows the reconstruction subprocess for ANOVA.

There are two different datasets (ExampleSets) connected to the ports of pre-processing. The first dataset is constructed for ANOVA analysis where the dataset with the original 70 attributes is split into two datasets with 14 attributes (attributes were selected by the expert). Those attributes are temperatures from different segments of hands. A subset of 14 attributes was selected by the SelectAttributes operator from the original dataset containing palmar and dorsal temperature values. In the next step the operator Rename was used for reconstructing those two sets into sets with the same attribute names. Figure 17.14 shows renaming definition for those attributes from the original datasets. Then description attribute Hand was defined for each of the subsets, where each of the subsets was labelled according to the side of the hand the temperature values were measured on. For this purpose, the *GenerateAttributes* operator was used, where $Hand = 0$ function was defined for the dorsal and $Hand = 1$ for the palmar side of the hand. Now both data subsets where joined together with an operator Append, and for the ANOVA analysis the numerical value of Hand was transformed into a binominal attribute with an operator Numerical to Binominal. That is how the first reconstructed dataset for ANOVA analysis was constructed. The second port presents another dataset which was defined for visualisation. In this case the operator Append is followed by an Aggregate operator, where we aggregate values (as average) to present the differences in the plot view. ANOVA calculation was implemented by an ANOVA operator and a plotter was used for visualisation of the hypothesis. For the plotter, the Aggregate operator was used and connected to the second port of the subprocess. The following steps correspond to Figure 17.13.

- FilterExample: filter the dataset on cases Severe and Without (1).
- RemoveUnusedValues: remove cases that have no value under label CTS state (2).
- Multiply: Generate two example sets for subsets processing (3).
  - SelectAttributes: select Palmar Attributes (4.1).
  - Rename: remove "Palmar" from the name of each attribute (4.2).
  - Generate Attributes: create new attribute Hand with value 0 for Palmar cases (4.3).
  - SelectAttributes: select Dorsal Attributes (5.1).

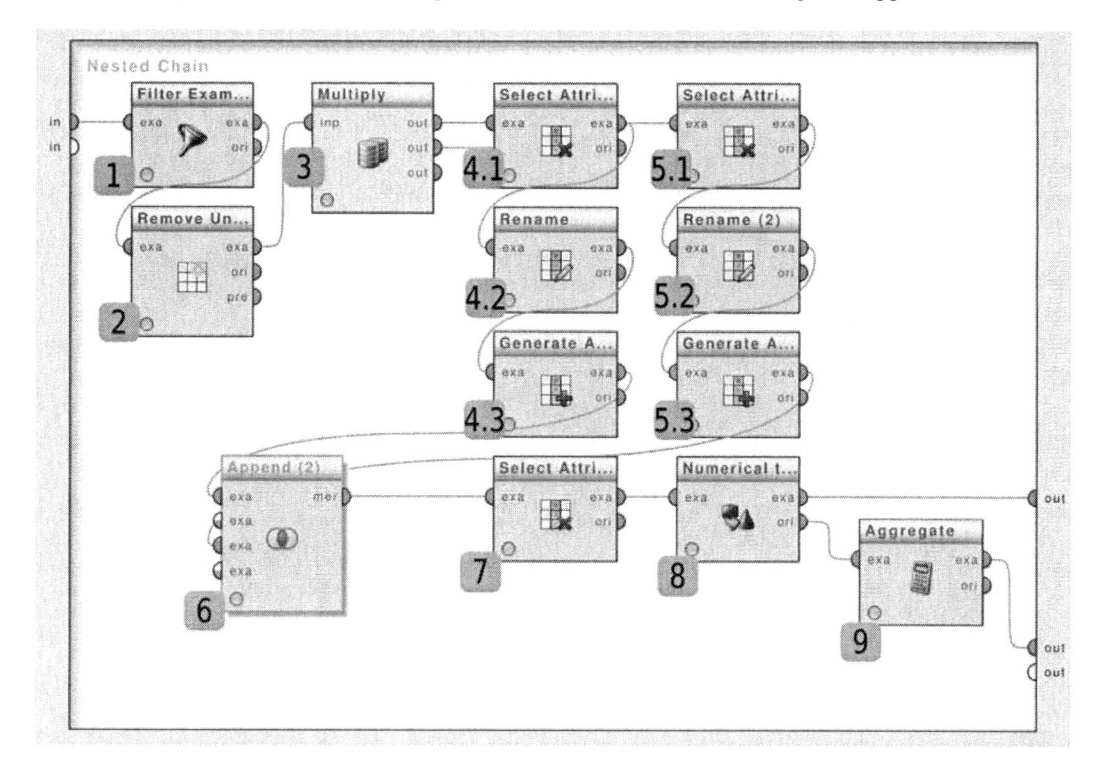

FIGURE 17.13: Reconstruction subprocess for ANOVA.

  - Rename: remove "Dorsal" from the name of each attribute (5.2).
  - Generate Attributes: create new attribute Hand with value 1 for Dorsal cases (5.3).

- SelectAttributes: select all attributes selected for Anova (6).
- Numerical2Nominal: convert type of attribute Hand to Binominal (7).
- Append: append two example sets with new attribute Hand in one example set (8).
- Aggregate: calculate the average values for attributes for aggregate plot (see Figures 17.15 and 17.16) (9).

New operators in the statistical analysis process are presented in the following subsections.

Rename Operator

This operator is used to change the name of specific attribute or list of attributes. In our case we renamed attributes from two splits to prepare the set for the Append operator, see Figure 17.14.

Generate Attributes Operator

This operator generates a new attribute with a value that is defined by user function. In our case we generate two attributes for each example set. In the example set containing temperatures from the palmar hand side, we generated a new attribute Hand with function $Hand = 0$ and for the dorsal example set $Hand = 1$.

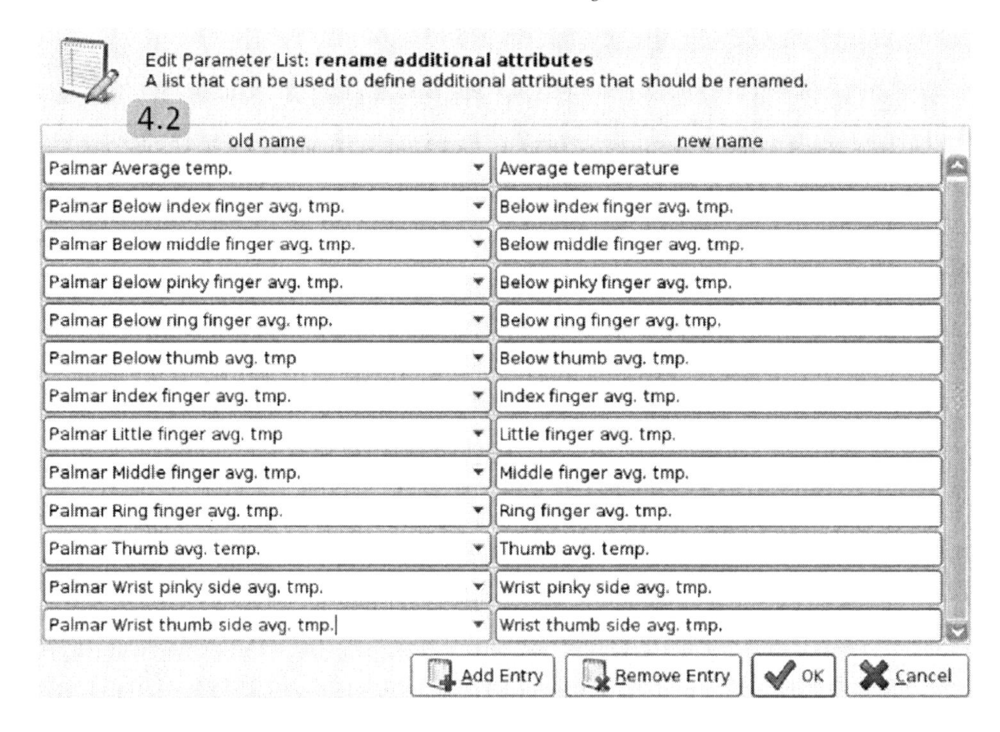

FIGURE 17.14: Renaming attributes in pre-processing.

Append Operator

This operator appends examples from one set to another. For this purpose, both sets must have an equal number of attributes of the same type.

NumericalToBinominal Operator

This operator converts numerical attributes to binomial. In our case the attribute Hand generated with the GenerateAttributes operator was converted to a binomial attribute with its values converted from 0 and 1 to false and true, respectively.

Aggregate Operator

This operator aggregates values of chosen attributes. In our example we used a typical aggregate scenario in which we are interested in average values across attributes. We obtained the average values of temperatures for all segments. Based on those values, different plots for evaluating the hypothesis are possible.

ANOVA Operator

This operator is a statistical operator that calculates the significance of the difference between average values for chosen numerical attributes belonging to different groups defined by a nominal attribute. In our scenario the average value of each attribute belonging to one side of the hand was statistically compared with the average value of a corresponding attribute from the other hand side. By doing this we can confirm or reject the null hypothesis that the average temperatures are more or less equal on both hand sides.

17.4.9 Results and Visualisation

The null hypothesis was not rejected. The ANOVA matrix with statistical probabilities is shown in Figure 17.15. We can notice the small differences between dorsal and palmar temperatures from Figure 17.16. We can see that our null hypothesis (there are no significant differences between dorsal and palmar temperatures) is true, as none of the tests resulted with a significance value under the threshold value of alpha = 0.05.

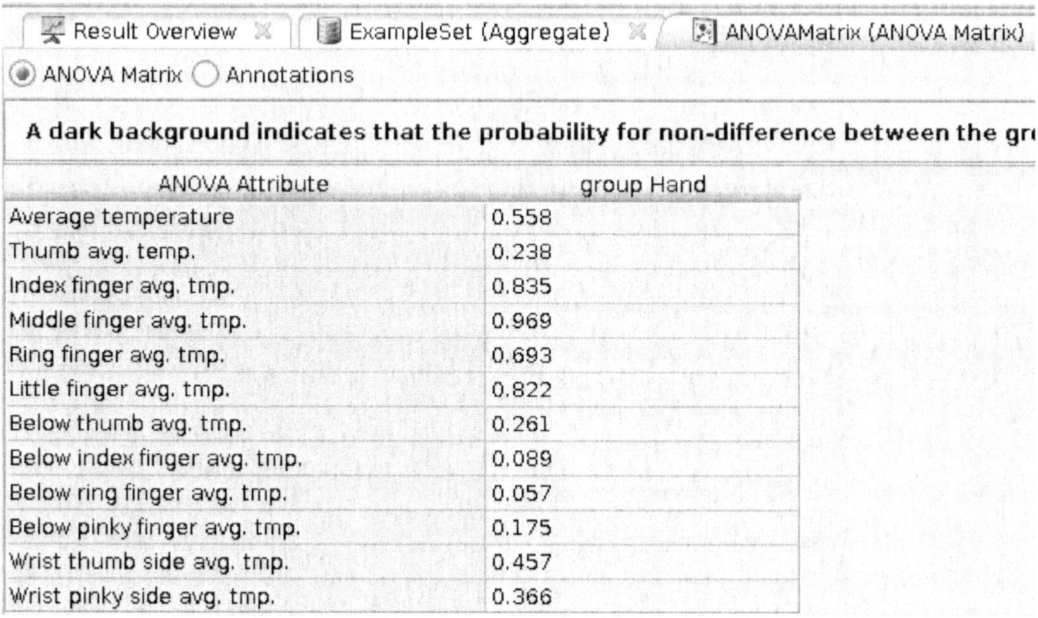

Result Overview | ExampleSet (Aggregate) | ANOVAMatrix (ANOVA Matrix)

ANOVA Matrix | Annotations

A dark background indicates that the probability for non-difference between the gr

ANOVA Attribute	group Hand
Average temperature	0.558
Thumb avg. temp.	0.238
Index finger avg. tmp.	0.835
Middle finger avg. tmp.	0.969
Ring finger avg. tmp.	0.693
Little finger avg. tmp.	0.822
Below thumb avg. tmp.	0.261
Below index finger avg. tmp.	0.089
Below ring finger avg. tmp.	0.057
Below pinky finger avg. tmp.	0.175
Wrist thumb side avg. tmp.	0.457
Wrist pinky side avg. tmp.	0.366

FIGURE 17.15: ANOVA matrix.

There are many different possibilities for displaying the achieved results. Figure 17.16 depicts the differences between segment temperatures obtained with an Aggregate operator in a matrix format. Figure 17.17 shows a Series plotter, where all average temperature values of different segments are presented for the palmar and dorsal hand sides.

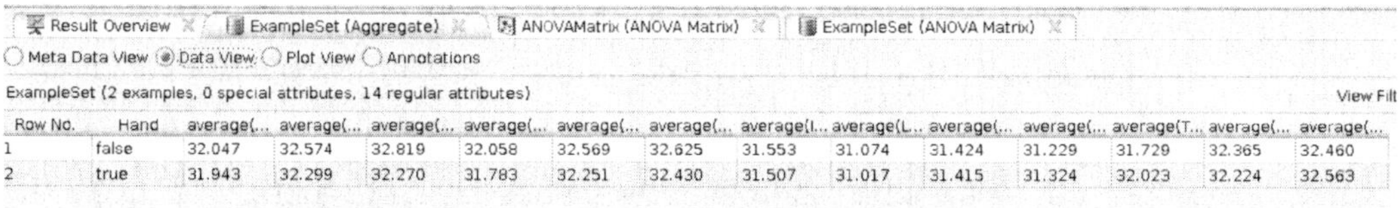

Result Overview | ExampleSet (Aggregate) | ANOVAMatrix (ANOVA Matrix) | ExampleSet (ANOVA Matrix)

Meta Data View | Data View | Plot View | Annotations

ExampleSet (2 examples, 0 special attributes, 14 regular attributes) View Filt

Row No.	Hand	average(...	average(...	average(...	average(...	average(...	average(...	average(I...	average(L...	average(...	average(...	average(T...	average(...	average(...
1	false	32.047	32.574	32.819	32.058	32.569	32.625	31.553	31.074	31.424	31.229	31.729	32.365	32.460
2	true	31.943	32.299	32.270	31.783	32.251	32.430	31.507	31.017	31.415	31.324	32.023	32.224	32.563

FIGURE 17.16: ExampleSet after Aggregate operator.

17.5 Knowledge Discovery Process in RapidMiner: Diabetes

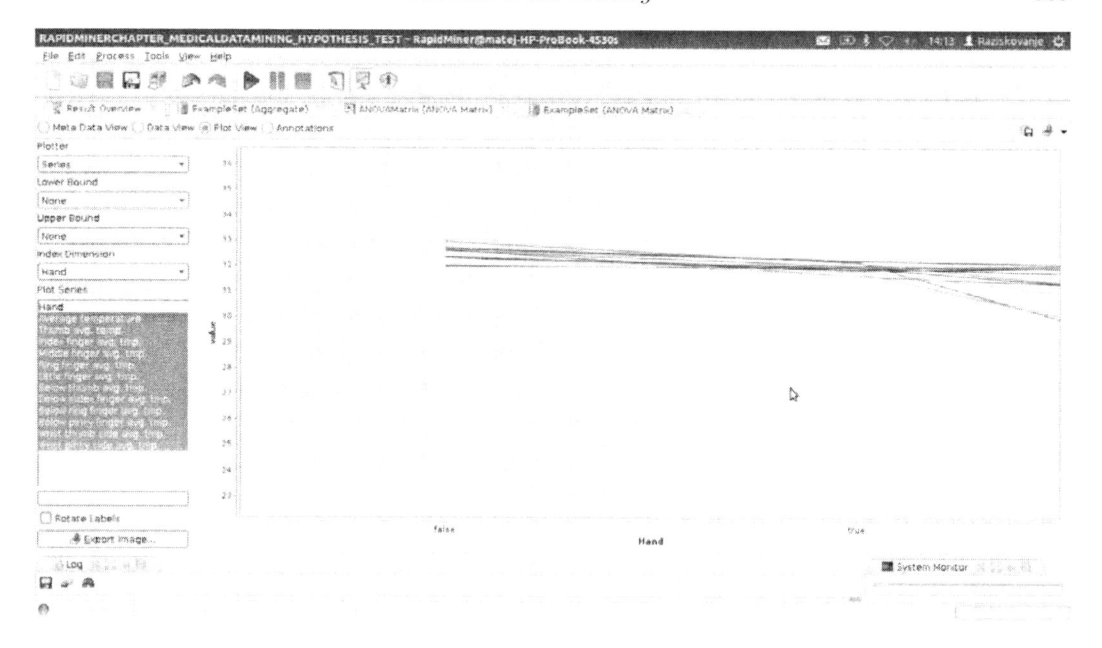

FIGURE 17.17: Plot for temperature differences.

17.5.1 Problem Definition, Setting the Goals

In the study of diabetes the various research questions were posed with the aim to evaluate the level of knowledge and overall perceptions of diabetes mellitus Type 2 (DM) within the older population in North-East Slovenia. The following differences in knowledge about diabetes with regard to the education and place of living were in focus of the study: (1) diet, (2) HbA1c, (3) hypoglycaemia management, (4) activity, (5) effect of illness and infection on blood sugar levels, and (6) foot care. The hypothesis about the different level of general knowledge of diabetes in an older population living in urban and rural areas was predefined within the study [12].

17.5.2 Data Preparation

Figure 17.18 presents the Meta Data View in RapidMiner on the diabetes dataset. There are 29 attributes of integer type as obtained from the questionnaire (scale from one to five). Figure 17.18 shows the characteristics of the data.

Figure 17.19 depicts the whole RapidMiner data mining process for diabetes. In the pre-processing phase we used similar operators to those we used in our previous example, however in this case, the MissingValues operator is introduced as we are dealing with missing data. The following operators are used in the whole process:

- ReplaceMissingValues: replace missing values in example set with average values (1).
- SelectAttributes: select attributes defined by experts (2).
- Clustering: search for clusters (3).
- SetRole: set new attribute cluster for label (4).
- Validation: make training model (5).
  - DecisionTree: use decision tree learner (5.1).

ExampleSet (225 examples, 0 special attributes, 30 regular attributes)

Role	Name	Type	Statistics	Range	Missings
regular	Serial number	integer	avg = 113 +/- 65.096	[1.000 ; 225.000]	0
regular	Gender	integer	avg = 1.401 +/- 0.491	[1.000 ; 2.000]	3
regular	Age	integer	avg = 69.714 +/- 10.366	[38.000 ; 92.000]	1
regular	Education	integer	avg = 1.818 +/- 0.789	[1.000 ; 5.000]	0
regular	Place of living	integer	avg = 1.476 +/- 0.501	[1.000 ; 2.000]	0
regular	Employment status	integer	avg = 2.733 +/- 0.641	[1.000 ; 3.000]	0
regular	Month income	integer	avg = 1.732 +/- 0.666	[1.000 ; 4.000]	5
regular	Duration of diabetes	integer	avg = 2.095 +/- 0.828	[1.000 ; 3.000]	4
regular	Number of medicament	integer	avg = 3.807 +/- 1.399	[1.000 ; 5.000]	2
regular	High blood presaure	integer	avg = 0.738 +/- 0.441	[0.000 ; 1.000]	4
regular	Kidney malfunction	integer	avg = 0.068 +/- 0.252	[0.000 ; 1.000]	4
regular	Heart disease	integer	avg = 0.353 +/- 0.479	[0.000 ; 1.000]	4
regular	Neural disease	integer	avg = 0.104 +/- 0.306	[0.000 ; 1.000]	4

FIGURE 17.18: Meta Data View of the diabetes dataset.

- ApplyModel: apply model to test set (5.2).
- Performance: measure performance (5.3).

Attributes selected for the experiment based on the expert's suggestions are presented in Figure 17.20.

Replace Missing Values Operator

This operator replaces missing values in a dataset. Different possibilities for replacing values are available (single, subset, regular expression, value type, block type, no missing values). We chose All since we wanted to replace all missing values in the data. An average value was chosen for replacing missing values (other possibilities are minimum, maximum, zero, and define value).

17.5.3 Modeling

Considering the goals of the study and the nature of diabetes data, descriptive data mining models were selected for discovering relationships in the data to form different hypotheses (see Figure 17.20). In RapidMiner we can choose many different descriptive operators and their variations under Clustering and Segmentation. The classical K-means cluster algorithm was selected for the task and a decision tree was applied on the cluster as a symbolic method for representation of attribute dependencies (see Table 17.4) We can see that two new operators were selected in the modelling phase: the cluster operator and the decision tree operator.

K-means Operator

This operator is an implementation of the k-means algorithm. The operator creates a cluster attribute in the dataset. The following algorithm parameters are required:

- K: the number of clusters which should be detected.
- Max runs: the maximal number of runs of k-means with random initialization.
- Max optimization steps: the maximal number of iterations performed for one run of k-means.

Following parameters were selected for our purpose.

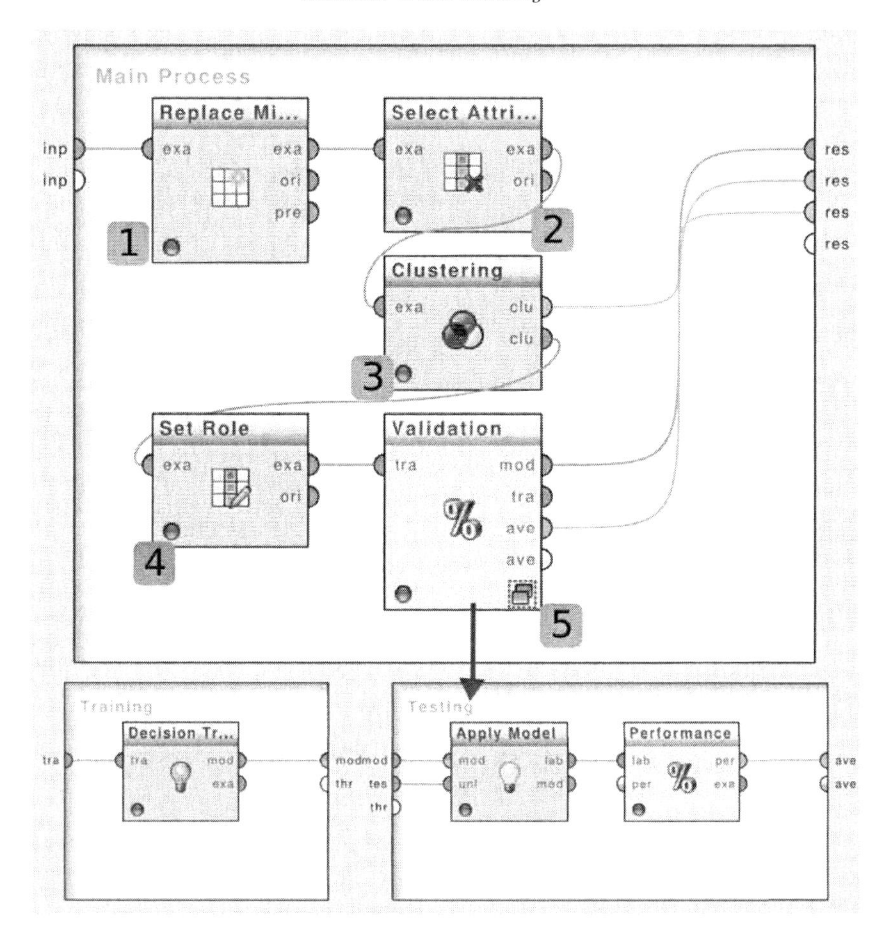

FIGURE 17.19: Descriptive data mining—searching for clusters.

- K: 3
- Max runs: 10
- Max optimization steps: 100

The Add Cluster attribute was selected to insert the cluster into a new dataset on which a decision tree was applied for representation of attribute dependencies.

Decision Tree Operator

Decision trees are powerful classifiers which are often easily understood. In order to classify an example, the tree is traversed bottom-down. Every node in a decision tree is labelled with an attribute. The value for this attribute determines which of the splits will be taken in next the iteration. The decision tree learner in RapidMiner works similar to Quinlan's C4.5 or CART [15] where different purity measures for creating the split are possible and can be selected by the user (information gain, gain ratio, gini index, etc.). In our example the parameters were set as follows:

- criterion: gain ratio
- minimal size for split: 4

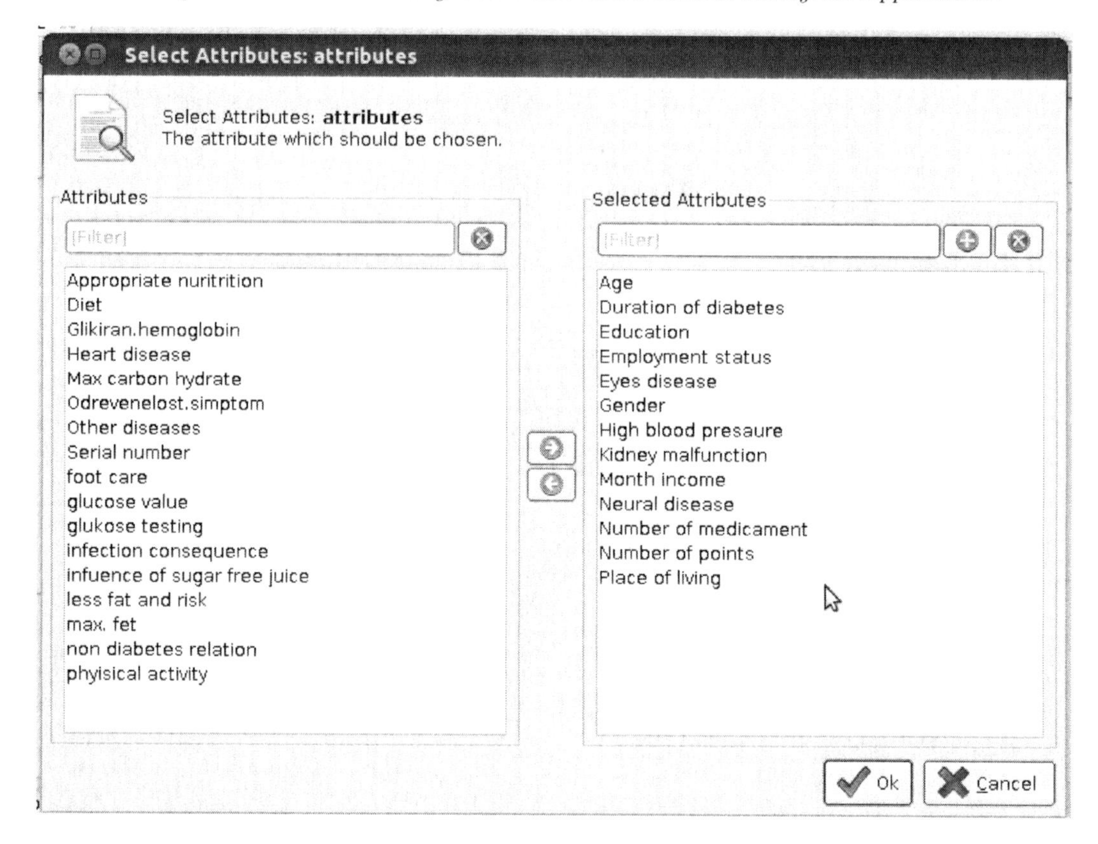

FIGURE 17.20: Diabetes—attribute selection.

- minimal leaf size: 2
- minimal gain: 0.1
- maximal depth: 20
- confidence: 0.25
- number of prepruning alternatives: 3

Pre-pruning and pruning were both selected for building the model.

17.5.4 Results and Data Visualization

In Figure 17.21 we can see the results of the first experiment. The following ports are presented in ResultView: ExampleSet (original), Cluster model, Decision Tree, and Performance Vector. There are 3 defined clusters with the following distributions: Cluster 0: 115, Cluster 1: 80, Cluster 2: 30. We can also see different dependencies between attributes that were selected by the SelectAttribute operator (see Figure 17.20). It can be seen that attributes Age and Number of points are the ones that had the greatest influence in defining the clusters. However, various representations of attributes can be produced for constructing the hypothesis for the study. Figure 17.22 presents various trees that were produced by removing the attributes that contributed to the splits from previous experiments. In other words, every additional tree was produced by removing the attributes from tree nodes of

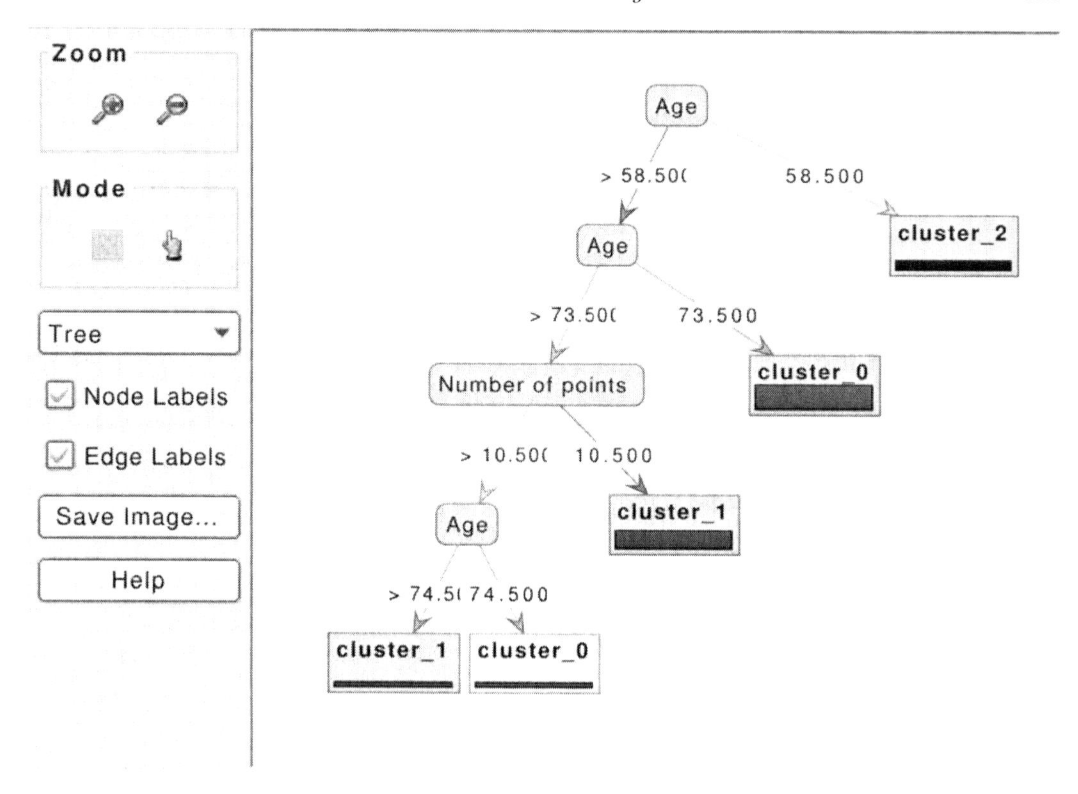

FIGURE 17.21: ResultViews for diabetes.

the parent experiment by using the SelectAttribute operator. Relationships between those attributes in different decision trees are appropriate for further statistical analysis that may be performed. An example of a simple statistical test for testing one of the study's hypothesis is performed in RapidMiner.

17.5.5 Hypothesis Testing

The hypothesis about the different level of general knowledge of diabetes in older populations living in urban and rural areas was defined before the experiments. We can see from the experiments above that the most important attributes for distinguishing individual clusters were age, number of points, and number of medicaments. The following statistical test was implemented in RapidMiner to test our hypothesis (see Figure 17.23). Operators ReplaceMissingValues (1), FilterExamples (2), SelectAttributes (3), Clustering (4), and NumericalToBinominal (5) were used. The Grouped Anova (6) operator was selected for this test where the number of points was selected for comparison and place of living was the attribute defining different groups. We can see in the figure that in this scenario there was no significant difference between the older people living in urban and rural areas. However further statistical analysis may be performed according to relationships between attributes in decision trees.

1st experiment

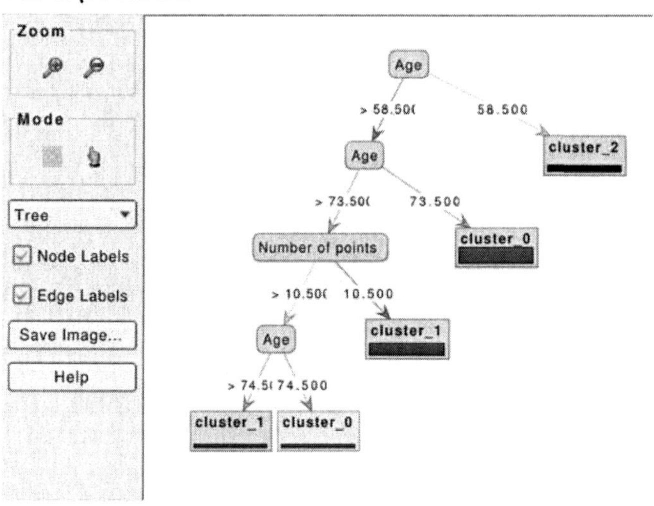

2nd experiment

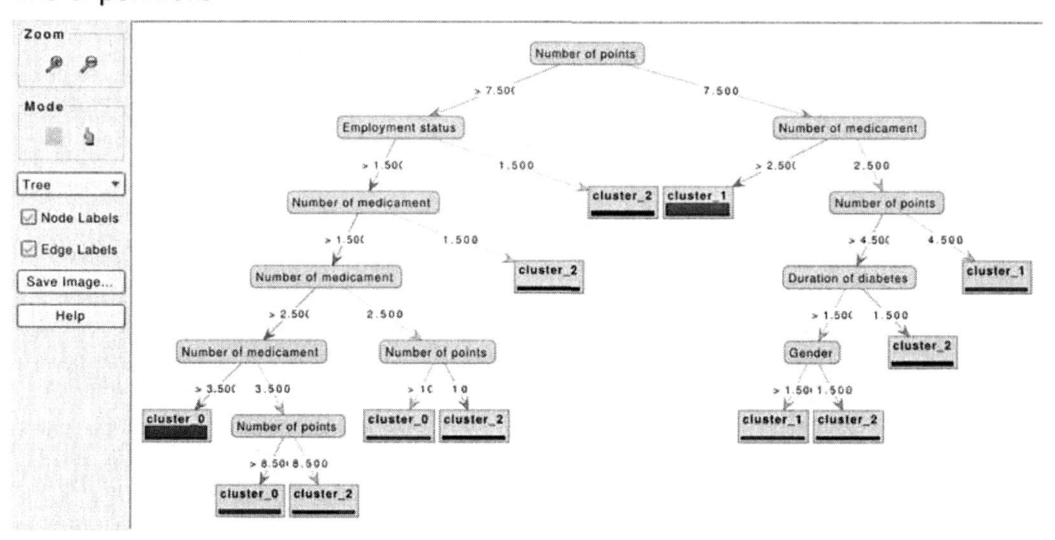

FIGURE 17.22: Various trees.

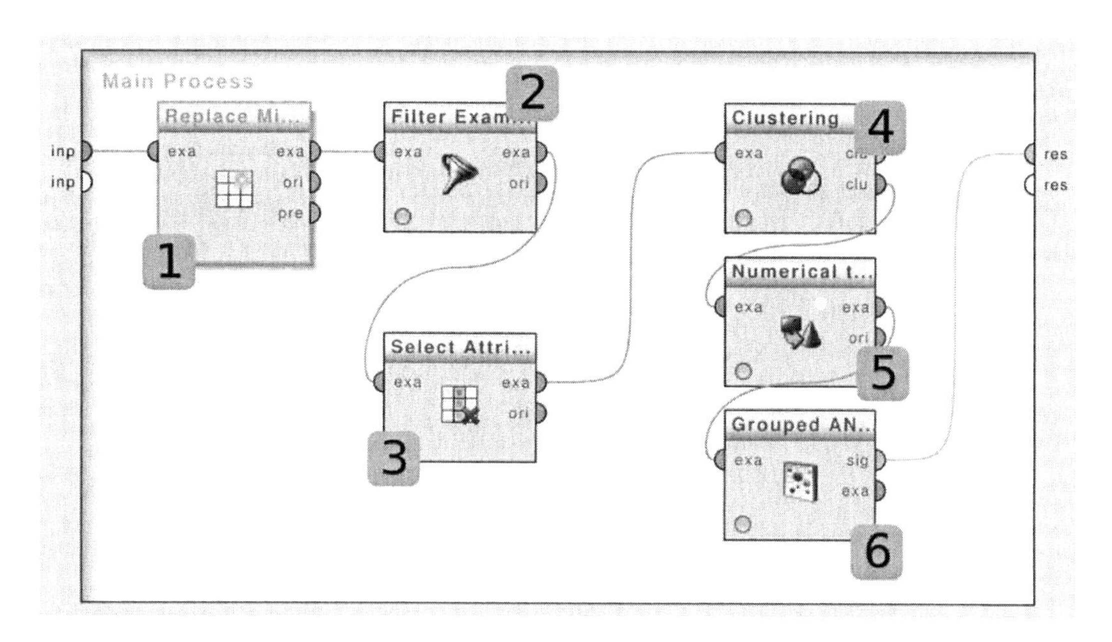

FIGURE 17.23: Hypothesis testing process for diabetes.

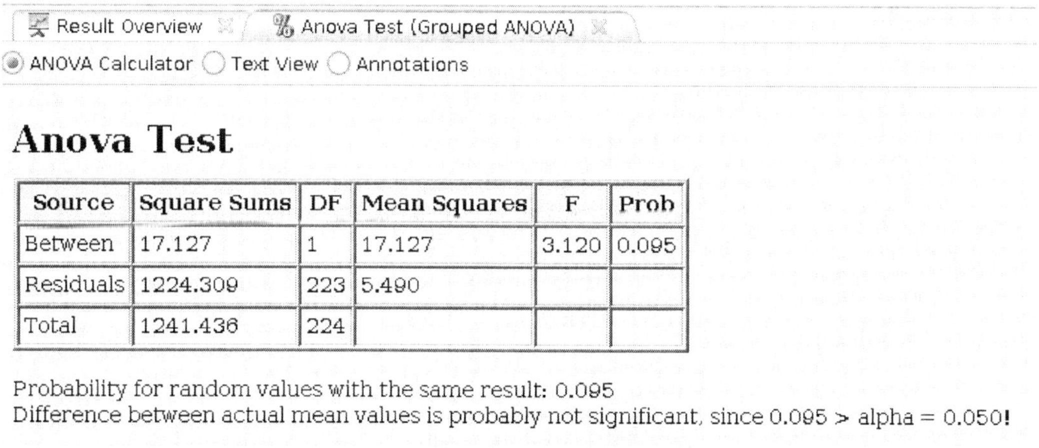

Result Overview | Anova Test (Grouped ANOVA)

ANOVA Calculator | Text View | Annotations

Anova Test

Source	Square Sums	DF	Mean Squares	F	Prob
Between	17.127	1	17.127	3.120	0.095
Residuals	1224.309	223	5.490		
Total	1241.436	224			

Probability for random values with the same result: 0.095
Difference between actual mean values is probably not significant, since 0.095 > alpha = 0.050!

FIGURE 17.24: Result of the hypothesis test.

17.6 Specifics in Medical Data Mining

In this section we summarize and outline some specifics concerning data and data mining in medicine.

- *Different types of data*
  When dealing with research problems in medicine, researchers collect different types of data from a vast range of procedures, ranging from simple, such as anthropometric measurements, to much more complicated, like acquiring complex images and signals, for example, by means of MRI, EEG, and EMG equipment.

- *Specific assumptions*
  The acquired data is more often used to test specific assumptions, and for this reason, it is usually quite reasonable to translate a medical research problem into a statistical problem with emphasis on formulation of a proper hypothesis which must be verified by means of an appropriate statistical test. Careful analysis of the test outcome is needed to properly interpret the research results.

- *Prone to measurement errors and bias*
  As with other data acquired by observation, medical data is also prone to measurement errors and bias, which usually stems from an inability to create a representative sample from a target population (as is the case when dealing with interesting but very rare medical conditions).

- *Ethics and privacy*
  Apart from the already mentioned statistical approach to medical data analysis, there are some other specific issues. One of them is the question of ethics and privacy of personal data, which puts constraints on availability of data for data mining. Another issue which shouldn't be overlooked is the importance of (wrong) conclusions derived from medical data—they can often mean the difference between life and death.

- *Heterogeneity*
  Maybe the most important issue, at least when looking strictly from the data mining point of view, is the heterogeneity of medical data. As already mentioned, the data pertaining to a specific diagnosis or research problem usually comes from many different sources and special data mining techniques or data pre-processing is needed to include all the necessary information in the analysis. It is also important and often difficult to include the interpretation of acquired data by different medical experts. They rely on years of experience and such additional information is hard to translate into quantifiable data, especially when it comes from different experts. This also emphasizes the need for unification of medical expressions when recorded in a database. Unfortunately there are many medical concepts for which there are yet no canonical forms or preferred notations [16].

17.7 Summary

In this chapter we presented data mining in the field of medicine and two examples of medical studies analysed with RapidMiner 5.3. Applying data mining to the medical field is a very challenging process due to the specifics of the medical profession. Data mining in medicine may differ from standard data mining practises in the process of knowledge discovery with its classical approach of predefining a hypothesis. The data mining results are obtained with the goal of rejecting or confirming this hypothesis. Another difference lies in the focus on data, where medicine is also interested in those minority cases that do not conform to the patterns and trends. We then presented some data mining techniques frequently documented in medicine that originated from predictive and descriptive mining and some of the classification and regression algorithms, production of association rules and clustering techniques. Two studies from medicine were presented: carpal tunnel syndrome and diabetes, where the knowledge discovery (data mining) process with RapidMiner is explained and designed. The appropriate RapidMiner operators were explained along with associated algorithms and the iterative process of constructing different models on those two scenarios. Readers can use RapidMiner and generate the models on the described datasets. The processes were described in detail, keeping in mind that knowledge discovery is a strictly iterative process with many possibilities and solutions. For this reason it is necessary to define the problem and its scope, and cooperation of experts from medicine and data analysis is crucial. We concluded and summarized the chapter with some specifics concerning data and data mining in the medical field as medical data is prone to measurement errors, bias, and heterogeneity, and must conform to strict ethics and privacy standards.

Bibliography

[1] Anna Shillabeer and John F Roddick. Establishing a lineage for medical knowledge discovery. In *Proceedings of the Sixth Australasian Conference on Data mining and Analytics. Volume 70*, pages 29–37. Australian Computer Society, Inc., 2007.

[2] Krzysztof J Cios and G William Moore. Uniqueness of medical data mining. *Artificial Intelligence in Medicine*, 26(1):1–24, 2002.

[3] Jeremy D. P. Bland. Carpal tunnel syndrome. *BMJ*, 335(7615):343–346, 8 2007.

[4] Richard Rosenbaum and José Ochoa. Carpal tunnel syndrome & other disorders of the meridian nerve. 2002.

[5] K Ammer. Diagnosis of nerve entrapment syndromes by thermal imaging. In *[Engineering in Medicine and Biology, 1999. 21st Annual Conf. and the 1999 Annual Fall Meeting of the Biomedical Engineering Soc.] BMES/EMBS Conference, 1999. Proceedings of the First Joint*, volume 2, pages 1117–vol. IEEE, 1999.

[6] K Ammer. Nerve entrapment and skin temperature of the human hand. *A Case Book of Infrared Imaging in Clinical Medicine*, pages 94–96, 2003.

[7] RT Herrick, SK Herrick, et al. Thermography in the detection of carpal tunnel syndrome and other compressive neuropathies. *The Journal of Hand Surgery*, 12(5 Pt 2):943, 1987.

[8] T Schartelmüller and K Ammer. Infrared thermography for the diagnosis of thoracic outlet syndrome. *Thermologie Österreich*, 6:130–134, 1996.

[9] S Tchou, JF Costich, RC Burgess, and CE Wexler. Thermographic observations in

unilateral carpal tunnel syndrome: report of 61 cases. *The Journal of Hand Surgery*, 17(4):631, 1992.

[10] American Academy of Neurology [AAN]. Report of the American Academy of Neurology and Therapeutics and Technology Assessment Subcommittee. Assessment: Thermography in neurologic practice. *Neurology*, 40(3):523–525, 1990.

[11] Hackett MEJ. The place of thermography in medicine. *Acta Thermographica*, 1:176–180, 1976.

[12] E Turk, M Palfy, V Prevolnik Rupel, and A Isola. General knowledge about diabetes in the elderly diabetic population in Slovenia. *Zdrav Vestn*, 81(7-8):517–525, 2012.

[13] Daniel T Larose. Discovering knowledge in data: An introduction to data mining. 2005.

[14] Nikola K Kasabov. *Foundations of Neural Networks, Fuzzy Systems, and Knowledge Engineering.* MIT press, 1996.

[15] J. Ross Quinlan. Induction of decision trees. *Machine Learning*, 1(1):81–106, 1986.

[16] K J Cios and Moore G W. Uniqueness of medical data mining. *Artif Intell Med*, 26(1–2):1–24, 2002.

Part VII

Molecular Structure- and Property-Activity Relationship Modeling in Biochemistry and Medicine

Chapter 18

Using PaDEL to Calculate Molecular Properties and Chemoinformatic Models

Markus Muehlbacher
Department of Psychiatry and Psychotherapy, University Hospital of Erlangen-Nuremberg, Friedrich-Alexander-University Erlangen, Germany; Computer Chemistry Center, Friedrich-Alexander-University Erlangen, Germany

Johannes Kornhuber
Department of Psychiatry and Psychotherapy, University Hospital of Erlangen-Nuremberg, Friedrich-Alexander-University Erlangen, Germany

18.1 Introduction

The Pharmaceutical Data Exploration Laboratory (PaDEL) is a software suite developed at the University of Singapore to calculate a variety of properties and fingerprints of molecules based on their molecular structure [1]. These properties are also named molecular descriptors in chemoinformatics, since they represent a molecular property by a numerical value. The aim of this use case is to demonstrate how easy a chemoinformatic prediction model, showing the relation between a number of calculated molecular descriptors and a biological property, can be established. Models based on simple molecular descriptors can be used to safe valuable resources and replace or support experiments.

18.2 Molecular Structure Formats for Chemoinformatics

The molecular structure is either given as a Simplified Molecular-Input Line-Entry System (SMILES) formatted data file, where each line represents the two-dimensional (2D) structure of one compound, or as an structure data format (SDF) formatted file, where the molecules are sequentially listed (either as 2D or 3D structures). Both SMILES and SDF structure files can be freely downloaded from the PubChem website [2]. Moreover, tools such as PubChemSR, allow the download of multiple structures in one process [3].

The attributes that can be calculated by PaDEL include a broad range of molecular properties, such as electrotopological state descriptors, topochemical atom descriptors, molecular volume descriptors, linear free energy relation descriptors, ring counts, and counts of chemical substructures. In total, PaDEL is able to calculate 861 different attributes. The 2D structures of the molecules (determined by their connectivity exclusively) are sufficient to compute 727 of these attributes, whereas 134 attributes are calculated on the basis of the molecules' 3D structure.

18.3 Installation of the PaDEL Extension for RapidMiner

Besides the standalone version, PaDEL is also available as an extension to the RapidMiner software suite. This extension enables the direct calculation of properties for a large number of molecules and the generation of a predictive model in a single process. The current version of the PaDEL extension can be downloaded directly from `http://padel.nus.edu.sg/software/padeldescriptor`. To integrate the PaDEL extension, the following steps must be performed:

1. Download the current version of the PaDEL extension from the website (version 2.10 as of May 2012). Unpack the archive.
2. Copy the file PaDEL-Descriptor.jar to the \lib\ plugins .
3. Copy all jar files into \ lib.

At the next program start, RapidMiner should show the folder "Chemistry" in the operators tree in the middle of the left side panel (see Figure 18.1).

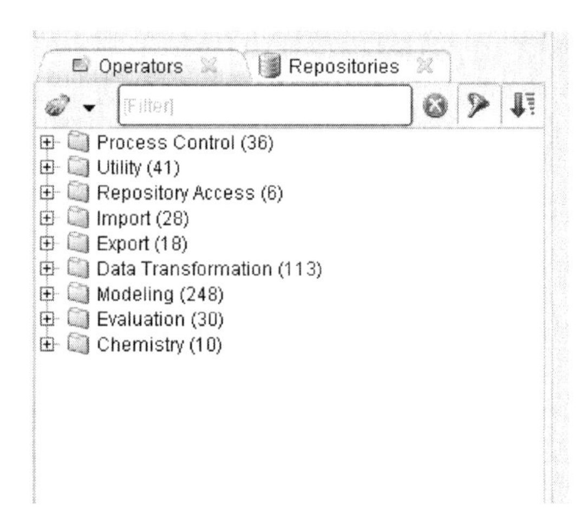

FIGURE 18.1: After successful installation the operator tree shows the *Chemistry* folder, which includes all 10 operators of the PaDEL extension, after installation.

This folder includes all 10 operators provided by the PaDEL extension and enables RapidMiner to directly handle molecular structures. All operators are outlined in the following table:

TABLE 18.1: Operator descriptions.

Folder	*Operator name*	*Description*
Import	Read Compounds	This operator is similar to all Read-Operators. It is used to import molecular compounds in any common molecular structure format, such as SDF or SMILES.
Export	Write Compounds	This operator is the counterpart to the Read Compounds operator. It exports a set of molecules in a corresponding format, such as SDF or SMILES, to an external file.
Basic Information	Count Substructure	This operator counts the number of substructures that are not covalently bound.
	Detect Dimensionality	This operator uses the coordinates of a molecule to identify its dimensionality (0D, 1D, 2D, or 3D).
Compound Transformation	Remove Salt	This operator removes counterions from a molecule by assuming that the largest fragment is the desired molecule.
	Convert to 3D	This operator converts a 2D (planar) molecular structure to 3D coordinates.
	Add Hydrogens	This operator adds hydrogen atoms to given structures.
	Detect Aromaticity	This operator detects aromaticity in a molecule.
	Rename Compound	This operator changes a compound name.
Descriptor	Calculate Descriptors	This operator does the main part of the computational work, as it calculates the selected molecular properties as attributes.

18.4 Applications and Capabilities of the PaDEL Extension

The combination of data mining (RapidMiner) and a tool to handle molecules (PaDEL) provides a convenient and user-friendly way to generate accurate relationships between chemical structures and any property that is supposed to be predicted, mostly biological activities. Relationships can be formulated as qualitative structure-property relationships (SPRs), qualitative structure-activity relationships (SARs), or quantitative structure-activity relationships (QSARs). SPR models aim to highlight associations between molecular structures and a target property, such as lipophilicity. SAR models correlate an activity with structural properties and QSAR models quantitatively predict an activity. Models are typically developed to predict properties that are difficult to obtain, expensive to measure, require time-consuming experiments, or are based on a variety of other complex properties. They may also be useful to predict complicated properties using several simple properties.

One of the major advantages of RapidMiner and PaDEL is that properties can be calculated quickly with little additional effort or cost in a single process. Another advantage is that properties can be calculated for any molecular structure, even if the compound is not physically accessible. An important feature of the presented solution is that all involved software, namely RapidMiner and PaDEL, are freely available and thus do not produce additional licensing costs.

18.5 Examples of Computer-aided Predictions

A large number of chemical properties and biological activities have been subjected to computer-aided studies aimed at developing models to be predicted *in silico* without the need to synthesize and test drug candidates. A chemoinformatic model bypasses this drawback by predicting the target property using a number of properties that are much easier to calculate. Conversely, models can be used to provide insight into properties influencing the target property.

These prediction systems can be split into qualitative and quantitative models. Quantitative models result in a metric value, whereas qualitative models yield a (mostly binary) classification, depending on the nature of the target value. A typical example of a metric target value would be the partition coefficient between 1-octanol and water (logP) [4][5], whereas P-glycoprotein transporter inhibition would be appropriate for a qualitative classification system [6]. As outlined before, many prominent studies have been conducted on a variety of target properties, which include the following fields:

- **Prediction of carcinogenicity** [7][8]: A lot of chemicals are known to increase the risk for cancer and therefore need to be handled carefully.
- **Blood-brain barrier permeability** [9][10][11]: The blood-brain barrier is essential to keep unwanted or toxic compounds out of the central nervous system (CNS), but allow CNS-active compounds to localize at their target.
- **Ion channel affinities** [12][13]: Ion channels are especially important for signal transduction and therefore of general importance.
- **Enzyme inhibition** [14][15]: Enzymes are vitally important to keep metabolism up. Inhibition of specific enzymes modulates metabolic pathways.
- **Transporter affinities** [16]: Active transporters are involved in metabolic pathways, whenever passive diffusion is not sufficient.
- **Intestinal drug absorption** [17]: Most drugs are applied in oral form. Therefore, oral drug absorption is a prerequisite for orally applied drugs.
- **Lipophilicity** [18][19][20]: Lipophilic environments such as lipid membranes play a key role in cellular biology. Thus it is important to assess and predict the lipophilic character of a molecule.

- **ADME properties** [21][22][23][24]: ADME is an acronym for absorption, distribution, metabolism, and excretion. It condenses several important fields of drug distribution and metabolism.
- **Aqueous solubility** [25][26][27]: The vast majority of relevant systems in biology and chemistry are based on aqueous solutions. It is an integral requirement to have knowledge about the aqueous solubility, even though when it is only predicted by a model.

18.6 Calculation of Molecular Properties

First, we need to import a set of chemical structures from a text file containing the SMILES- or SDF-formatted data, using the "Read Compounds" operator. When running the import process, the numbers of atoms, bonds, fragments, and hydrogen atoms are calculated by default. Next, we highly recommend that the molecules should be passed to the "Remove Salt" operator to remove counterions, if included. In case any 3D descriptors are calculated afterward, the set of molecules must be converted into 3D structures, as the SMILES data format only contains the 2D structure. This can be easily accomplished using the "Convert to 3D" operator. To apply this process to all compounds of an example set, the corresponding checkbox must be activated. Obviously, if only 1D and 2D descriptors are calculated, this step can be skipped. Subsequently, the set of molecules can be passed over to the "Calculate Descriptors" operator, which does the main work of calculating the molecular descriptors. Please note that a log file must be specified. By default, only 1D/2D descriptors are calculated; additionally, 3D descriptors can be calculated by activating the checkbox. This operator also provides an option to generate the 3D structures, in case the corresponding operator has not been applied before. Please note that the output of the "Calculate Descriptors" operator is an example set containing the calculated properties; in contrast, the input has been a set of molecules.

18.7 Generation of a Linear Regression Model

In the following we want to establish a multiple linear regression model as a prime example of a simple chemoinformatic prediction model. The model generation process begins with the import of an example set, to which a learner is applied to generate a model and predict the label attribute (y-value). The dataset can be either imported from an external file (CSV, Excel, etc.) or obtained directly from the repository. An appropriate example set includes an "id" attribute, which is used to identify the examples. By default this attribute is ignored by learner operators to construct the model. Every example set also requires a "label" attribute, which represents the attribute to be predicted and a number of regular attributes (the properties of the molecules). With such an example set, a learner is able to generate a model predicting the label attribute using the regular attributes. RapidMiner offers a large variety of operators to generate prediction models appropriate for different types of example sets. All learner operators can be found in the repository folder "Modeling". The choice of the actual learner is crucial for the performance of the resulting model. Additionally, the value type of the label attribute already suggests a group of learners. Binominal labels, for example, are appropriate for decision tree or random forest learners, whereas numerical labels are appropriate for multiple linear regression, support vector machines, or artificial neural network learners. A model must be applied, independent of the used learner, to an example set before the performance is evaluated by the corresponding performance operator. The "Apply Model" operator generates the "prediction" attribute based on the supplied model. As long as the example set contains all attributes included in the model it is able to calculate a prediction attribute. Thus, this workflow can also be used to predict properties or activities for compounds with an unknown label attribute.

18.8 Example Workflow

As listed above, aqueous solubility has been subjected to QSAR studies several times. In chemistry, it is vitally important to accurately determine the aqueous solubility of a compound. This property is given as the aqueous solubility (in gram per liter) or as logarithm (logS). We selected this important parameter as a prime example of how to create a simple prediction system using RapidMiner and the PaDEL extension.

Initially, we need a set of compounds with their experimental aqueous solubility measurements to train our model. The quality of the training set and the number of examples are crucial to the quality of the generated model. Obviously, training sets should be as accurate and as large as possible. The dataset used as the example set here was originally published by Kramer and colleagues [28]. This set contains measured aqueous solubility in g/l for 789 compounds as well as their corresponding structures in the SMILES format. As shown in Figure 18.2, we separated the SMILES data from the numerical data, as they are processed differently. Initially, we multiplied the imported example set (Figure 18.2 #**2**); the first copy consists of the SMILES attribute only and is saved again as a ".smi" file (Figure 18.2 #**4**). The second copy contains all other attributes (most importantly, the measured solubility) (Figure 18.2 #**5**). Next, the .smi file is imported with the "Read Compounds" operator (Figure 18.2 #**7**) as the set of molecules. The molecules are then preprocessed ("Remove Salt" Figure 18.2 #**8**) and forwarded to the "Calculate Descriptors" operator (Figure 18.2 #**9**), where the main calculation process is performed. In this example, we only calculate 1D/2D properties. Please note that for 3D properties, the structures must have been converted to 3D previously. The "Calculate Descriptor" output (Figure 18.2 #**9**) is an example set (not a set of molecules) with the selected calculated properties. To join them to the original dataset, we must generate an ID attribute (Figure 18.2 #**10**). Finally, the complete example set is stored in the repository (Figure 18.2 #**12**). We suggest saving the import and descriptor calculation processes separately, as the fully prepared example set can then be easily exchanged with other analogous example sets for model generation.

With the stored example set, we can start building a prediction model (see Figure 18.3). Before we can hand the example set over to the learner operator, we must ensure that there are no missing values within the example set, as this can cause the learner to stop. For this purpose, the corresponding operator is applied (Figure 18.3 #**3**). The example set, consisting of id, label, and 789 regular attributes without missing values, is then passed over to the learner operator. Since we want to demonstrate the basic assembly of a model generation process, we use the simple "Linear Regression" operator (Figure 18.3 #**7**) as learner. A multiple linear regression, constructs a linear combination of several regular

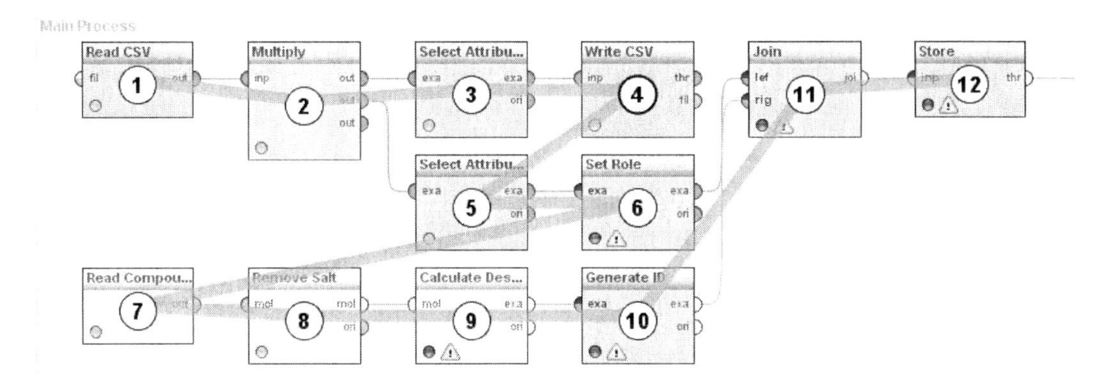

FIGURE 18.2: The workflow for the complete import, preprocessing, and descriptor calculation process. The operators 7, 8, and 9 are provided by the PaDEL extension.

attributes to predict the label attribute. The feature selection parameter determines the algorithm used to find the optimal (best performing while smallest) number of attributes for the regression which is "M5 Prime", by default [29]. A linear regression without any kind of feature selection would use all of the supplied attributes to generate the model. Example sets with few examples and many attributes would entail the risk of over-fitting. Since this workflow should only exemplify the basic components of the process, we use the default settings. We want to emphasize that the choice of an appropriate learner and the optimization of the parameters are crucial to obtain a high performance. In the next step, the regression model is applied to the example set ("Apply Model", Figure 18.3 #**8**) to calculate and add the prediction of the label to the example set. Finally, the "Performance (Regression)" (Figure. 18.3 #**9**) operator calculates the selected performance criteria, such as the squared correlation, from the labeled example set. For the performance operator (Figure 18.3 #**9**), the example set must include a label and a prediction attribute from which the performance can be calculated.

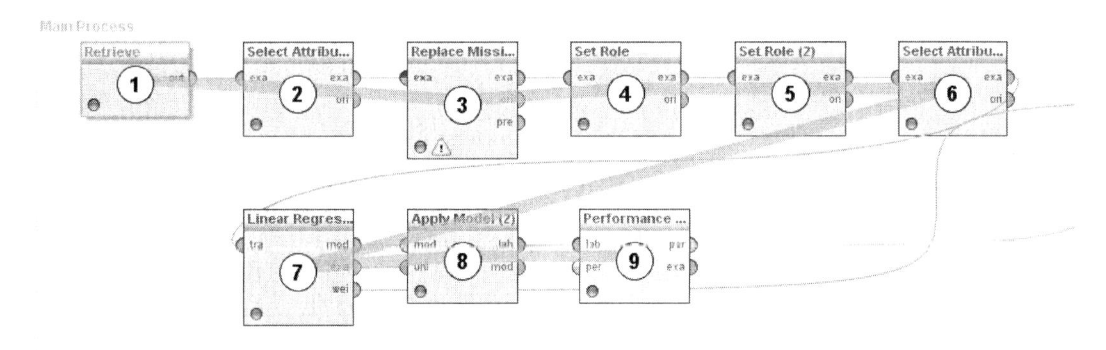

FIGURE 18.3: Example workflow to generate a simple quantitative model for the prediction of aqueous solubility.

To summarize the second process, the "Linear Regression" operator (Figure 18.3 #**7**) selected 312 out of 723 regular attributes building a linear combination to predict the label attribute. This linear regression model achieved a squared correlation of 0.621 and a root mean squared error of 58.4. Figure 18.4 shows the result of this multiple linear regression attempt. Although the performance using this simple workflow is already reasonable, it clearly leaves lots of space for improvement and optimization.

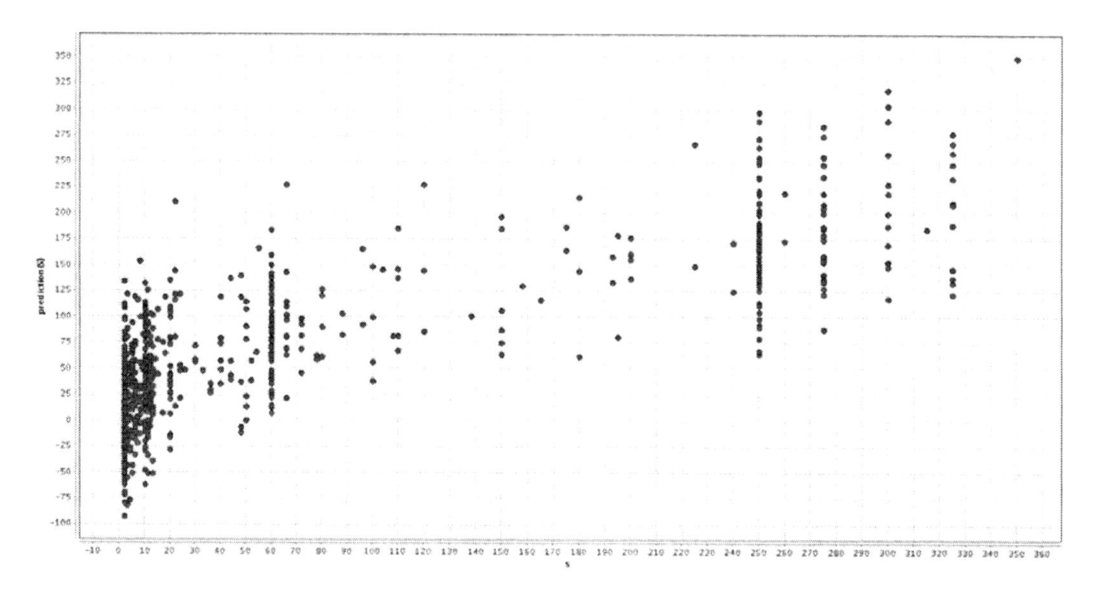

FIGURE 18.4: This scatterplot shows the experimental solubility (S) and the prediction of S.

In many cases a validation is performed to avoid chance correlations and to separate the training and test process. For this example workflow we did not want to complicate the process by including a validation, although we highly recommend it for productive workflows.

18.9 Summary

In this chapter, we explain the usability of the PaDEL extension for RapidMiner and highlight its application to generate predictive models for chemical, biochemical, or biological properties based on molecular properties, which is a frequently encountered task in the chemistry and pharmaceutical industries. A combination of RapidMiner and PaDEL provides an Open Source solution to generate prediction systems for a broad range of biological properties and effects.

Acknowledgment

A portion of this work was supported by funding from the German Ministry of Education and Research (BMBF Grant 01EX1015B).

Bibliography

[1] C.W. Yap. PaDEL-descriptor: An open source software to calculate molecular descriptors and fingerprints. *Journal of Computational Chemistry*, 32(7):1466–1474, 2011.

[2] Q. Li, T. Cheng, Y. Wang, and S.H. Bryant. PubChem as a public resource for drug discovery. *Drug Discovery Today*, 15(23-24):1052–1057, 2010.

[3] J. Hur and D. Wild. PubChemSR: A search and retrieval tool for PubChem. *Chemistry Central Journal*, 2(1):11, 2008.

[4] C. Kramer, B. Beck, and T. Clark. A surface-integral model for log POW. *Journal of Chemical Information and Modeling*, 50(3):429–436, 2010.

[5] M. Muehlbacher, A. El Kerdawy, C. Kramer, B. Hudson, and T. Clark. Conformation-dependent QSPR models: logPOW. *Journal of Chemical Information and Modeling*, 51(9):2408–2416, 2011.

[6] P. de Cerqueira Lima, A. Golbraikh, S. Oloff, Y. Xiao, and A. Tropsha. Combinatorial QSAR modeling of P-glycoprotein substrates. *Journal of Chemical Information and Modeling*, 46(3):1245–1254, 2006.

[7] N.X. Tan, H.B. Rao, Z.R. Li, and X.Y. Li. Prediction of chemical carcinogenicity by machine learning approaches. *SAR and QSAR in Environmental Research*, 20(1-2):27–75, 2009.

[8] N.C.Y. Wang, R. Venkatapathy, R.M. Bruce, and C. Moudgal. Development of quantitative structure-activity relationship (QSAR) models to predict the carcinogenic potency of chemicals. II. Using oral slope factor as a measure of carcinogenic potency. *Regulatory Toxicology and Pharmacology*, 59(2):215–226, 2011.

[9] H. Chen, S. Winiwarter, M. Fridin, M. Antonsson, and O. Engkvist. In silico prediction of unbound brain-to-plasma concentration ratio using machine learning algorithms. *Journal of Molecular Graphics and Modelling*, 29(8):985–995, 2011.

[10] G. Prabha and V. Jitender. In silico prediction of blood brain barrier permeability: An Artificial Neural Network model. *Journal of Chemical Information Modeling*, 46(1):289–97, 2006.

[11] M. Muehlbacher, G.M. Spitzer, K.R. Liedl, and J. Kornhuber. Qualitative prediction of blood-brain barrier permeability on a large and refined dataset. *Journal of Computer-Aided Molecular Design*, 25:1095–1106, 2011.

[12] O.S. Smart, J. Breed, G.R. Smith, and M.S. Sansom. A novel method for structure-based prediction of ion channel conductance properties. *Biophysical Journal*, 72(3):1109 – 1126, 1997.

[13] P. Willett, D. Wilton, B. Hartzoulakis, R. Tang, J. Ford, and D. Madge. Prediction of ion channel activity using binary kernel discrimination. *Journal of Chemical Information and Modeling*, 47(5):1961–1966, 2007.

[14] J. Kornhuber, L. Terfloth, S. Bleich, J. Wiltfang, and R. Rupprecht. Molecular properties of psychopharmacological drugs determining non-competitive inhibition of 5-HT_{3A} receptors. *European Journal of Medicinal Chemistry*, 44(6):2667–2672, 2009.

[15] J. Kornhuber, M. Muehlbacher, S. Trapp, S. Pechmann, A. Friedl, M. Reichel, C. Mühle, L. Terfloth, T.W. Groemer, and G.M. Spitzer. Identification of novel functional inhibitors of acid sphingomyelinase. *PloS ONE*, 6(8):e23852, 2011.

[16] L. Zhang, P.V. Balimane, S.R. Johnson, and S. Chong. Development of an in silico model for predicting efflux substrates in Caco-2 cells. *International Journal of Pharmaceutics*, 343(1):98–105, 2007.

[17] F. Johansson and R. Paterson. Physiologically Based in Silico Models for the Prediction of Oral Drug Absorption. In Carsten Ehrhardt and Kwang-Jin Kim, editors, *Drug Absorption Studies*, volume VII of *Biotechnology: Pharmaceutical Aspects*, pages 486–509. Springer US, 2008.

[18] H.F. Chen. In silico log P prediction for a large data set with support vector machines, radial basis neural networks and multiple linear regression. *Chemical Biology & Drug Design*, 74(2):142–147, 2009.

[19] J. Ghasemi and S. Saaidpour. Quantitative structure-property relationship study of n-octanol-water partition coefficients of some of diverse drugs using multiple linear regression. *Analytica Chimica Acta*, 604(2):99–106, 2007.

[20] M. Ikuo. Simple method of calculating octanol/water partition coefficient. *Chemical & Pharmaceutical Bulletin*, 40(1):127–130, 1992.

[21] D. Butina, M.D. Segall, and K. Frankcombe. Predicting ADME properties in silicon: Methods and models. *Drug Discovery Today*, 7(11):S83–S88, 2002.

[22] P.M. Gleeson. Generation of a set of simple, interpretable ADMET rules of thumb. *Journal of Medicinal Chemistry*, 51(4):817–834, 2008.

[23] R. Narayanan and S.B. Gunturi. In silico ADME modelling: Prediction models for blood–brain barrier permeation using a systematic variable selection method. *Bioorganic & Medicinal Chemistry*, 13(8):3017–3028, 2005.

[24] J. Shen, F. Cheng, Y. Xu, W. Li, and Y. Tang. Estimation of ADME properties with substructure pattern recognition. *Journal of Chemical Information and Modeling*, 50(6):1034–1041, 2010.

[25] D.S. Cao, Q.S. Xu, Y.Z. Liang, X. Chen, and H.D. Li. Prediction of aqueous solubility of druglike organic compounds using partial least squares, back-propagation network and support vector machine. *Journal of Chemometrics*, 24(9):584–595, 2010.

[26] J. Huuskonen. Estimation of aqueous solubility for a diverse set of organic compounds based on molecular topology. *Journal of Chemical Information and Computer Sciences*, 40(3):773–777, 2000.

[27] C. Kramer, B. Beck, and T. Clark. Insolubility classification with accurate prediction probabilities using a MetaClassifier. *Journal of Chemical Information and Modeling*, 50(3):404–414, 2010.

[28] C. Kramer, T. Heinisch, T. Fligge, B. Beck, and T. Clark. A consistent dataset of kinetic solubilities for early-phase drug discovery. *Chem Med Chem*, 4(9):1529–1536, 2009.

[29] Y. Wang and I.H. Witten. Inducing model trees for continuous classes. In *Hamilton, New Zealand: University of Waikato, Department of Computer Science*, 1996.

Chapter 19

Chemoinformatics: Structure- and Property-activity Relationship Development

Markus Muehlbacher
Department of Psychiatry and Psychotherapy, University Hospital of Erlangen-Nuremberg, Friedrich-Alexander-University, Erlangen; Computer Chemistry Center, Friedrich-Alexander-University Erlangen, Germany Germany

Johannes Kornhuber
Department of Psychiatry and Psychotherapy, University Hospital of Erlangen-Nuremberg Erlangen, Germany

19.1 Introduction

The process of drug design from the biological target to the drug candidate, and subsequently the approved drug has become increasingly expensive. Therefore, strategies and tools that reduce costs have been investigated to improve the effectiveness of drug design [1] [2]. Among the time-consuming and cost-intensive steps are the selection, synthesis and experimental testing of drug candidates. Numerous attempts have been made to reduce the number of potential drug candidates for experimental testing. Several methods that rank compounds with respect to their likelihood to act as an active drug have been developed and applied with variable success [3]. *In silico* methods that support the drug design process by reducing the number of promising drug candidates are collectively known as virtual screening methods [4]. If the structure of the target protein has been discovered and is available, this information can be used for the screening process. Docking is an appropriate method for structure-based virtual screening [5][6][7]. Ligand-based virtual screening employs several known active ligands to rank other drug candidates using physiochemical property pre-filtering [8][9], pharmacophore screening [10][11], and similarity search [12] methods. All methods share the common goal to reduce the number of drug candidates subjected to biological testing and increase the efficacy of the drug design process [13][14]. In this chapter, we want to demonstrate an *in silico* method to predict biological activity based on RapidMiner workflows. This chapter is based on our previous experience using RapidMiner for chemoinformatic predictions [15][16][17]. In a companion chapter, we demonstrate how to process the chemoinformatic descriptors generated in PaDEL.

19.2 Example Workflow

We want to demonstrate the basic components of an accurate and validated model to predict inhibition of acid sphingomyelinase (ASM), which has been previously published by our group [15]. The ASM enzyme is located at the inner lysosomal membrane, where it cleaves sphingomyelin into ceramide and phosphorylcholine. The activity of ASM is correlated to several disease patterns, such as major depression [18], multiple sclerosis [19][20], and respiratory disease [21][22], which accounts for the interest in modulation and prediction of ASM activity [15][23]. Moreover, the balance between ceramide and its metabolite sphingosine-1-phosphate is assumed to be involved in the signaling pathways of cell proliferation and cell death [24][25][26]. Therefore, drugs modulating the activity of ASM have been closely studied for a wide range of drug candidates and drugs commonly used in clinical treatment. Functional inhibitors of the sphingomyelinase acid (FIASMAs) are frequently found among antidepressants and neuroleptics [27].

Recently, we tested a large set of compounds with respect to their ability to inhibit ASM [15]. Based on these data, we developed an *in silico* model to predict inhibitors of ASM using RapidMiner. Most of the steps can be easily modified for application to a broad range of similar chemoinformatic tasks.

19.3 Importing the Example Set

First, we need an example set to train the model. Each example represents a molecular compound with a corresponding "id" attribute that identifies the example. This attribute should be excluded from the modeling process (as it is for most operators a per-default settings). The "label" represents the target property that ought to be predicted (in this case, the experimentally measured inhibition of the ASM). Finally, our example sets contains a large number of regular attributes used for model generation. The regular attributes in our case were provided by MOE 2009.10 [28] and ACD/Labs [29], two software packages dedicated to calculating molecular properties from chemical structures. In total, we started with 296 compounds (examples) with measured activities and 462 regular attributes for each compound, imported from the repository (Fig. 19.1 #1). The properties can contain non-numerical values and some properties are only filled with numerical values for specific compounds (i.e., acidity constants).

19.4 Preprocessing of the Data

In most cases, the raw example set needs to be pre-processed to avoid errors and warnings in the workflow. We want to highlight some of the prerequisites that should to be considered when preparing an example set for RapidMiner. We will only explain the preprocessing steps that are important to our example workflow. The actual arrangement of the operators might differ or be insufficient when another example set is employed. Nevertheless, several common issues and things to keep in mind are addressed in this example workflow.

- **Retrieve (Figure 19.1 #1):** The data import requires special treatment, especially when working with molecular structures. The default character signaling a comment is "#". In chemistry, this character usually represents a triple bond and should be replaced by some other character that is not used in chemistry (i.e., "§"). Moreover, the "name" attribute was automatically assigned as a "polynomial" type. Although this is not wrong, we suggest the "nominal" type for this attribute. In addition, we highly recommend manually checking the imported data, for example, by setting a breakpoint after the example set is loaded to determine whether the import was executed as expected.

- **Rename attributes (Figure 19.1 #2):** In general, attribute names are chosen to be intuitive and easy to understand for the reader. They frequently contain characters, such as "+", "-", "(",")" and "/". These attribute names can cause problems when selecting attributes with a regular expression or when generating additional attributes from existing attributes because some symbols represent mathematical operations. Therefore, we highly recommend replacing characters that are also used by RapidMiner as signaling characters.

- **Label type conversion:** For our example set, a qualitative prediction is more appropriate as the label attribute is bimodal distributed [15]. Therefore, we converted the numerical activities into binominal values (positive or negative) using the "Dichomatize" (Figure 19.1 #3) and "Nominal to Binominal" (Figure 19.1 #4) operators. Binominal labels are required for qualitative models (i.e., decision trees, random forests, etc.), whereas numerical labels are appropriate for quantitative models (i.e., linear regression, neural network, support vector machines, etc.). Remap Binominals is applied to set more intuitive classifications for the label column (Figure 19.1 #5).

- **Changes in regular attributes**: For regular attributes, missing values can cause the learner to stop. Therefore, all missing values need to be replenished. Finally, we applied "Select attributes" (Figure 19.1 #6) to retain only numerical regular attributes. In addition, all nominal and polynominal attributes (such as the compound name) are removed (Figure 19.1 #7). Special attributes (such as the "id" and "label" attribute) can be retained, if they are excluded from the modeling process, which is the default setting for most operators.

- **Generation of additional attributes**: RapidMiner is able to calculate linear combinations and mathematical derivatives from given attributes using the "Generate Attributes" operator (Figure 19.1 #8, #9). In the example workflows, we calculate size-intensive variations for all of the attributes (Figure 19.1 #9). The idea of size-intensive attributes is to normalize the attributes based on the molecular weight of a compound [30]. In other words, each attribute is divided by its molecular weight. This is an operation that can be automatically performed for all of the attributes using the "Loop Attributes" (Figure 19.1 #9)) operator.

- **No missing values (Figure 19.1 #10):** For some operators it is necessary to have example sets without any missing values. For this reason the "Replenish Values" operator can be used.

Finally, we sort (Figure 19.1 #11) and store (Figure 19.1 #12) the preprocessed example set consisting of 910 attributes in the repository to run the model generation on this prepared example set. The complete workflow is shown in Figure 19.1.

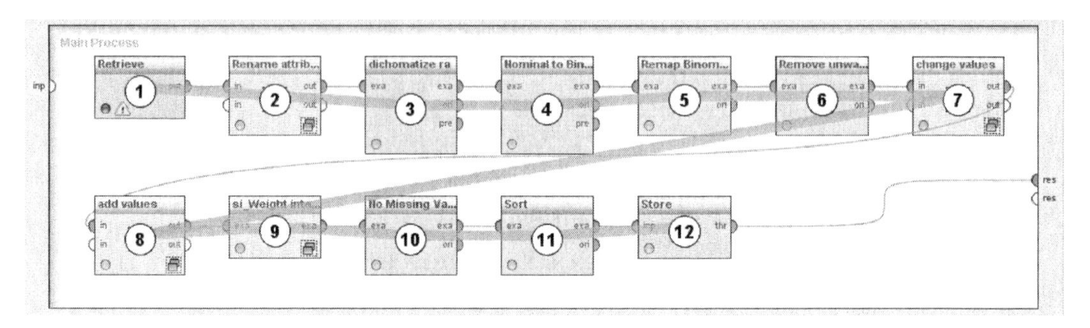

FIGURE 19.1: This workflow outlines the import and preprocessing of the example dataset. It includes several operators for the preparation of the example set.

19.5 Feature Selection

There are several reasons to reduce the number of attributes prior to model generation:

- A reduction of attributes dramatically reduces the computational time required to calculate and validate models because the number of attribute combinations increases exponentially with the number of given attributes.
- A large number of attributes can lead to over-fitting, especially when the number of examples is low. In addition, a low number of attributes enables more sophisticated and complex algorithms to search for predictive models.
- A high number of correlated attributes indicates redundancy, while it requires more computational time.
- A high number of attributes always increases the likelihood of chance correlations. Whenever the number of attributes is nearly equal or even larger than the number of examples, chance correlations might be obtained.

These points highlight the benefits of feature selection. RapidMiner offers a broad range of operators to weight attributes, which can be used for feature selection, according to correlation (Figure 19.2 (#4)), tree importance (Figure 19.2 #3), information gain (Figure 19.2 #2), relief (Figure 19.2 #5) and several more. In this case we use 4 filters based on the criteria named before. In combination with the "Select Attributes by Weight" operator, the 10 best scoring attributes with respect to the corresponding criterion are retained. After that, the 4 sets consisting of top 10 attributes each are joined (Figure 19.2 #8) to a dataset including potentially important attributes exclusively. In other words, we reduced the number of attributes and therefore the dimensionality. Attributes that have not shown to be in relation to the label attribute with respect to any of the used criteria have been removed from the set (e.g., if an attribute has no variance, it might not be useful for the modeling process, as this is done by the "Remove Useless Attributes" operator). Simultaneously, our workflow monitors the attribute names of the selected features, which captures multiple selection of an attribute (the second output of each filter process contains the names and numbers of the selected attributes) (Figure 19.2 #9). Please note that this workflow only uses some examples of the operators for feature selection. Altogether, this reduces the number of attributes from 910 to 41 attributes (40 regular and 1 label attributes). The number of regular attributes can be lower in case the same attribute is selected by two or more filtering algorithms.

19.6 Model Generation

After the preprocessing and feature selection is completed, the example set is ready for the model generation process. Because our dataset is built on a binominal label attribute, we can choose from a variety of appropriate learner operators, such as decision trees, logistic regression and random forest operators. We decided on the random forest operator (Figure 19.3 #6), since it is known to be robust, fast, and effective [31]. To avoid over-fitting and chance correlations, we aimed for a simple model based only on a handful of attributes. All random forest operators were limited to a depth of 3.

From the filtered example set consisting of 41 attributes, we had to find combinations of a small number of attributes that were able to generate models with high performance. The most important step here is to stop adding attributes when the performance does not significantly improve.

The fastest and most straightforward way to select features for model generation is to iteratively select and include the single attribute that gives the highest improvement in terms of performance. In the first step of this algorithm, the best attribute is selected with respect to the chosen performance criterion. Next, the second attribute is added to the model and so on. The search is terminated when no significant increase in the performance criterion is achieved by addition of another attribute. This so-called forward-stepping feature selection bears the risk that the combination of two alternative attributes might be superior to the combination of the first and second best performing attributes. However, the "Forward Selection" operator provides a simple and fast way to perform this feature selection algorithm in RapidMiner. The "Backward Elimination" operator is very similar to the forward feature selection. It starts with the complete attribute set and iteratively removes attributes until the performance falls by specified value. This algorithm provides the benefit that beneficial combinations of descriptors are not eliminated. On the other hand, complex models are more complicated to calculate and therefore require more computational time.

In contrast to the fast iterative selection of the single best attribute per generation, there is also the option of performing an exhaustive search. In this case, all possible combinations of attributes are calculated and evaluated by means of the selected performance criterion. Obviously, this algorithm is the most time-consuming, and it can be applied only for a very low number of attributes because the number of combinations scales exponentially with the

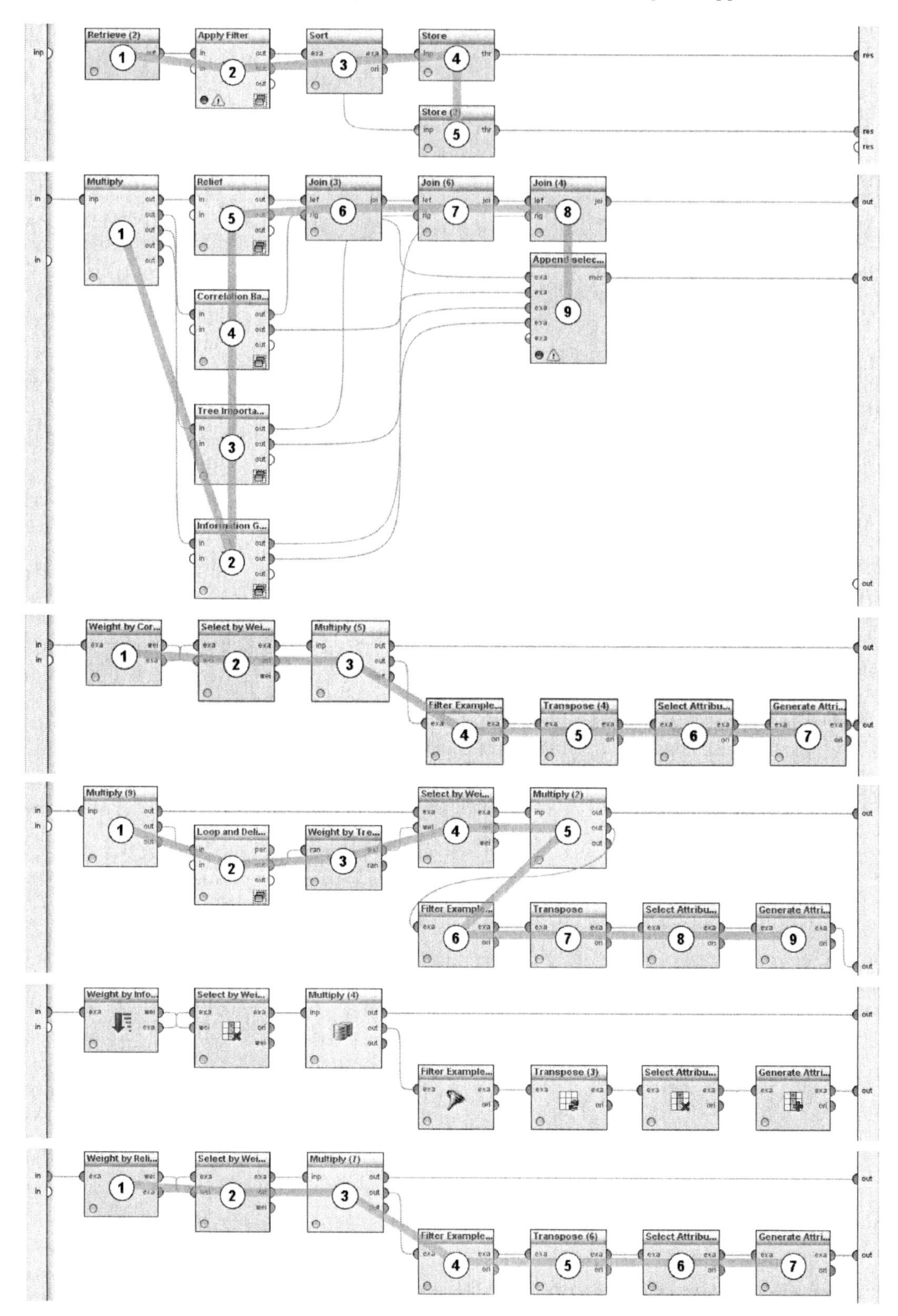

FIGURE 19.2: The figure illustrates the complete feature selection process. It is split into different nested processes for each selection process (according to a specific criterion) that are joined together afterward. Simultaneously, the names of the top 10 attributes from each weighting operator are captured.

number of attributes. RapidMiner has the ability to perform an exhaustive search using the "Loop Subset" operator. This operator generates every possible combination using a defined maximum and minimum number of attributes.

As a compromise between these two approaches (forward selection, exhaustive search), we employed the beam search algorithm [32]. Instead of using only the best or all of the attributes, it maintains the best N models per generation. In general, the first attributes selected have the highest impact on the performance. Therefore, those are the most important to be chosen correctly. When the width is set equal to the number of attributes, a beam search provides the benefit of an exhaustive search for the first two generations and limits the number of combinations thereafter. By this we limit the number of combinations to be tested and avoid the exponential growth and take advantage of including more than a single attribute in each generation. We calculated and validated the models for the N=41 best models per generation (note that 41 is the number of attributes in our example set). In RapidMiner, the beam search is performed using the "Optimize Search" operator (see Figure 19.3 #3).

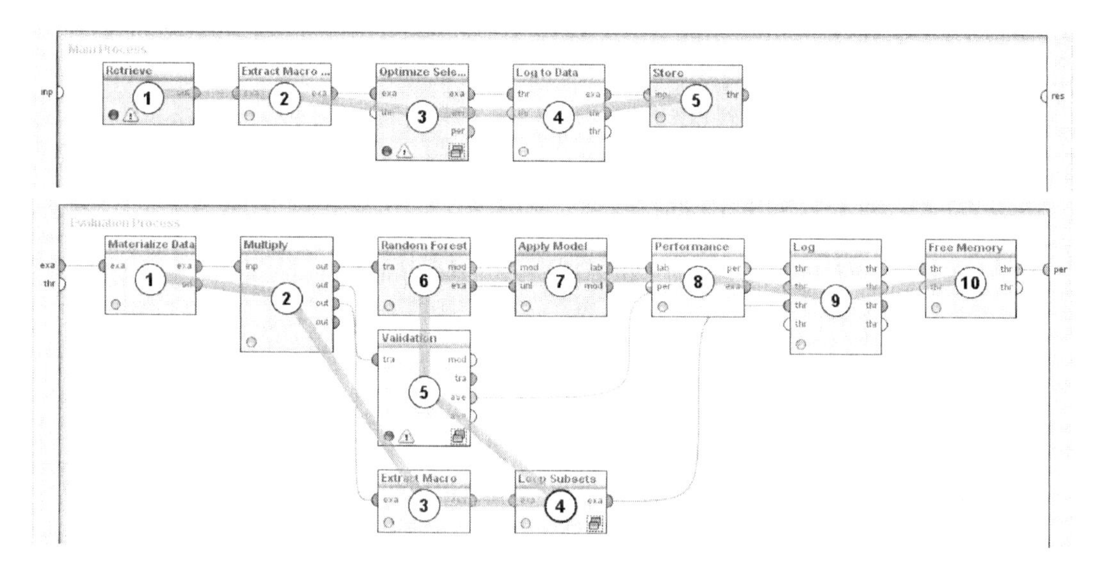

FIGURE 19.3: The workflow illustrates the nested process of the "Optimize Selection" operator performing a beam search.

19.7 Validation

Validation is one of the most important steps in generating predictive models [33][34]. It ensures that a model is not only predictive for the dataset it was trained on, but also for external data on which the model has not been used for model generation previously. Essentially, the model generation is a mathematical process of minimizing the prediction error regardless of the method or example set. Over-fitting occurs when the model is specifically predicting the example set employed for training rather than representing a general relationship between the regular attributes and the label attribute. Validation is highly useful to avoid over-fitting. For example, if the number of examples and attributes is of the same magnitude, the learner might return a model that is fitted to the trained example set and only predictive for this specific example set. Validation is of utmost importance when working with a small example set, which is frequently the case in medicine or chemistry, where a limited amount of experimental data is a common issue.

Several algorithms are available to validate a model [35][36], such as leave-one-out, cross-validation [37], or bootstrapping. All of the validation processes evaluate the performance of a model on an example set (i.e., test set), which has not been used for the training. The validation methods differ in how they split the example set into a training and test set. To perform a validation, a learner needs to be nested into the validation operators. The validation process consists of two nested chains: one for the training set and the other for the test set. The learner is applied to the training set, and the model is connected to the output of the first chain (Figure 19.4, left side). The second chain applies this model to the test set (Figure 19.4, right side). Then, the labeled example set is employed to generate a validated performance measure. The whole validation process is repeated several times to even out the effects of different splits in the training and test set. Figure 19.4 shows the composition needed to assemble a simple validation process (i.e., "Bootstrap Validation"). This composition can be employed for any type of validation operator (optionally with minor modifications).

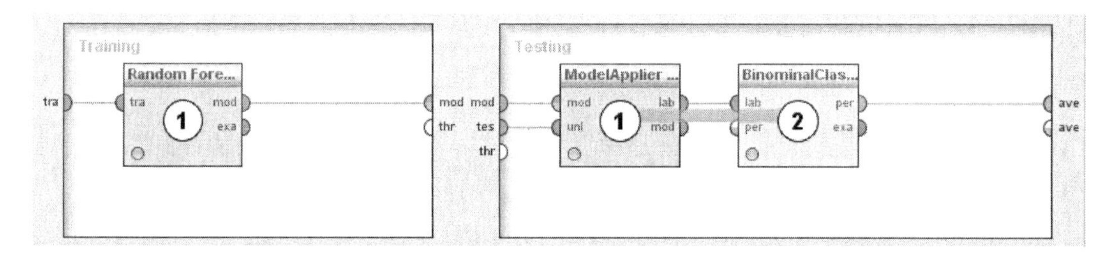

FIGURE 19.4: A validation consists of a nested training and testing process. The training process is used to train the model using an appropriate learner. In the testing process, the model is applied to the test example set and the corresponding performance is calculated.

19.8 Y-Randomization

The performance criterion is calculated by a mathematical equation, which has a defined maximum and minimum value, such as 0 and 1 for the squared correlation, and should assess the quality of a prediction model. In practice, both extreme values are not typically obtained. Therefore, it is unclear if the obtained performance also stands for relationship that is performing significantly better than a zero hypothesis, which is the relationship between randomized label and regular attributes. When the labels are randomly permutated, the learner should not be able to generate a model with a reasonable performance (in contrast to the original performance).

In RapidMiner, a Y-randomization can be easily assembled using the "Permutation" operator (Figure 19.5 #4) on the label attribute. First, the example set is split into the label attribute and regular attributes because the permutation randomly shuffles the order of the examples with all their attributes. After the permutation, the regular attributes and the label attribute are joined together (Figure 19.5 #9). With the shuffled example set, model generation can be performed in the same way as it was performed for the original set. For a valid and predictive model, the performance for the permutated example sets should be considerably lower.

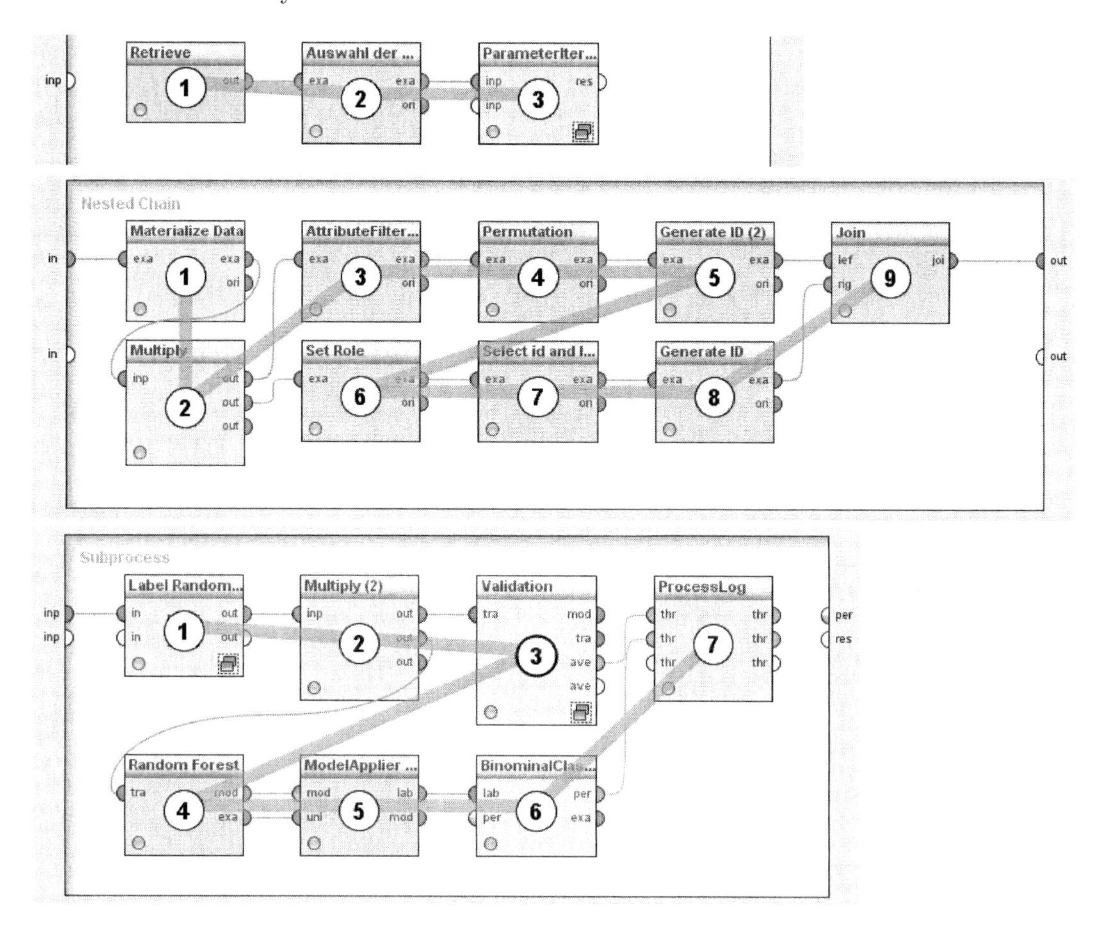

FIGURE 19.5: The Y-randomization workflow generates models based on randomly shuffled label values. These models should not be as predictive as the original model.

19.9 Results

Overall, this workflow has been used to predict functional inhibition of ASM [15]. The data used for the workflow described in this chapter can be downloaded from the supplemental material.

Please note that we describe a simplified workflow for this use case. For the workflow presented here we obtained the following results shown in Table 19.1. The zero-hypothesis performance obtained from the Y-randomization is below all other obtained models. Moreover the accuracy is improving from 1 to 4 attributes, but stays constant for 5 attributes.

TABLE 19.1: Summary of the results achieved using the workflow from this use case.

generation	feature names	Mean Ac racy
5	XLogP, si_prot_vsa_pol, si_TPSA, ACD_pKa1, si_prot_PEOE_VSA_POL	0.917
4	ACD_LogP, XLogP, si_prot_vsa_pol, prot_Qamines	0.917
3	XLogP, ACD_pKa1, neutral_PEOE_VSA_POL	0.902
2	ACD_LogP, si_neutral_vsa_pol	0.877
1	si_neutral_TPSA	0.841
-	Y-randomization	0.743

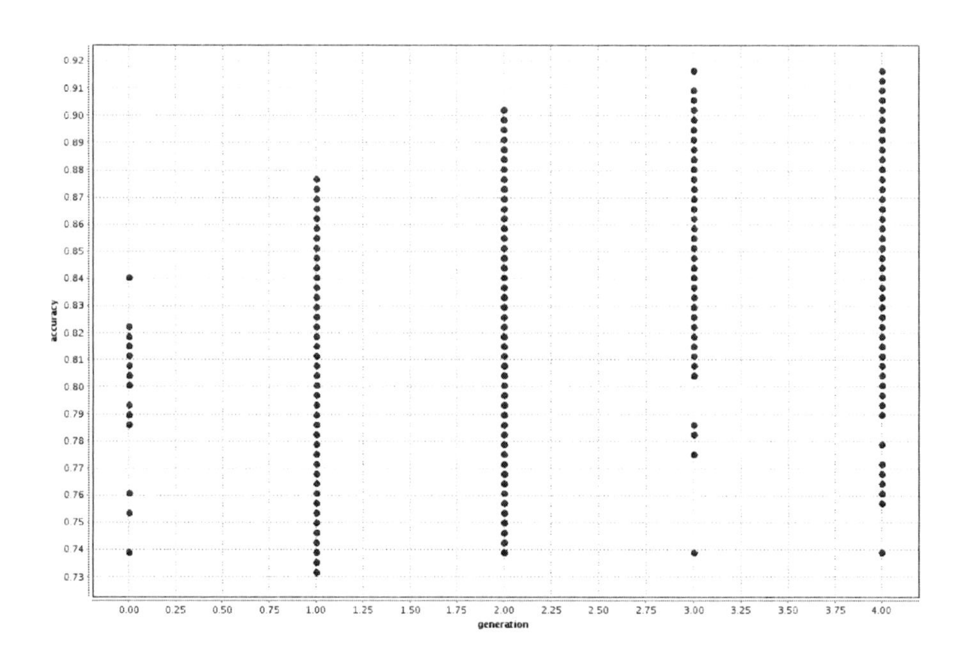

FIGURE 19.6: The plotted evaluation of the beam search shows that there is an increase of the accuracy up to 4 attributes. Please note that generation 0 represents the case of 1 selected attribute in Table 19.1.

19.10 Conclusion/Summary

In this chapter, we explain how to use RapidMiner to generate a model to predict biological activities, such as the inhibition of ASM. Starting with a large number of properties of the example set, we highlight several important preprocessing steps, followed by a feature selection to vastly reduce the number of attributes before performing a systematic search to find the most predictive model for the feature generation. Finally, we discuss why validation is necessary to avoid over-fitting. In addition, we show the usability and benefits of Y-randomization.

Acknowledgment

A portion of this work was supported by funding from the German Ministry of Education and Research (BMBF Grant 01EX1015B).

Bibliography

[1] C.P. Adams and V.V. Brantner. Spending on new drug development. *Health Economics*, 19(2):130–141, 2010.

[2] S.M. Paul, D.S. Mytelka, C.T. Dunwiddie, C.C. Persinger, B.H. Munos, S.R. Lindborg, and A.L. Schacht. How to improve R&D productivity: the pharmaceutical industry's grand challenge. *Nature Reviews Drug Discovery*, 9(3):203–214, 2010.

[3] Nature Publishing Group. A decade of drug-likeness. *Nature Reviews Drug Discovery*, 6(11):853–853, 2007.

[4] T.I. Oprea and H. Matter. Integrating virtual screening in lead discovery. *Current Opinion in Chemical Biology*, 8(4):349–358, 2004.

[5] D.B. Kitchen, H. Decornez, J.R. Furr, and J. Bajorath. Docking and scoring in virtual screening for drug discovery: methods and applications. *Nature Reviews Drug Discovery*, 3(11):935–949, 2004.

[6] U.H. Zaheer, B. Wellenzohn, K.R. Liedl, and B. Rode. Molecular docking studies of natural cholinesterase-inhibiting steroidal alkaloids from Sarcococca aligna. *Journal of Medicinal Chemistry*, 46(23):5087–5090, 2003.

[7] G. Schneider and H.J. Böhm. Virtual screening and fast automated docking methods. *Drug Discovery Today*, 7(1):64–70, 2002.

[8] P.D. Leeson and A.M. Davis. Time-related differences in the physical property profiles of oral drugs. *Journal of Medicinal Chemistry*, 47(25):6338–6348, 2004.

[9] C.A. Lipinski, F. Lombardo, B.W. Dominy, and P.J. Feeney. Experimental and computational approaches to estimate solubility and permeability in drug discovery and development settings. *Advanced Drug Delivery Reviews*, 2012.

[10] G. Wolber. 3D pharmacophore elucidation and virtual screening. *Drug Discovery Today: Technologies*, 7(4):e203–e204, 2010.

[11] J. Kirchmair, S. Ristic, K. Eder, P. Markt, G. Wolber, C. Laggner, and T. Langer. Fast and efficient in silico 3D screening: toward maximum computational efficiency of pharmacophore-based and shape-based approaches. *Journal of Chemical Information and Modeling*, 47(6):2182–2196, 2007.

[12] P. Willett. Similarity-based virtual screening using 2D fingerprints. *Drug Discovery Today*, 11(23):1046–1053, 2006.

[13] W.P. Walters, M.T. Stahl, and M.A. Murcko. Virtual screening - an overview. *Drug Discovery Today*, 3(4):160–178, 1998.

[14] T. Lengauer, C. Lemmen, M. Rarey, and M. Zimmermann. Novel technologies for virtual screening. *Drug Discovery Today*, 9(1):27–34, 2004.

[15] J. Kornhuber, M. Muehlbacher, S. Trapp, S. Pechmann, A. Friedl, M. Reichel, C. Mhle, L. Terfloth, T.W. Groemer, G.M. Spitzer, K.R. Liedl, E. Gulbins, and P. Tripal. Identification of novel functional inhibitors of acid sphingomyelinase. *PLoS ONE*, 6(8):e23852, 2011.

[16] J. Kornhuber, L. Terfloth, S. Bleich, J. Wiltfang, and R. Rupprecht. Molecular properties of psychopharmacological drugs determining non-competitive inhibition of $5 - HT_{3A}$ receptors. *European Journal of Medicinal Chemistry*, 44(6):2667–2672, 2009.

[17] M. Muehlbacher, G.M. Spitzer, K.R. Liedl, and J. Kornhuber. Qualitative prediction of blood–brain barrier permeability on a large and refined dataset. *Journal of Computer-Aided Molecular Design*, 25(12).

[18] J. Kornhuber, A. Medlin, S. Bleich, V. Jendrossek, A.W. Henkel, J. Wiltfang, and E. Gulbins. High activity of acid sphingomyelinase in major depression. *Journal of Neural Transmission*, 112(11):1583–1590, 2005.

[19] A. Jana and K. Pahan. Sphingolipids in multiple sclerosis. *Neuromolecular Medicine*, 12(4):351–361, 2010.

[20] S. Walter and K. Faßbender. Sphingolipids in multiple sclerosis. *Cellular Physiology and Biochemistry*, 26(1):49–56, 2010.

[21] Y. Yang and S. Uhlig. The role of sphingolipids in respiratory disease. *Therapeutic Advances in Respiratory Disease*, 5(5):325–344, 2011.

[22] K.A. Becker, J. Riethmüller, A. Lüth, G. Döring, B. Kleuser, and E. Gulbins. Acid sphingomyelinase inhibitors normalize pulmonary ceramide and inflammation in cystic fibrosis. *American Journal of Respiratory Cell and Molecular Biology*, 42(6):716, 2010.

[23] J. Kornhuber, P. Tripal, M. Reichel, L. Terfloth, S. Bleich, J. Wiltfang, and E. Gulbins. Identification of new functional inhibitors of acid sphingomyelinase using a structure property activity relation model. *Journal of Medicinal Chemistry*, 51(2):219–237, 2007.

[24] E.L. Smith and E.H. Schuchman. The unexpected role of acid sphingomyelinase in cell death and the pathophysiology of common diseases. *The FASEB Journal*, 22(10):3419–3431, 2008.

[25] A. Haimovitz-Friedman, S. Martin, D. Green, M. McLoughlin, C. Cordon-Cardo, E.H. Schuchman, Z. Fuks, and R. Kolesnick. Acid sphingomyelinase–deficient human lymphoblasts and mice are defective in radiation-induced apoptosis. *Cell*, 86:189–199, 1996.

[26] V. Teichgräber, M. Ulrich, N. Endlich, J. Riethmüller, B. Wilker, C. De Oliveira-Munding, A.M. van Heeckeren, M.L. Barr, G. von Kürthy, K.W. Schmid, B. Tmmler F. Lang H. Grassme G. Döring , M. Weller, and E. Gulbins. Ceramide accumulation mediates inflammation, cell death and infection susceptibility in cystic fibrosis. *Nature Medicine*, 14(4):382–391, 2008.

[27] J. Kornhuber, P. Tripal, M. Reichel, C. Mühle, C. Rhein, M. Muehlbacher, T.W. Groemer, and E. Gulbins. Functional Inhibitors of Acid Sphingomyelinase (FIASMAs): a novel pharmacological group of drugs with broad clinical applications. *Cellular Physiology and Biochemistry*, 26(1):9–20, 2010.

[28] Chemical Computing Group. MOE (Molecular Operating Environment). 2010: Montreal, Quebec, Canada, mar 2013.

[29] A.C.D Inc. ACD PhysChem: Toronto, ON, Canada, mar 2013.

[30] G.D. Purvis. Size-intensive descriptors. *Journal of Computer-Aided Molecular Design*, 22(6):461–468, 2008.

[31] V. Svetnik, A. Liaw, C. Tong, C.J. Culberson, R.P. Sheridan, and B.P. Feuston. Random forest: a classification and regression tool for compound classification and QSAR modeling. *Journal of Chemical Information and Computer Sciences*, 43(6):1947–1958, 2003.

[32] D. Furcy and S. Koenig. Limited discrepancy beam search. In *International Joint Conference on Artificial Intelligence*, volume 19, page 125. Lawrence Erlbaum Associates Ltd, 2005.

[33] A. Tropsha, P. Gramatica, and V.K. Gombar. The importance of being earnest: validation is the absolute essential for successful application and interpretation of QSPR models. *QSAR & Combinatorial Science*, 22(1):69–77, 2003.

[34] D.M. Hawkins, S.C. Basak, and D. Mills. The problem of overfitting. *Journal of Chemical Information and Computer Sciences*, 44(1):1–12, 2004.

[35] P. Burman. A comparative study of ordinary cross-validation, v-fold cross-validation and the repeated learning-testing methods. *Biometrika*, 76(3):503–514, 1989.

[36] S. Borra and A. Di Ciaccio. Measuring the prediction error. A comparison of cross-validation, bootstrap and covariance penalty methods. *Computational Statistics & Data Analysis*, 54(12):2976–2989, 2010.

[37] D.M. Hawkins, S.C. Basak, and D. Mills. Assessing model fit by cross-validation. *Journal of Chemical Information and Computer Sciences*, 43(2):579–586, 2003.

Part VIII

Image Mining: Feature Extraction, Segmentation, and Classification

Chapter 20

Image Mining Extension for RapidMiner (Introductory)

Radim Burget

Brno University of Technology, Czech Republic

Václav Uher

Brno University of Technology, Czech Republic

Jan Masek

Brno University of Technology, Czech Republic

Acronyms

RGB - Red, Green, Blue

IMMI - IMage MIning Extension

HSV - High Saturated Values

20.1 Introduction

In general, data mining is the process of analysing data from different perspectives and summarizing it into useful information—information that, for example, can be used to increase revenue, cuts costs, or both, and so on. The most common approach in data mining is to mine information from structured data, i.e., the data is structured into tabular form, where each column has its own given meaning. In RapidMiner, most of the learning algorithms expect tabular data (Example Set) as an input.

Unfortunately, many data available, not only on the Internet but generally, are in unstructured form. One case of the unstructured data is image data. Of course, it is possible to connect each pixel intensity of an image to each input of, e.g., a neural network, support vector machine, or other learning algorithm. However, results will not be very good in most of the cases and a more advanced approach is needed—the pixels have to be given some meaning. This process of information retrieval from the image data is often called "image mining".

To enable image mining capabilities in RapidMiner, you need to install the IMage MIning extension (IMMI) first. You can install the extension directly from the RapidMiner extension manager, which you can download directly from a website of the extension maintainer[1] or alternatively, you can switch to a new update URL[2] and install the extension directly from RapidMiner's menu at: *Help > Manage Extensions*. The extension is written in pure JAVA, therefore it is platform independent. So far it has been tested on Windows, Linux, and Mac platforms. If you want to work with image mining, it is strongly recommended to turn the expert mode on. Otherwise parameter settings of IMMI operators will be strongly limited. How to enable expert mode is depicted in Figure 20.1.

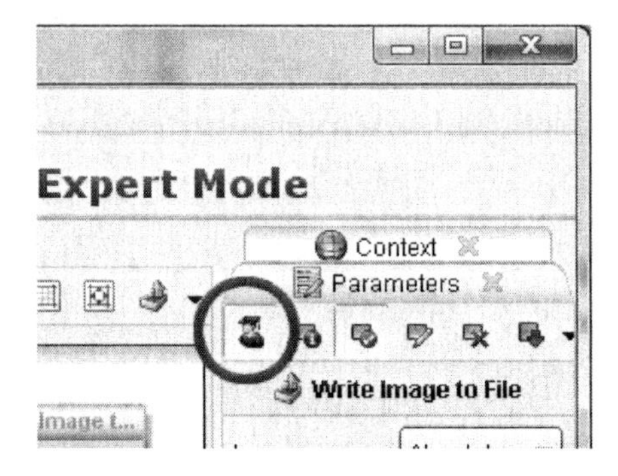

FIGURE 20.1: How to enable expert mode in RapidMiner.

The main idea behind IMMI extension and its functionality in RapidMiner is depicted in Figure 20.2. The extension provides various methods for transformation of the image data into the tabular data. This process is called feature extraction. When the extracted features are obtained in the structured form (RapidMiner uses *Example Set* structure for representation of the structured data), any learning algorithm or data mining algorithm can be utilized and work with the data is the same as in general data mining tasks and therefore the required knowledge is mined (in the figure it is depicted as smiley). The IMMI supports

[1] Signal processing laboratory, website: http://splab.cz/immi.

[2] In the menu, go to Tools > Preferences > Update and set rapidminer.update.url to http://rapidupdate.de:8180/UpdateServer.

many different operators supporting variety of algorithms for image transforms and feature extraction. The feature extraction algorithms by their nature can be divided into four basic classes: local-level, segment-level, global-level, and similarity measurement between images. More information about the category's meaning will be provided in Section 20.4.

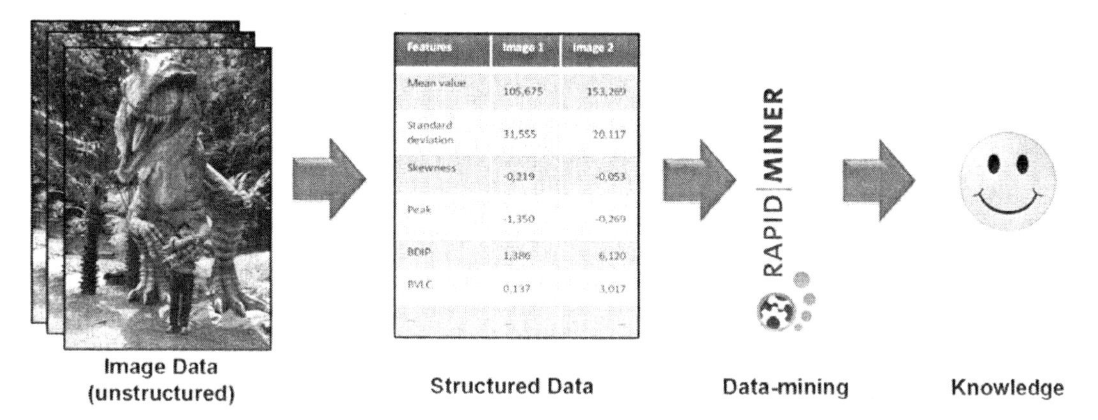

FIGURE 20.2: Concept of image mining extension (IMMI) integration into RapidMiner.

This chapter covers issues related to local-level, segment-level, and global-level feature extraction which are demonstrated by examples. The Chapter 21 Image Mining Extension for RapidMiner (Advanced) covers local, segment and global-level feature extraction in more detail and also covers the similarity measurement and object detection topics.

The rest of this chapter is structured as follows. In the next section several examples for reading and writing images are given. It also gives some examples of how to convert between colour and gray-scale representation of images and how to combine two images into one resulting image. Section 20.4 is devoted to feature extraction, i.e., conversion of the image data to structured tabular data. Local, segment, and global-level feature extraction are covered and are demonstrated in several examples.

20.2 Image Reading/Writing

The IMMI extension works internally with 8-bit gray-scale images or with RGB colour images with 8-bit depth for each colour component (i.e., 24-bit depth). Different operators in IMMI expect different image formats and therefore IMMI internally distinguishes between the image representation types. IMMI provides two operators for reading images: one for reading colour images and one for reading gray-scale images.

In IMMI it is possible to read a single gray-scale or colour image or to iterate over a set of images (gray-scale, colour, or even a mixture of gray-scale and colour images). In case of iterating over set of images IMMI does not hold images in memory, but opens and closes the image in each iteration. Therefore, it is possible to iterate over big databases of images without any memory problems. A basic demonstration how to read and write an image is given in the following example.

Example 1—I/O operation with images: We need to read an image, blur the image, and save the blurred result to a disk. Let's say we decide to work with gray-scale images. For this purpose we need three operators: "Open Grayscale Image", "Gaussian blur", and

"Write Image to File". The first image just reads the image from the given file, the second applies Gaussian blur transform[3] to the image, and the last saves the result to the given file. For colour images you should use the "Open Color Image" operator instead. If you use the gray-scale variant of the operators and the read image is not in the required format, it is internally converted to the required internal representation (see Figure 20.3).

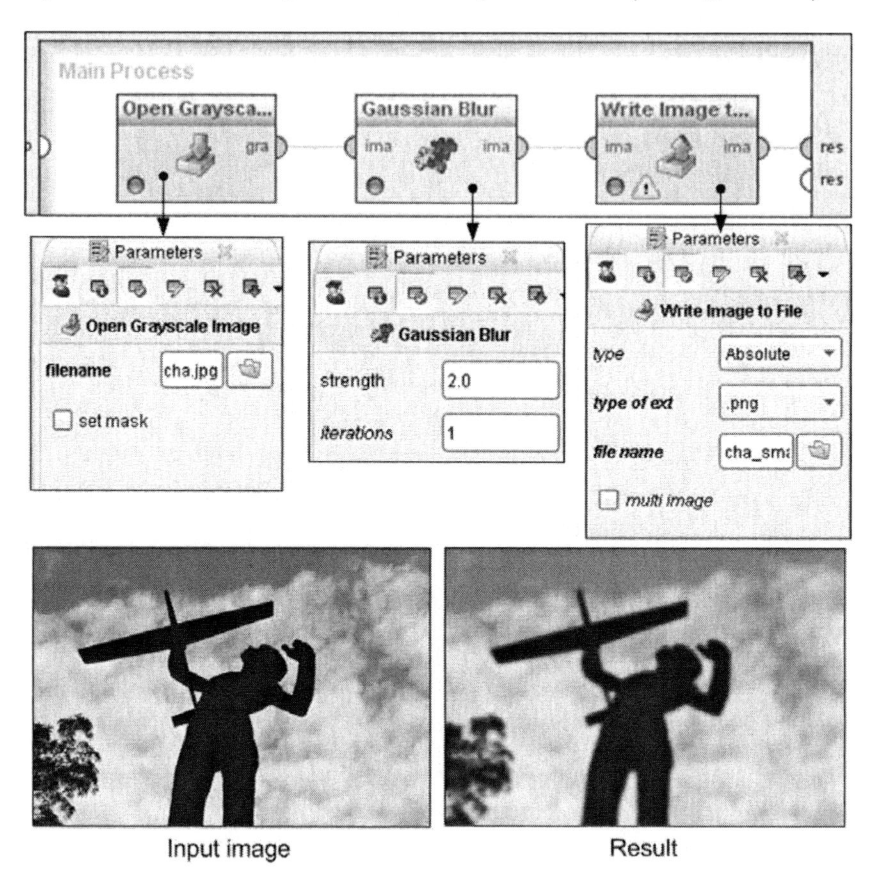

FIGURE 20.3: Example demonstrating reading, blurring, and saving image to disk.

You can experiment with the process by changing the parameter "strength" of the "Gaussian Blur" operator. You can also try to replace the "Open Grayscale Image" operator with the "Open Color Image". The resulting image after this change should be a color image. So as you can see, reading and writing images in RapidMiner is quite easy.

Unfortunately, in many cases a single image does not contain enough information for training a reliable model. In such cases utilization of information from several images is necessary. In IMMI there are several ways to easily iterate over set of images.

For example, if you need to classify images, you will need to extract global features from a set of images. For this purpose there are the "Multiple Color Image Opener" and "Multiple Grayscale Image Opener" operators. In these operators you can choose any number of folders. An example demonstrating the use of the colour variant of the operators is in

[3]Gaussian blur is an image transform that smoothes images with the use of the Gaussian function.

Section 20.4.3. An alternative way to iterate over set of images is with the use of the "Loop" or "Loop Files" operators in combination with macros.[4]

Example 2—Iterate over set of images: Consider the following problem: We have a set of images and we need to adjust the contrast of these images (for example, using the histogram equalization method) in a folder and store result to another folder, but the file names must be kept the same. For this purpose we will use the "Loop Files" operator with a combination of macros. Other operators we need are "Open color image", "Histogram equalization" and "Write image to file". The "Loop Files" operator iterates over a set of images in the given folder (see Figure 20.4). In each iteration it sets macro "file_path" to the path of the current file and "file_name" to the current file name. In the subprocess of the operator there are placed operators for reading, equalizing, and writing the images. As an source image, the macro "%{file_path}" is given and as an output, a folder with "%{file_name}" is set.

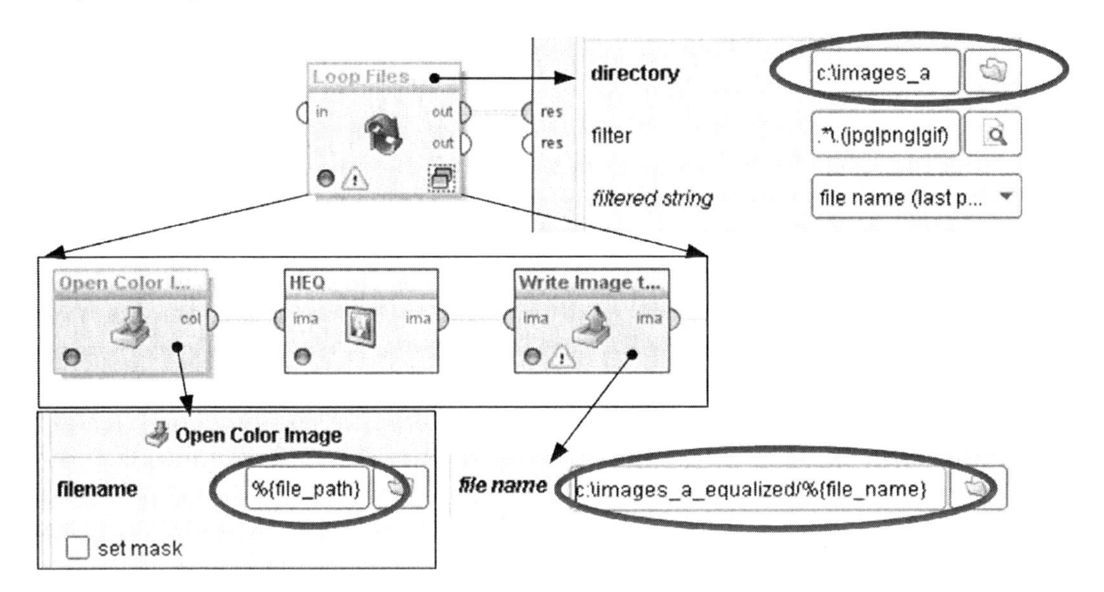

FIGURE 20.4: Example: Iterating over set of images.

Notice that the colour and gray-scale variants of the open image operators also have the parameter "mask". Since RapidMiner is a data mining tool, image data often needs to be labeled. Labeling is a process of marking which part of an image belongs to what class (e.g., background vs. foreground, healthy vs. affected tissue, etc.). In order to assign classes to image data, so-called masks exist. The mask is not mandatory, however many image mining algorithms need it, just as it is with classical machine learning. According to the the mask, a label value is often assigned to extracted data. More information will be provided in Sections 20.4.1 and 20.4.2.

[4]Macro stands for variable. Since in data mining the term variable denotes something different, in order to avoid this ambiguity, the variables are called macros in RapidMiner.

20.3 Conversion between Colour and Grayscale Images

As mentioned before, in IMMI there are two types of internal image representation: gray-scale or colour. For the purpose of conversion between these types of images, there are several operators. Two of the most important are the "Color to gray-scale" and "Grayscale to color" operators. It is not only possible to convert from colour to gray-scale, but it is also possible to create colour image from a gray-scale image by adding so-called pseudo colors. Of course correct assigned colours cannot be expected, but especially in biomedicine it can be very useful. With the use of pseudo-colors, the human eye can more easily recognize identical gray-scale levels. The pseudo-colours can be applied using the "Look up table applier" (LUT) operator. In many cases red-green-blue (RGB) image representation is not very suitable and a different point of view can reveal interesting features. For example, for skin detection, the Hue-Saturation-Value (HSV) color model can be more suitable than the RGB colour model. For this purpose the "Color Component Extractor" operator exists, which can extract from a color image a variety of components from several models such as RGB, HSV, YUV, IHLS, YIQ, CIE L*a*b*, and others.

Example 3—Conversion between gray-scale and RGB images: We would like to analyze a colour image, i.e., an image composed of red, green, and blue (RGB) colour components. This is how IMMI (and also most of the computer programs) represent colour images. Therefore we need the original RGB image decomposed into the RGB. Further, to help analyze gray-scale levels with the human eye, each colour component should be pseudo-colorized. This can often give a hint, if the approach based on gray level can work.

The process to accomplish this task is depicted in Figure 20.5. First "Open Color Image" reads, input image. In this case we do not plan any data mining, therefore the mask is not used in this case. The "Multiply" operator creates three copies of the original image. The "Color Component Extractor" operator can extract the color component. In the first case the extracted color component is red, in the second it is green and in the third it is blue. In order to enhance differences between the images, the "LUT" operator is used. In this case the predefined fire LUT table was used.

Component extraction and image transforms are the key work, which can obtain valuable features from images. There are plenty of ways to extract different features from an image. Another task may be how to combine several transformations back into a single image. A quite handy operator for this purpose is "Image Combinator". Use of this operator is demonstrated in the next example.

Example 4—Combining images into one result: We have two images—the original image and the image dividing the image into two parts—foreground and background. In order to avoid confusion, the term "mask" is not used here, since masks have a different purpose in the IMMI. Of course, the segmenting image does not necessarily have to be an object from a file but it can be a result of some trainable segmentation or object selection. The task is to remove the background of the image (denoted in the background/foreground image with black/white colour) and keep the foreground of the image.

The process for solving this task is depicted in Figure 20.3. First the original image and the background/foreground image are read by two "Open Grayscale Image" operators. Their parameter filename was set to the path of the file on disk. The "action" parameter of the "Image Combinator" operator is set to max, which stands for taking the maximum for each pixel in the resulting image.

This combination can be especially interesting for post-processing, when a trained result should be somehow applied to the original input image, e.g., to emphasize or suppress some part of image.

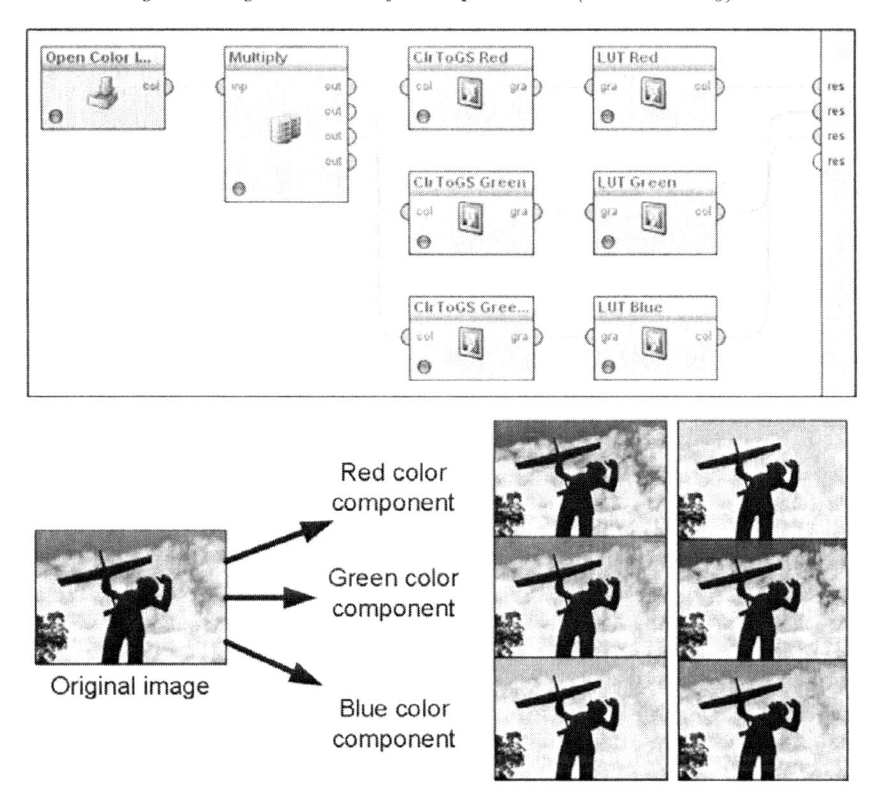

FIGURE 20.5 **(see color insert)**: Example: Extraction of color component from image.

In IMMI there are currently many operators for image transforms and image manipulation. Although it is often good and sometimes may be necessary to know the theory behind the operators, these topics are well covered in many other books devoted to image processing, and much information can be found on the Internet. Since this book is focused on data mining, the rest of the chapter will be focused on feature extraction and image mining.

20.4 Feature Extraction

Feature extraction is a special form of dimensionality reduction, which is often applied to data that are complex, unstructured, or suspected to be highly redundant (much data, but not much information). In the context of image mining it is conversion of image data (i.e., also often highly redundant) in unstructured form into a tabular form. Feature extraction algorithms for images can be divided into three basic categories: local-level, segment-level and global-level.

The term "local-level" denotes that the information is mined from given points (locations) in the image. Local-level feature extraction is suitable for segmentation, object detection, or area detection. From each point in the image it is possible to extract information such as pixel gray value, minimal or maximal gray value in a specified radius, value

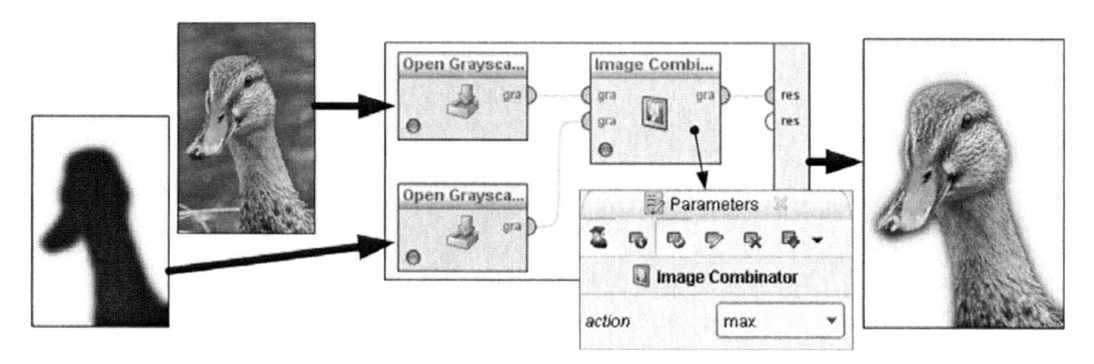

FIGURE 20.6: Example: How to combine two images into a single result.

after applying kernel function (blurring, edge enhancements), and so on. An example of utilization of such data can be trainable segmentation of an image, point of interest detection, and so on. Object detection is also a related field.

The term "segment-level" denotes segmentation from (as is obvious) segments. A large number of segmentation algorithms exist, such as k-means [1], watershed or statistical region merging [2]. By the use of segment-level feature extraction algorithms you can extract information from the whole segments. Examples of such features can be, e.g., mean, median, lowest and highest gray value, circularity, eccentricity and many others. In contrast to local-level features, it does not take into consideration only a single point and its neighborhood; it considers the whole segment and its shape. With the use of such knowledge, it is possible to select or remove objects according to their size, shape, or merge over-segmented segments into well segmented image and so on.

The "global-level" denotes feature extraction from the whole image. For example, mean colour, dominant colour, maximal gray value, minimal gray value, variance of pixels, number of edges etc. Unlike the local or segment level, global-level feature extraction is not suitable for points or area identification or segmentation. Rather, it is suitable for classification of images and determining properties of the image as a whole.

In the following text, examples demonstrating use of local-level, segment-level, and global-level image mining techniques are presented.

20.4.1 Local Level Feature Extraction

The basic concept of how the local feature extraction works is depicted in Figure 20.7. The figure is split to two parts: the training part, where a model is being trained, and application part, where the trained model is being applied to some new, usually previously unseen data. The training can often be very time-consuming and also demand a lot of computer resources, where supercomputers can be also used. On the other hand, the resulting model can often work in real time with low demands on hardware.

In order to train a model, some training images are needed (i.e., original images) together with specification what the results should look like. The desired results are also common images, which are manually prepared, usually by a human domain expert and they specify to the objective to which the resulting model should be trained. In order to provide the learning algorithm with a sufficient amount of information, the original images should be preprocessed by several selected transformations (e.g., blurring, edge detection, convolution with different kernels and so on) in order to enhance image structures at different scales and retrieve more information from each pixel. These transformations can include simple and

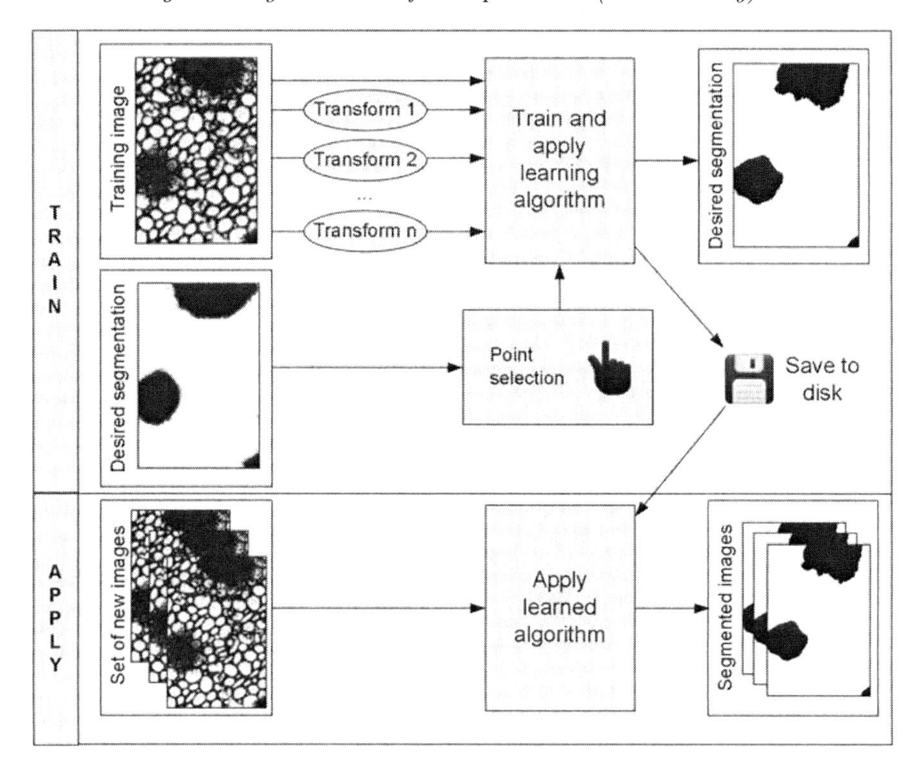

FIGURE 20.7: Local-level feature extraction, training, and application of the model.

also some more advanced transformations. When different transformations are prepared, the training of a learning algorithm can begin. For this purpose any learning algorithm available in RapidMiner can be used and you can choose from many different algorithms. Proven algorithms which usually give good results are Random Forests, Decision Trees, and Support Vector Machines.

The way the image data are transformed from image representation into the structured table form is depicted in the Figure 20.8. At first, training points are selected. From each transform pixel, values at the given point positions are taken and each value is stored in the given column of table. Each line in the table represents a single point in the image. Each column represents a different image transform.

As mentioned before, the training is often a relatively time and resources demanding process. Therefore it is suggested to store the trained model to a disk and when the model is about to be used, it can be loaded and used quickly. This is very fast and does not require any significant resources. An example demonstrating use of local level features for trainable segmentation is demonstrated in following example.

Example 5—Trainable image segmentation: Consider the following problem. An input image is given (see left part of Figure 20.9) and we want to transform it into an image as shown in right part of Figure 20.9. Note that some small objects that should be removed are connected to bigger objects that should be kept (emphasized with red circles in the figure). This is the reason why a segment-level approach (will be discussed in the next section) is not suitable for this problem, and rather a trainable segmentation based on local-level features is preferred here.

The process that solves this task is depicted in Figure 20.10. First the image is read

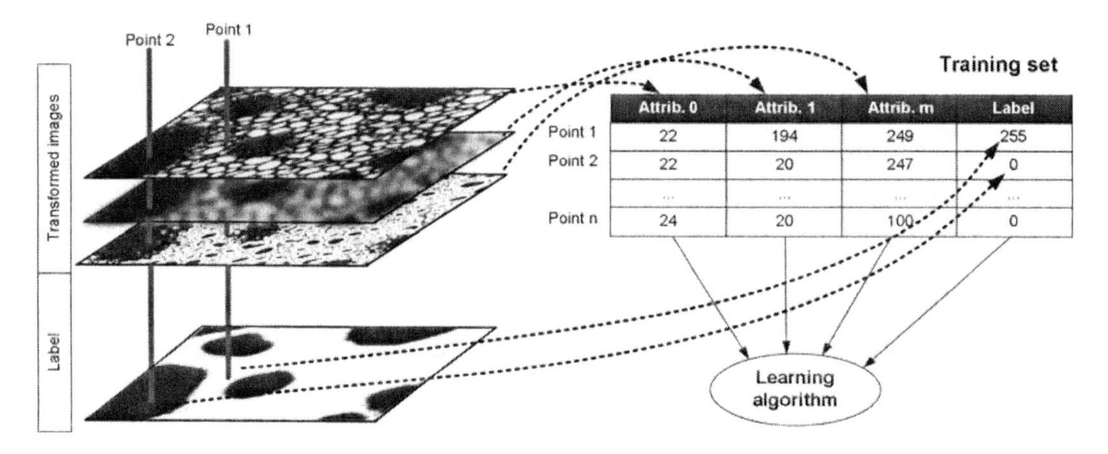

FIGURE 20.8: Principle how image data are transformed into tabular data.

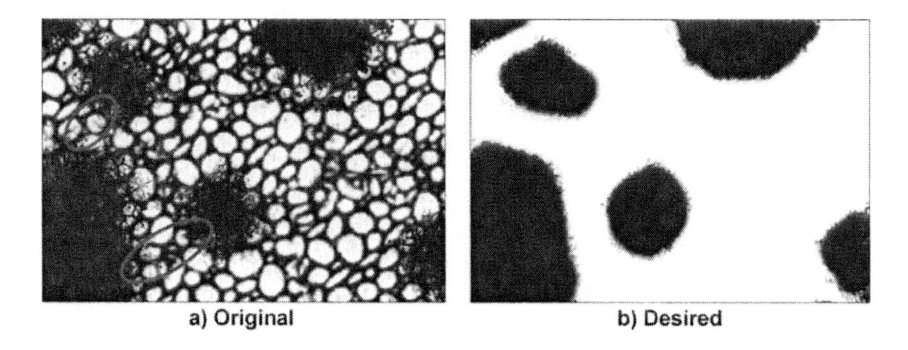

FIGURE 20.9: Example: Trainable image segmentation based on low-level features.

by using the "Open Grayscale Image" operator. Using the dimension of the image, 2500 random points are generated. The key operator which does the work is the "Trainable segmentation" operator. As an input the operator gets the following data: set of generated points, the original image and another three images, which are a result of Gaussian blur with kernel size 8, 16, and 24 applied the original image. As a learning algorithm, a Support Vector Machine with radial kernel was used.

One of the outputs of the "Trainable segmentation" operator is the trained model, which can be saved to disk by the use of the "Store" operator. If parameter "operation" of the operator "Trainable segmentation" is set to "Apply" mode and the stored model is retrieved by "mod" input of the operator, it does not train a model. It often saves considerable processing time.

20.4.2 Segment-Level Feature Extraction

In the previous case, there was a problem to segment objects in an image because some of the unwanted parts were connected to the wanted objects. However, in some cases the image can be segmented easily. In this case it is preferable to use segment-level feature extraction and trainable segment selection, because the segment features can contain many

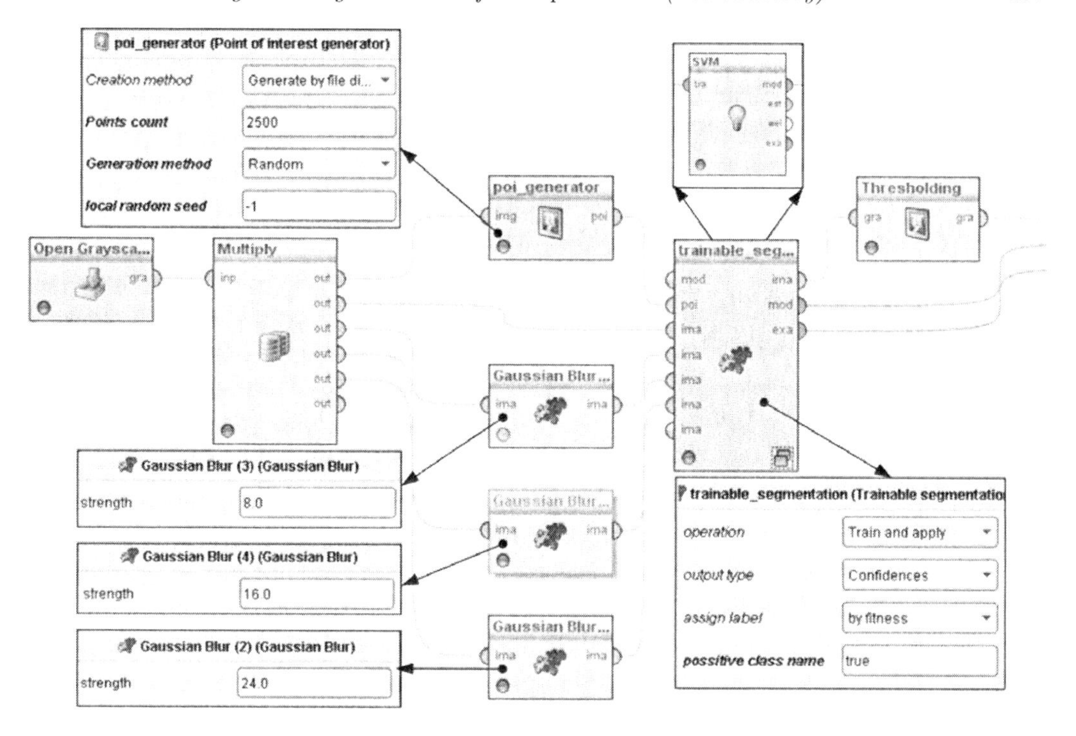

FIGURE 20.10: Example: Trainable image segmentation based on low-level features.

interesting properties such as size, shape, roundness, etc., which can be used for image analysis. How to use trainable segment selection is demonstrated in the following example.

Example 6—Trainable segment selection: Consider the following problem, which is depicted in Figure 20.11. At the left side of the figure, an original input image is depicted. The task of this example is to remove all the objects of the figure except the stars (see right part of Figure 20.11).

The process, that solves this problem is depicted in Figure 20.12. At first the input image is read by the "Open Color Image" operator. After that, the image is segmented using Statistical region merging method. In the body of the "Subprocess" operator, the most significant work is done and (since the result returned by the subprocess operator is inverted) some post-processing using the "Inverse" operator is performed. The result is exactly the same as depicted in the right part of the Figure 20.11.

The "Subprocess" operator at first extracts segment-level features from each segment obtained from the "statistical region merging" operator. Which features will be extracted is determined by the inner operator. In this case an "ROI Statistics" operator was used. Output of this operator are segments (unchanged) and an example set containing all the extracted features obtained from the figure. The operator "Segment Filter by Example Set" removes all the segment, which are not present in the Example set. For this reason the Example set is preprocessed by the "Filter examples" operator, where condition class is set to "attribute value filter" and the condition is to filter all the examples, where the parameter "roi_statistics_circularity" (this parameter was extracted from the previous "Segment Feature Extraction" operator) is lower than 0.1. Of course, any other parameter can be used, however, as is obvious from the original image, the stars are less circular than circles and also as rectangles and squares. Then the segments need to be converted to a single

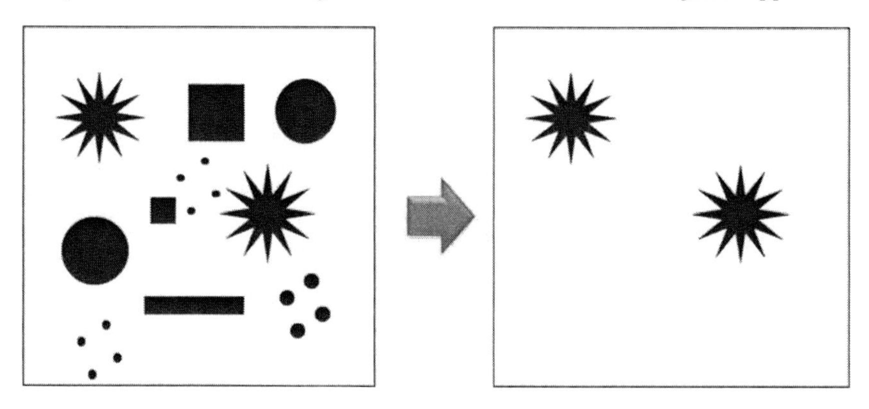

FIGURE 20.11: Example of segment manipulation according to the segment shape.

image. This is performed by the "Merge Segments to Image" operator, where parameter "operation" is set to "Set all white", i.e., to the black image all the filtered segments are denoted as white.

20.4.3 Global-Level Feature Extraction

The third level of feature is the global-level. The global-level feature extraction is especially useful in cases where we have many images and we need to classify these images into some classes. The global features are computed from the whole image. Some of the features are dominant colour, minimal intensity, maximal intensity, percent of edges, etc. The next example shows how to distinguish between two classes: images containing birds and images containing sunsets.

Please note that this example is only for demonstration of how to use RapidMiner with global-level features. Understanding of images is still an open research problem and for some real-world deployment, a more advanced training and validation set would be needed.

Example 7—Image classification: Consider that we need to automatically classify images into two classes, "Birds" and "Sunsets". We have already a set of classified images in the folders "images_a" and "images_b". Please note that each folder contains only 4 images (i.e., 8 in total), which is certainly not enough for real production. However for demonstration purposes, this should suffice. What we need is to iterate over a set of images. For each image in the folders we need to extract some features and assign a label according to its folder.

Since the images are colour images, for this purpose we can use the "Multiple Color Image Opener" operator. For its parameter "images" you need to enter two folders "images_a" and "images_b" and assign their classes, for example, "MY_CLASS_A" and "MY_CLASS_B" (see Figure 20.13). You can choose any names but they must differ from each other, otherwise it would not be possible to classify images since all the images would be members of one class. Since we need to assign a class to each example, don't forget to check the "assign label" check-box. Double-click on the operator and enter the operator's sub-process area. In the sub-process it is expected to transform colour image into an example set, i.e., extract features from images. For this purpose you can use the "Global Feature Extractor from a Single Image" operator can be used (see Figure 20.13). Again, double-click on the operator and enter its sub-process. You can use any global feature extraction algorithm, for example, "Global statistics".

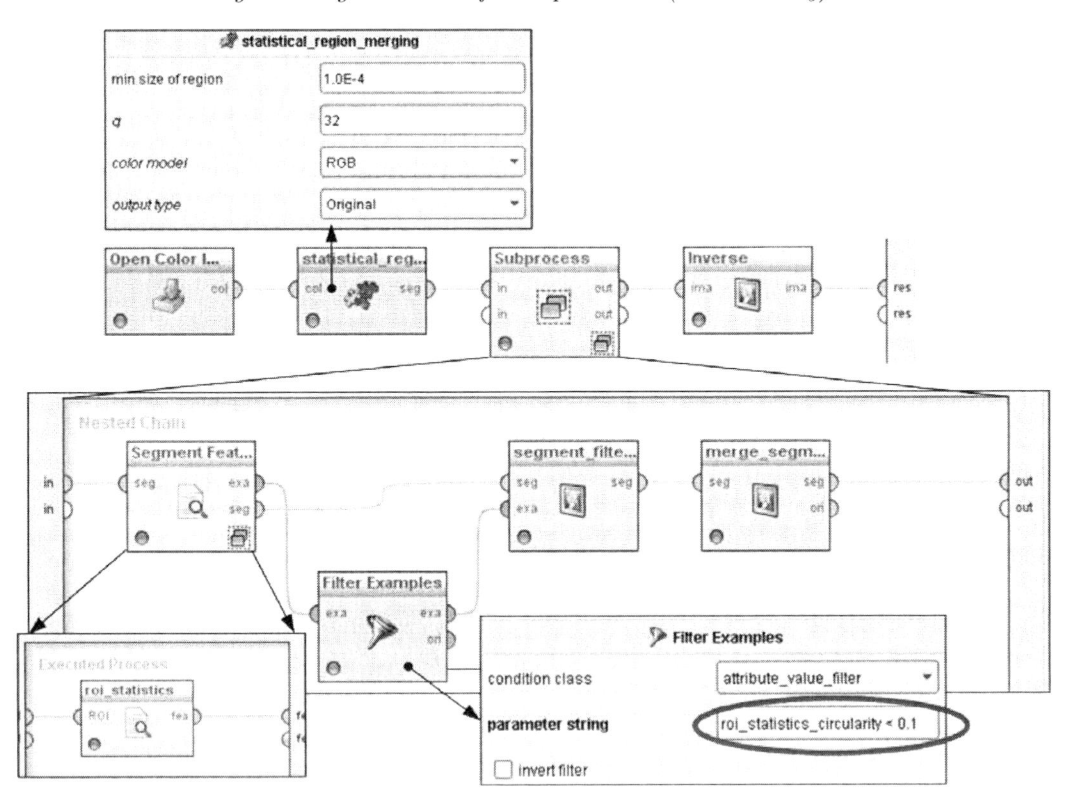

FIGURE 20.12: Example of segment manipulation according to the segment shape.

If you apply, e.g., the decision tree learner on the resulting example set you can get a tree as depicted in the Figure 20.13.

20.5 Summary

Although understanding of images by computers is in most cases far below the capabilities of humans, this area of science has seen significant progress in recent years. This chapter has given some introductory insight into image mining using RapidMiner, and IMage MIning extension (IMMI) tool.

In this chapter some basics about the IMMI were covered including reading single or multiple colour and gray-scale images and conversion between them. Further, local, segment, and global-level feature extraction were presented in selected examples.

Exercises

1. Consider Example 1, where some basic transformation of the image is demonstrated.

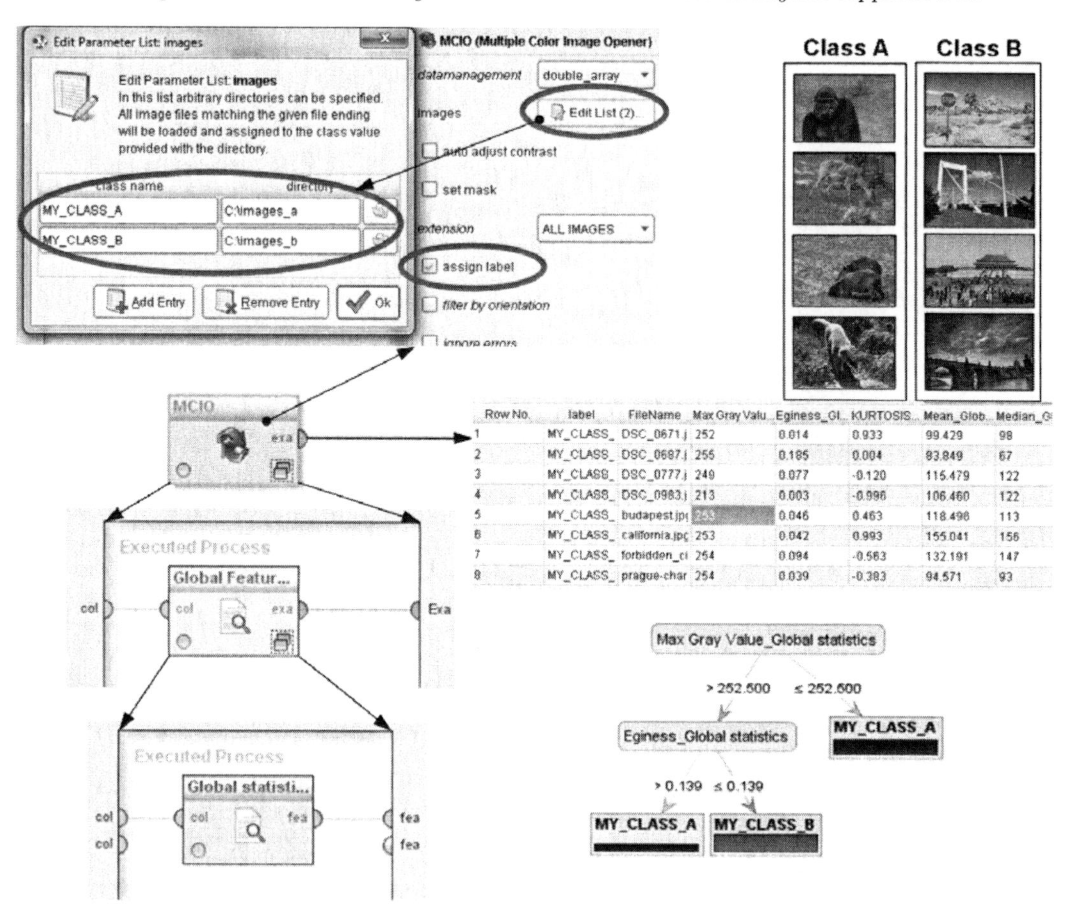

FIGURE 20.13: Example of a process for training an image classifier.

(a) What is the reason that the resulting image is gray-scale and not color?

(b) Replace the gray-scale image read operator by its color variant. What is the effect?

(c) Find the operator for image inversion and connect the operator to the process.

2. Example 3 demonstrated how to decompose an image into several color components.

(a) Examine different predefined LUT tables and notice the effect on the resulting image.

(b) The YUV color model defines a color space in terms of one luma (Y') and two chrominance (UV) components. Extract the Y, U, and V components from the original image and display the components.

3. Consider Example 4, where the foreground of the original image is selected.

(a) Modify the example, so that the background is selected and the foreground will be removed.

(b) Modify the example, so that the transition between background and foreground is smoother.

   (c) Modify the example, so that only borders between background and foreground is selected (TIP: for edge detection you can use the "convolution filter" with matrix type for edge detection and Laplacian of Gaussian 9x9.)

4. Consider Example 5, where segmenting algorithm is trained by selected points in the training image.

   (a) Add a breakpoint to the "poi_generator" operator and start the process. When the process stops on the "poi_generator" operator, click by left and right mouse buttons to add positive and negative points. Click on a point in order to remove the point.

   (b) Add several red points so the red will dominate over the green points (you can reduce number of points generated by the "poi_generator" operator). What is the effect of overall distribution? What is the ideal ratio between red and green points?

   (c) Modify the example, so that the "Gaussian Blur" operators will have set strengths (i.e., sizes of kernels) of 2, 4, 8. What is the effect? What is the effect when some "Gaussian Blur" operator is added/removed?

   (d) Replace the "SVM" operator with different learning algorithms such as Decision Tree, Random Forest, k-Nearest Neighbour, etc. Watch the effect of changes on resulting image.

   (e) Store the resulting model from the "Trainable segmentation" operator to the disk using the "Write model" operator. Then switch the "operation" parameter of the "Trainable segmentation" operator to "Apply" and load the stored model to the operator using the "Read Model" operator. What is the effect? Does the operator train a classifier? Is it faster or slower?

5. Consider Example 6, where segments are selected according to the shape of objects.

   (a) Add a breakpoint before execution of the "Filter Examples" operator and examine the input example set. What is the name of the parameter that describes size of the segment? What is the minimal size of segment and what is the maximal size of segment?

   (b) Modify the "parameter string" parameter in the "Filter Examples" operator, so that only objects of bigger sizes will be selected. Experiment with different values.

   (c) See the parameters of the "roi_statistics" operator and its parameters. Experiment with checking / unchecking of these parameters and watch the resulting example set, which is produced by the "Segment Feature Extractor" operator. Is there any effect on performance?

Glossary

Image Mining is the process of analyzing data from different perspectives and summarizing it into useful information in order to obtain some knowledge from the data.

RGB is an additive colour model where red, green, and blue colour components are mixed in order to reproduce a broad array of colours.

Local-level features are features obtained from a single point of an image.

Segment-level features are features obtained from a single segment of an image.

Global-level features are features obtained from the whole image.

Example set is a tabular representation of data containing several attributes (columns) and examples (rows).

Bibliography

[1] D E Ilea and P F Whelan. Color image segmentation using a spatial k-means clustering algorithm. *10th International Machine Vision and Image Processing Conference*, 1, 2006.

[2] Richard Nock and Frank Nielsen. Statistical region merging. *IEEE Transactions on Pattern Analysis and Machine Intelligence*, 26(11):1452–8, 2004.

Chapter 21

Image Mining Extension for RapidMiner (Advanced)

Václav Uher

Brno University of Technology, Czech Republic

Radim Burget

Brno University of Technology, Czech Republic

Acronyms

IMMI - Image Mining

OBCF - Order-based Block Color

ROI - Region Of Interest

21.1 Introduction

This chapter demonstrates several examples of image processing and feature extraction using the image mining plug-in for RapidMiner. Firstly, the process of global feature extraction from multiple images with some advanced operators will be described. These operators are briefly described. Secondly, an example of the Viola-Jones algorithm for pattern detection will be demonstrated. The final example will demonstrate the process of image segmentation and mask processing.

21.2 Image Classification

The classification of an image is used to identify which group of images this image belongs to. This might concern, for example, distinction between scene types like nature, urban environment (only urban in the following), interior, images with and without people, etc. Global features are usually used for this purpose. These features are calculated from the whole image. The key to a correct classification is to find the features that differentiate one class from other classes. Such a feature can be, for example, the dominant colour in the image. These features can be calculated from the original image or image after pre-processing like Gaussian blur or edge detection.

This example describes the procedure of creating a process which distinguishes two image categories (urban and nature). The whole process is quite simple and has three major steps:

- Read images.
- Calculate features.
- Convert features to example set.

FIGURE 21.1: Example of two image classes.

The entire process is demonstrated in Figure 21.2. Images in the folder are iterated one by one. Defined features are calculated separately for each image.

21.2.1 Load Images and Assign Labels

- Add *Multiple Color Image Opener* operator to the process.
- Click on the *Edit list* button, insert two rows (or more, one for each class), and set up fields according to Figure 21.3. Choose the paths to folders depending on your current folder tree.
- Check the *assign labels* box (Figure 21.4). It will cause labels to be assigned to images according to settings.

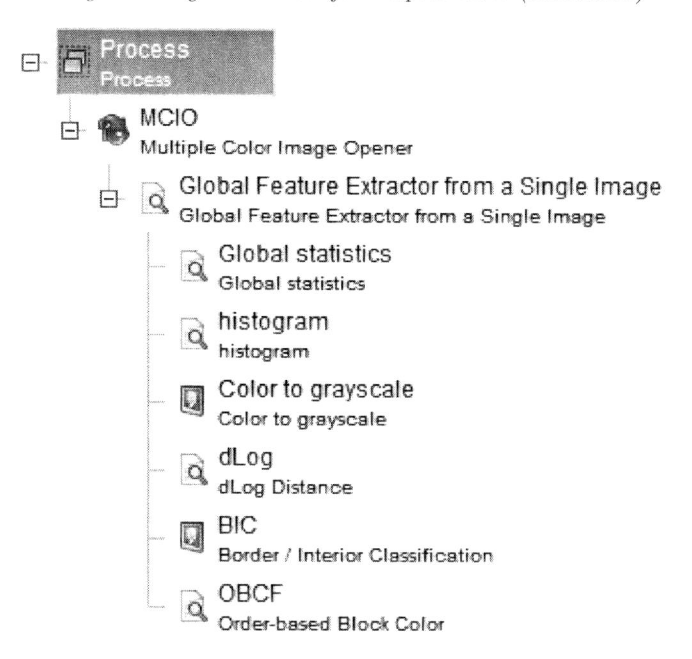

FIGURE 21.2: Process tree.

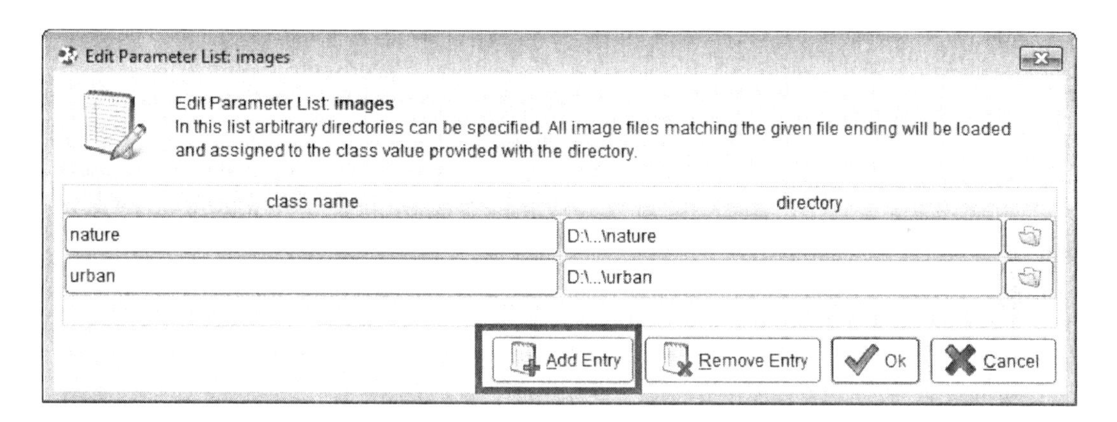

FIGURE 21.3: Choosing the class.

21.2.2 Global Feature Extraction

- Paste the *Global Feature Extractor from a Single Image* operator inside the *Multiple Color Image Opener* operator. This operator converts features from inside operators to ExampleSet with one row.

- Paste the *Global statistics*, *histogram*, *Order-based Block Color* (*OBCF*), and *dLog Distance* inside the *Global Feature Extractor from a Single Image*. These operators calculate global-level features.

  – *Global statistics* – This operator calculates several statistics from the whole im-

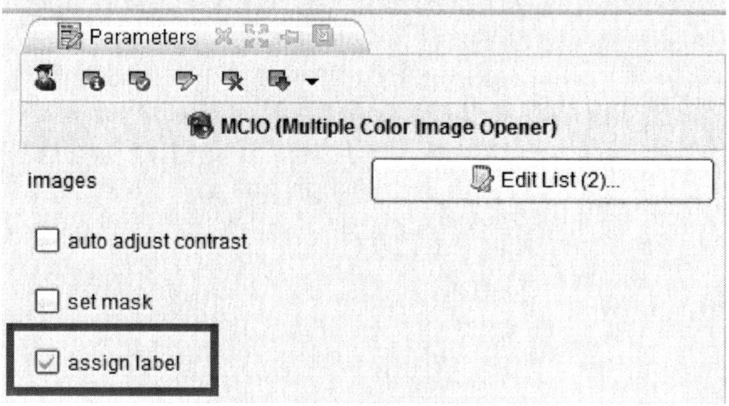

Parameters

MCIO (Multiple Color Image Opener)

images Edit List (2)...

auto adjust contrast

set mask

assign label

FIGURE 21.4: Image loading settings.

age. Pixels converted from colour to 8-bit gray-scale are used as the input data. Therefore, the processed values are in the range from 0 to 255. The available features are mean, median, standard deviation, skewness, kurtosis, minimum and maximum grey value, peak, etc.

- *Histogram* – A histogram is a graph of the distribution of the pixel values. Number of bins can be pre-defined because the default value can be quite high. The pixel values are divided by scaling factor if the number of bins is smaller than 255. If an input image is colour, then it is converted to the gray scale.
- *Order-based Block Color* – An image is divided into the grid when these features are calculated. After splitting the image into cells, statistics (mean, minimum, and maximum) are calculated for each cell. The resulting set of features has rows times columns times statistics attributes (Figure 21.5). This operator can work only with gray-scale images. Therefore, it is necessary to prepend the *Color to gray-scale* operator. As previously mentioned, this operator converts a colour image to gray-scale. You can select from several colour spaces and test which colour channel produces the best performance.

FIGURE 21.5: OBCF operator function.

– dLog Distance – This operator computes the logarithmic distance between interior and border pixels. The image with marked interior and border pixels must be provided as an input image. This image is generated by the second output of the *Border / Interior Classification* operator. This operator works as follows: the input image colour space is quantized to 64 levels. Each colour channel has 4 levels. After the quantization, the image pixels are classified as border or interior. A pixel is classified as interior if its 4 neighbours (top, bottom, left, and right) have the same quantized colour. Otherwise, it is classified as a border. The example of the image with pixels marked as border / interior is in Figure 21.6.

FIGURE 21.6: Pre-processed image by Border / Interior Classification operator.

- The connection of the features extraction operators is outlined in Figure 21.7.

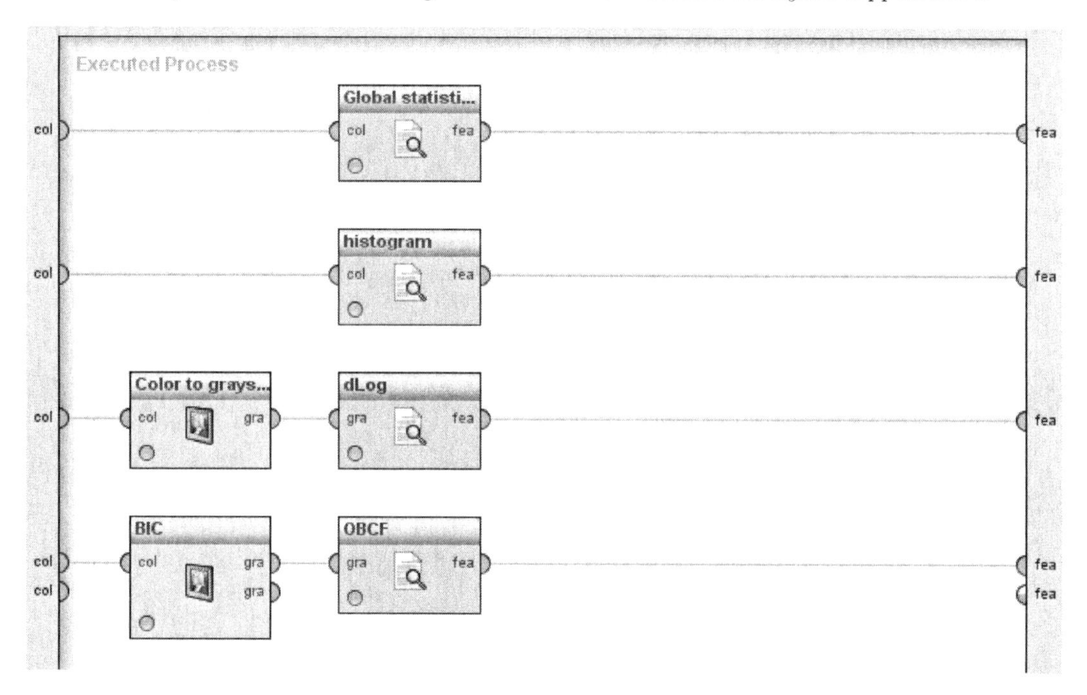

FIGURE 21.7: Operators for global features extraction.

21.3 Pattern Detection

Pattern detection is a task which searches in the image for a pattern known in advance, but only approximately, what it will look like. A good algorithm for detection should be sensitive to neither the size of the image nor the rotation.

One possible approach is to use a histogram. This approach compares the histogram of the pattern with the histogram of a selected area in the image. In this way, the algorithm passes step by step through the whole image, and if the match of histograms is larger than a certain threshold, the area is declared to be the sought pattern.

Another algorithm, which is described in the following example, is the Viola-Jones algorithm. The algorithm works with a set of features that are shown in Figure 21.8. The classifier is trained based on the positive and negative sets. Appropriate features are selected using the Adaboost algorithm. An image is iterated during pattern detection using a window with increasing size. Positive detections are then marked with a square area of the same size as the window.

Next example that we will show is the process used to detect the cross-sectional artery from the ultrasound image. An example of such an image is shown in Figure 21.9. After detection, it can be used to measure the patient's pulse.

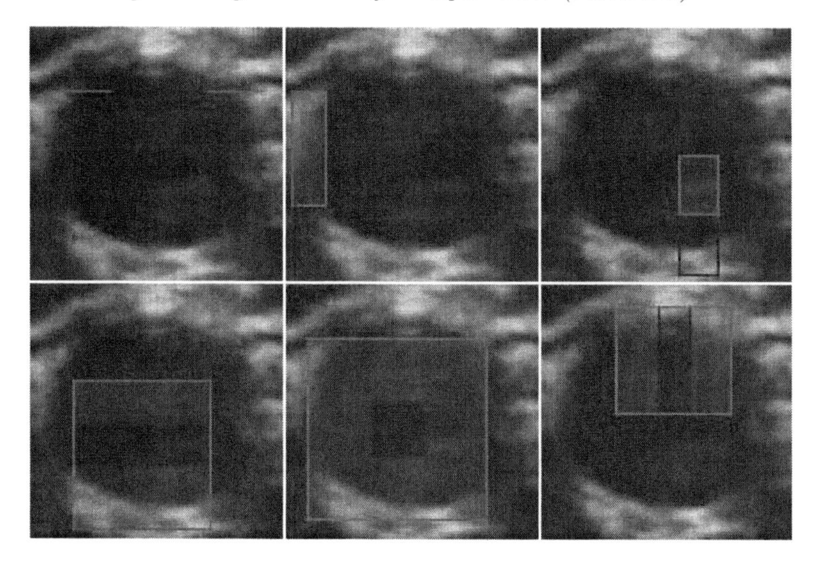

FIGURE 21.8 **(see color insert)**: Haar-like features.

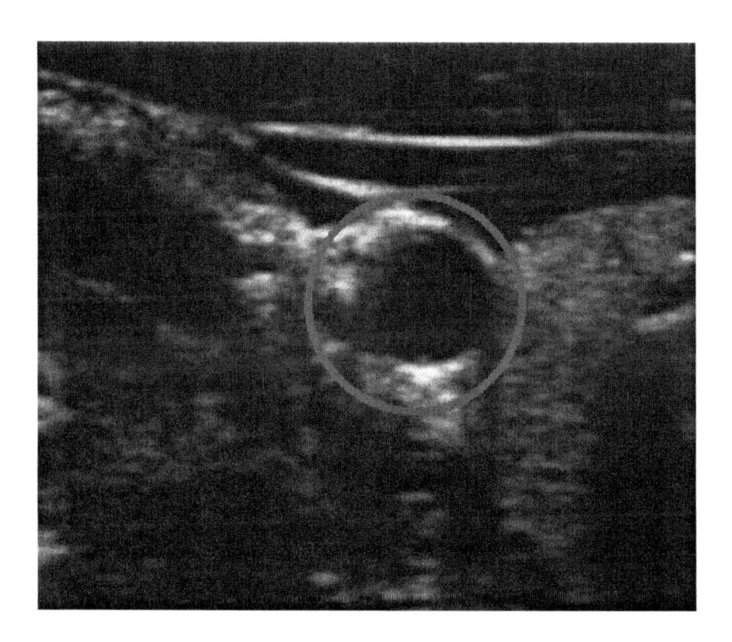

FIGURE 21.9: Ultrasound artery image.

As previously mentioned, the Viola-Jones algorithm is used for detection. For its proper function it is necessary to create a positive and a negative set of examples. Examples of positive samples are shown in Figure 21.10(a). Negative samples are shown in Figure 21.10(b).

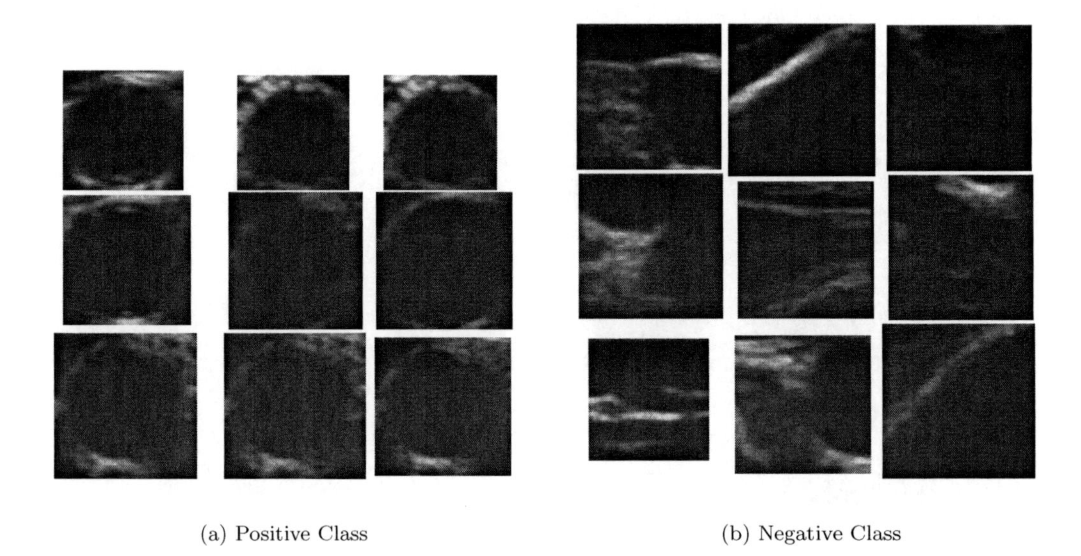

(a) Positive Class (b) Negative Class

FIGURE 21.10: Samples for artery detection.

21.3.1 Process Creation

- Paste the *Read Image Set* operator to the process and select input folders in a similar way to Figure 21.11.

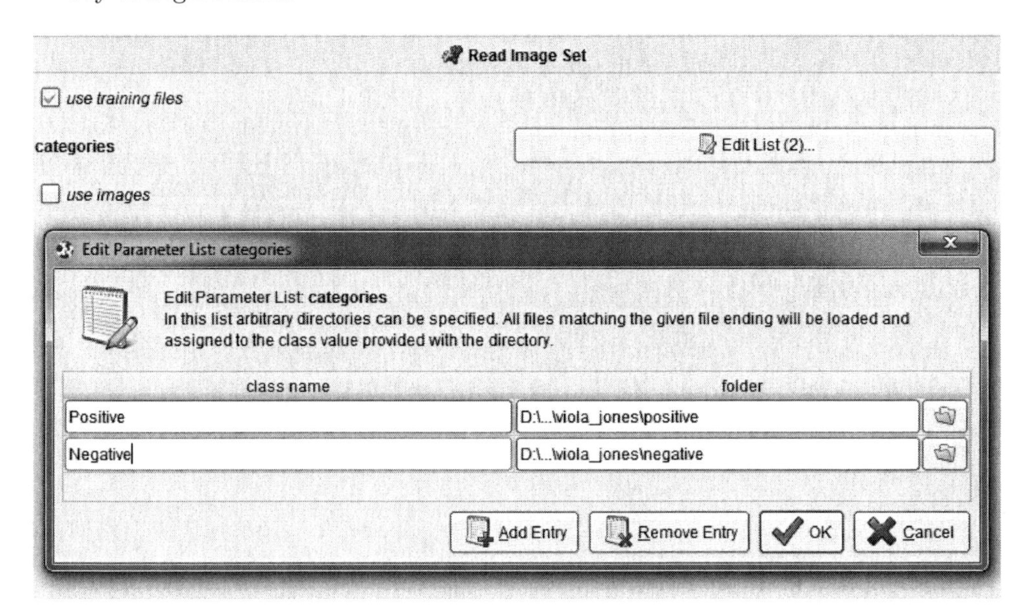

FIGURE 21.11: Read Image Set operator settings.

- Connect the *Viola-Jones* operator after the *Read Image Set* operator. You can experiment with the *evaluation* type and other settings. Default settings are in Figure 21.12.

This operator generates the model for detecting patterns according to the positive class.

Viola-Jones

evaluation type	Matthews correlation coefficient
create cascade	evolutionary
window width	24
window height	24

- [ ] *rotated 45*
- [ ] *add false positive*
- [ ] *save cascade to xml*
- [x] *parallel process*

FIGURE 21.12: Viola-Jones operator settings.

- Used for detection is the *Detector* operator. This operator loads the model and detects the pattern in an input image according to this model to defined scales. Connect the output port from the *Open Grayscale Image* operator to the second input port. The output port of the *Detector* provides only coordinates and the size of the area of interest. To check the visual results, the *Interest Points Visualizer* operator is needed. This operator draws the generated areas of interest onto an input image.
- The whole process should look like that in Figure 21.13.

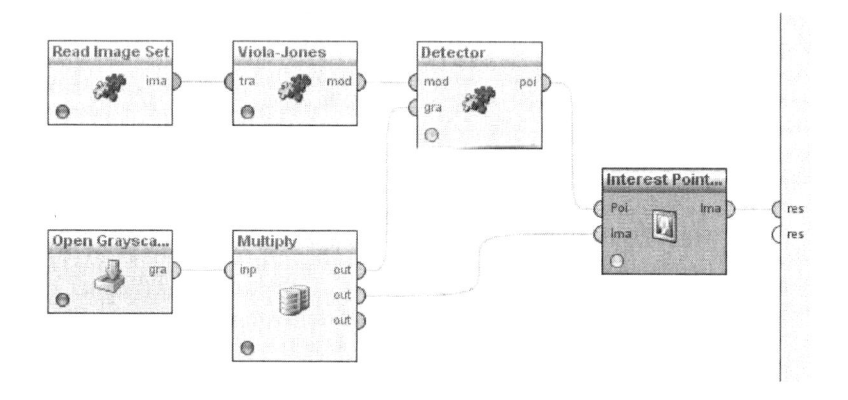

FIGURE 21.13: Process for pattern detection.

21.4 Image Segmentation and Feature Extraction

Segmentation is often used for the detection of different objects in the image. Its task is to split the image into parts so that the individual segments correspond to objects in the image. Unfortunately there is no universal algorithm that can be applied to all types of images. Therefore, there are many algorithms which differ in their properties. These properties can be, for example, the coverage of the entire image by segments and their overlap, the capability to work with colour channels, or monolithic segments. The functionality of algorithms can be based on thresholding, edge detection, statistical models, boundary tracking, active contour, watershed, etc.

In this example we will show a process which works with segmentation and segmented images. The process has several steps:

- Segment the image into parts.
- Process each segment mask separately.
- Extract features to ExampleSet.

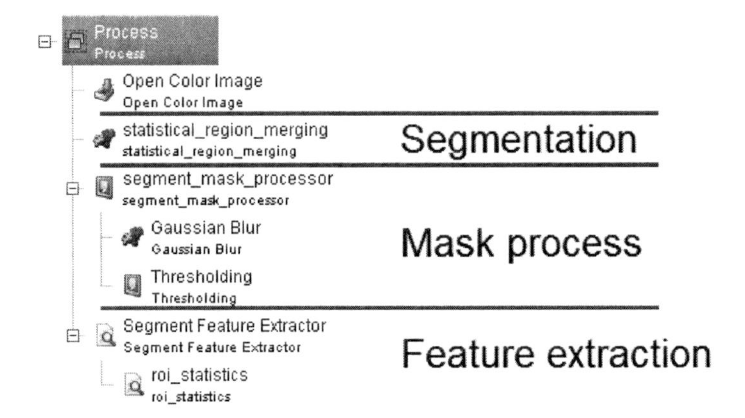

FIGURE 21.14: Process overview.

Segmentation is the process of splitting an image into parts. These parts usually cover the whole image and sometimes overlap each other. The overlap depends on the segmentation algorithm. There are several segmentation algorithms like Statistical Region Merging or Markov Random Field. The first one works with colour images and the second one with gray-scale images. After an image is segmented into parts, it is ready for further processing.

Each mask of segment can be processed as any other grey-scale image using the *Segment Mask Processor* operator. This operator iterates over each segment and sends the mask as *GSImagePlus*. The processing of an image by inner operators differs depending on the concrete example. For example, if we want to increase the size of each region, we will use the *Dilate* operator. The *Gaussian Blur* operator with the *Thresholding* operator can be used for smoothing the boundaries. There are many ways to process the mask. But keep in mind that the final image should always be thresholded if other than binary operators are used.

FIGURE 21.15 (**see color insert**): Example of segmentation.

The *Segment Feature Extractor* operator sends to the inner operator the image with the Region of Interest (ROI) defined. The *Roi_statistics* operator calculates features like median, mean, and shape-based features and sends them to the parent operator.

21.5 Summary

This chapter describes several examples of the use of the IMMI extension for segmentation, feature extraction, and image classification. Image processing is a challenging task and it is not possible to use one universal procedure for all types of tasks. For this reason it is necessary to gain practical experience in creating new processes. It is therefore appropriate to start with simple examples and test the image processing operators, and their effect on the resulting image. The above-mentioned examples can serve as an inspiration for the design of new processes in individual cases. Better results can be achieved with the use of evolutionary algorithms for tuning the parameters of operators.

Bibliography

[1] M D Abramoff, P J Magelhaes, and S J Ram. Image processing with imagej. *Biophotonics International*, 11(7):36–42, 2004.

[2] Hong Qin, Shoujue Wang, Huaxiang Lu, and Xinliang Chen. Human-inspired order-based block feature in the HSI color space for image retrieval. In *Proceedings of the 2009 International Conference on Robotics and Biomimetics*, ROBIO'09, pages 1978–1982, Piscataway, NJ, USA, 2009. IEEE Press.

[3] Renato O. Stehling, Mario A. Nascimento, and Alexandre X. Falcão. A compact and

efficient image retrieval approach based on border/interior pixel classification. In *Proceedings of the Eleventh International Conference on Information and Knowledge Management*, CIKM '02, pages 102–109, New York, NY, USA, 2002. ACM.

Part IX

Anomaly Detection, Instance Selection, and Prototype Construction

Chapter 22

Instance Selection in RapidMiner

Marcin Blachnik
Silesian University of Technology, Department of Management and Informatics, Poland

Miroslaw Kordos
University of Bielsko-Biala, Department of Mathematics and Computer Science, Bielsko-Biala, Poland

Acronyms

ISPR - Instance Selection and Prototype-based Rules

22.1 Introduction

In this chapter we discuss instance selection algorithms and show how to use them in RapidMiner with the ISPR plug-in.

Instance selection is an important pre-processing procedure, which provides us with many benefits. In the past, instance selection was primarily used to improve the accuracy of the nearest neighbor classifier or to speed up the decision-making process. Another significant problem that has appeared in recent years is the large volume of training data that needs to be processed. This involves very high computational complexity and presents a real challenge for most learning algorithms. When examining possible solutions to this problem, instance selection and construction algorithms deserve special attention.

Another application of instance selection can be found in prototype-based rules, where the selected instances help us understand the data properties and the decision made by the model (for example, a nearest neighbor model). In the case of prototype-based rules the goal of the training process is to reduce, as much as possible, the size of data and keep the smallest possible set of representative examples (prototypes) without significant accuracy reduction. The selected examples become the representatives of the population and preserve the representation of all the distinctive features of all class categories.

One example of the prototype-based rules are the prototype threshold rules, where each of the prototypes is associated with a threshold, so that the prototype-threshold pair represents a reception field. The prototype-based rules are typically constructed by the sequential covering algorithms or by distance-based decision trees.

The examples of instance selection and construction methods used in RapidMiner are presented in the subsequent sections using popular datasets from the UCI repository [1] (the last dataset is the *SpamBase* dataset 5 times multiplied with additive noise). The datasets are listed in Table 22.1. The examples require the Instance Selection and Prototype-based Rule plug-in (ISPR), which is available at the RapidMiner Marketplace.

TABLE 22.1: List of datasets used in the experiments.

Dataset Name	# Instances	# Attributes	#Classes
Iris	150	4	3
Diabetes	768	8	2
Heart Disease	297	13	2
Breast Cancer	699	9	2
SpamBase	4601	57	2
5xSpamBase + noise	23005	57	2

22.2 Instance Selection and Prototype-Based Rule Extension

The ISPR plug-in is an extension to RapidMiner that can be obtained from the Rapid-Miner Markeplace. To install the ISPR plug-in go to *Help → Updates and Extensions (Marketplace)* and on the search tab enter the *prototype* keyword. Locate the ISPR extension, select it, and press the *Select for installation* button. The final installation starts by pressing the *Install packages* button in the bottom of the window.

Another option to install the ISPR extension is to manually download the ISPR extension jar file from the `www.prules.org` and place the downloaded file into the *lib/plug-in* folder of the RapidMiner installation directory. On that web page there are also more details about this extension including the source files, access to the SVN repository, and JavaDoc.

Installation of the Instance Selection and Prototype-based Rule extension adds a new group of operators called **ISPR**, which contains all operators implemented in the ISPR extension. The main **ISPR** group includes eight sub-groups:

- **Classifiers.** Classifiers contain classification and regression operators. Currently, there is only one operator called *My k-NN*. This is an alternative implementation of the k-NN classifier in RapidMiner. The main difference is that it can operate on a modified distance metric, for example, including a covariance matrix that can be calculated by some of the instance selection or construction algorithms.

- **Optimizers.** A group of supervised instance construction operators. This group also includes a single operator called LVQ which implements 7 different LVQ training algorithms.
- **Selection.** It is a group of instance selection operators, where each operator uses the nearest neighbor rule for the evaluation.
- **Generalized selection.** This group includes similar algorithms to the ones which can be found in the *Selection* group, however, the operators are generalized so that they can handle both nominal and numerical labels, and instead of k-NN the quality evaluation function, any prediction operators can be used to evaluate the quality of instances.
- **Feature selection.** This is an extra plug-in to the ISPR extension, which wraps the Infosel++ [2] library for feature selection.
- **Clustering.** This group contains various clustering algorithms like Fuzzy C-means, Conditional Fuzzy C-Means, and the vector quantization (VQ) clustering algorithm.
- **Misc.** The Misc Group consists of utility operators like the class iterator, which iterates over class labels of the ExampleSet, and allows for independent clustering of each class. Another operator allows for assigning a class label for prototypes obtained from clustering.
- **Weighting.** The last group includes operators for instance weighting. The obtained weights can be used for conditional clustering and the WLVQ algorithm.

The subsequent sections present examples of operators from the eight groups.

22.3 Instance Selection

As described in the introduction, there are several purposes of instance selection. One is to speed up the k-NN classification process for online prediction tasks. This is especially important for a large training dataset. The computational complexity of the training process of the kNN classifier is $O(1)$ (hence this group of methods is called lazy learning), but the prediction process scales linearly with the number of training samples $O(n)$ (where n is the number of samples in the training set). So you must calculate the distance between the query vector and all the training samples, which makes the prediction process very expensive (time consuming). For example, if the training set consists of $n = 100.000$ samples, to make the prediction for each input query, the system has to calculate $n = 100.000$ distances. So to speed up the prediction process an instance selection is performed. Now the question is which algorithm or combination of algorithms to choose, and to answer this question a property called compression C can be used. This property is defined as

$$C = \frac{\text{instances after selection}}{\text{instances before selection}}. \tag{22.1}$$

A high C value (close to 1) of a given instance selection process means that the algorithm did not successfully reduce the dataset size. A small C value indicates that the set of selected instances is a small subset of original dataset, thus the prediction process will be C times faster than before the selection.

Another purpose is to help us understand the data by selecting a small reference subset as in Prototype Based Rules [3]. In both cases, the goal is to reduce the maximum number of selected instances without affecting the accuracy of the k-NN classifier.

Yet another purpose is to remove the outliers from the training set. An outlier can be defined as a training instance, which significantly differs from the rest of samples in the training set. For instance, in classification problems the miss-labelled samples will (almost) surely be outliers. Pruning such instances improves the convergence of a training process and usually increases the generalization ability of the prediction system.

The instance selection group in the ISPR extension includes several operators. Each operator implements a single instance selection algorithm. All the operators in the group have a single input port—an ExampleSet and three output ports: the set of selected instances, the original ExampleSet, and the trained k-NN prediction model. In most cases, the model is equivalent to the process, which starts by selecting instances and then trains the k-NN model. However, some of the instance selection algorithms calculate some extra parameters like covariance matrix or local feature weights. Although the k-NN prediction model obtained from the instance selection operator can use the extra parameters, currently none of the implemented operators provide this functionality. Thus both solutions presented in Figure 22.1 are equivalent, except that using the model obtained in the output of the instance selection operator we save memory and Java Garbage Collector time used to create new instances of the k-NN model and free up the memory used by the internal nearest neighbour data structures. Additionally, all of the instance selection operators provide in-

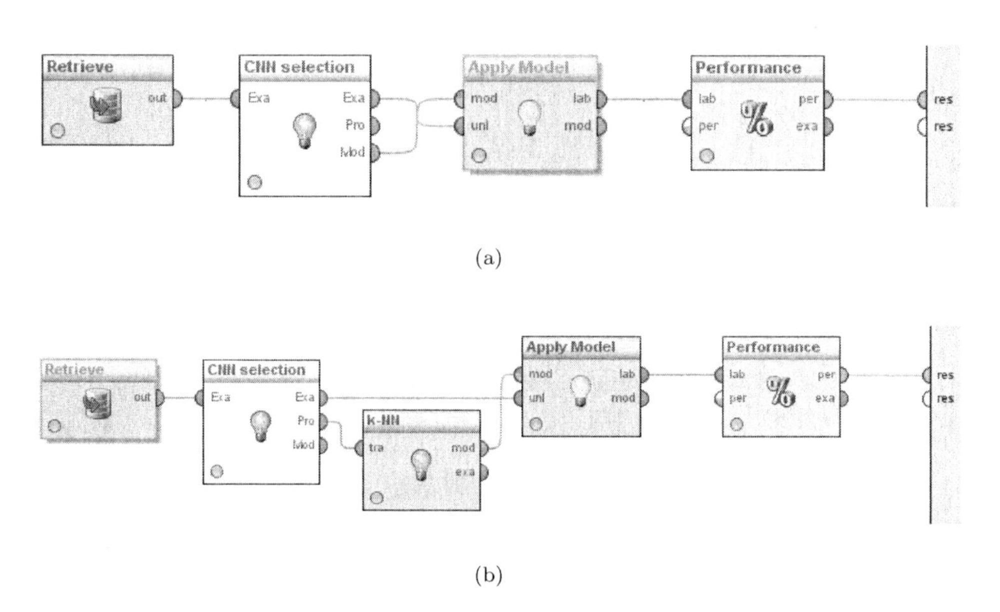

FIGURE 22.1: Equivalent examples of applications of the Instance Selection operators (a) use of internal k-NN model (b) use of selected instances to train external k-NN model.

ternal parameters (properties) such as *instances_after_selection*, *instances_before_selection*, and *compression*. These parameters can be used to determine the quality of the instance selection process.

22.3.1 Description of the Implemented Algorithms

At the time of writing this chapter the following 11 algorithms are implemented:

Random selection It is the simplest instance selection operator, which randomly drawns instances from the example set. The instances can be selected independently or stratified, so that the ratio of instances from different class labels is preserved [4].

MC selection This is a simple extension of the random selection, also known as the Monte-Carlo algorithm [4]. It repeats the random selection given a number of times (the number of iterations) and selects the best subset. In this algorithm the quality of the set of selected instances is determined by the accuracy of the 1-NN classifier. See Algorithm 22.3.1.

Algorithm MC algorithm.

Require: $\mathbf{T}$, z, $maxit$
  for $i = 1 \ldots maxit$ **do**
    $\mathbf{P}^* \leftarrow Rand(\mathbf{T}, z)$
    $tacc \leftarrow Acc(1NN(\mathbf{P}^*, \mathbf{T}))$
    if $tacc > acc$ **then**
      $acc \leftarrow tacc$
      $\mathbf{P} \leftarrow \mathbf{P}^*$
    end if
  end for
  return P

RMHC selection This operator also belongs to random selection-based operators. It implements the Random Mutation Hill Climbing algorithm defined by Skala [5]. The algorithm has two parameters—the number of prototypes and the number of iterations. In addition, it uses the classification accuracy of the 1-NN classifier as a cost function. The algorithm is based on coding prototypes into a binary string. To encode

---

Algorithm RMHC algorithm.

Require: $\mathbf{T}$, z, $maxit$
  $m \leftarrow |\mathbf{T}|$
  $st \leftarrow Bin(l \log_2 m)$
  $st^* \leftarrow SetBit(st, z)$
  for $i = 1 \ldots maxit$ **do**
    $\mathbf{P}^* \leftarrow GetProto(\mathbf{T}, st^*)$
    $tacc \leftarrow Acc(1NN(\mathbf{P}^*, \mathbf{T}))$
    if $tacc > acc$ **then**
      $acc \leftarrow tacc$
      $\mathbf{P} \leftarrow \mathbf{P}^*$
      $st \leftarrow st^*$
    end if
    $st^* \leftarrow PermuteBit(st)$
  end for
  return P

---

a single prototype the $\lceil \log_2(n) \rceil$ bits are required, where n is the number of vectors in the training example set. $c \cdot \lceil \log_2(n) \rceil$ bits are required to encode c prototypes. The algorithm in each iteration randomly mutates a single bit and evaluates the predictive accuracy of the actual set of instances. If the mutation improves results, the new solution is kept, otherwise the change is rolled back. This step is repeated until the defined number of iterations is reached—see Algorithm 22.3.1. The number of prototypes is predefined, so the *compression* is manually defined by the user. The algorithm it is also insensitive to outliers, which are pruned by default.

CNN selection CNN is one of the oldest instance selection algorithms [6]. CNN starts by adding a single randomly selected instance from the original set $\mathbf{T}$ to the subset of selected examples $\mathbf{P}$, and then tries to classify all other examples from $\mathbf{T}$ using k-NN (for the purpose of the classification $\mathbf{P}$ now becomes the training set and $\mathbf{T}$ the test set). If any instance from $\mathbf{T}$ is incorrectly classified, it is moved to the selected set $\mathbf{P}$. The algorithm is rather fast and allows for small compression (the number of training samples which remain after selection), so it is useful for reducing the size of the training set. However, the quality of the compression depends on the level of noise in the data. Moreover, it is sensitive to outliers, so its usage should be preceded by a noise and outlier filter. The CNN algorithm is shown on algorithm 22.3.1, where $\mathbf{T}$ is a training set, $\mathbf{x}_i$ is an i'th instance from the training set, $\mathbf{P}$ is the selected subset of examples, and $\mathbf{p}_j$ is a single j'th example from the set $\mathbf{P}$.

Algorithm CNN algorithm.

Require: $\mathbf{T}$

$m \leftarrow |\mathbf{T}|$
$\mathbf{p}_1 \leftarrow \mathbf{x}_1$
$flaga \leftarrow$ **true**
while flaga **do**
  $flaga \leftarrow$ **false**
  for $i = 1 \ldots m$ **do**
    $\bar{C}(\mathbf{x}_i) = \text{kNN}(\mathbf{P}, \mathbf{x}_i)$
    if $\bar{C}(\mathbf{x}_i) \neq C(\mathbf{x}_i)$ **then**
      $\mathbf{P} \leftarrow \mathbf{P} \cup \mathbf{x}_i$;
      $\mathbf{T} \leftarrow \mathbf{T} \setminus \mathbf{x}_i$
      $flaga \leftarrow$ **true**
    end if
  end for
end while
return $\mathbf{P}$

ENN selection ENN is a base for a group of algorithms [7]. The ENN operator uses a single parameter k, which is the number of nearest neighbors in the k-NN algorithm used by the ENN. ENN iterates over all vectors in the example set $\mathbf{T}$ and tries to predict with k-NN their class label. If the predicted label is incorrect the vector is marked for removal, so this algorithm is useful for outlier detection and noise reduction. Its sketch is presented in 22.3.1

RENN selection RENN repeats *ENN selection* until no vector is removed (see algorithm 22.3.1). [8] It has similar noise-reduction properties as the ENN algorithm.

All-kNN selection is also based on the ENN algorithm, so it shares the same properties. This operator repeats the ENN procedure for a range of different k values [8].

GE selection is a Gabriel Editing proximity graph-based algorithm [9]. It uses a simple

Algorithm ENN algorithm.

Require: $\mathbf{T}$
  $m \leftarrow |\mathbf{T}|$;
  $rem_i \leftarrow 0$;
  for $i = 1 \ldots m$ **do**
    $\bar{C}(\mathbf{x}_i) =$kNN$((\mathbf{T} \setminus \mathbf{x}_i), \mathbf{x}_i)$;
    if $C(\mathbf{x}_i) \neq \bar{C}(\mathbf{x}_i)$ **then**
      $rem_i = 1$;
    end if
  end for
  for $i = 1 \ldots m$ **do**
    if $rem_i == 1$ **then**
      $\mathbf{T} = \mathbf{T} \setminus \mathbf{x}_i$
    end if
  end for
  return P

Algorithm RENN algorithm.

Require: $\mathbf{T}$ $flag \leftarrow$ **true**
  while falg **do**
    $flag \leftarrow$ **false**
    $\bar{\mathbf{P}} =$ENN$(\mathbf{P}, \mathbf{T})$
    if $\bar{\mathbf{P}} \neq \mathbf{P}$ **then**
      $flag \leftarrow$ **true**
    end if
  end while
  return P

evaluation defined as:

$$\underset{a \neq b \neq c}{\forall} D^2(\mathbf{p}_a, \mathbf{p}_b) > D^2(\mathbf{p}_a, \mathbf{p}_c) + D^2(\mathbf{p}_b, \mathbf{p}_c) \tag{22.2}$$

to determine if an example $\mathbf{p}_b$ is a neighbor of an example $\mathbf{p}_a$. If they are neighbors according to formula 22.2 and the class label of $\mathbf{p}_a$ and all its neighbors is equal, then $\mathbf{p}_a$ is marked for removal. This algorithm is sensitive to noise and outliers. However, it usually allows for obtaining a low compression ratio.

RNG selection is an implementation of the *Relative neighbor Graph* algorithm. It has similar properties to the GE method. It differs in the cost function and the function used to evaluate the neighbors is modified as follows:

$$\underset{a \neq b \neq c}{\forall} D(\mathbf{p}_a, \mathbf{p}_b) \geq \max(D(\mathbf{p}_a, \mathbf{p}_c), D(\mathbf{p}_b, \mathbf{p}_c)). \tag{22.3}$$

For more details see [9].

IB3 selection is a very popular algorithm defined by D. Aha [10]. It is a member of the IBL (Instance Based Learning) methods. The first two algorithms in the IBl group (IB1 and IB2) are just small modifications of 1-NN introducing the normalization step and rejection of samples with unknown attributes (IB1) and a simplified one-pass CNN algorithm (IB2). The IB3 algorithm extends this family by applying the *wait and see* principle of evaluating samples according to the following formula:

$$AC_{up|low} = \frac{p + z^2 \big/ 2n \pm z\sqrt{p(1-p)/n + z^2/4n^2}}{1 + z^2 \big/ n} \tag{22.4}$$

where p denotes the probability of success in n trials and z is a confidence factor.

The prototype is accepted if the lower bound of its accuracy (with the confidence level 0.9) is greater than the upper bound of the frequency of its class label and a prototype is removed if the upper bound of its accuracy is lower (with the confidence level 0.7) than the lower bound of the frequency of its class. It is insensitive to noise and outliers and also achieves good compression.

ELH selection is an implementation of the instance selection algorithm proposed by Cameron Jones [11]. The algorithm is based on Encoding Length Heuristic (ELH) evaluates the influence of an instance rejection. This heuristic is defined as:

$$J(n, z, n^{err}) = F(z, n) + z\log_2(c) + F(n^{err}, n - z) + n^{err}\log_2(c-1) \tag{22.5}$$

where $F(z, n)$ is a cost of coding n examples by z examples.

$$F(l, n) = \log^*\left(\sum_{j=0}^{l} \frac{n!}{j!(n-j)!}\right) \tag{22.6}$$

where $\log^*$ is a cumulative sum of positive factors $\log_2$.
The algorithm starts by rejecting an instance and then evaluates the influence of the rejection on the ELH heuristic. If the rejection does not decrease the value of the heuristic, the vector is removed and the whole procedure is repeated. This algorithm has very good properties such as low compression, usually high accuracy, and it is insensitive to outliers.

22.3.2 Accelerating 1-NN Classification

To speed up the classification of the 1-NN algorithm, the family of proximity graph algorithms can be used. This group is represented by the *GE Selection* and *RNG Selection* operators. Both of them attempt to preserve the decision border identical to the original 1-NN classifier border. Also the CNN, ELH, and IB3 algorithms can be used as members of this group. Because particular algorithms may produce different accuracy and different compression, it is worth comparing them. The best practice to compare the quality of different learning algorithms is to use the *X-validation* operator as shown in Figure 22.2.

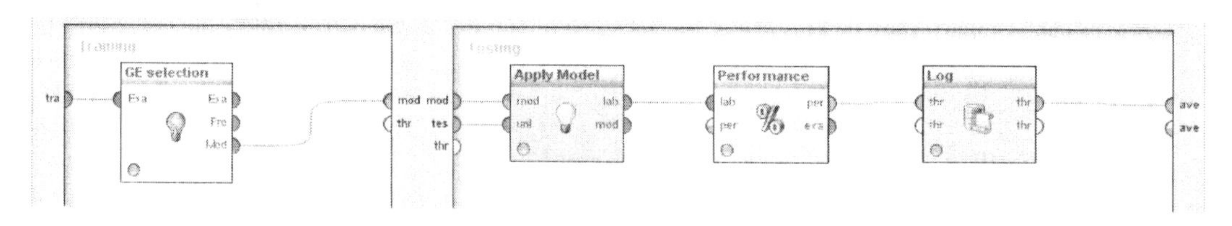

FIGURE 22.2: Validation of the accuracy and compression of the *GE Selection* operator.

This example does not provide any information about the compression. The compression can be obtained with the *Log* operator. The *Log* operator allows for extracting internal information provided by any of the operators in the process. The logging options include two types of values: the configuration parameters called *parameter* of an operator, such as the value of k in kNN or the type of the distance function, and the *value* parameter, which is an internal property returned by a given operator. Sample configuration settings of the *Log* operator for the process presented in Figure 22.2 is shown in Figure 22.3. In this example three parameters are logged: *Compression* of the *GE selection* operator, *accuracy* of the *Performance* operator, and the type of the distance measure *mixed_measure* of the *GE selection* operator. The *Log* operator logs all the values whenever it is executed, so if it is placed inside the *X-Validation* operator, it is executed in each validation iteration. Finally, all of the logged values can be stored in a file and are available when the process finishes in the *Results Perspective* in the form of a plot or a Table 22.3.

For large datasets where the selection process may be time-consuming it is more efficient to compare different instance selection techniques within a single process. The model used for that should provide an identical testing environment for all instance selection operators. This can be accomplished by embedding all the instance selection algorithms into a validation loop as shown in Figure 22.4. In this example, we start by explaining the validation procedure, which is equivalent to the one shown in Figure 22.1, except that instead of a single instance selection algorithm, a family of different algorithms is used. The trained 1NN models are then passed to the testing process where each of them is applied, and for each of them the performance is calculated. In the last step the *Log* operator is applied to log all the values of accuracy and the compression in each validation loop. The logged values are further used outside of the *Validation* operator. Now let us see the main process presented in Figure 22.5. First, the data is retrieved from the repository (here we use the Iris dataset for the simplicity) and then after reading the data, all attributes are normalized using the *Normalize* operator (this operator can be found in *Data Transformation* → *Value Modification* → *Numerical Value Modification*). This step is very important because attributes in the dataset may be differently scaled, which affects the distance calculations. After that, we start the validation operator.

When it finishes, we can access the extracted *Log* values. To perform further calculations the logged values are transformed into an *ExampleSet* using the *Log to Data* operator (it requires defining the name of the Log that we want to transform into data; the Log name is the name of the Log operator). This allows for further manipulations of the values. The logged values contain the accuracy and the compression logged in each validation step, which should be averaged. For that purpose, *Aggregate* operator is used. The configuration settings of the *Log* operator and the *Aggregation* operator are shown in Figure 22.6.

TABLE 22.2: Comparison of various instance selection methods. *acc* denotes classification accuracy, *compr* denotes compression.

Dataset Name	kNN acc	GE acc	RNG acc	CNN acc	GE compr.	RNG compr.	CNN compr.
Iris	94.66	94.66	90.66	91.33	39.55	18.22	16.96
Diabetes	70.97	70.97	67.59	66.67	94.06	59.57	49.16
Heart Disease	77.90	77.90	76.94	73.61	99.89	57.86	44.29
Breast Cancer	94.85	94.70	92.13	92.71	38.07	12.82	13.54

The results for some datasets are presented in Table 22.2.

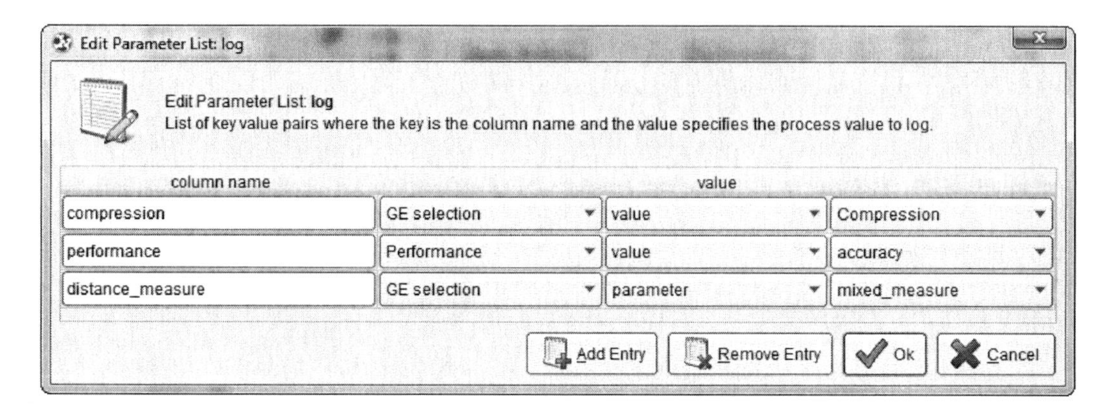

(a) *Log* operator configuration.

Log | Result Overview

Table View | Plot View

Log (10 rows, 3 columns)

compression	performance	distance_mea...
0.281	0.933	MixedEuclidean
0.252	0.933	MixedEuclidean
0.289	1	MixedEuclidean
0.304	1	MixedEuclidean
0.289	0.933	MixedEuclidean
0.289	0.933	MixedEuclidean
0.259	1	MixedEuclidean
0.267	1	MixedEuclidean
0.274	0.933	MixedEuclidean
0.252	0.933	MixedEuclidean

(b) *Log* view in the results perspective.

FIGURE 22.3: Configuration of *Log* operator, and the table view of the results.

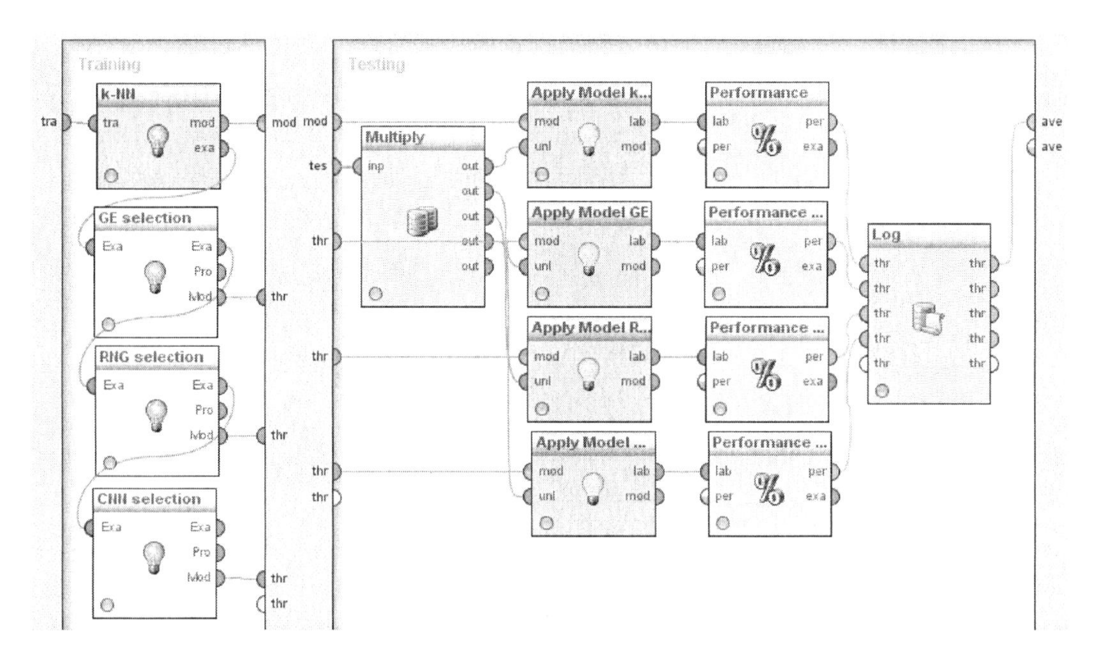

FIGURE 22.4: Selecting the most useful instance selection operator. Validation inner process configuration.

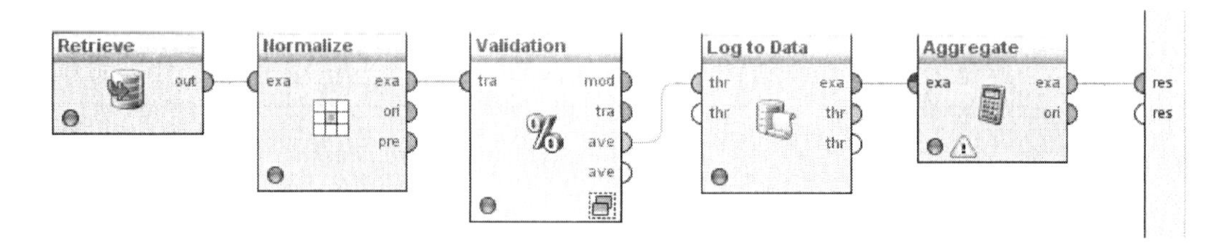

FIGURE 22.5: Selecting the most useful instance selection operator Main process.

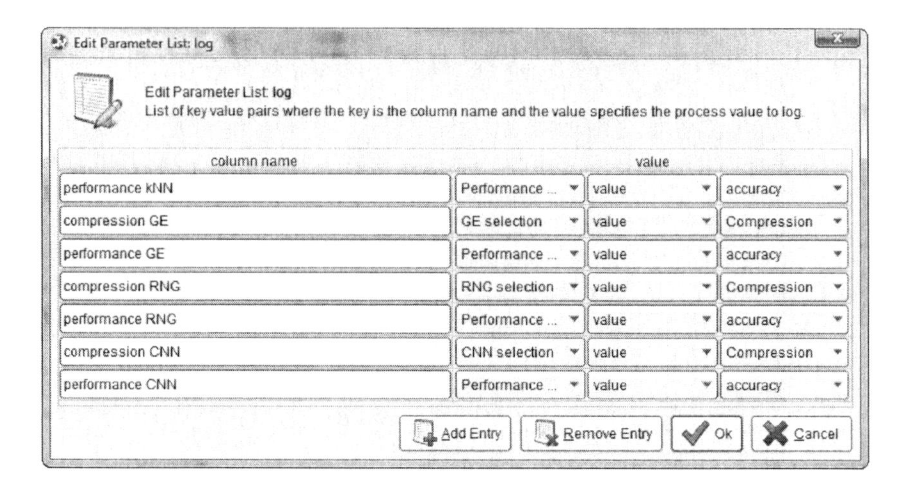

(a) *Log* configuration.

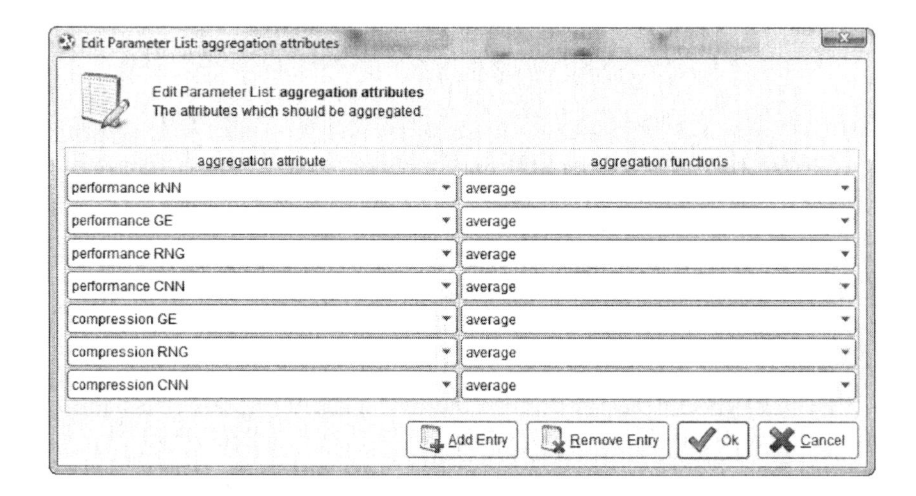

(b) *Aggregate* configuration.

FIGURE 22.6: Selecting the most useful instance selection operator Validation inner process configuration.

The results were obtained by extending the process with some extra operators, which allows for automatic result acquisition. In this example the *Loop Datasets* operator was used to iterate over several datasets as shown in Figure 22.7(a). The inner process was equipped with a mechanism that allows combining results. This was obtained by the *Recall* and *Remember* operators. These two operators are designed to store and cache any types of objects like example sets, trained models, etc. in an operational memory. In our example they are used to cache the results already obtained for each of the datasets connected to the *Loop Dataset* operator. The *Recall* operator is used to retrieve the previous results and the *Remember* stores the new, combined results in a memory cache. Note: While calling the *Remember* and *Recall* operators, we have to specify the type of input which will be stored. Here we store ExampleSet. The old results and the new ones obtained at the output of the *Aggregate* operator (see Figure 22.7(b)) are combined using the *Append* operator inside the *Handle Exception* operator (see Figure 22.7(c)). Since there are no initial results in the first run, the *Recall* operator will pop up an exception. For that reason, the *Handle Exception* operator in case of an exception calls the *Catch* subprocess as shown in Figures 22.7(b) and 22.7(c), where it just stores the initial results into the memory cache.

22.3.3 Outlier Elimination and Noise Reduction

While mining datasets, we often encounter the problem of noise and outliers. The meaning of outliers was already described, while the noise complicates the shape of the decision border of the prediction system. It usually drives from imprecise measurements, and in many learning systems its influence is reduced by introducing Tihonov regularization or by over-sampling with a manually added small level of noise.

Especially important are outlier examples, which may affect the quality of the learning process. Their influence usually depends on the cost function used by the learner and on the size of the data (for smaller datasets their influence is much higher than for larger ones). Some of the instance selection operators can be used to eliminate the outliers and reduce the noise, for example, the *ENN selection* operator and others in that family like *RENN selection* or *All kNN selection*.

To analyse the influence of the level of noise on the model accuracy, a process in RapidMiner can be built. In the sample process we will consider the influence of different levels of noise on the accuracy of the LDA classifier applied to the Iris dataset. The main process is presented in Figure 22.8(a).

The process starts with retrieving the Iris dataset from the repository (*Retrieve* operator), then the data is normalized using the *Normalize* operator (this operator is set to normalize in range 0–1) followed by the *Loop Parameters*. The *Loop Parameters* operator is a loop, which iterates over different subsets of parameters of any of sub-process models. In the process the final results are combined into a single set of examples using the *Recall* and *Remember* operators wrapped by the *Handle Exception* operator, as described in the previous section. The sub-process of the *Loop Parameters* is presented in Figure 22.8(b). The process is similar to the one described in Figure 22.7(b), except that the *Add Noise* operator is added. This operator allows adding noise into variables or labels of the input example set, but for our test we will only add noise to the label attribute. The sub-process of the validation step is presented in Figure 22.8(c). It uses two *LDA* operators. The first one uses the dataset without outlier elimination and noise reduction, and the second one uses the prototypes obtained from the *ENN selection* operator, so the second one uses the dataset after noise filtering obtained with the ENN algorithm. The second LDA is renamed and called *LDA Pruned*. The remaining operation required to execute the process is the *Loop Parameters* operator configuration. For the experiment, the *Add Noise* operator has to be selected and then from the list of parameters the *label_noise* parameter. We set the

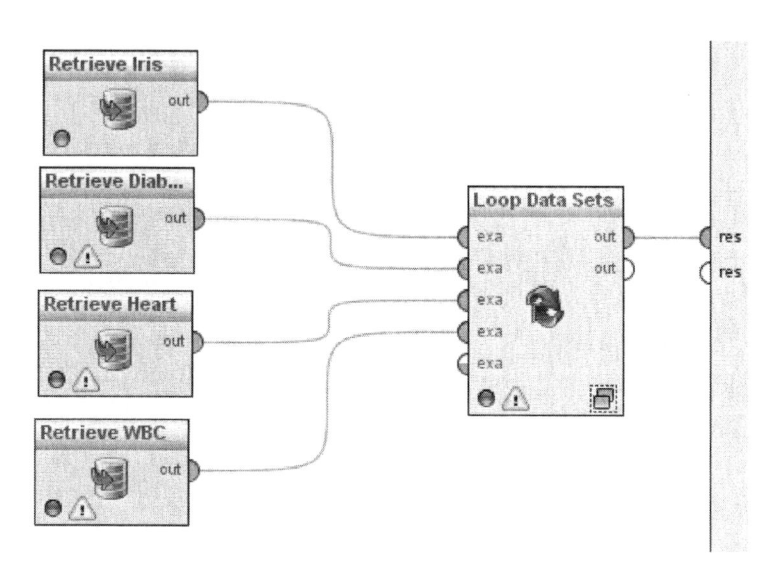

(a) *Main process* configuration.

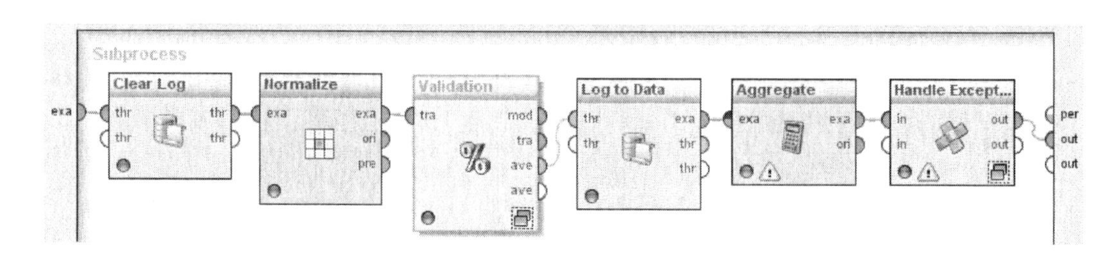

(b) *Inner process of Loop Datasets* configuration.

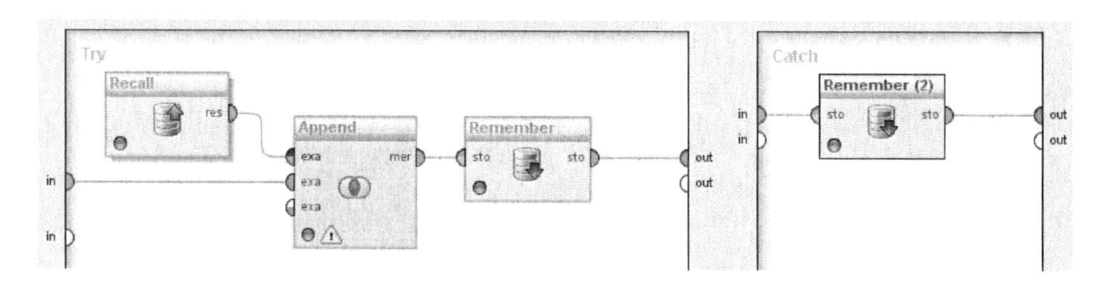

(c) *Handling exception* configuration.

FIGURE 22.7: Processing many datasets at once.

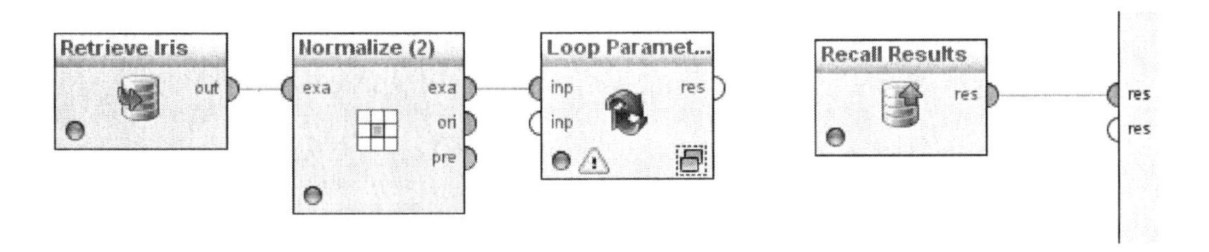

(a) The main process.

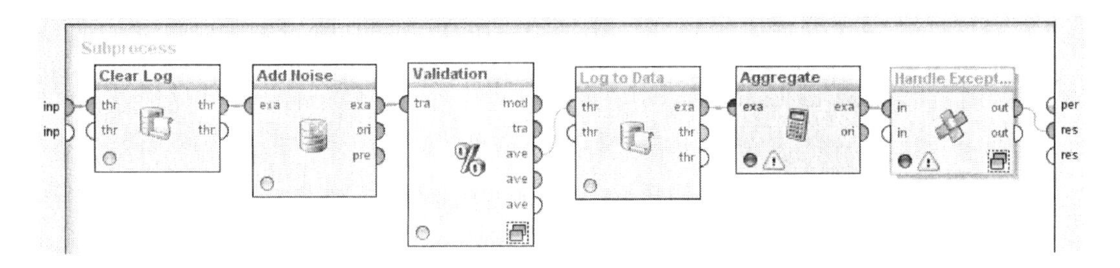

(b) The subprocess of the *Loop parameters* operator.

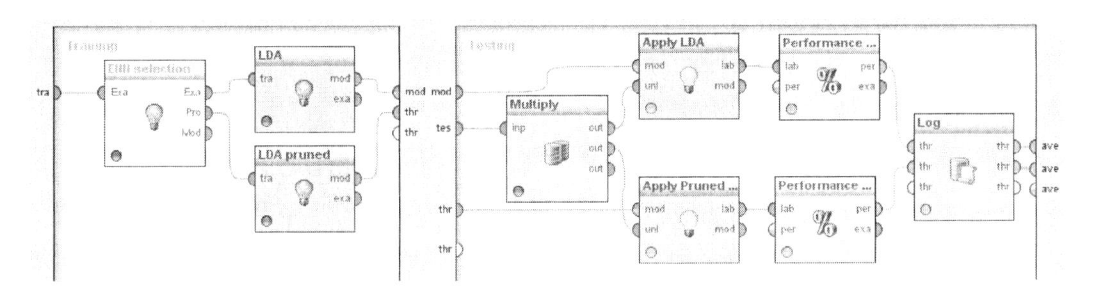

(c) The subprocess of *Validation* operator.

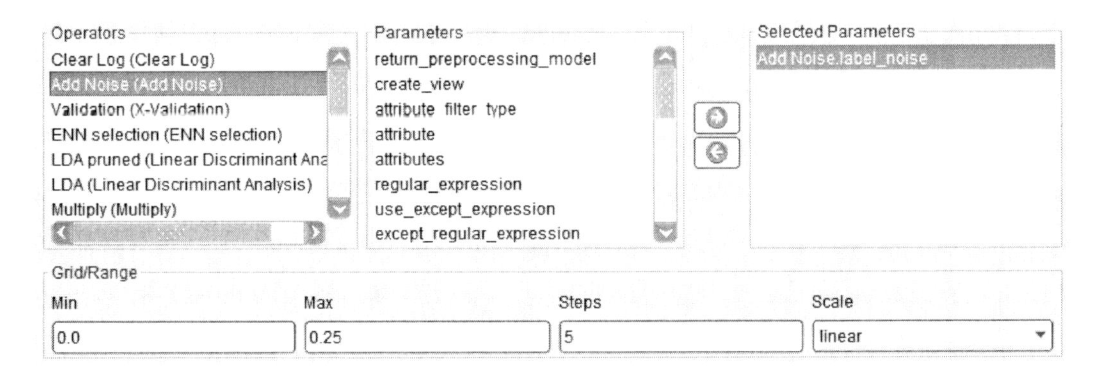

(d) *Loop Parameters* configuration.

FIGURE 22.8: Test of the outlier elimination task.

minimum level of noise to 0 and the maximum to 0.25 in 5 steps (see Figure 22.8(d)). The obtained results are presented in Table 22.3. From these results we can see that when the level of noise is increasing the *LDA Pruned* improves the results. However, in the case of absence of noise or for low levels of noise, the accuracy may decrease.

TABLE 22.3: Comparison of the influence of noise on the accuracy with and without ENN noise filtering.

Pruned LDA	LDA	Noise Level
84.00	86.66	0.0
81.33	80.00	0.05
73.33	74.66	0.10
64.00	62.66	0.15
55.33	50.66	0.2
45.33	42.66	0.25

22.3.4 Advances in Instance Selection

All the examples presented above assume that only a single instance selection algorithm was applied. In practice, instance selection algorithms can be combined into a series of data transformations. This is useful for obtaining prototype-based rules as well as when we are interested in high compression.

A good example of such processing is a combination of outlier and noise elimination algorithms and algorithms preserving the 1-NN decision border (described in Section 22.3.2). The algorithms, which preserve the 1-NN decision boundary are very sensitive to any outliers and noise examples, so it is desired to eliminate the outliers before applying the *CNN, GE,* or *RNG* algorithms. An example of such a process is presented below. In this example we compare the accuracy and compression of three combinations of algorithms. Here we use the *Select Subprocess* operator, rather than the processes already presented, which will be placed between different instance selection schemes. It is also important to set the *Use local random seed* for the *Validation* operator, which ensures that each model will have identical training and testing datasets. The main process is presented in Figure 22.9(a). At the beginning, all the datasets are loaded and then the already presented *Loop Datasets* operator is used. However, we add here an extra operator called *Set Macro* to define a variable *Dataset_Iterator*, which identifies datasets used in the experiments. *Macros* are process variables which can be used to represent a value, which is changing during the process execution. In this case the macro is used to enumerate the ExampleSets used in the experiments, such that in each iteration of the *Loop Dataset* the value of *Dataset_Iterator* macro is incremented. When the *Loop Datasets* operator finishes, the *Log* used inside the process is transformed into anExampleSet using the *Log to Data* operator. Then the *Parse Numbers* operator is applied to convert nominal values into numerics. We need to perform the nominals-to-numeric conversion, because RapidMiner stores macro values as nominals (Strings) even when they represent numbers like the *Compression.* Then the ExampleSet is aggregated. However, in this example the *Aggregate* operator uses two attributes to group results before averaging: the *Dataset_Iterator* and the *Model.* The final three operators are used to assign appropriate nominal values. The *Numerical to Nominal* operator transforms the DataSet (the number logged into the *Log* from the *Dataset_Iterator* macro) and Model (similar macro value) attributes into the appropriate names. Two *Map* operators assign

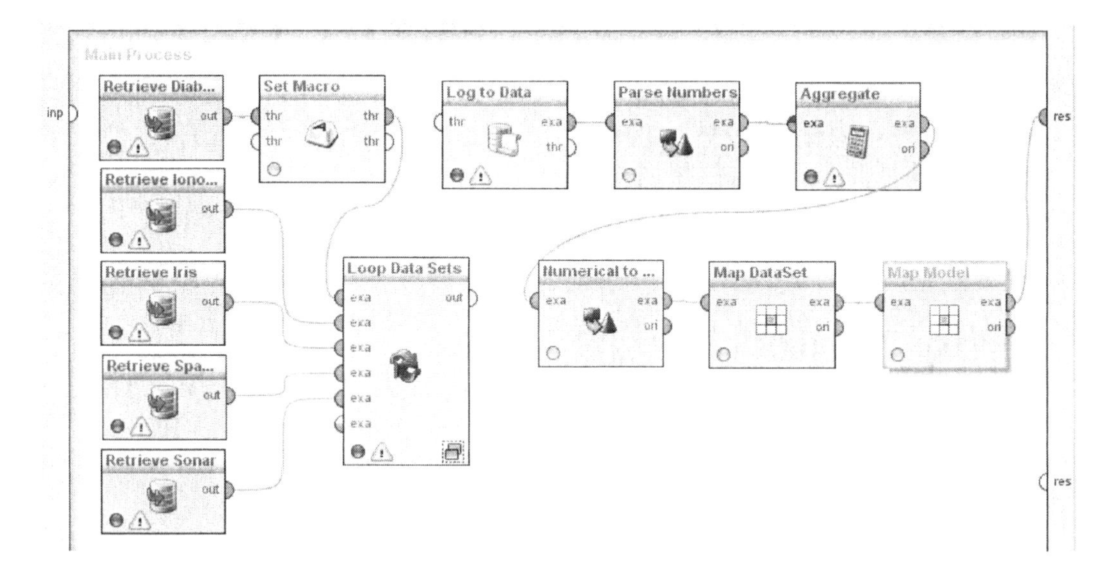

(a) *Main process* configuration.

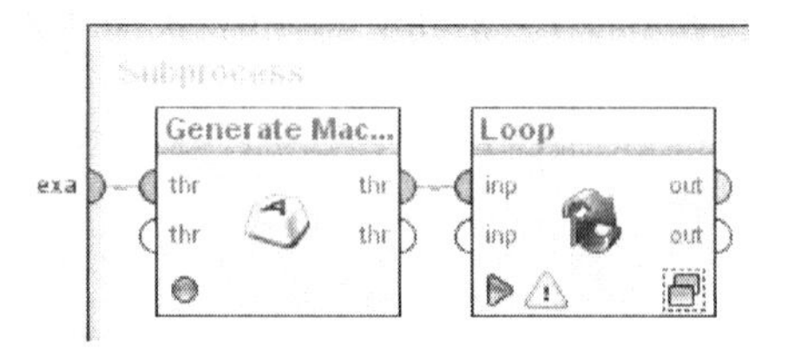

(b) Subprocess of the *Loop Datasets*.

FIGURE 22.9: Process configuration used to compare different instance selection scenarios (1).

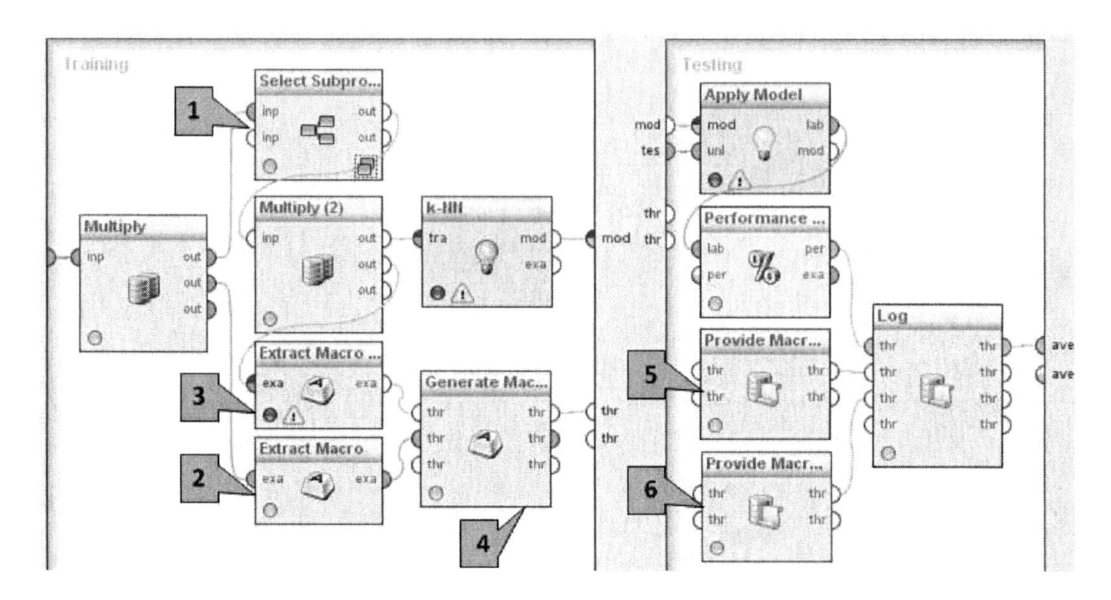

(a) Subprocess of the *Validation* operator.

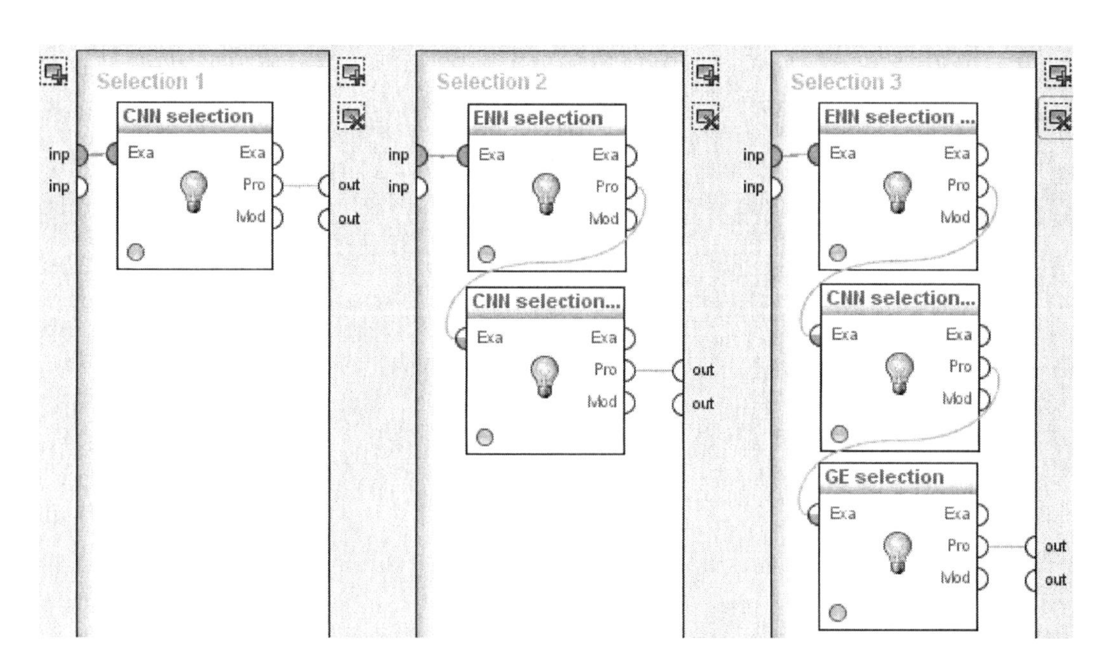

(b) Configuration of the *Select Subprocess* operator.

FIGURE 22.10: Process configuration used to compare different instance selection scenarios (2).

appropriate names to the numbers like *DataSet* value equal 1 is mapped into *Diabetes*, value 2 into *Ionosphere*, etc.

As it was described, inside the *Loop Datasets* operator the value of the *Dataset_Iterator* macro is incremented using *Generate Macro* and then the *Loop* operator is used to iterate over different combinations of instance selection chains (Figure 22.9(b)). To implement this, the *iteration* macro is used and the number of iterations is set to the number of different instance selection configurations. The sub-process of the *Loop* operator includes just the *X-Validation* operator. Its sub-process is presented in Figure 22.10(a). The main part of this sub-process consists of the *Select Subprocess* and *k-NN* operators. The first one (marked as 1) is equivalent to the *switch* statement and allows switching among different sub-processes. The problem with the process is the need to manually calculate the *Compression* of the whole system, because the compression provided as the property of the operator describes just compression of a single operator and not of the operators chain. Moreover, because we use the *Select Subprocess* we don't know from which the *Compression* property should be read. For that purpose, the *Extract Macro* operators are used. They extract the number of instances before (operator marked as 2) and after instance selection (operator marked as 3). Then the *Generate Macro* (marked as 4) is used to calculate the *Compression*, which is passed to the testing part. The test sub-process is relatively simple just with two extra *Provide Macro* operators. They are necessary to provide the *Macros Values* in a form that can record the log. Here, using an operator marked as 5, we transform the *Compression* macro into a value which can be logged, and similarly operator 6 is used to provide *Dataset_Iterator* macro as a log value.

The performance, *Compression*, and *iteration* of the *Loop* operator equivalent *select_which* parameter of the *Select Subprocess* operator and the value of *Dataset_Iterator* macro are logged. The final part of the process is the *Select Sub-process*, which is visualized in Figure 22.10(b). It shows different configuration schemes. In the first one, only the *CNN* algorithm is used; in the second one it is preceded by the *ENN selection*, while in the last one the two former operators are followed by the *RNG selection*.

The results obtained for several datasets are visually presented in Figure 22.11(a) using the *Advanced Charts* data visualization. To do so, we need to switch to the *Results perspective* and on the returned example set switch to the *Advanced Charts* view. In that view we have to adjust attributes of the dataset to appropriate configuration settings using a drag-and-drop technique. For example, we have to move the *Average(compression)* attribute to the *Domain dimension*, *Average(accuracy)* to the *Empty axis*, and the remaining attributes to the *Color dimension* and *Shape dimension*. The *Advanced charts* configuration settings are shown in Figure 22.11(b).

22.4 Prototype Construction Methods

The family of prototype construction methods includes all algorithms that produce a set of instances at the output. The family contains all prototype-based clustering methods like *k-means*, *Fuzzy C-Means* (FCM), or *Vector Quantization* (VQ). The family also includes the Learning Vector Quantization (LVQ) set of algorithms. The main difference between these algorithms and the instance selection ones is the way of obtaining the resulting set. In instance selection algorithms the resulting set is a subset of the original instances, but in the case of instance construction methods, the resulting set is a set of new instances.

Next, we would like to show different approaches to prototype construction. In our

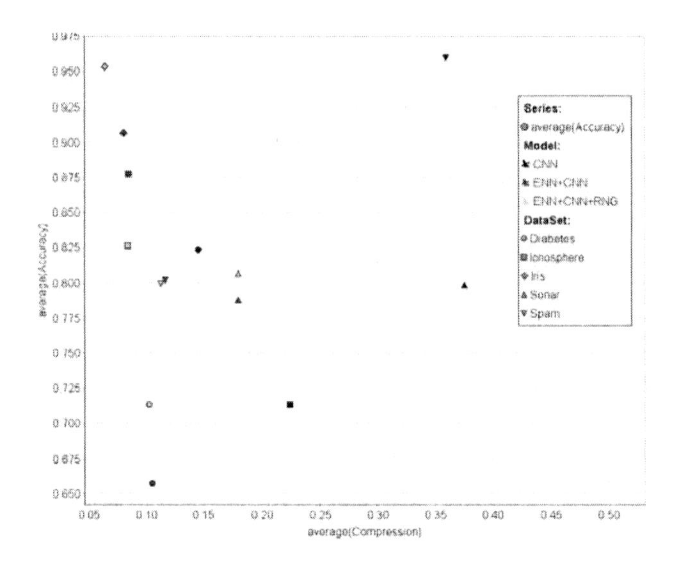

(a) Results.

(b) Configuration settings.

FIGURE 22.11: Comparison of three different combinations of instance selection algorithms (CNN, ENN+CN, ENN+CNN+RNG, accuracy as a function of Compression plot, and configuration settings required to generate it.

examples, we will use them to train k-NN classifier. However, the obtained instances can also be used to train other models. According to the paper of L. Kucheva [12], there are many approaches to use clustering algorithms to obtain the labeled dataset. The easiest one is to cluster the dataset and then label the obtained prototypes. This kind of process is shown in Figure 22.12. It commences with loading the dataset, and next the data is clustered using the FCM algorithm. The obtained cluster centers are then labeled using the *Class assigner* algorithm. After re-labeling, the prototypes are used to train the k-NN model and the model is applied to the input dataset.

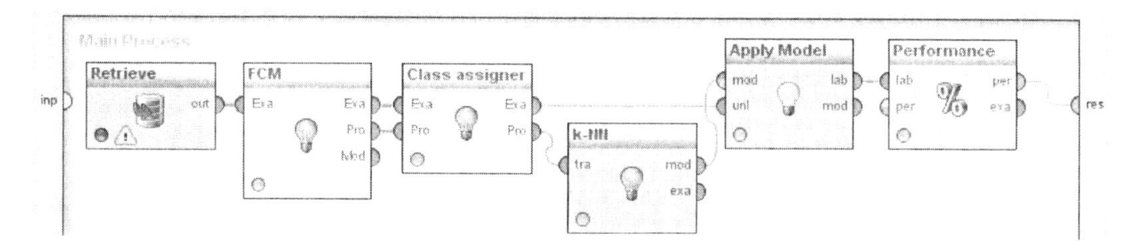

FIGURE 22.12: Process of prototype construction based on clustering and re-labeling.

Another approach is to cluster each class independently. In this approach we need to iterate over class labels and in each iteration we select instances from one single class label which will be clustered. The prototypes obtained after re-labeling should be concatenated. This example is shown in Figure 22.13(a). In this example the *Class iterator* is used to

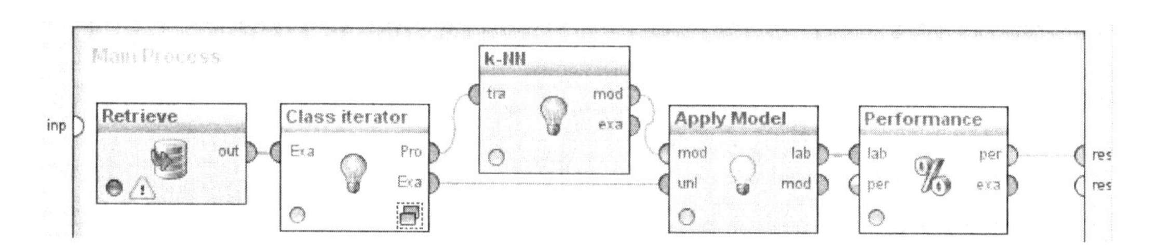

(a) *Main process* configuration.

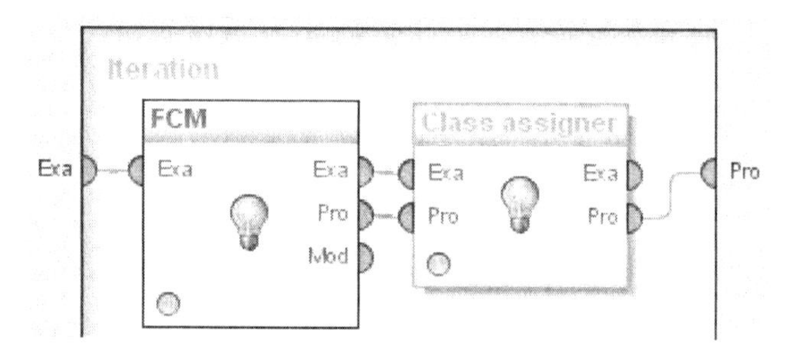

(b) Subprocess of the *Class Iterator*.

FIGURE 22.13: Process of prototype construction based on clustering of each class and combing obtained cluster centres.

iterate over classes. Instances from each single class are clustered and delivered to the inner process in a loop (Figure 22.13(b)). At the end the obtained prototypes are automatically combined into a final set of prototypes. This task can also be solved using the *Loop labels* operator, however this doesn't concatenate prototypes obtained in each iteration, so in that case it must be done manually.

In the next example an LVQ network is used to select the prototypes. The LVQ algorithm has many advantages and the most important one is the ability to use optimized codebooks (prototypes) according to the classification performance. The use of the LVQ network is equivalent to the previous examples except that the LVQ network requires initialization. For the initialization, any instance selection or construction algorithm can be used. The typical initialization is based on random codebooks selection so the *Random selection* operator can be used as shown in Figure 22.14(a).

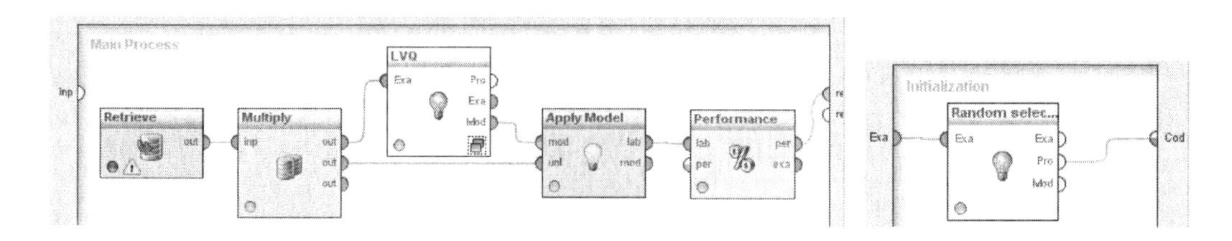

(a) *Main process* configuration.

(b) Random initialization of *LVQ* operator.

FIGURE 22.14: Process of prototype construction based on clustering of each class and combing obtained cluster centres.

In a more complex scenario, we can use the series of LVQ processing with prototypes initialized using VQ clustering. The tree structure of this process is shown in Figure 22.15. The process starts with loading the data and then the *Validation* operator is executed to estimate the accuracy of the model. In the training part of the *Validation* the k-NN operator uses the prototypes obtained with the *LVQ2.1* algorithm. The obtained prototypes are then filtered by the *GE selection* algorithm because some of the determined codebooks may be useless. The *LVQ2.1* algorithm is initialized by the prototypes obtained from the *LVQ1* algorithm, which in turn is initialized by the prototypes from the *Class Iterator* where vectors from each class were clustered using the *VQ* algorithm starting at random. Always the number of prototypes is determined by the initialization operator. So if the *Random selection* is set to determine 10 prototypes and the problem is a three-class classification task, the resulting set of codebooks obtained with the LVQ algorithm will include 30 prototypes.

In our final example, we will compare three different LVQ algorithms and show how to speed up the execution process caching sub-results. In this example we use the *Weighted LVQ* algorithm [13], which assigns example weights during the training. The weights can be determined in different ways, but if we are interested in the model accuracy we can use the *ENN Weighting* operator, which assigns weights according to the mutual relations between classes. This operator generates a new *weights* attribute, which is used by the *Weighted LVQ* algorithm. The weighting scheme assigns weights according to the confidence of the *ENN* algorithm, which is also used for instance selection. An instance weight takes a value close to 1 when all neighbor instances are from the same class and very small values for instances which lie on the border. The weights should be then transformed with the *Weight Transformation* operator, for example, using the formula $exp(-pow(weight - 0.5, 2) * 0.5)$.

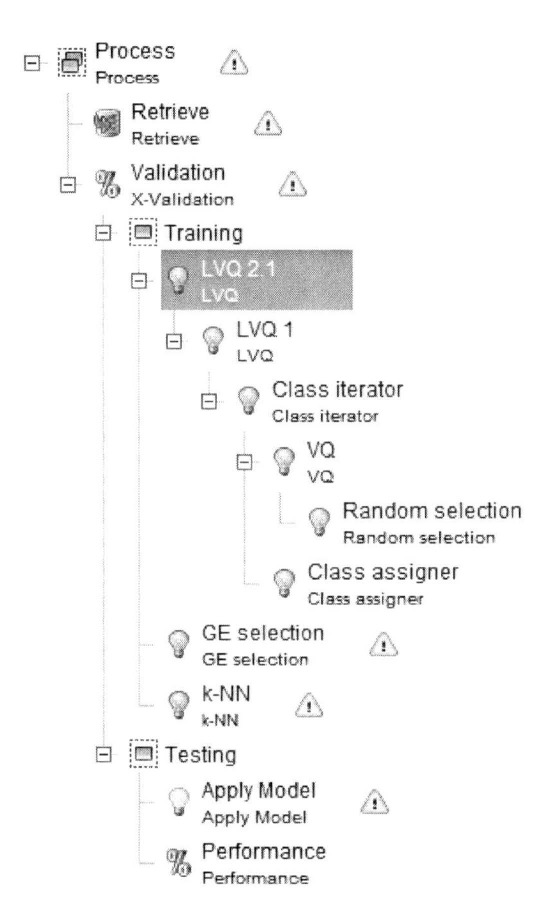

FIGURE 22.15: Process tree of prototype construction based on the LVQ algorithm.

We perform a comparison of the *WLVQ* algorithm, the *LVQ2.1*, and the *LVQ1* algorithm

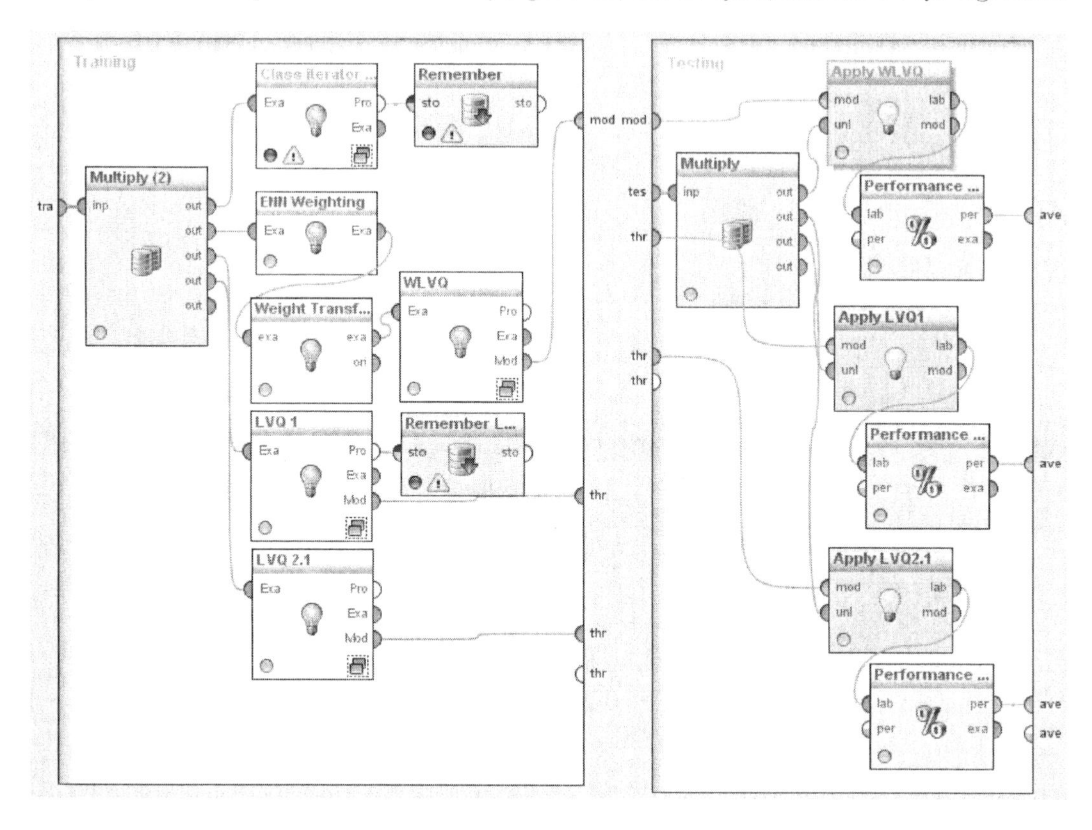

FIGURE 22.16: Accelerating the comparison of three LVQ algorithms using subresults caching.

for the SpamBase dataset using the process in Figure 22.16. In this example all the *LVQ* algorithms share a common part used to initialize *LVQ* codebooks. To accelerate the process, instead of recalculating it, we can store it in a cache using the already described *Remember* operator and retrieve it by the *Recall* operator. To analyze the process we focus on the training sub-process of the *Validation* step. First, the codebooks initialization is calculated using the *Class iterator* operator and the *VQ* clustering algorithm. The obtained results are cached using the *Remember* operator, subsequently. Then the trainings with the LVQ algorithms begin. The first one starts with determining weights by the *ENN Weighting*, then weights are transformed as discussed earlier, by the *Weight Transformation* operator. The training set prepared in this way is then used to train the *WLVQ* model. The initialization of the *WLVQ* algorithm is carried out with the *Recall* operator using the previously stored values. The final k-NN model is obtained directly from the *mod* LVQ output port. Identically, the next *LVQ1* model is trained and its k-NN model is delivered to the output of the subprocess. The obtained prototypes are also stored in the cache as in the case of initialization of the LVQ2.1 algorithm. Finally all the *k-NN* models are delivered to the *Testing* part of the process, where all of the algorithms are applied to the test data. An important element of the process is the execution order, which should follow the order described in this paragraph, otherwise it may happen that the *Recall* operator may try to retrieve data which is not yet available in the cache. To set the appropriate execution order, we should go to the menu *Process →Operator Execution Order → Order Execution* and then set right execution order with a mouse. The correct execution order is shown in Figure 22.16.

22.5 Mining Large Datasets

Today more and more data is logged into databases and mining such large datasets becomes really a challenge. Most of the prediction models have polynomial computational (quadratic or cubic) complexity, for example, similarity-based methods like kNN, SVM, etc. Therefore, a question arises how to deal with such large datasets and how to process them with current computers. One of the solutions is to re-distribute the mining process between many computing nodes like in Machout, a sub-project of Apache Hadoop. We can take advantage of it using the Radoop extension to RapidMiner. Another approach is to use updateable models, which update the model parameters according to the newly presented example, so they scale linearly with the size of the dataset. A disadvantage of the updated models is usually poor generalization abilities. Perhaps the simplest solution is resampling the data to a given number of samples, but then we may lose some important information. The last method is similar to re-sampling but the samples are not random; instead they are obtained with instance selection and construction methods. A state-of-the-art approach is based on initial clustering of the data, for example, using the *k-means* algorithm. In this approach, instead of the original data, the prediction model is trained using cluster centers. However, this is also not the best solution. Clustering does not use the label information, so examples which lie close to the decision border may be grouped into a single cluster even when they are from different classes. In addition, vectors which are far from the decision border are less important for the prediction algorithms, because they don't bring any useful information. These instances should be pruned, but clustering algorithms can't do that and consequently they usually produce many prototypes, which are useless from the prediction model perspective, reducing the information gain of the pruned datasets.

Instance selection algorithms are dedicated to prune useless instances using the label information, so the ISPR extension can be applied for mining big datasets. The idea of using instance selection methods to train prediction models on the pre-selected instances was already presented in [14, 15], but the aim was to improve the quality of the learning process. Now we can also use them as filters, similarly to the feature selection filters, and prune some of the useless data thus reducing the size of the data and filtering noise and outliers.

The only drawback of the instance selection methods is the computational complexity, which in most of the cases is quadratic, so the gain of using them is obtained afterwards by precisely optimizing the prediction model. Sample results are presented in Figure 22.19, where the dataset containing over 20,000 instances is used to show the dependencies between the computational complexity and the obtained accuracy.

In the process presented in Figure 22.18, first the dataset is loaded into the memory, then normalized to fit the range $[0, 1]$. Then the main loop *Loop Parameters* starts and sets the number of examples sampled from the dataset. Inside the operator (Figure 22.17(b)) first the data is sampled using the *Sample (stratified)* operator. The number of samples is set by the *Loop Parameters*. Next, the subset is re-built in memory using the *Materialize Data* operator. This operator creates a duplicate copy of the input example set. In RapidMiner the ExampleSet is a complex data structure, which can be treated as a view on some original so-called *Example Table*. If we sample or filter the examples or do any complex

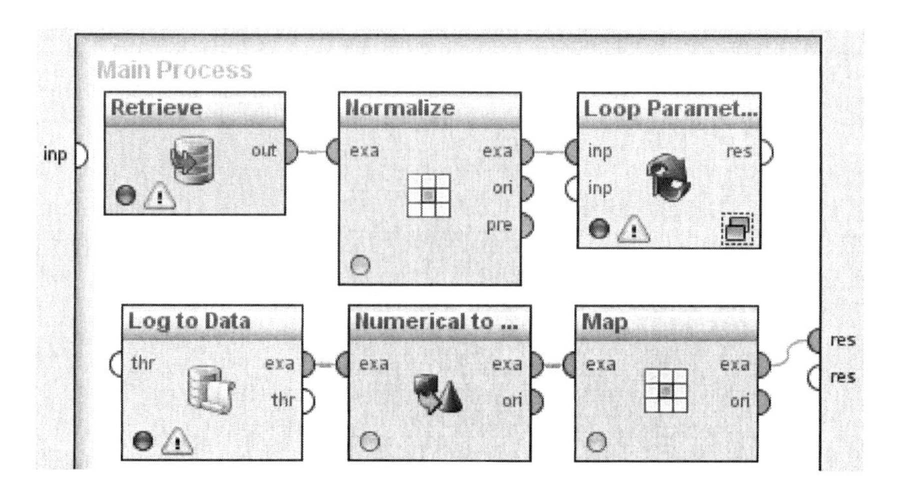

(a) Main process.

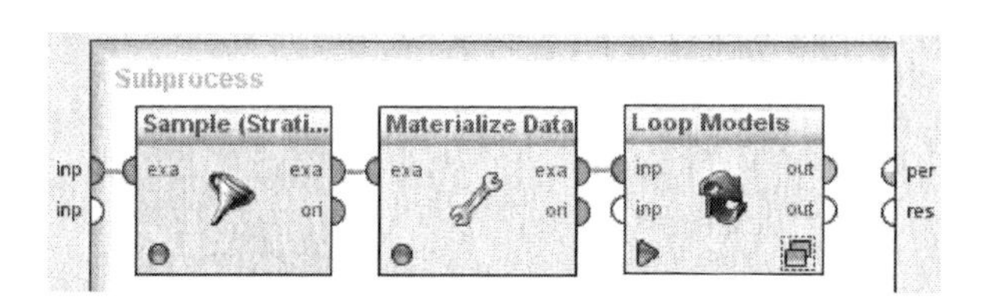

(b) *Loop Parameters* subprocess.

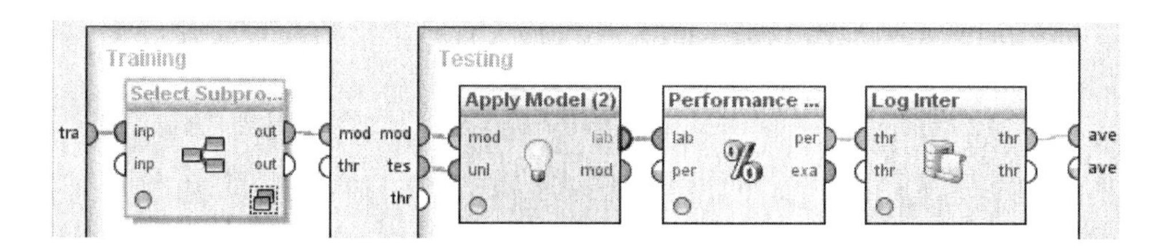

(c) *Validation* subprocess.

FIGURE 22.17: Evaluation process of the time complexity of the instance selection methods (1).

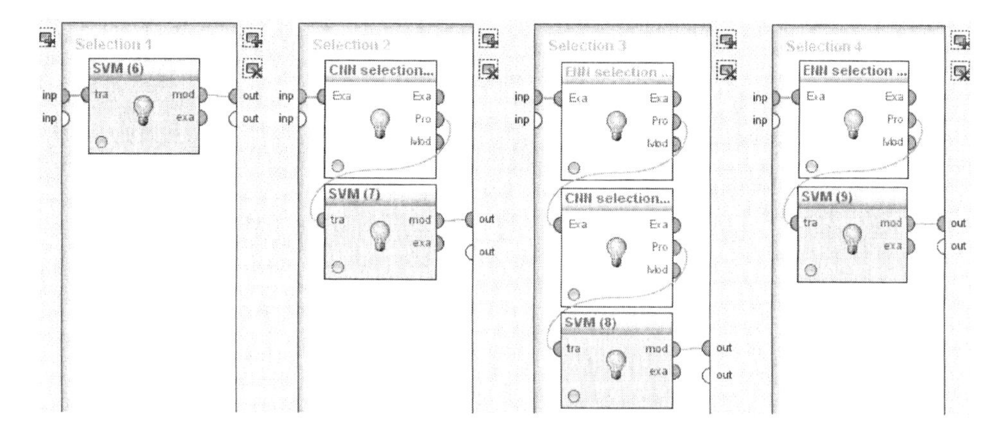

(a) *Select Sub-process* operators configuration.

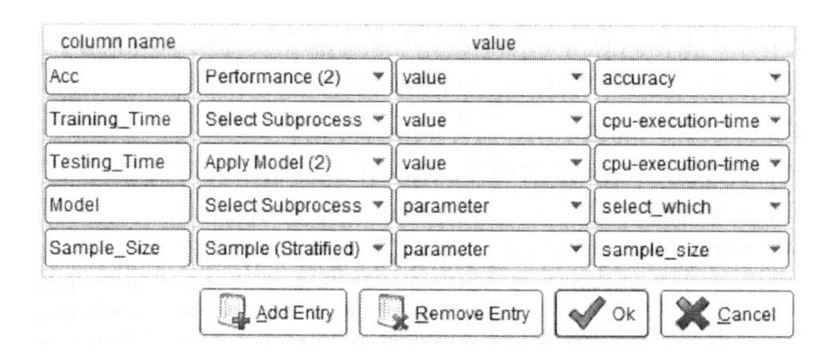

(b) *Log* operator configuration.

FIGURE 22.18: Evaluation process of the time complexity of the instance selection methods (2).

preprocessing steps, new views over views are created, increasing the overhead to access the data (but it pays off by reducing the memory usage, especially when dealing with large datasets). Then the final loop iterates over different combinations of models. The models are evaluated with the *X-Validation* operator. Inside the validation loop (Figure 22.17(c)) the *Select Sub-process* operator is used to switch between different combinations of models, which are shown in Figure 22.18(a). The evaluated algorithms include a pure SVM (in all cases, for simplicity, the LibSVM implementation was used with the default settings), CNN instance selection followed by the SVM, combination of ENN and CNN as a pruning step before applying an SVM model, and lastly, a pure ENN algorithm as a filter for the SVM model. In the testing sub-process of the validation the obtained model was applied and the performance was measured. The *Log* operator was used to measure the execution time of the whole training process, including both example filtering and training of the SVM model, and the execution time of the *Apply Model* operator. In addition, the size of the dataset and the identifier of selected processing scheme were logged. Finally, after all the calculations were finished in the *Main Process*, the logged values were converted into a dataset. For the details of the log configuration settings, see Figure 22.18(b). To analyze the results we can use the *Advanced Charts* to plot the relations. However, to create the visualization we need to convert the *Model* identifier, which is a number, into the nominal value using the *Numerical to Polynomial* operator and finally map the obtained values to appropriate names like SVM, CNN+SVM, ENN, CNN, SVM, etc. The last operation which remains is configuration of *Advanced Charts*. First we need to set the *Sample size* as *Domain dimension* and set the *Grouping* option to *Distinct Values*. Then we need to set *Model* attribute as *Color dimension*. It is also important to set the *Grouping* option in the *Color dimension* to *Distinct values*. Finally, we set the *Acc* attribute (or alternatively *Training_Time* or *Testing_Time* depending on what we want to show) as the *Empty axis*. We also need to set the *Aggregation* function to *Average*. Finally, to link the points with a line, we need to configure the *Format* of the *Series* setting the *Line style* to *Solid*.

From the results we can see that the pre-processing based on the *CNN* algorithm has the lowest computational complexity, but, unfortunately, also the lowest accuracy. From the perspective of the quality of the prediction model, the combination of *ENN+CNN* leads to the best accuracy but it also has the highest time complexity. The *ENN* pre-processing has similar complexity to the original *SVM*, but the average accuracy is a little higher than the one of the pure *SVM* model. In many applications, the prediction time is very important. The *CNN* and a combination of *ENN+CNN* algorithms decrease the time significantly. This is due to their high selection ratio, which results in a much lower number of support vectors needed to define the separating hyperplane.

A disadvantage of the instance selection methods is the computational complexity. This weakness is currently under development for a future release of the ISPR extension, but it can be also overcome by instance construction methods. As already discussed, the clustering algorithms are not a good solution for the classification tasks, but the *LVQ* algorithm is. It has a linear complexity as a function of the number of samples and attributes for a fixed number of codebooks and it takes advantage of the label information. The comparison of the pre-processing based on the *LVQ* and the *k-means* clustering is shown in Figure 22.20.

The obtained results presented in Figure 22.21 show linear time complexity. This was obtained for both the *LVQ* and the *k-means* algorithm, while the pure SVM scales quadratically. Similarly, the prediction time due to the fixed number of prototypes obtained after both prototype construction methods is linear or even constant. The plot showing the accuracy as a function of the size of the data visualizes the gain of using the LVQ instance construction over the clustering-based dataset size reduction.

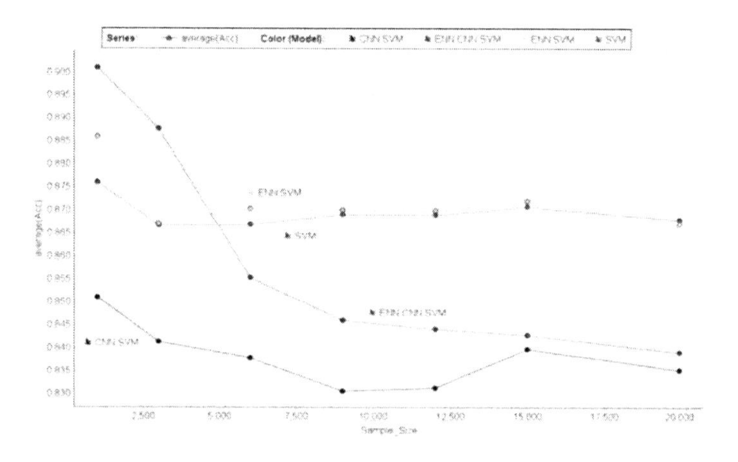

(a) Accuracy.

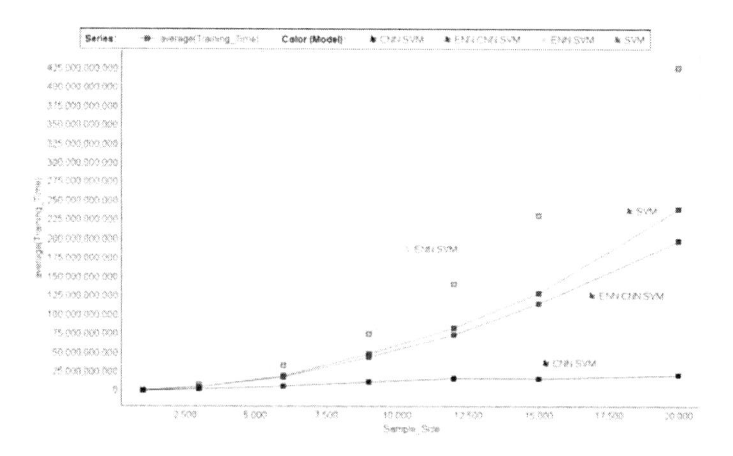

(b) Training Time.

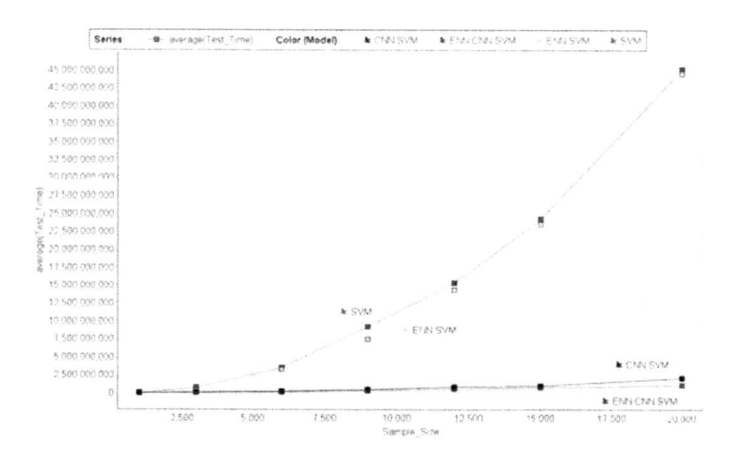

(c) Testing Time.

FIGURE 22.19: Accuracy, training time, and testing time in a function of the sample size using different instance selection methods followed by the SVM (no instance selection, CNN, ENN, and ENN+CNN).

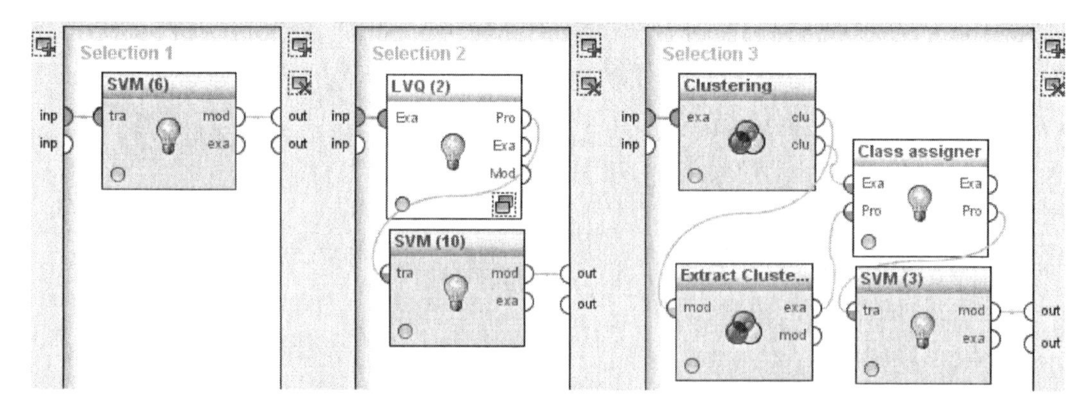

FIGURE 22.20: *Select Subprocess* operators configuration used to evaluate the time complexity of the instance construction methods.

22.6 Summary

In this chapter, we have shown usage examples of instance selection and construction methods, which are contained in the ISPR extension. The presented examples also show some advances in building RapidMiner processes, which can be used as a baseline to evaluate the quality of these and other prediction methods. We hope that this will be useful for users involved in distance or similarity-based learning methods but also for those who are interested in noise and outlier filtering. In addition, the set of available operators can be useful for mining large datasets for transforming of changing big data into small data. This makes mining large datasets with the state-of-the-art techniques feasible. Most of the available operators implement the algorithms, which were not designed to minimize the computational complexity, so the challenge is to create even better and more efficient instance filters. We plan to constantly develop the ISPR extension and release its future versions.

Bibliography

[1] A. Asuncion and D.J. Newman. UCI machine learning repository. http://www.ics.uci.edu/~mlearn/MLRepository.html, 2007.

[2] A. Kachel, W. Duch, M. Blachnik, and J. Biesiada. Infosel++: Information based feature selection C++ library. *Lecture Notes in Computer Science*, (in print), 2010.

[3] W. Duch and K. Grudziński. Prototype based rules - new way to understand the data. In *IEEE International Joint Conference on Neural Networks*, pages 1858–1863, Washington D.C, 2001. IEEE Press.

[4] L. Kuncheva and J.C. Bezdek. Nearest prototype classification: Clustering, genetic

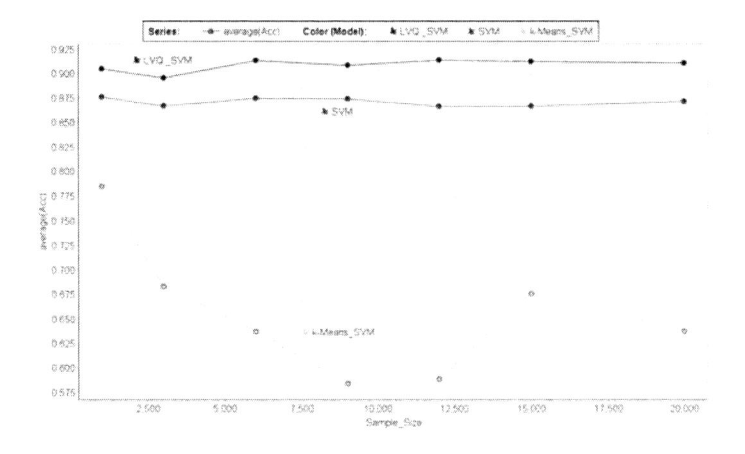

(a) Accuracy.

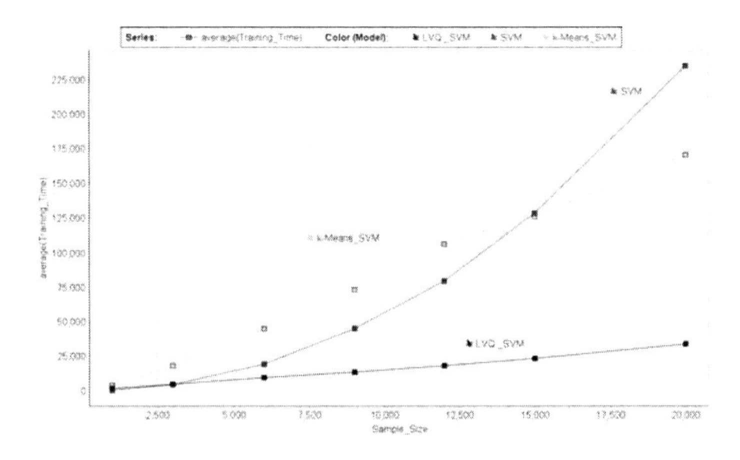

(b) Training Time.

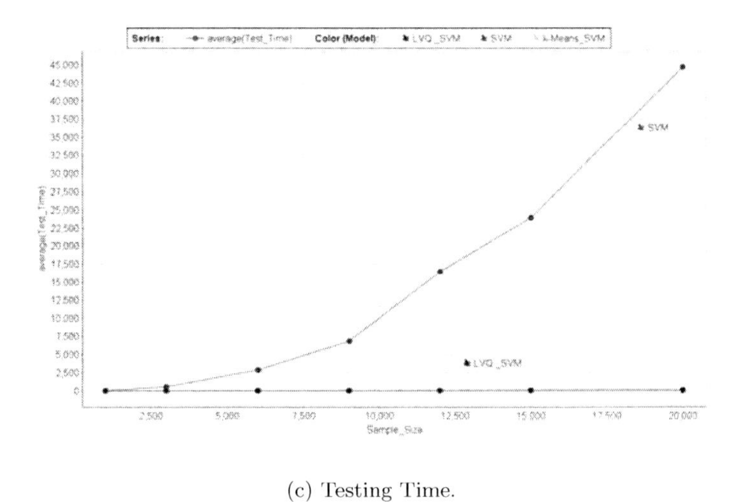

(c) Testing Time.

FIGURE 22.21: Accuracy, training time, and testing time in a function the sample size using SVM and two instance construction algorithms LVQ and k-means.

algorithms or random search? *IEEE Transactions on Systems, Man, and Cybernetics*, C28(1):160–164, 1998.

[5] D.B. Skalak. *Prototype selection for composite nearest neighbor classifiers*. PhD thesis, University of Massachusetts Amherst, 1997.

[6] P.E. Hart. The condensed nearest neighbor rule. *IEEE Transactions on Information Theory*, 114:515–516, 1968.

[7] D.L. Wilson. Asymptotic properties of nearest neighbor rules using edited data. *IEEE Trans. Systems, Man and Cybernetics*, 2:408–421, 1972.

[8] I. Tomek. An experiment with the edited nearest-neighbor rule. *IEEE Trans. on Systems, Man, and Cybernetics*, 6:448–452, 1976.

[9] J.S. Sánchez, F. Pla, and F.J. Ferri. Prototype selection for the nearest neighbour rule through proximity graphs. *Pattern Recognition Letters*, 18(6):507–513, 1997.

[10] D. Aha, D. Kibler, and M.K. Albert. Instance-based learning algorithms. *Machine Learning*, 6:37–66, 1991.

[11] R. Cameron-Jones. Instance selection by encoding length heuristic with random mutation hill climbing. *In Proc. of the Eighth Australian Joint Conference on Artificial Intelligence*, pages 99–106, 1995.

[12] L. Kuncheva and J.C. Bezdek. Presupervised and postsupervised prototype classifier design. *IEEE Transactions on Neural Networks*, 10(5):1142–1152, 1999.

[13] M. Blachnik and W. Duch. Lvq algorithm with instance weighting for generation of prototype-based rules. *Neural Networks*, 2011.

[14] N. Jankowski and M. Grochowski. Comparison of instance selection algorithms. i. algorithms survey. *Lecture Notes in Computer Science*, 3070:598–603, 2004.

[15] M. Grochowski and N. Jankowski. Comparison of instance selection algorithms. ii. results and comments. *LNCS*, 3070:580–585, 2004.

Chapter 23

Anomaly Detection

Markus Goldstein
German Research Center for Artificial Intelligence, Kaiserslautern, Germany

Acronyms

AD - Anomaly Detection

CBLOF - Cluster-based Local Outlier Factor

COF - Connectivity-based Outlier Factor

CSV - Comma-Separated Values

DLP - Data Leakage Prevention

IDS - Intrusion Detection System

INFLO - Influenced Outlierness

LDCOF - Local Density Cluster-based Outlier Factor

LOCI - Local Correlation Integral

LOF - Local Outlier Factor

LoOP - Local Outlier Probability

LRD - Local Reachability Density

NASA - National Aeronautics and Space Administration

NBA - National Basketball Association

NN - Nearest-Neighbor

SVM - Support Vector Machine

23.1 Introduction

Anomaly detection is the process of finding patterns in a given dataset which deviate from the characteristics of the majority. These outstanding patterns are also known as anomalies, outliers, intrusions, exceptions, misuses, or fraud. The name usually refers to a specific application domain, thus, we are using the generic term anomaly in the following. Anomaly detection can basically be classified as a sub-area of data mining and machine learning. However, the term anomaly detection is not well defined from a mathematical point of view, which makes it necessary to give a more detailed overview in the following section. Even if the reader is very familiar with classification and machine learning, in general, we recommend reading Section 23.2 completely since anomaly detection is fundamentally different and a well-working RapidMiner process needs in almost all cases a deeper understanding of the nature of the given anomaly detection problem.

The first attempts at anomaly detection go back to the 1970s, where researchers attempted to remove noise or incorrect measurements from their data in order to ensure that the data fits best to their proposed models. In this context, the Grubbs' outlier test [1] is most likely the best-known algorithm. It takes univariate (one-dimensional) input data assuming it is normally distributed and determines whether the minimum and maximum of the data is an outlier based on a significance level α. Once an outlier has been found, it is removed from the data and the procedure is repeated. It is obvious to see the shortcomings of this algorithm: it works only on univariate data and it fails if the data is not normally distributed. However, it illustrates the first attempts at removing anomalies, also known as *data cleansing*. In this case, one is interested in an anomaly-free dataset and not in the outliers itself. They are only flagged as outliers using a binary classification and then removed. RapidMiner contains operators for data cleansing on multivariate data for marking outliers, for example, the *Detect Outlier (Distances)* and the *Detect Outlier (Densities)* operators. These operators require as a parameter an estimation of outliers, either the concrete number or the expected proportion.

Exercise 1 - Data cleansing
As a first very simple example, we want to remove errors from the *iris* dataset, which contains four manual measurements of sepal and pedal width and length of the iris flower as well as the class label describing the species of the iris. It is already shipped with RapidMiner in the Samples/data repository. Use the *Detect Outlier (Distances)* operator with the number of neighbors set to 10 and detect 2 outliers. Use the *Scatter 3D Color* plotter to visualize the results so that the color represents the outlier. Note that the data contains four regular attributes, but you can only plot three dimensions at a time while the color represents the outlier prediction.

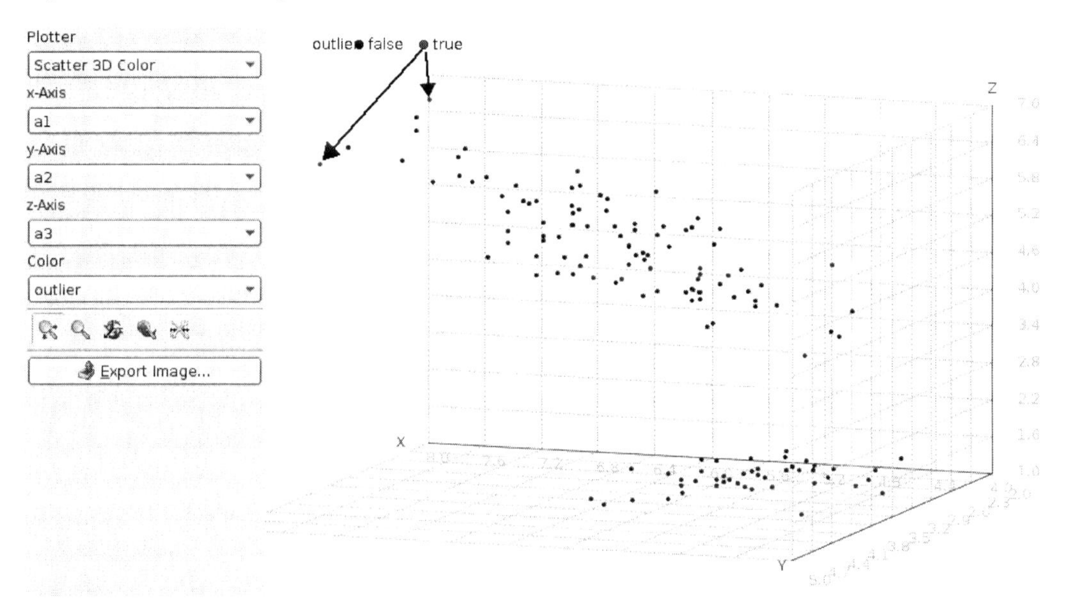

FIGURE 23.1: The result of data cleansing using the *Detect Outlier (Distances)* operator on the *iris* dataset. Two outliers at the top-left are identified based on their distance to their 10 nearest neighbors.

If we take a closer look at the Figure 23.1, which visualizes the results of the exercise above, a number of questions arise: why are these two points chosen as outliers? There is one point at the top right, which also seems to be an anomaly (at least within the visualization). Which is the more significant outlier of these two and how significant are both compared to a possible third anomaly?

To answer these questions, a more modern approach of anomaly detection has become increasingly more popular in the last number of years and has attracted a growing and active research community, which began in 2000, approximately. This was attributed to the fact that focus changed; in particular, people were not primarily interested in data cleansing any longer but in detecting anomalies instead. The rapid growth of databases and the availability of massive log data led to the desire to discover anomalous records, deviating from the norm. In addition, the output of such algorithms evolved. Instead of having a binary label indicating an anomaly, algorithms which assign *scores* to all instances indicating their extent of being an outlier, gained attention. Of course, this new view could also be used for data cleansing, for example, instances are ordered by their outlier score and a simple threshold then removes the top n anomalies.

Today, anomaly detection is applied in many application domains. In the area of network security, it is used in network intrusion detection systems [2]. Suspicious behavior of malicious software, unusual network activity, and break-in attempts should be detected without prior knowledge of the specific incident. In modern anti-virus applications, behavioral analysis also complements the traditional pattern matching detection engines [3]. In addition, this example describes the difference of traditional classification with the prior knowledge of the classes (malware patterns) and anomaly detection, dealing with the question if there is new previously unseen activity. In data leakage prevention (DLP) systems, researchers also try to incorporate anomaly detection algorithms for detecting abnormal data access patterns in order to avoid data to be stolen. A second core application area is misuse detection in transactional data. In this context, anomaly detection is used to detect fraudulent credit card transactions caused by stolen credit cards [4], fraud in Internet payments, or suspicious transactions in financial accounting data [5]. In the medical domain, anomaly detection is also used, for example, for detecting tumors in medical images or monitoring patient data (electrocardiogram) to get early warnings in case of life-threatening situations [6]. Furthermore, a variety of other specific applications exist such as anomaly detection in surveillance camera data [7], fault detection in complex systems, or detecting forgeries in the document forensics domain [8]. If anomaly detection should be applied on image data as shown in some of the above examples, representative features need to be extracted from that data first. This is a non-trivial task and, in some cases, for example document analysis, the pre-processing evolved to its own research domain.

23.2 Categorizing an Anomaly Detection Problem

As already mentioned, no concrete mathematical definition of anomaly detection exists and the term is used by a variety of systems and algorithms. This is the reason why some categorization attempts have been conducted in the past trying to clarify the differences. It is extremely important to understand the differences in order to choose the correct algorithm for a given problem and get an understanding of the required pre-processing of raw data to be analyzed. The following subsections are based on the definitions of Chandola et al. [9], the most likely most detailed survey paper in the anomaly detection area.

23.2.1 Type of Anomaly Detection Problem (Pre-processing)

For anomaly detection, the nature of the input data plays a very important role. In the following, we expect the data to be in a tabularized format, where an instance (*example*) has multiple attributes. The attributes can be either categorical or numerical. For almost all anomaly detection algorithms, distances between instances are computed to determine their anomaly score. Therefore, it is very important to be aware of the scales of the attributes. For categorical attributes, an appropriate distance measure needs to be selected, for example, the *MixedEuclideanDistance* measure type, which assumes a distance of one between non-identical categories and zero if the instances share the same categorical value. For numerical values, it is often a good idea to normalize the attributes in order to have a fair distance computation.

Let us examine an example. Suppose your dataset describes items you are selling in your company, consisting of only two attributes, (1) the price of a product and (2) the profit of that product in percent. Theoretically, the attribute in percent ranges between zero and one, whereas the absolute price might have a range of zero up to several thousand. If you directly apply anomaly detection on that dataset, the distances on the price attribute are much larger on average than the distances in the attribute measured in percent. This means that the anomaly detection results will be more influenced by the price instead of the percentage attribute. In this context, it is a good idea to normalize both attributes into the same range, for example, between zero and one using the *Normalize* operator using the *range transformation* method. Another possibility for pre-processing this example would be to remove the percentage feature by making it absolute. In RapidMiner, this can be done using the *Generate Attribute* operator, generating a new attribute being computed by multiplying both original attributes. Please note that this pre-processing will generate different anomaly scores.

This very simple example already demonstrates that pre-processing is extremely important in anomaly detection. In contrast to supervised machine learning, it has a larger influence on the results. In addition, the above example demonstrates that there is no "correct" way of pre-processing the data—it always depends on the anomalies you are searching for. While designing anomaly detection processes, it is also a good idea to simply try out different pre-processing methods.

Besides normalization, another pre-processing step which is very important for anomaly detection is generating *data views*. Data views are generated from the raw data and are different representations of the data using aggregation, filtering, or attribute removal. In contrast to most supervised machine learning, anomaly detection algorithms are very sensitive to attributes containing no information and they need to be removed. In the example above, the computation of the absolute profit value can also be seen as generating a data view. Data views are also used for removing the context or dependency between instances as explained in the following sections.

Point Anomalies

Point anomalies are the simplest type of anomalies. A single instance differs from the others according to their attribute values. If the dataset is visualized, these anomalies can be seen immediately, at least when not having more than three dimensions (attributes). In Figure 23.2, the points p_1 and p_2 are such anomalies, having a large distance to their neighbors. Nearly all anomaly detection algorithms detect this kind of anomaly, which is why they are sometimes called point anomaly detection algorithms. In the remainder of this chapter, all presented algorithms are of this type. The above example which demonstrated the sales prices of products and their profit is also an example of a point anomaly detection problem.

Contextual Anomalies

Contextual anomalies can only be identified with respect to a specific context in which they are appearing. Since this is best explained using an example, please have a look at Figure 23.3, which shows the average monthly temperatures in Germany over the years 2001–2010. If a point anomaly detection algorithm was directly applied on the univariate temperature values, only the extreme temperatures could be found, which might be the peak in July 2006 or the extreme low temperatures in December 2010. However, it is obvious to a person that the mild winter in December 2006/ January 2007 is an anomaly with respect to the *context time*. In order to detect such contextual anomalies with point anomaly detection algorithms, the context needs to be integrated when generating an appropriate data view. This must be done in a pre-processing step. For example, the month can be added as an additional attribute ranging from 1 to 12 to the data (since we assume the data should

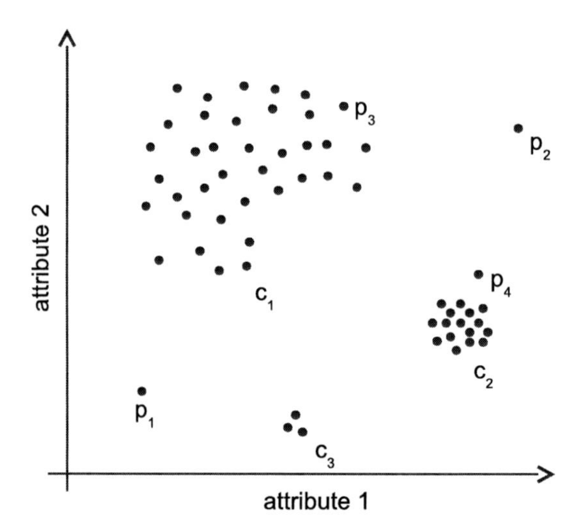

FIGURE 23.2: A point anomaly detection problem. p_1 and p_2 are identifiable as anomalies, p_4 also seems to be an anomaly with respect to its (local) neighborhood whereas p_3 is one of the normal examples.

have a yearly periodic behavior). Another possibility would be to use the difference to the monthly average temperature instead of using the absolute value for removing the context. Time information is not the only source for causing contextual anomalies, they may also occur within other contexts, for example, spatial (geographic) information, user IDs in log files, and many more.

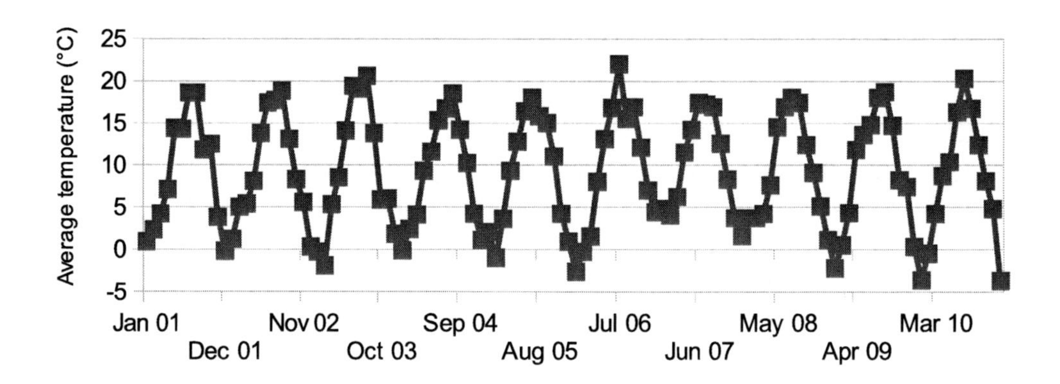

FIGURE 23.3: A contextual anomaly detection problem: The average monthly temperature in Germany from Jan 2001 to Dec 2010. The univariate temperature data is plotted here within its context time.

The generic way of integrating the context is to add appropriate attributes. In the case of having a continuous context, for example, time, it can be aggregated into meaningful bins, which are shown in the above example on monthly bins. For log file analysis, usually a per-user data view can be generated to integrate the context. For example, assume we are having a log file of a database server containing the information time-stamp, user-id, and the query. A possible way of transferring the problem into a point anomaly detection problem is counting how many accesses to the database each user performed in a particular hour. This means we are binning over time and user-id resulting in a two-dimensional dataset (hour, number of accesses), which we can analyze using a point anomaly detection algorithm. This procedure is illustrated in Figure 23.4, where the original dataset is used to generate a new data view, only containing the aggregated information in the orange bars (number of accesses per user per hour).

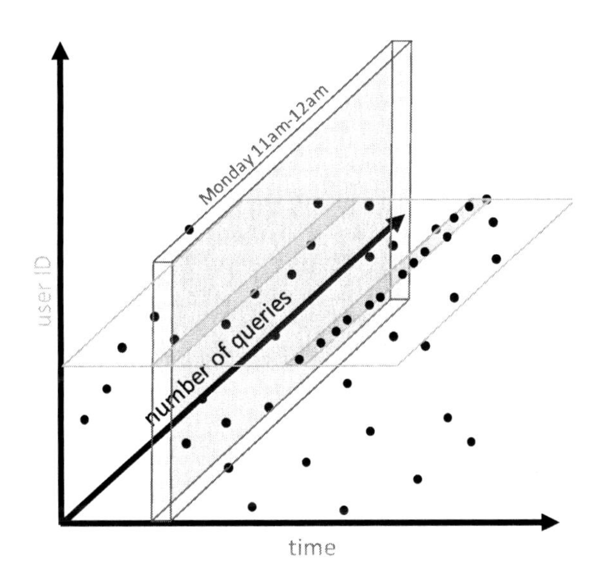

FIGURE 23.4: A contextual anomaly detection problem having two contexts: The time and the user ID. To remove the context, the number of database queries are aggregated over time and over each user. In this example, an hourly time bin per user contains the number of queries (dark orange plane).

For the special case of having univariate data within a time context, time series analysis can also be used for anomaly detection (cf. the "Series Extension" of RapidMiner). In this case, a forecast based on the previous values can be used to determine whether a value is within an expected range or not. Since this technique is a very special case requiring periodic behavior, training data only works with univariate data, which is not discussed in detail in this chapter.

Collective Anomalies

If a certain combination of instances defines an anomaly and not only a single instance, we have a collective anomaly detection problem. A good example for this type of anomalies is a network intrusion detection system (IDS). Imagine an IDS logs all system calls of a computer. If a hacker wants to break into a system, he usually uses an unpatched security vulnerability causing a buffer overflow and then injecting his own code. Thus, a buffer overflow logged event is very suspicious, but this might also occur due to poorly implemented software or other defects. So an observed buffer overflow is only supposed to be an anomaly within a series of other events indicating compromised systems, for example, if a new user account is created immediately afterwards. The sequence "http, buffer_overflow, newuser" could thus indicate a used web server vulnerability with an exploit being run creating a backdoor account for the hacker.

Of course, there is no generic way of handling such collective anomalies since the knowledge of what defines these anomalies needs to be modeled by hand, generating an appropriate data view from the input data in order to apply point anomaly detection algorithms. In the example above, it could make sense to generate a data view consisting of triples of the original sequence data.

23.2.2 Local versus Global Problems

Let us again have a closer look at Figure 23.2. We already identified p_1 and p_2 as anomalies. These kind of outliers, which are obviously far away from *all* other data points, are called *global outliers*. For example, the distance from p_1 to its next neighbor is significantly larger than the distance from p_3 to its nearest neighbor. Global anomaly detection algorithms are able to find such outliers. But what about p_4? Compared to cluster c_2, it appears to be an anomaly. On the other hand, if we compare the distance of p_4 to its next neighbor, it is not further away than the next neighbor for the normal instance p_3. Thus, p_4 is not a global anomaly but a *local outlier* with respect to its local neighborhood. These local anomalies can be found in datasets having different local densities (compare the density of clusters c_1 and c_2) or simply in datasets with multiple clusters "between" the clusters. Local outlier detection algorithms can find local and global outliers. One might conclude that it is always a good idea to use a local method, but this is not correct. It strongly depends on the anomaly detection problem you are trying to solve. In some cases you might not be interested in local outliers at all, and finding them will corrupt your results. A detailed investigation showed [10, 11] that results are best if you choose the correct algorithm (local algorithm for local problems) and that results are slightly worse if you apply a global algorithm on a local anomaly detection problem. In contrast, if you use a local anomaly detection algorithm on a global problem, it will fail.

In Figure 23.2, the cluster c_3 is also a phenomena worth pointing out. One might dispute whether this is an anomaly ("only three instances") or a normal cluster ("the instances are pretty close together"). These borderline clusters are called *microclusters* and it usually depends on the parameter setting of the used algorithms if these instances are considered to be anomalies or not.

23.2.3 Availability of Labels

Depending on the availability of labels, we can define three different anomaly detection modes.

Supervised Anomaly Detection
Supervised anomaly detection is a synonym for supervised learning. In this mode, a labeled training set is used to train a classifier, which is afterwards applied on a (labeled) test set. The only difference between supervised anomaly detection and supervised machine learning is the fact that the prior of the anomalous class is low (times less outliers compared with the normal instances). In general, any supervised learning algorithm, such as Support Vector Machines, Neural Networks, Decision Trees, or k-nearest-neighbors can be used in this mode. However, this scenario is extremely rare in practice since the anomalies are in most cases unknown in advance.

Semi-Supervised Anomaly Detection
Semi-supervised anomaly detection is also divided into a training and a test phase. In contrast to supervised anomaly detection, the training dataset does not contain any anomalies but only examples of the normal class. During the training phase, a model for the "normal" behavior is learned and afterwards evaluated using the test set, which then contains normal

records and anomalies. A showcase of this mode is the network intrusion detection use case. In this context, normal behavior of a system can be learned, but the attack patterns (the anomalies) are unknown in advance. During the operation of the IDS, unusual situations can then be reported as suspicious activity to an administrator.

One traditional idea in semi-supervised anomaly detection is that the normal data is clustered during the training phase and during the testing phase simply the distance to the cluster centroids is computed. Since there is no RapidMiner operator doing this, it is not straightforward to implement such a process, but we will have a look at this issue later. Furthermore, a one-class support vector machine can be used in this mode, already available in the *Support Vector Machine (LibSVM)* RapidMiner operator.

Unsupervised Aanomaly Detection
Unsupervised anomaly detection is the most difficult mode. In this case, no assumption about the data is made and no training of a model is performed. The data is analyzed based on statistical measures only, for example, the distances to the nearest neighbors, and an anomaly score rates the single instances. Of course, this is the most powerful and "magic" mode of anomaly detection, however, it comes at the price that a correct pre-processing and data view generation is the key of success. In many cases the term anomaly detection is implicitly used for the this unsupervised mode, which is also the focus of this chapter. RapidMiner does not ship with these kind of algorithms, but the *Anomaly Detection Extension*[1] offers a large variety. Please download and install this extension into the */lib/plugins/* directory of your RapidMiner installation since we are using it in the following examples.

Once again, it is very important to understand your anomaly detection problem. Many people are mixing up semi-supervised and unsupervised anomaly detection problems, even in scientific literature.

23.3 A Simple Artificial Unsupervised Anomaly Detection Example

A meaningful visualization of the unsupervised anomaly detection results makes it easier to understand the algorithms, their differences, and also the basic concepts, such as local and global anomalies. In general, real-world datasets have a high number of dimensions, which makes an easy visualization difficult or even impossible. For this reason, we start with a two-dimensional artificial dataset, which we generate by ourselves.

Exercise 2 - Data Generation
Generate a mixture of Gaussians using the *Generate Data* operator. As parameters, use Gaussian mixture clusters as a target function, and set the number of attributes to 2 (dimensions). The resulting data will have a label attribute naming the 4 clusters. Since this is our normal data, we set all labels to "normal" using the *Map* operator. Choose single attribute "label", use a regular expression, and remember to tick the include special attributes checkbox. In order to generate anomalies, we use the *Generate Data* operator again, but this time with the random classification as target function. Use the *Map* operator again to map the label values to "outlier". Finally, use the *Append* operator to merge the two data

[1] For source code, binary downloads and installation instructions, see http://madm.dfki.de/rapidminer/anomalydetection

sets and *Store* the result in your repository. Figure 23.8 shows a visualization using the "Scatter" plotter and the color corresponds to the label.

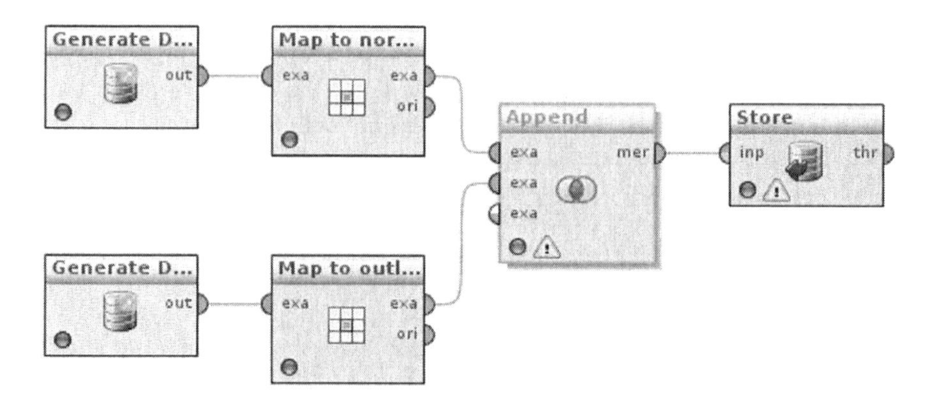

FIGURE 23.5: The process for generating an artificial 2D anomaly detection data set.

target function	gaussian mixture clusters
number examples	10000
number of attributes	2
attributes lower bound	-10.0
attributes upper bound	10.0
☑ use local random seed	
local random seed	3594
datamanagement	double_array

FIGURE 23.6: The parameter settings of the *Generate Data* operator. 10,000 examples are randomly generated as normal data using a mixture of Gaussians.

attribute filter type	single
attribute	label
☐ invert selection	
☑ include special attributes	
value mappings	Edit List (0)...
replace what	.*
replace by	normal
☑ consider regular expressions	
☐ add default mapping	

FIGURE 23.7: The parameter settings of the *Map* operator. A regular expression is used to map the cluster names in the label attribute to a single label entitled "normal".

Exercise 3 - Unsupervised Anomaly Detection

Firstly, we apply a global anomaly detection method on the generated dataset. After ensuring you have installed the *Anomaly Detection Extension*, as described in Section 23.2.3, the *k-NN Global Anomaly Score* operator is available. Now, retrieve the previously generated dataset from the repository, apply the operator, and visualize the results using a bubble plot as illustrated in Figure 23.9, where the bubble size corresponds to the outlier score.

In the resulting visualization (cf. Figure 23.9) of the example above, we can already observe that the local outliers, especially the two at the right side of the top-left cluster, do not get very high scores. You might want to repeat the experiment using a local detector, for example, the *Local Outlier Factor (LOF)*, which will be able to score these outliers higher. If using LOF, please make sure to use the *Local Outlier Factor (LOF)* from the extension, not the *Detect Outlier (LOF)* from RapidMiner, which computes something different.

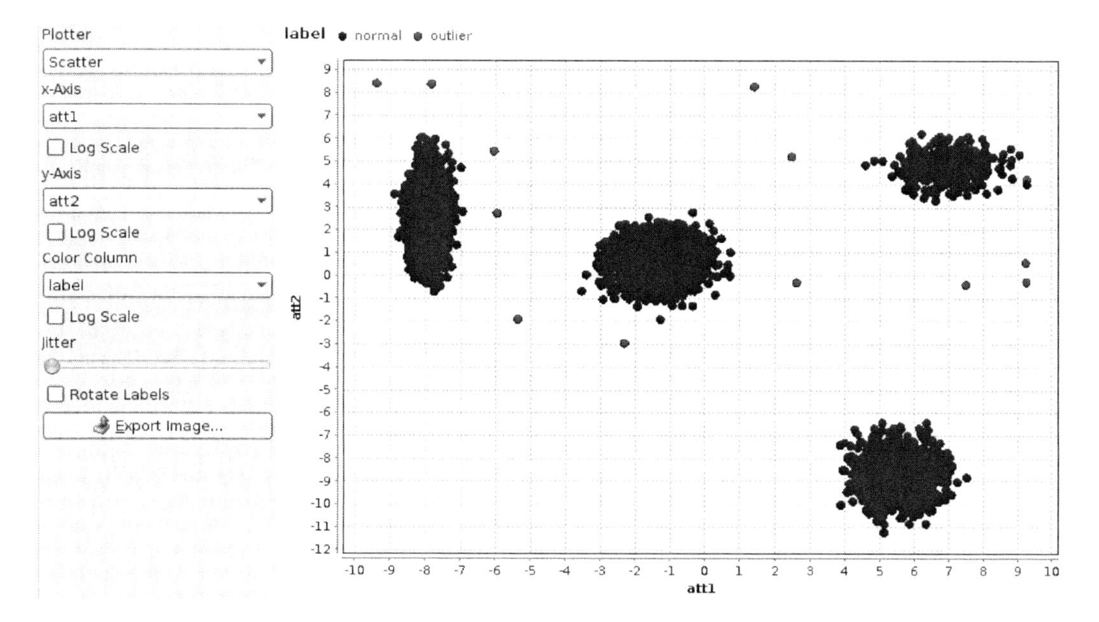

FIGURE 23.8 **(see color insert)**: The scatter plot of the generated artificial 2D dataset for a unsupervised anomaly detection process. Blue color indicates normal data instances, red color the sampled outliers.

23.4 Unsupervised Anomaly Detection Algorithms

In this section, we briefly address the available unsupervised anomaly detection algorithms of the RapidMiner extension. Since the *Local Outlier Factor (LOF)* is one of the most important algorithms, we examine it in detail. For other algorithms, their main ideas are illustrated and mathematical details are avoided. For interested readers, a link to the original publication is given.

23.4.1 k-NN Global Anomaly Score

The k-nearest-neighbor algorithm is a global anomaly detection algorithm. In each instance, the k nearest neighbors are located first. Then, the score can be computed in two ways: (1) the distance to the kth nearest-neighbor (only one) is used [12] or (2) the average to all k-nearest-neighbors is used as a score [13]. In practice, the second method performs generally better due to the fact that statistical fluctuations are averaged out. The RapidMiner operator can compute both, where the second method is the default and the first can be enabled by ticking the corresponding checkbox in the operator's parameter settings. Furthermore, it is also possible to choose the distance measure, whereas it is a good idea to use the *MixedEuclideanDistance* measure type. It can deal with categorical and numerical attributes and if no categorical attributes are present it is identical to the *EuclideanDistance*. The parameter for how many nearest neighbors are taken into account, k, also needs to be set. Unfortunately, we cannot optimize this parameter because we usually do not have a test set in unsupervised anomaly detection. In practice, a setting between 10 and 50 is a good choice, depending on the dataset. You might want to try out multiple settings and manually check the ordering of the detected anomalies for meaningfulness.

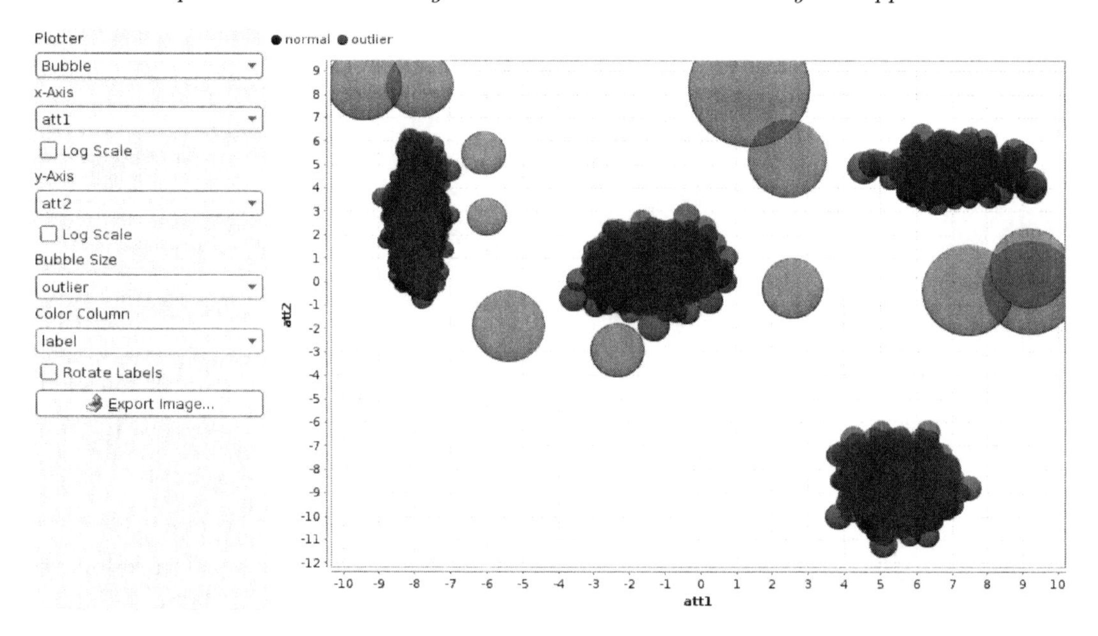

FIGURE 23.9 **(see color insert)**: The result of the global k-NN anomaly detection.

23.4.2 Local Outlier Factor (LOF)

The LOF algorithm [14] was the first algorithm proposed for dealing with local outliers. We explain it in more detail in the following example since it is also the basis for most of the following nearest-neighbor-based algorithms, which are variants of LOF. Basically, the algorithm has three main steps:

1. Find the k-nearest-neighbors for each instance.
2. For each instance, estimate the local density using the k neighbors.
3. Compute the ratio of the density of an instance compared to the density of its neighbors.

Algorithmically speaking, the following values are computed for each instance p:

reach_dist(p,o) The reachability distance is the maximum of the Euclidean distance between p and o and the k-distance of o. In most cases, the maximum is the Euclidean distance and the k-distance is only used to avoid extreme small distances in highly dense clusters.

Local Reachability Density (LRD) The LRD is computed for all data points p based on the set of k neighbors $N_{min}(p)$, using

$$LRD_{min}(p) = 1/\left(\frac{\sum_{o \in N_{min}(p)} reach\_dist_{min}(p,o)}{|N_{min}(p)|}\right). \quad (23.1)$$

In other words, the LRD is (the inverse of) the average reach_dist of its neighbors. The LRD is illustrated as a circle in Figure 23.10.

Local Outlier Factor (LOF) Finally, the LOF is the ratio between the average LRDs of the neighborhood to the LRD of p itself:

$$LOF_{min}(p) = \frac{\sum_{o \in N_{min}(p)} \frac{LRD_{min}(o)}{LRD_{min}(p)}}{|N_{min}(p)|}. \tag{23.2}$$

Computing the ratio of the LRDs results in a nice characteristic of LOF: Normal instances get LOF scores close to one (their LRD is the same as the one of their neighbors), whereas anomalies get higher scores (their LRD is larger). A rule of thumb says that LOF values larger than 1.2 (or 2.0) may indicate anomalies, depending on the dataset. This easy interpretation of the results is most likely also responsible for the success of LOF in practical applications.

For illustrating the principle, a zoomed view of the cluster c_2 is outlined in Figure 23.10. In this example, the $k = 3$ nearest neighbors for the instance p_4 are used and the circles around the instances represent the estimated LRDs. It can be seen that the LRD of p_4 is larger compared to the average LRDs of its neighbors (dashed lines), resulting in a higher LOF score.

In contrast to k-NN, the operator for computing LOF accepts a minimum and a maximum value for k. If both are set, multiple LOF scores using different values for k are computed and the maximum is used for the final score, which was proposed in the original publication. If you want to use a fixed value for k, simply set the minimum and maximum to the same number.

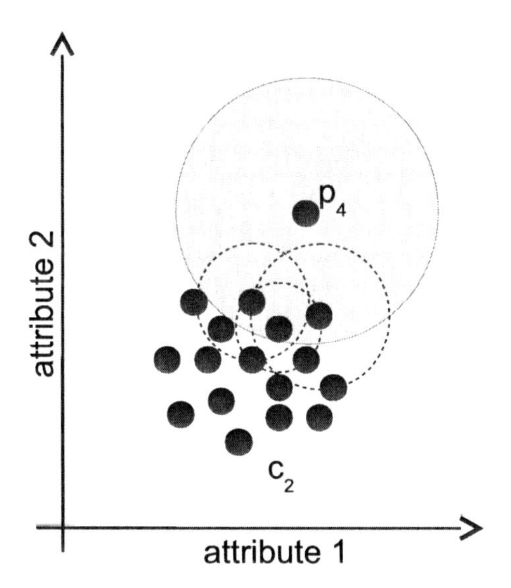

FIGURE 23.10: A zoomed view of the c_2 cluster of Figure 23.2 illustrating the local outlier p_4 detected by LOF. The circles represent the LRDs for $k = 3$. The LOF score is the ratio of the light grey circle to the average of the dashed circles.

23.4.3 Connectivity-Based Outlier Factor (COF)

COF [15] is very similar to LOF, however, the difference is that instead of the LRD, a minimum spanning tree-like approach is used for the density estimation (average chaining distance). This results in the fact that COF focuses more on "connected" structures in the data instead of assuming normal data to be spherical. Figure 23.11 and 23.12 show the difference of LOF and COF on an example having line-shaped data with two outliers.

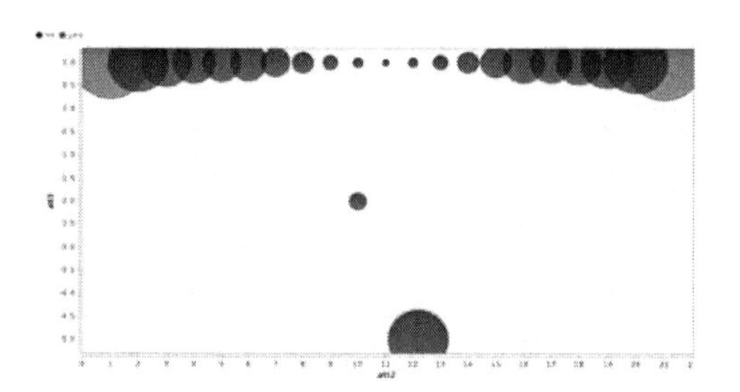

FIGURE 23.11 (**see color insert**): Results of a straight line dataset with two outliers using LOF. The top two outlier scores are marked with red color.

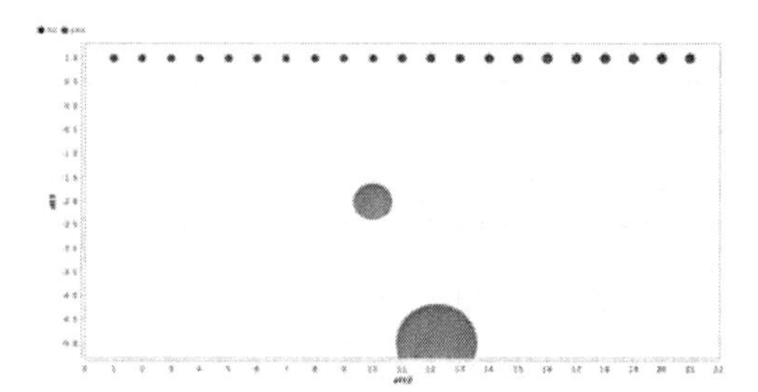

FIGURE 23.12 (**see color insert**): Results of a straight line dataset with two outliers using COF. The top two outlier scores are marked with red color.

23.4.4 Influenced Outlierness (INFLO)

The INFLO [16] algorithm is another extension of LOF. In contrast to LOF, it not only takes the k-neighborhood into account, but also the reverse-k-neighborhood. Theoretically, this performs better if clusters of varying densities are very close to each other. The interested reader is referred to the publication for more details. In practice, the results of INFLO are not as good as the results of LOF and COF [10, 11], leading to the hypothesis that the assumption of the algorithm is often not fulfilled on real-world datasets.

23.4.5 Local Outlier Probability (LoOP)

Inspired by the success of LOF, the authors developed LoOP [17], a probability-based outlier detection algorithm. In theory, the idea of having probabilities as an output comes with a lot of advantages, especially for interpreting the results and for further processing. Unfortunately, probabilities always need to refer to a specific reference point, which is not available in practice. Thus, the algorithm is, in fact, a scoring algorithm in the range of [0,1]. Nevertheless, LoOP performs average in the area and if a probability-like output is needed, it might be the algorithm of choice. LoOP estimates the local density by computing the deviation of a normal distribution of the local neighborhood. Then, similar to LOF, ratios are computed, normalized, and mapped to a [0,1] range.

In contrast to other algorithms, an additional parameter needs to be set in the RapidMiner operator. A normalization factor λ (default: 3.0) corresponds to the confidence level of the used normal distribution ("three sigma rule": $\lambda = 3 \Rightarrow\approx 0.3\%$ outlier in the data).

23.4.6 Local Correlation Integral (LOCI) and aLOCI

The LOCI algorithm [18] avoids the critical parameter k being specified by the user. The idea is that all possible k values are tested for each single data instance and the maximum score is taken. In contrast to LOF, the density is estimated by comparing two different sized neighborhoods with each other. In theory, this algorithm should work on local and global anomaly detection problems since a small value of k represents a local approach and a large value of k a global approach. Experiments demonstrated that it is a good choice to use LOCI if it is absolutely unclear how to set k and it is unknown if the problem is local or global. However, if k is optimized, in respect of a usually unavailable labeled test set, other algorithms perform better. Unfortunately, this comes at the price that the run-time complexity is $O(n^3)$ (standard NN algorithms: $O(n^2)$) and the space requirement also increases dramatically to $O(n^2)$ (standard NN algorithms: $O(nk)$). This leads to a very limited usage of the algorithm, which can only be applied on very small datasets (at most 3,000 examples). The authors tried to avoid this shortcoming by developing an approximate version of the algorithm, called aLOCI [18], which is also available as a RapidMiner operator. However, the performance on simple datasets is already very poor, which you might want to try on our artificial dataset from Exercise 2.

The two parameters α and n_{min} are responsible for the size of the neighborhoods used for computation and need not to be changed, in general. However, the interested reader is referred to the original publication. The parameter settings of the aLOCI algorithm are very complex and most settings refer to the used search tree. We cannot go into details here, but the output of the algorithm is very sensitive to the settings and general advice on a good parameter set cannot be given.

23.4.7 Cluster-Based Local Outlier Factor (CBLOF)

Until now, only nearest-neighbor methods have been discussed. The anomaly detection extension also contains clustering-based algorithms which work fundamentally different. The basic idea is that the data is clustered first and then an anomaly score is computed based on the cluster model and the data. This means that for building a RapidMiner process, a clustering operator such as k-means, X-means, or k-medoids needs to be applied first on the data. Subsequently, both the ClusterModel and the ExampleSet are passed to the CBLOF operator.

CBLOF [19] uses the cluster model and separates clusters into small and large clusters with a heuristic using the parameters α and β. α should be set to the amount of expected normal instances, whereas β can be left to its default—it does not have any influence on most cases. After the division into small and large clusters, the distance of each instance to the nearest largest cluster center is used as an anomaly score. In the original publication, the size of the cluster is also used as a weighting factor, but it has been shown that this should be ignored [10]. Therefore, we always recommend to untick the checkbox "use cluster size as weighting factor". Figure 23.13 shows an example process and the parameter settings of the CBLOF operator. The process consists of the *Retrieve* operator, which retrieves the data generated in Exercise 2 from a repository, the *X-Means* clustering algorithm, and the *Cluster-Based Local Outlier Factor (CBLOF)* operator.

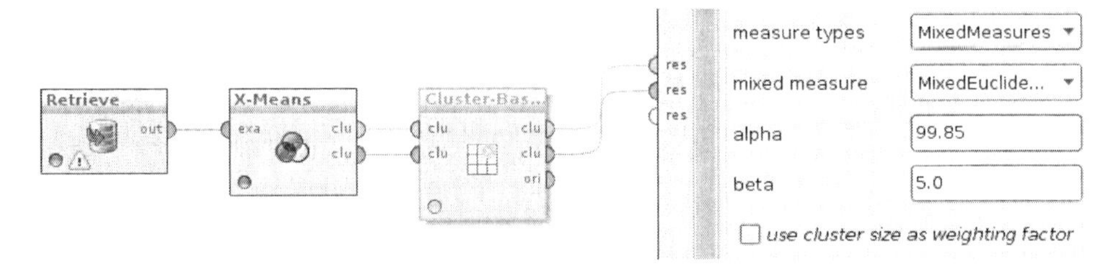

FIGURE 23.13: A sample process for applying a cluster-based anomaly detection algorithm on the data. On the right side, the parameter settings for CBLOF on the artificial dataset are shown.

23.4.8 Local Density Cluster-Based Outlier Factor (LDCOF)

LDCOF [11] is an advancement of CBLOF. It also uses the separation into small and large clusters but then not only the distance of an instance to the nearest large cluster center is used, but the distance as a ratio to the average of all instances associated with that cluster. This has two main advantages: (1) local densities of the clusters are taken into the anomaly score and (2) the outlier score is (as LOF scores) easily interpretable: a value of 1 refers to a normal instance and outliers have much larger values. In addition, only one parameter γ is used to determine the separation into small and large clusters. It represents a minimum boundary in percent for the average number of elements in a cluster. In other words, if $\gamma = 0.05$, a cluster is considered to be a small cluster if it contains less then 5% of instances compared to the average of all clusters.

Exercise 4 - LDCOF on Artificial Data
Use the data generated in Exercise 2 and apply the *X-Means* clustering algorithm using its default parameters. X-means is a clustering algorithm, which determines the number of clusters automatically from the data. You can verify its result since we know that our artificial data has four clusters. Then, apply LDCOF, also with the default parameter settings. For a representative visualization, we want to color all instances describing a binary decision (outlier or normal) based on a threshold. Use the *Generate Attribute* operator to generate a new attribute "outlier_binary" being set to "yes" if the outlier score exceeds 4.0 and to "no" otherwise (expression: if(outlier>4,"yes","no")).

The result is shown in Figure 23.14. It can be seen that one anomaly close to the top left cluster was not found. This is due to the fact that the distance to the cluster center is used as an anomaly score regardless of the shape of the cluster. This is also the reason why instances of elongated clusters get higher scores at the far-away ends.

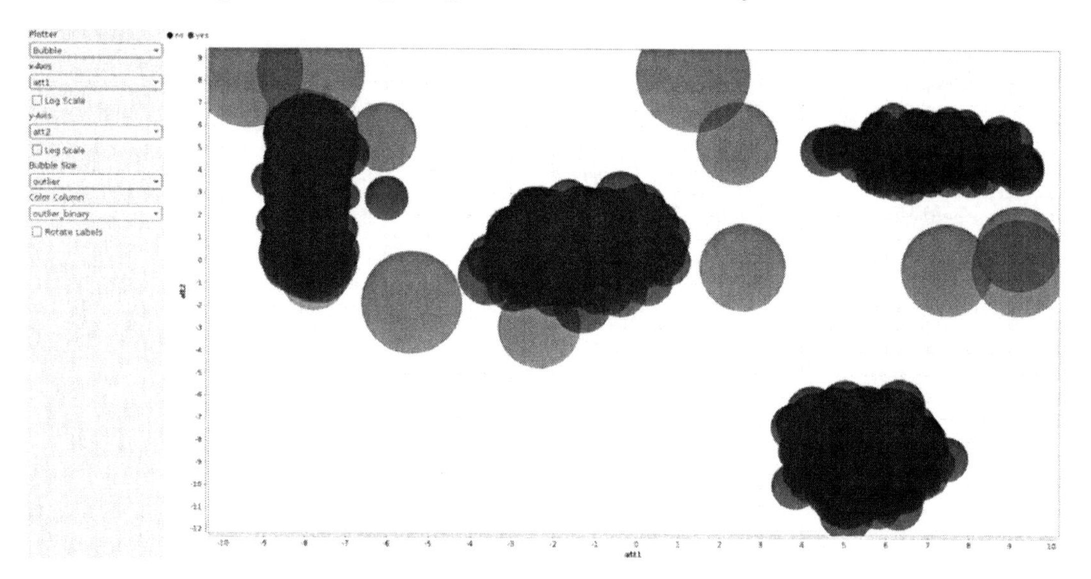

FIGURE 23.14 **(see color insert)**: The result of the LDCOF anomaly detection algorithm using X-means clustering. The bubble size indicates the outlier score and the color indicates the binary outlier decision.

In general, clustering-based unsupervised anomaly detection algorithms came with the advantage that they are faster than the nearest-neighbor based methods. Typically, a nearest-neighbor search has a complexity of $O(n^2)$, whereas clustering approaches, for example k-means, only have a complexity of $O(n \cdot log(n))$ making them much more appropriate for large datasets. However, clustering-based approaches seem to be not as accurate on (at least some) real-world datasets. Summarizing the experience on an experimental study, here are some advice you should consider when constructing your anomaly detection RapidMiner processes:

- If possible, with respect to computation time,use a nearest-neighbor-based method for better performance,
- As a clustering algorithm, X-means is a good choice on low-dimensional data (rule of thumb: $2^{dimensions} << examples$).
- For high-dimensional data, use k-means for nominal data and k-medoids for categorical/mixed datasets.
- When dealing with higher dimensions, it is usually a good idea to overestimate the number of clusters. Since data is usually not Gaussian anymore, this helps to reduce false positives. Do not worry about too small clusters since they are ignored by the CBLOF/LDCOF anyway.
- Be aware of the fact that clustering is not deterministic (at least when using the fast EM-based algorithms as we do) leading to different results/ outliers on multiple executions of the process.

23.5 An Advanced Unsupervised Anomaly Detection Example

Our artificial example explains well the differences of the algorithms, but now it is time to apply unsupervised anomaly detection on a real-world data set, including an appropriate pre-processing.

As a real-world dataset we use the freely available NBA regular-season basketball player statistics[2]. This dataset contains statistics of NBA players from 1946 to 2009 (version 2.1). The data has a lot of attributes, including the player name, the season, the team, the number of played games, and their statistics such as minutes played, points scored, blocks, and many more. In total, the dataset has 23 attributes and 21,961 examples. The dataset does not contain any label information.

Exercise 5 - Anomaly Detection on the NBA Dataset
Our task is now to find outstanding players using unsupervised anomaly detection without any prior knowledge of the data. First, we need to identify contextual attributes, then we need to generate an appropriate data view, normalize the data, and then we can apply an anomaly detection algorithm. All details are described in the following example.

Firstly, we need to import the CSV data using the *Read CSV* operator and its *Import Configuration Wizard.* Make sure to use the comma "," as the column separator and skip comments beginning with "#". In step 3 of the wizard it is wise to associate the annotation "Name" to the first data row so that we have meaningful names of our attributes. During step 4 of the wizard it is not required to define any role now since we do that later anyway.

The data has basically two contexts, the time information "year" and the player's name.

[2]Available at http://www.databasebasketball.com/stats_download.htm.

The data is already a statistic aggregation of many games into these two contextual bins, which is perfect for us, because it makes sense to find outstanding players in a particular season. To this end, we can now remove the "ilkid" and "leag" columns from the data using the *Select Attributes* operator because this is unnecessary information for us. Hint: use a subset as an attribute filter, select both attributes, and tick the "invert selection" box.

Next, we need to identify the attributes which define our ID. Since the role ID can only be used once in an example set, we need to generate this attribute by combining the player's first and last name, the season, and the team into one new attribute. For this task, we use the *Generate Attributes* operator. We set the name of the new column to "id" and use the expression "concat(lastname,firstname,"_",team,"_",str(year))", which generates a new string with useful identifier information, for example, "BarrJohn_ST1_1946". Afterwards, we use the *Set Role* operator in order to set the role of this attribute to "id". This ensures that this information is only used for identification and is ignored as an attribute by the anomaly detection algorithm. Of course, the old attributes lastname, firstname, and year need to be removed now using the *Select Attributes* operator again.

A quick look into the *meta data view* of our dataset reveals that we have missing values in our data, which anomaly detection algorithms cannot deal with (at least not the ones available in the RapidMiner extension). We use the *Filter Examples* operator with the condition class "missing_attributes" and invert the filter so that all examples with missing values are removed. Missing values are, in fact, a problem in anomaly detection and dealing with them is not simple. When removing them, you might also remove some (interesting) anomalies, but on the other hand, this might be better than replacing missing values (e.g., with the average) and thus generating incorrect anomalies by incident. Figure 23.15 illustrates the process to this point.

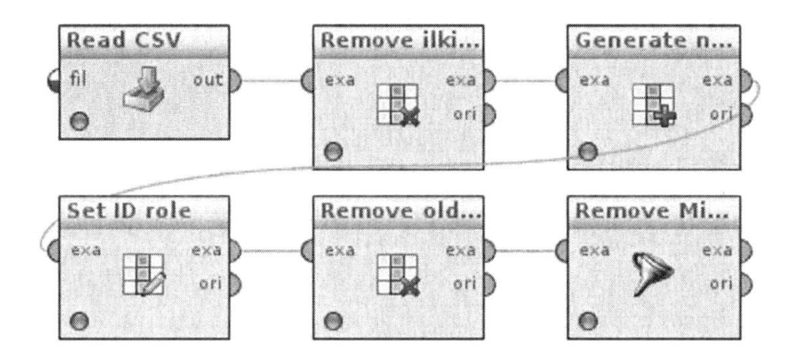

FIGURE 23.15: The pre-processing of the NBA data: Removing attributes, generating an appropriate ID, and removing missing values.

Now we need to select examples with the attributes that are semantically important for defining a good NBA player. Therefore, we select the attributes id, games played (gp), points scored (pts), rebounds (reb), and assists (asts). Other values are omitted due to the fact that they are either redundant, for example, "points per game" or not available for the complete dataset, for example, "blocks" due to not gathering these statistics in the past. For more information about the detailed statistical attributes, refer to the website databasebasketball.com.[3] Of course, this is done using a subset attribute filter type of the *Select Attribute* operator. If you have a look at the resulting data, you will notice that rebounds are always set to "0" before 1950. This fact is not stated in the database

[3]See http://www.databasebasketball.com/about/aboutstats.htm

description, but we are not concerned about it at this point. Since there are multiple zero entries in the dataset, an anomaly detection algorithm will notice that as a normal behavior.

Before continuing with normalization, please duplicate the current dataset using the *Multiply* operator; we need the original example set later on.

All absolute performance measurements of a player, such as the number of scored points, depend, of course, on the number of played games. Since we want to find outstanding players based on their actual effort, we divide the measurements points, rebounds, and assists by the number of played games. This can be achieved by using the *Generate Attributes* operator and the corresponding expression, i.e., generating the attribute avg_pts calculating "pts/gp". Using three entries in the list of the operator, you can generate three new attributes avg_pts, avg_reb, and avg_asts at once. Do not forget to remove the "old" attributes again from the dataset, when completed.

Finally, we need to normalize all attributes to the range of [0,1] so that the anomaly detection algorithm treats all distances equally (cf. Section 23.2.1). To do so, use the *Normalize* operator, select all attributes, and apply a "range transformation".

Now we are finished with all the pre-processing and we can finally apply the anomaly detection algorithm. Most likely the given problem is a global problem, therefore, we try the *k-NN Global Anomaly Score* for now with a parameter setting of $k = 50$. After the anomaly detection has been applied, a new "outlier" column is added to the data. If we now examine the results, the normalized attributes are not easily readable to the human eye. For this reason, we combine the outlier output with the previously multiplied original dataset again. To do this, we use the *Join* operator, which maps the two example sets back together based on their ID. Afterwards, we can remove the avg_ columns, which we are not interested in anymore. This procedure is illustrated in Figure 23.16.

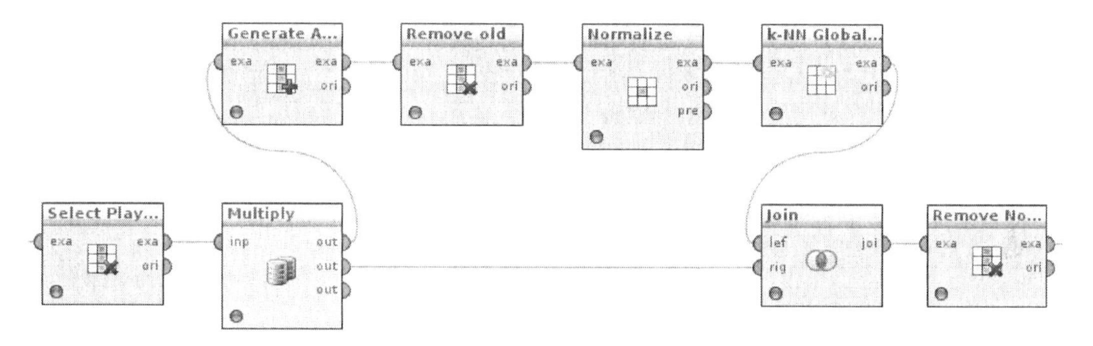

FIGURE 23.16: Processing the NBA data: Selecting important attributes, keeping a copy of the (not yet normalized data), normalizing the data and generating a data view, applying k-NN anomaly detection, and finally merging the outlier score back to the original data.

In the RapidMiner results perspective we can now order the data based on the outlier score and get the top outliers as shown in Table 23.1. Obviously, Wilt Chamberlain seems to be an outstanding player, because he shows up 9 times among the top-10 outliers. Investigation demonstrated that he was, until now, the best offensive player in the NBA. Furthermore, the anomaly with the highest score refers to his 1961 season, which is still considered today to be the most legendary season in NBA history. In addition, Elgin Baylor was rated as one of the top offensive players in the NBA. Figure 23.17 visualizes three of the four dimensions such that the color indicates the anomaly score. It can be seen that the high scores are associated with outstanding examples with respect to all dimensions.

Please feel free to try out other anomaly detection algorithms by yourself. Also local anomaly detection methods reveal new insights. If you apply LOF, for example, you will

TABLE 23.1: Top-10 anomalies found in the NBA dataset using the k-NN global outlier detection algorithm.

Outlier Score	Player	Team	Season	Games Played	Points Scored	Rebounds	Assists
0.5387	Chamberlain, Wilt	PH1	1961	80	4029	2052	192
0.4991	Chamberlain, Wilt	SFW	1964	38	1480	893	117
0.4296	Chamberlain, Wilt	SFW	1962	80	3586	1946	275
0.4282	Chamberlain, Wilt	PH1	1960	79	3033	2149	148
0.4282	Chamberlain, Wilt	PHI	1964	35	1054	780	133
0.4094	Chamberlain, Wilt	PH1	1959	72	2707	1941	168
0.3903	Chamberlain, Wilt	PHI	1967	82	1992	1952	702
0.3900	Chamberlain, Wilt	LAL	1969	12	328	221	49
0.3801	Baylor, Elgin	LAL	1961	48	1836	892	222
0.3578	Chamberlain, Wilt	PHI	1966	81	1956	1957	630

find, together with Wilt Chamberlain, George Lee in the top-5 outliers. He only played one game in the season of 1966 but scored 12 points. However, this is not much, but usually all players playing only one game scored less times (in most cases 0 points). This is a typical local outlier because it is not outstanding with respect to all attributes but only due to its direct local neighborhood consisting of players playing only one season in their lifetime.

23.6 Semi-supervised Anomaly Detection

Even if the focus of this chapter is on unsupervised anomaly detection, many applications require semi-supervised algorithms. Unfortunately, the anomaly detection extension does not (yet) offer such algorithms. However, the *LibSVM* operator can be used as a one-class classifier for semi-supervised anomaly detection.

As a real-world dataset, we choose the Shuttle dataset from the UCI machine learning repository.[4] This data is also freely available and its purpose is to find suspicious states during a NASA shuttle mission. However, the dataset is originally used as a classification dataset, so that we need to modify it before it can be used for anomaly detection. In the training set, we choose "1" as the normal class and in the testing dataset, the (rare) classes 2, 3, 6, and 7 are used as anomalies. This leads to a training dataset with a size of 34,108 instances, which we sample down to a size of 5000. The test set contains 11,536 instances (sampled down to 2558) with 58 being labeled as outliers (2,3%). The reduction of the size is only done to reduce the computation time, but if you want you can also use the complete dataset for the experiments.

23.6.1 Using a One-Class Support Vector Machine (SVM)

Exercise 6 - Semi-Supervised Anomaly Detection on the Shuttle Dataset
A one-class SVM is now used to perform semi-supervised anomaly detection on the shuttle dataset. The data needs to be imported, normalized, and the SVM needs to be trained and applied. The process is described in detail in the following.

[4] http://archive.ics.uci.edu/ml/datasets/Statlog+%28Shuttle%29

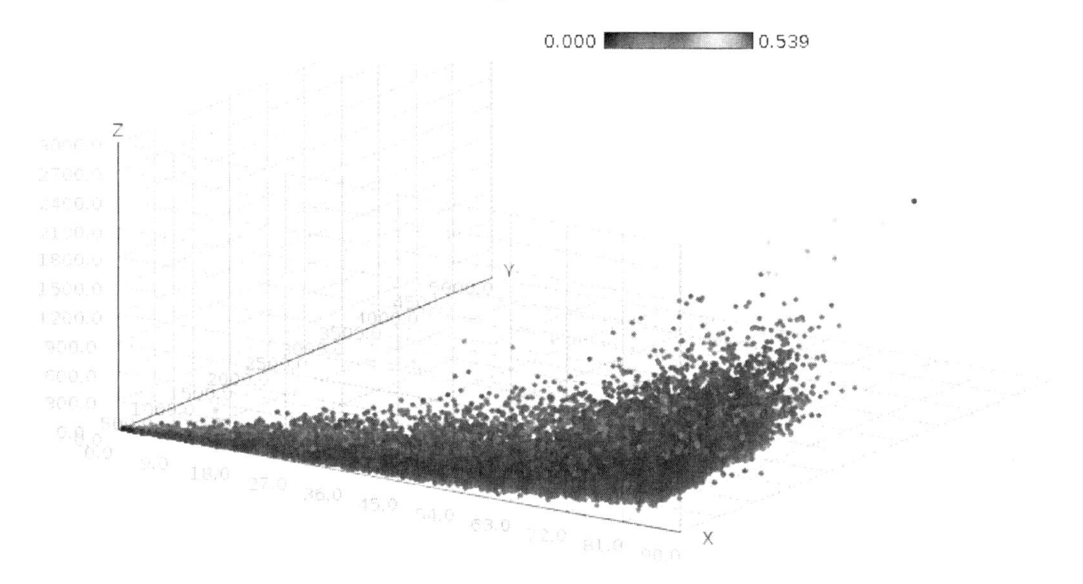

FIGURE 23.17 **(see color insert)**: Visualizing the anomaly scores of the NBA dataset. It can be seen that instances have higher scores which are outstanding in all dimensions—a property of a global anomaly detection algorithm.

Use the *Read CSV* operator twice to read both datasets—the training and test set. Again, use the import wizard, make sure no annotation is set (the dataset does not contain any attribute names), and assign the special attribute "label" to the last column (att10). Important: Set the type of the label to nominal; the default setting of binominal will not work for unknown reasons with the LibSVM operator! In case of the training set, there exist only "normal" instances, whereas both labels "normal" and "outlier" occur in the test set. You might want to verify this in the *meta data view* of the result perspective. Before training the one-class SVM, the data should be normalized in the range of [0,1] using the *Normalize* operator and the range transformation method. Also use the pre-processing model of the *Normalize* operator output and an *Apply Model* operator to do the same pre-processing on the test set. The simple way of using the *Normalize* operator twice might cause a different mapping since it is very unlikely that the training and test set have the same minimum and maximum values. Next, apply the *Support Vector Machine (LibSVM)* operator on the normalized training data. Set the parameters of the operator as follows: svm type = "one-class", kernel-type = "rbf", gamma = "0", nu = "0.05" and epsilon = "0.001". You might later want to try changing the gamma, nu, and epsilon settings to see the effect on the performance (or even optimize the parameters using the *Optimize Parameters (Grid)* operator). In order to do the actual prediction of our test data, the model of the one-class SVM is applied on the normalized test data using the *Apply Model* operator. The output of the prediction has the name "prediction(att10)" and has the values "inside" or "outside" by default. Since our data uses "normal" and "outlier" as labels, the *Map* operator is used to map "inside" to "normal" and "outside" to "outlier". Finally, you can either look at the results in the results perspective or apply the *Performance (Classification)* operator to get a qualitative measure of the anomaly detection success. Figure 23.18 shows the complete process and Table 23.2 shows the results of the predicted accuracy with the parameter setting from above. Since the parameters are not optimized, you might want to try to tweak the performance by yourself.

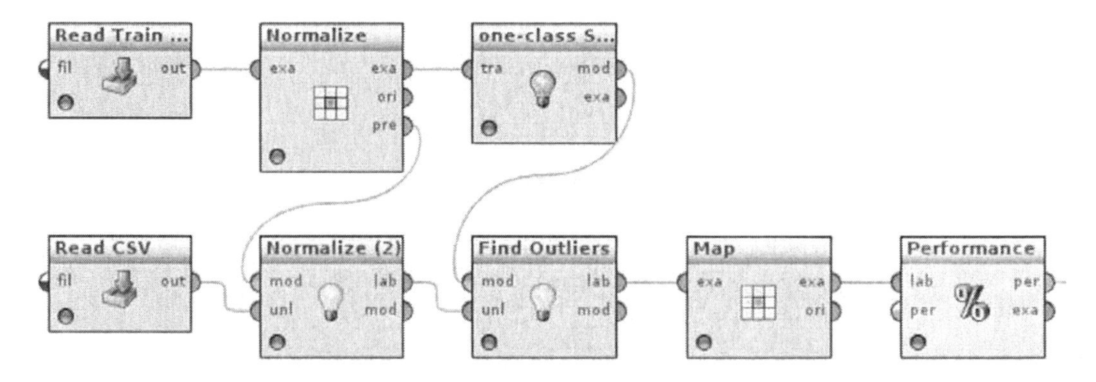

FIGURE 23.18: A sample process for semi-supervised anomaly detection on the shuttle dataset. Training and test data are normalized with the same transformation, a one-class SVM is trained, outliers are detected by applying the SVM model, and finally the performance is measured.

TABLE 23.2: The confusion matrix of the ground truth and the semi-supervised predictions for the shuttle dataset using a one-class SVM.

	True normal	True outlier	Class precision
predicted normal	**2309**	26	98.89%
predicted outlier	191	**32**	14.35%
Class recall	92.36%	55.17%	

23.6.2 Clustering and Distance Computations for Detecting Anomalies

If non-experts are asked how to apply semi-supervised anomaly detection, the answer is always as follows. Cluster the training data and compute for the test data the distances to the nearest cluster center. Then, a distance threshold can separate anomalies from normal instances. The bad news is that there is no operator in RapidMiner for such an approach. Recently, the author was asked how this could be achieved in RapidMiner and it emerged that this is not a simple process. Actually, there is a *Cluster Distance Performance* measure which can be abused to compute the distance of instances to a cluster centroid. However this involves looping over all examples, *Remember*ing/ *Recall*ing the cluster model as well as *Log*ging the distance to the cluster centroid within the loop. Since the process is really complex it is not explained here in detail. However, if you are interested, the process is illustrated in Figure 23.19 and 23.20 and ships with this book, too.

It is obvious that the approach above is not ideal. Firstly, the complexity and the high usage of logging, the macros, and remembering/ recalling make it error prone and hard to maintain. Secondly, looping over single examples is very slow in RapidMiner. Fortunately, RapidMiner offers an *Execute Script* operator so that we can implement the required functionality easily using Groovy[5], a language for executing Java code dynamically. The functionality we want to implement is really easy and it can be done in a few lines of code as shown in Algorithm 1.

After retrieving the example set and the cluster model as input (lines 6 and 7), a new attribute "outlier" with the role outlier is added to the example set (lines 9 to 15). Then we loop over all examples (lines 19 to 32) and we compute the distance of each example to

[5]See http://groovy.codehaus.org/

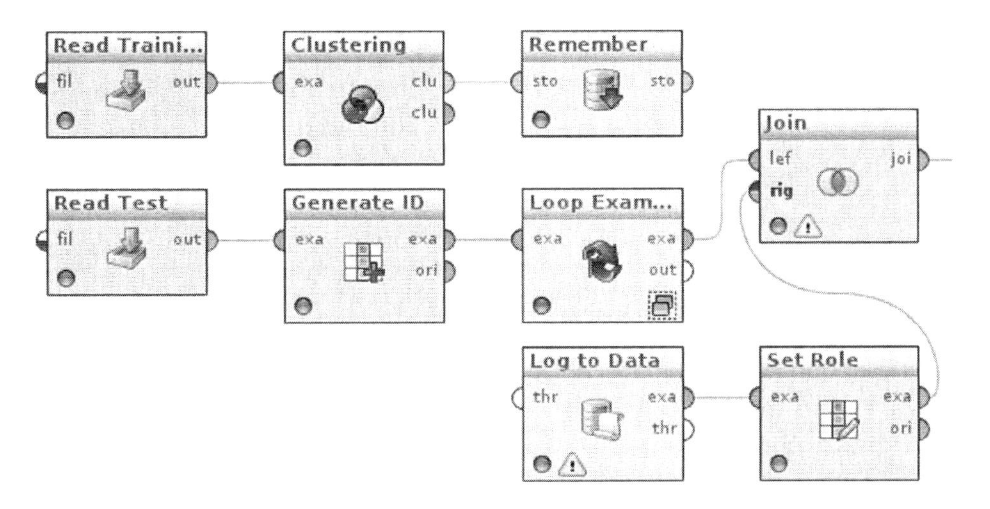

FIGURE 23.19: A complex process for semi-supervised cluster-based anomaly detection. At the top, the training data is clustered and the CentroidClusterModel is remembered. At the bottom, the test data is loaded and it is looped over each single example using the *Loop Examples* operator, as shown in Figure 23.20. The *Log to Data* operator reads the logged data from the inner loop (the result of the distance to the next cluster) and joins it again with the original test data.

each of the cluster centroids using the Euclidean distance (lines 25 to 30). The minimum distance, which is the distance to the nearest cluster center, is set as the outlier score (line 31) and the modified example set is returned (line 33).

Exercise 7 - A Groovy Script for Semi-Supervised Cluster-Based Anomaly detection

Implement Algorithm 1 in an *Execute Script* operator. Then load the training and test data from the shuttle dataset, cluster the training data using the *k-means* operator, apply the cluster model to the test data, and connect its output (the cluster model and the labeled training set) to the *Execute Script* operator.

The operator with the Groovy script can simply be called as illustrated in Figure 23.21. The parameter for k-means for the shuttle dataset was set to $k = 10$. The results are very promising. You might want to try to evaluate the performance by yourself, which can either be done by defining a threshold or selecting the first n samples as outliers if ordered by the outlier score. The latter was applied and it results in the confusion matrix as shown in Table 23.3, which reveals that our simple method performs even better than the one-class SVM. The performance evaluation is not shown in Figure 23.21 but it can be found in the processes which ship with this book.

TABLE 23.3: The confusion matrix of the ground truth and the semi-supervised predictions for the shuttle dataset based on our simple cluster-based anomaly score.

	True normal	True outlier	Class precision
predicted normal	**2478**	16	99.36%
predicted outlier	22	**42**	65.62%
Class recall	99.12%	72.41%	

Algorithm 1: Cluster-based Anomaly Score.

```
01: import com.rapidminer.operator.clustering.CentroidClusterModel;
02: import com.rapidminer.operator.clustering.Centroid
03: import com.rapidminer.tools.math.similarity.numerical.EuclideanDistance
04: import com.rapidminer.tools.Ontology;
05:
06: ExampleSet exampleSet = operator.getInput(ExampleSet.class)
07: CentroidClusterModel cmodel = operator.getInput(CentroidClusterModel.class)
08: EuclideanDistance m = new EuclideanDistance();
09: Attributes attributes = exampleSet.getAttributes();
10:
11: // add an outlier attribute to the ExampleSet
12: Attribute o = AttributeFactory.createAttribute("outlier",Ontology.NUMERICAL);
13: exampleSet.getExampleTable().addAttribute(o);
14: exampleSet.getAttributes().addRegular(o);
15: exampleSet.getAttributes().setOutlier(o);
16:
17: Centroid[] centroids = cmodel.getCentroids();
18: double[] exampleValues = new double[attributes.size()];
19: for (Example example : exampleSet) {
20:   int i = 0;
21:   for (Attribute attribute : attributes) {
22:     exampleValues[i++] = example.getValue(attribute);
23:   }
24:   double minDist = Double.MAX_VALUE;
25:   for (Centroid c : centroids) {
26:     double d = m.calculateDistance(c.getCentroid(), exampleValues);
27:     if (d < minDist) {
28:       minDist = d;
29:     }
30:   }
31:   example.setValue(o,minDist);
32: }
33: return exampleSet;
```

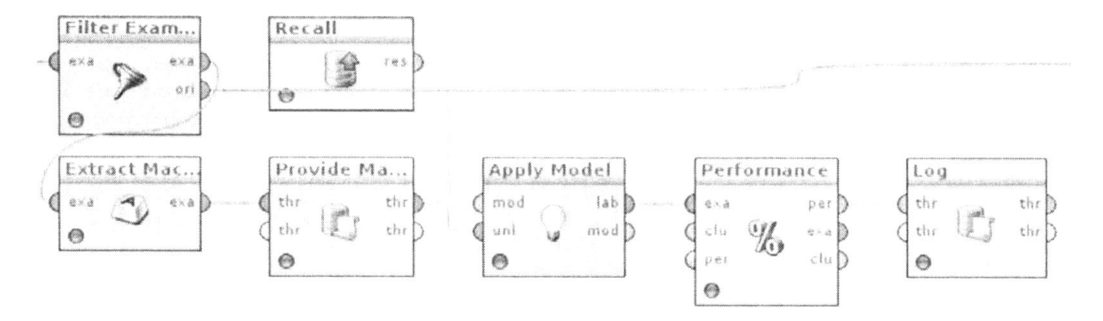

FIGURE 23.20: The inner loop of the process: The *Cluster Distance Performance* operator is misused to compute the distance of each single instance to its cluster center and the result is logged. Besides the distance, also the ID of the example needs to be logged, which is done by extracting a macro.

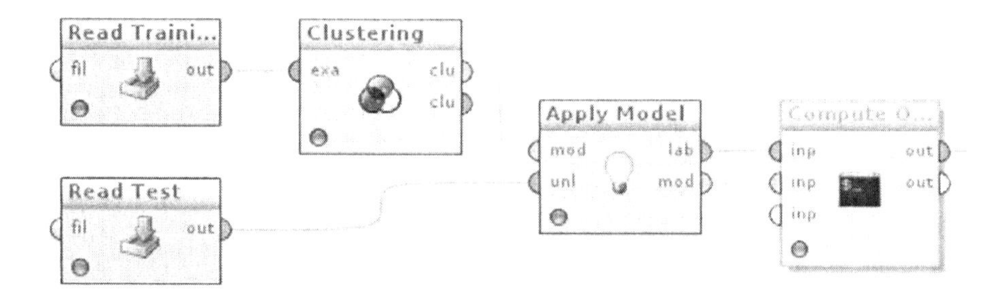

FIGURE 23.21: Using the *Execute Script* operator for computing anomaly scores based on the distances to the nearest cluster center. The CentroidClusterModel and the example set is handed over to the operator. The resulting example set at the output then has an additional outlier attribute.

23.7 Summary

Anomaly detection is the process of finding single records in datasets which significantly deviate from the normal data. Application domains are, among others, network security (intrusion detection), fraud detection, misuse detection, complex system supervision, or finding suspicious records in medical data.

Despite the differences of the various application domains, the basic principle remains the same. Multivariate normal data needs to be modeled and the few deviations need to be detected, preferably with a score to indicate their "outlierliness".

The relatively new research area of anomaly detection is not extremely well structured, such that the chapter starts off with a categorization attempt of the different approaches. Anomalies can occur in three different types: (1) point anomalies are single data records deviating from others, (2) contextual anomalies with occur with respect to their context only, for example, with respect to time), and (3) collective anomalies where a bunch of data points cause the anomaly. Most anomaly detection algorithms detect point anomalies only, which leads to the requirement of transforming contextual and collective anomalies to point anomaly problems using an appropriate pre-processing and thus generating processable data views.

Furthermore, anomaly detection algorithms can be categorized with respect to their operation mode: (1) supervised algorithms with training and test data as used in traditional machine learning, (2) semi-supervised algorithms with the need of anomaly-free training data, and (3) unsupervised approaches without the requirement of any labeled data. Anomaly detection is, in most cases, associated with an unsupervised setup, which is also the focus of this chapter. In this context, all available unsupervised algorithms from the RapidMiner anomaly detection extension are described and the most well-known algorithm, the Local Outlier Factor (LOF), is explained in detail in order to get a deeper understanding of the approaches itself.

Besides a simple example consisting of a two-dimensional mixture of Gaussians, which is ideal for first experiments, two real-world datasets are analyzed. For the unsupervised anomaly detection, the player statistics of the NBA are analyzed for outstanding players, including all necessary pre-processing. The UCI shuttle dataset is used for illustrating how semi-supervised anomaly detection can be performed in RapidMiner. In this context, a Groovy script is implemented for a simple cluster-distance-based anomaly detection approach.

Glossary

Anomaly Detection - Anomaly detection is the process of finding rare and outstanding instances in a dataset.

Collective Anomalies - Collective anomalies are anomalies which are defined by the occurrence of multiple instances in a specific combination. In order to detect such anomalies, a data view needs to be generated for mapping these anomalies to point anomalies.

Contextual Anomalies - Contextual anomalies always occur within a specific context, such as time. In order to apply anomaly detection, the context needs to be integrated by using an appropriate data view.

Data Cleansing - Data cleansing is the opposite of anomaly detection. In anomaly detection the anomalies are of interest, whereas in data cleansing the goal is to remove these outliers in order to have noise-free data.

Data View - A data view is used as a pre-processing method in anomaly detection. It guides the anomaly detection algorithm in the desired direction by aggregating, removing, or modifying attributes in an appropriate way. It is also used for integrating contexts.

Global Outlier - An outlier which deviates from the norm with respect to all instances within a dataset.

Groovy - A scripting language with a Java syntax. In RapidMiner it can be used for an easy implementation of one's own code accessing the native Java classes of RapidMiner.

Local Outlier - An outlier which deviates from the norm with respect to its direct neighborhood.

One-class SVM - A one-class support vector machine is a semi-supervised classifier. It is trained only with a dataset containing a single class and can afterwards predict if instances are inside or outside of that learned class.

Outlier - An instance in a dataset which deviates from the norm and only occurs rarely.

Outlier Score - A score estimated by an anomaly detection algorithm indicating the extend of an instance being an outlier. Usually instances are ordered by their outlier score in order to detect the top outliers.

Point Anomalies - Point anomalies are instances in a multidimensional space which can be identified using anomaly detection algorithms.

Semi-Supervised Anomaly Detection - Semi-supervised anomaly detection requires a training set containing only normal instances. During testing phase, anomalies can be identified which deviate from the normal model.

Unsupervised Anomaly Detection - Unsupervised anomaly detection does not require any labeled dataset. Usually instances are scored according to their likelihood to be an outlier.

Bibliography

[1] F. E. Grubbs. Procedures for detecting outlying observations in samples. *Technometrics*, 11:1–21, 1969.

[2] Leonid Portnoy, Eleazar Eskin, and Sal Stolfo. Intrusion detection with unlabeled data using clustering. In *In Proceedings of ACM CSS Workshop on Data Mining Applied to Security (DMSA-2001*, pages 5–8, 2001.

[3] Aubrey-Derrick Schmidt, Frank Peters, Florian Lamour, and Sahin Albayrak. Monitoring smartphones for anomaly detection. In *Proceedings of the 1st International Conference on MOBILe Wireless MiddleWARE, Operating Systems, and Applications*, MOBILWARE '08, pages 40:1–40:6, ICST, Brussels, Belgium, Belgium, 2007. ICST (Institute for Computer Sciences, Social-Informatics and Telecommunications Engineering).

[4] Richard J. Bolton and David J. Hand. Unsupervised profiling methods for fraud detection. *Statistical Science*, 17(3):235–255, 2002.

[5] Sutapat Thiprungsri. Cluster analysis for anomaly detection in accounting data. *19th Annual Research Workshop on Strategic and Emerging Technologies*, 2010.

[6] Jessica Lin, Eamonn Keogh, Ada Fu, and Helga Van Herle. Approximations to magic: Finding unusual medical time series. In *In 18th IEEE Symp. on Computer-Based Medical Systems (CBMS)*, pages 23–24, 2005.

[7] Arslan Basharat, Alexei Gritai, and Mubarak Shah. Learning object motion patterns for anomaly detection and improved object detection. *CVPR*, pages 1–8, 01 2008.

[8] Joost van Beusekom and Faisal Shafait. Distortion measurement for automatic document verification. In *Proceedings of the 11th International Conference on Document Analysis and Recognition*. IEEE, 9 2011.

[9] Varun Chandola, Arindam Banerjee, and Vipin Kumar. Anomaly detection: A survey. *ACM Comput. Surv.*, 41(3):1–58, 2009.

[10] Mennatallah Amer. Comparison of unsupervised anomaly detection techniques. Bachelor's Thesis, 2011. http://madm.dfki.de/_media/theses/bachelorthesis-amer_2011.pdf.

[11] Mennatallah Amer and Markus Goldstein. Nearest-neighbor and clustering based anomaly detection algorithms for RapidMiner. In Ingo Mierswa Simon Fischer, editor, *Proceedings of the 3rd RapidMiner Community Meeting and Conferernce (RCOMM 2012)*, pages 1–12. Shaker Verlag GmbH, 8 2012.

[12] Sridhar Ramaswamy, Rajeev Rastogi, and Kyuseok Shim. Efficient algorithms for mining outliers from large data sets. In *Proceedings of the 2000 ACM SIGMOD International Conference on Management of Data*, SIGMOD '00, pages 427–438, New York, NY, USA, 2000. ACM.

[13] Fabrizio Angiulli and Clara Pizzuti. Fast outlier detection in high dimensional spaces. In Tapio Elomaa, Heikki Mannila, and Hannu Toivonen, editors, *Principles of Data Mining and Knowledge Discovery*, volume 2431 of *Lecture Notes in Computer Science*, pages 43–78. Springer Berlin / Heidelberg, 2002.

[14] Markus M. Breunig, Hans-Peter Kriegel, Raymond T. Ng, and Jrg Sander. LOF: Identifying density-based local outliers. In *Proceedings of the 2000 ACM SIGMOD International Conference on Management of Data*, pages 93–104, Dallas, Texas, USA, May 2000. ACM.

[15] Jian Tang, Zhixiang Chen, Ada Fu, and David Cheung. Enhancing effectiveness of outlier detections for low density patterns. In Ming-Syan Chen, Philip Yu, and Bing Liu, editors, *Advances in Knowledge Discovery and Data Mining*, volume 2336 of *Lecture Notes in Computer Science*, pages 535–548. Springer Berlin / Heidelberg, 2002.

[16] Wen Jin, Anthony Tung, Jiawei Han, and Wei Wang. Ranking outliers using symmetric neighborhood relationship. In Wee-Keong Ng, Masaru Kitsuregawa, Jianzhong Li, and Kuiyu Chang, editors, *Advances in Knowledge Discovery and Data Mining*, volume 3918 of *Lecture Notes in Computer Science*, pages 577–593. Springer Berlin / Heidelberg, 2006.

[17] Hans-Peter Kriegel, Peer Kröger, Erich Schubert, and Arthur Zimek. LoOP: Local Outlier Probabilities. In *Proceeding of the 18th ACM Conference on Information and Knowledge Management*, CIKM '09, pages 1649–1652, New York, NY, USA, 2009. ACM.

[18] Spiros Papadimitriou, Hiroyuki Kitagawa, Phillip B. Gibbons, and Christos Faloutsos. Loci: Fast outlier detection using the local correlation integral. *Data Engineering, International Conference on*, 0:315, 2003.

[19] Zengyou He, Xiaofei Xu, and Shengchun Deng. Discovering cluster-based local outliers. *Pattern Recognition Letters*, 24(9–10):1641–1650, 2003.

Part X

Meta-Learning, Automated Learner Selection, Feature Selection, and Parameter Optimization

Chapter 24

Using RapidMiner for Research: Experimental Evaluation of Learners

Jovanović Miloš

Faculty of Organizational Sciences, University of Belgrade, Serbia

Vukićević Milan

Faculty of Organizational Sciences, University of Belgrade, Serbia

Delibašić Boris

Faculty of Organizational Sciences, University of Belgrade, Serbia

Suknović Milija

Faculty of Organizational Sciences, University of Belgrade, Serbia

24.1 Introduction

RapidMiner is a tool for conducting data mining workflows for various tasks, ranging from different areas of data mining applications to different parameter optimization schemes [1]. One of the main traits of RapidMiner is its advanced ability to program execution of complex workflows, all done within a visual user interface, without the need for traditional programming skills.

This paper provides a tutorial on how to use RapidMiner for research purposes. It presents a way to experimentally evaluate effectiveness of different learning algorithms, parameter settings, or other choices that influence data mining results.

One way research workflows differ from application workflows is the methodology, which is aimed at discovering the properties of a method, for example, an algorithm, a parameter setting, attribute selection, etc. under study. Since more factors influence the overall results, the discovery process needs to be controlled, meaning that we want to isolate the influence of one factor on overall performance. Therefore, the research methodology is aimed at removing biases and noise from other sources of variation, which is discussed more in Section 24.2 of the tutorial.

There are many research questions which require a framework for experimental evaluation, however, this tutorial will focus on experimental evaluation of learning algorithms. Similar problems will also be discussed, but not shown through RapidMiner processes. Several RapidMiner workflows will be given to enhance the explanation of the issues, and also to serve as templates for similar research tasks.

24.2 Research of Learning Algorithms

Research is a scientific and systematic search for pertinent information on a specific topic [2]. It can also be defined as a "systematized effort to gain new knowledge" [3]. Scientific research is not any insight, but one that is, as far as possible, controlled, rigorous, systematic, valid, verifiable, empirical, and critical. Throughout this chapter, we demonstrate how such efforts could be made using tools in RapidMiner, in order to answer questions related to the use of learning algorithms.

The main objective will be to analyze the effect of a learning algorithm on the quality of the prediction model, outputted by an algorithm. So our research hypothesis is that the **selection of the learning algorithm greatly influences the prediction ability of the model, for the task of bank credit scoring.**[1] Furthermore, we would like to also learn **which algorithms would perform better, and a way to define when to prefer one algorithm over another**.

We are going to measure the quality of the prediction model by its accuracy of predictions, which we will measure in each experimental trial. Research methodology in this case could be classified as quantitative, empirical, applied, experimental, and analytical [3].

24.2.1 Sources of Variation and Control

An experiment trial is one run of the learning algorithm over given data, which produces a prediction model that can be evaluated. We limit this study to classification algorithms, but the same methodology would apply to other types of learners.

In order for research to be systematic and controlled, we need to be able to detect and control all sources of variation in our measurable results (i.e., accuracy of the predictions). For this experimental setting we could intuitively identify several sources of variation, which would affect the output measurement if changed. These are:

1. the choice of a particular **dataset** chosen for training and testing;
2. the choice of a **learning algorithm** that produces a model;
3. the choice of **algorithm options** available for tweaking;

[1]Note that this hypothesis would not constitute an interesting research direction today, since it has been explored so many times in the past. Thus is should be considered merely as a guideline to the way of exploring other research questions.

4. the choice of **pre-processing steps** applied to the dataset before the learning algorithm;
5. the choice of a **method of validating results** and calculating performance; and
6. the choice of **random seeds** used for the stochastic part (if exists) of the algorithm.

All these choices (we will call "factors") are determining the output we measure, so we want to control them in one of these ways:

- vary different values, so to see the effect on the output;
- fix single values, to exclude the effect on the output; or
- randomize, by trying different random values, to exclude the effect on the output.

By controlling factors that have an effect on the output, we want to isolate the effect of one of the factors, so that we can draw conclusions on how it influences the results.

The important thing to note is the difference between fixing and randomizing values, for factors we wish to exclude. By fixing a single value, e.g., fixed way to pre-process data or fixed dataset, we are narrowing down the experiment to that context, and all conclusions are valid only if interpreted in that context, i.e., under the conditions fixed in the experiment. In this way, we are making the research more specific. If we want to generalize the conclusions, we would need to try different (random) values for the factor, and if the results are statistically consistent, we could argue that the influence of this factor (for which we tried many random values) is opted out. Fixing factors, on the other hand, are simpler and also valid, but then the conclusions must also be narrowed down to these fixed factors.

24.2.2 Example of an Experimental Setup

Considering our example goal defined in Section 24.2, we will illustrate the experimental setup for that purpose, which will be used throughout the chapter.

In our research of learning algorithms, our main source of interest is the second factor (the choice of a learning algorithm), so we want to vary this factor, switching it between more learning algorithms which we want to compare experimentally. We know that algorithms have biases, meaning that they have implicit assumptions, which make them perform better in some contexts than others. This is why we want to compare the algorithms, and hopefully figure out which one is the best for our purpose. We will use several well-known algorithms for classification, which will be mentioned in later sections.

The first factor (dataset) could be fixed, but then we would have to discuss results only for this single dataset. Sometimes this is appropriate; however, in our example we would want to draw a somewhat general conclusion on what algorithm to use in a credit scoring setting. This is why we will use more datasets and see if the results are consistent. Each of the datasets originates from different banks; therefore, the more datasets there are, and the more representative they are for this area, the more the conclusions will hold.

Another thing concerning the data is the concrete sample of data we have from each bank. We could assume that the given samples are representative, but it is more appropriate not to assume this. Thus we would have to somehow randomize the sample, to exclude the possibility that the conclusions will hold only on this sample, as explained in the previous section. We do this by sampling techniques, like the most widely used, cross-validation technique. This way, we will repeat the experiment on various random sub-samples. Generally, if the dataset is quite big, there is no need for these repetitions, as one could only use the separate test set for evaluation, assuming that the large dataset should be representative for the given problem.

Other sources of variation we need to control are different options that algorithms have, and pre-processing steps that could be taken before the running of algorithms. For the sake of simplicity, we will fix these factors to default values, which will partially limit our conclusions, but still be useful for this exercise.

In addition, some of the algorithms have stochastic parts, which means that they have some randomness in calculation which makes results hard to replicate. This is usually inappropriate for the research community, so we will fix the *random seed* to a pre-defined random value, making the experiment results reproducible.

24.3 Experimental Evaluation in RapidMiner

In this section we will showcase how the research experiment can be conducted in RapidMiner, which we set up in the previous section. The example will be gradually built, to dissect the overall complex process. It will include the following parts:

1. setting up of the evaluation scheme;
2. looping through a collection of datasets;
3. looping through a collection of learning algorithms;
4. logging and visualizing the results;
5. statistical analysis of results;
6. exception handling and distribution of experiments; and
7. setup for meta-learning.

24.3.1 Setting Up the Evaluation Scheme

Firstly, we will set up a simple way to measure how good a single algorithm on a single dataset is. This means that we will fix almost all the factors discussed in the previous section. Figure 24.1 shows the simple process for this purpose. We import a single dataset with the *Read AML* operator, since the dataset is stored on a local hard drive, with the AML description format, native to RapidMiner. The *Read AML* operator is set to point to a local file path where the dataset is stored (attributes file: "*australian.aml*"). The dataset describes clients of a bank by their demographic attributes, financial state, usage of products, and previous behavior. Each client is labeled as "good" or "bad", whether he repaid the previous credit, or had problems with it. The goal is to build a model that would be able to predict if some new client is suitable for crediting.

The way we intend to measure how good is the algorithm, is by measuring its accuracy, which is a percentage of correctly classified bank clients into one of the two classes ("good for credit", "bad for credit"). This is a widely used metric for measuring the performance of an algorithm; however, one should use accuracy with caution, because there are circumstances when it is not appropriate. Typically, accuracy is misleading when the dataset has a huge class imbalance (skewed class distribution), and when different errors that the classifier makes do not have an equal cost. The example would be a disease detection classifier, when there are a lot more healthy people in the dataset than those with the disease. Also, the cost of an error of not detecting the disease is not the same as the cost of falsely detecting a non-existing one. We will not go through all evaluation metrics available in RapidMiner (and there are a quite a number available), since the evaluation scheme will not differ much if we select another metric.

For estimating accuracy in an unbiased way, we will use the cross-validation (*X-Validation*) operator that conducts the evaluation by re-evaluating the algorithm on more subsets (folds). Another alternative is to use the operator *Bootstrapping Validation*, which is similar because it also repeats the evaluation on many random sub-samples. The differences are subtle, but for smaller datasets, *X-Validation* tends to be a good choice.

Inside the *X-Validation* operator, we chose to evaluate a *Decision tree* learning algorithm, and we define that we want to measure accuracy inside a *Performance* operator (Figure 24.1). All parameters for operators are set to default values. Running this process, we get a single number[2] 84.06%, depicting the estimated accuracy for this algorithm.[3]

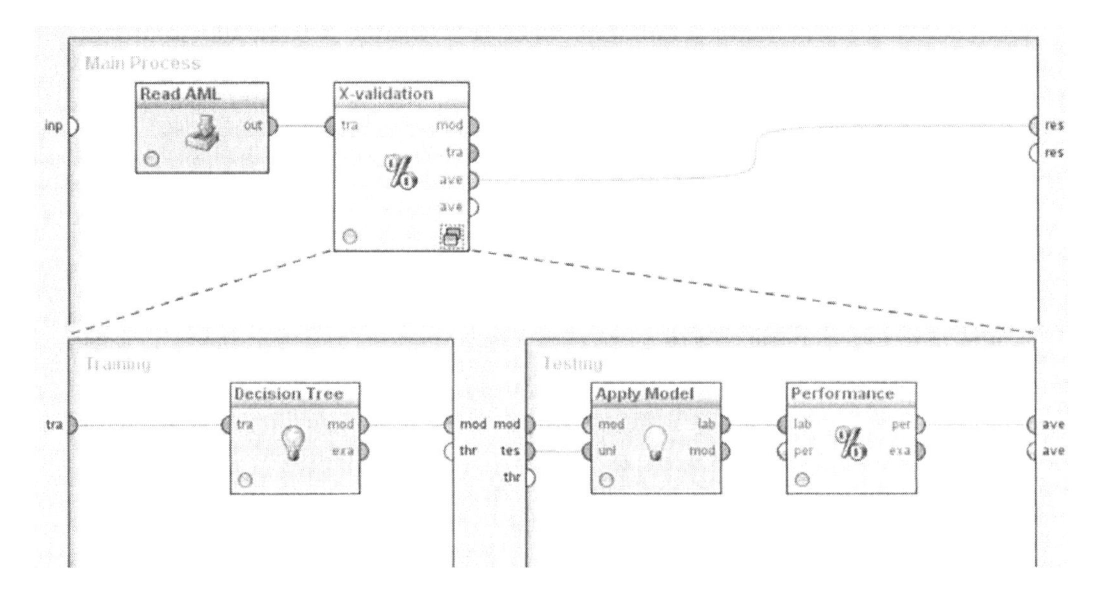

FIGURE 24.1: The first setting for the evaluation of learning algorithms.

24.3.2 Looping Through a Collection of Datasets

Now that we have our evaluation procedure set, we want to evaluate the algorithm on more datasets, to check the consistency of the results on other sources of similar data. In this example, we are using the problem of credit scoring, and will be using two datasets that are publicly available for this purpose [4]. Although two datasets are not enough to generalize conclusions sufficiently, the setup we create will be able to scale to as many datasets as one collects. For this example, two datasets, *German* and *Australian* credit scoring, are used mainly because of their availability.

To run the evaluation multiple times for each dataset, we will wrap the previously built evaluation setup into a *Loop Files* operator, renamed to "Loop Datasets" (Figure 24.2).

This operator will iterate over all files in a selected folder, and we will set the folder

[2]Actually, the results also renders a confusion matrix, which we will ignore, since we don't analyze it in this chapter.

[3]Throughout this chapter, we will use classification algorithms, evaluated by accuracy, but the same general process could be applied for other supervised tasks, like regression.

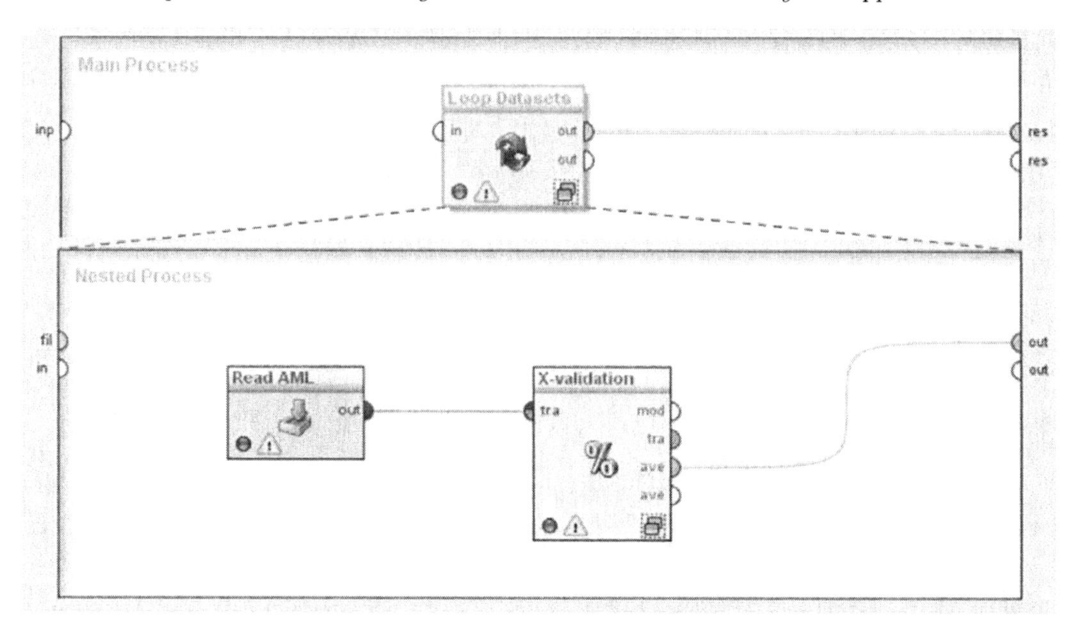

FIGURE 24.2: Looping through different datasets.

parameter to point to our local folder that contains all the datasets. Since the folder contains both ".aml" files as dataset descriptions, as well as ".dat" files with raw data, we will set the filter parameter of the Loop operator to iterate only over ".aml" files (Figure 24.3). The *filter* parameter defines the pattern of file names it will iterate over, and uses *regular expressions* to define the pattern. Other examples of using the filter could be `[Data].*`, for iterating over files whose names start with "Data"; or `.*[0-9]+.*`, for iterating over files whose name contains a number.

Loop Dataset (Loop Files)

directory	RapidChapter\Datasets
filter	.*[.]aml
filtered string	file name (last part of the ...
file name macro	dataset_name
file path macro	dataset_path
parent path macro	parent_path
☑ recursive	
☑ iterate over files	

FIGURE 24.3: Parameter setting for the *Loop Files* operator.

In each iteration this operator will set the value of the *dataset_path* macro (a RapidMiner variable) to a dataset path in current iteration. Inside this operator, we need to reset the *attributes* parameter of the *Read AML* operator to this macro *dataset_path*, so that the different dataset is loaded in each iteration. The *attributes* parameter would then have the value `%{dataset_path}`.

One thing to note is that because the file path will be determined in run-time, RapidMiner will report errors for these operators (as seen on Figure 24.2). Nevertheless, the errors should be ignored and the process would still work, with the drawback that we cannot see dataset meta-information in design time.

When we run this process, RapidMiner will output a collection of *PerformanceVectors*, and we can read the average accuracy for every dataset. So we get 84.06% for dataset *Australian*, and 67.90% for dataset *German*, all evaluating the accuracy of *Decision tree* learning operator.

This setup is made for the case when datasets are stored on a local folder, and would work for any number of datasets, as long as they are converted to ".aml" format. It is also possible to replicate this setup when the datasets are located in the RapidMiner repository, either local or remote. The difference would only be in the used *Loop* operator, having to switch from *Loop Files*, to *Loop Repository*, and discard the *ReadAML* operator. The *Loop Repository* operator would be set to point to a folder within the repository, and the entry type parameter should be set to *IOObject*, in order to loop through dataset objects.

24.3.3 Looping Through a Collection of Learning Algorithms

Now that we can repeat the validation for each dataset, we would also need to iterate over all learning algorithms that we want to compare. We will use a set of algorithms, originating from different fields, like statistics and machine learning. The algorithms are: *Decision Tree*, *k-Nearest Neighbors*, *Naive Bayes*, *Support Vector Machines (SVM)*, and *Random Forests*. Since we cannot tell in advance which one of these approaches would be the best for our credit scoring datasets, we would like to evaluate them all and somehow reason which one we should prefer over another.

One problem is that these algorithms could not be represented as files, so we cannot reuse the *Loop Files* operator. Since all these algorithms are actually operators, we could use *Select Sub-process* operator, which can define more disjoint sub-processes, and allow the user to pick one with the parameter, as shown in Figure 24.4.

This selection of algorithms is actually put inside the cross-validation operator, as a learner for training (Figure 24.4). The main problem is that the user needs to manually pick one of the algorithms, by setting the parameter to one value. We will override this by placing the whole cross-validation operator inside a simple *Loop* operator (renamed to "Loop Algorithm"), which executes the validation a fixed number of times. In each iteration of this *Loop* operator, we will have a macro (named "algorithm") which will point to the algorithm currently selected, and will use this macro to select the sub-process in each turn, as shown in Figure 24.5.

Now we have two loops, one inside of another, which will iterate over all datasets, and all selected algorithms, shown in Figure 24.6.

One final thing: the *Select Subprocess* operator does not restrict us in using any combination of preprocessing and algorithms. This is the case with algorithm *SVM*, for which we needed to convert the attributes to numerical, since it is one of its requirements. This also means that we could use the same setup to test different preprocessing options, different parameter settings, or any combination of operators that produce some model at the end.

24.3.4 Logging and Visualizing the Results

If we run the previous process, we would get a collection of results for accuracy, for each algorithm over each dataset. However, these results are not easy to look at and compare,

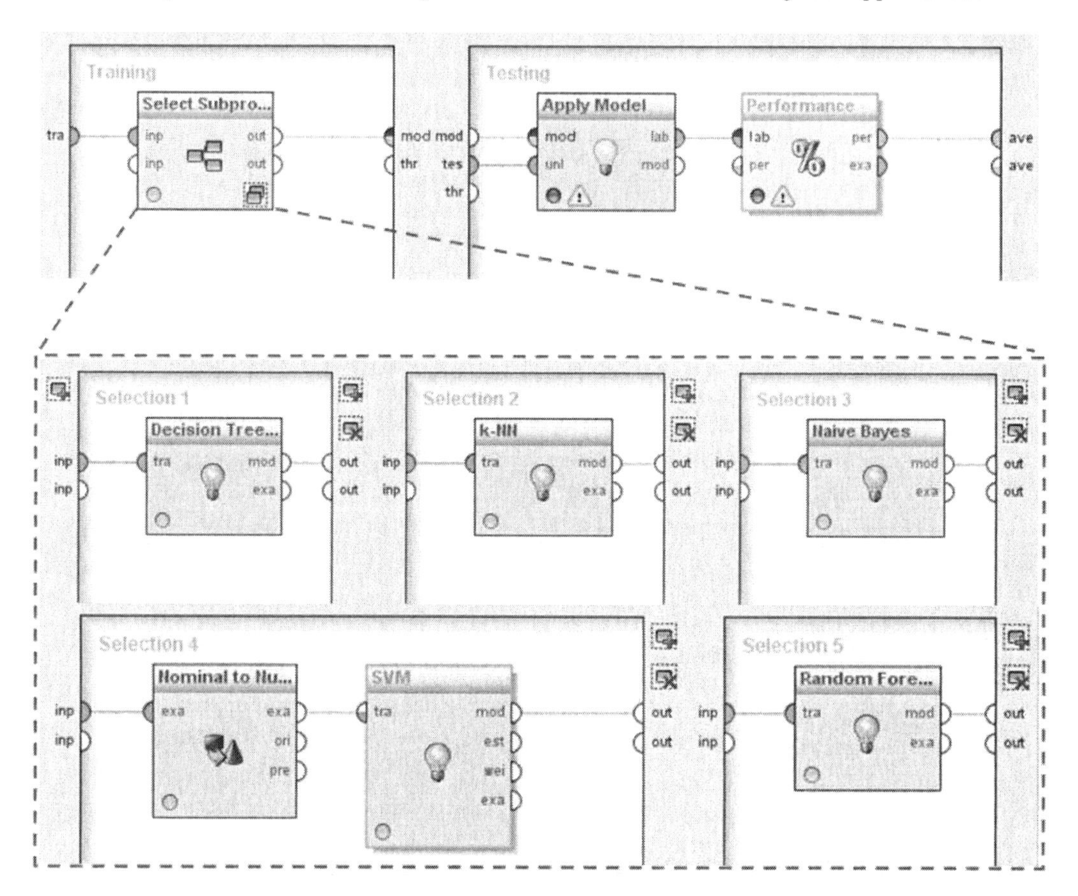

FIGURE 24.4: Selecting one of the learning algorithms.

since we would need to manually collect all the numbers and form a table, for inspection. This is another thing that we can nicely automate in RapidMiner, using the *Log* operator.

Firstly, we add a sub-process for all of our logging operators directly after the cross-validation operator, so that it can "catch" the accuracy after each evaluation. The *Subprocess* operator is renamed to "Logging". This is shown in Figure 24.7, along with the content of this sub-process, holding our operators for logging.

Secondly, we insert the operators *Provide Macro as Log Value*, which will enable us to log values of macros. One of these operators will catch the value of *dataset_name* macro (defined in Figure 24.3), and the other will catch the value of *algorithm*, macro, used for iterating over our algorithms. These operators are renamed into "LogDataset" and "LogAlgorithm", respectively.

Finally, we use the *Log* operator to define what information we want to log. We would log dataset, algorithm and the accuracy outputted by the cross-validation operator, also shown in Figure 24.7.

Now when we start the process, RapidMiner outputs a log, which we can see as a table (Figure 24.8) or draw some basic plots for visualization. As seen in Figure 24.8, we now have all the results in tabular form, which we can export and use for any further analysis of the algorithms.

One thing to remember is that the same process would run even if we add more datasets or algorithms, giving us these results in a more usable format. Furthermore, RapidMiner can

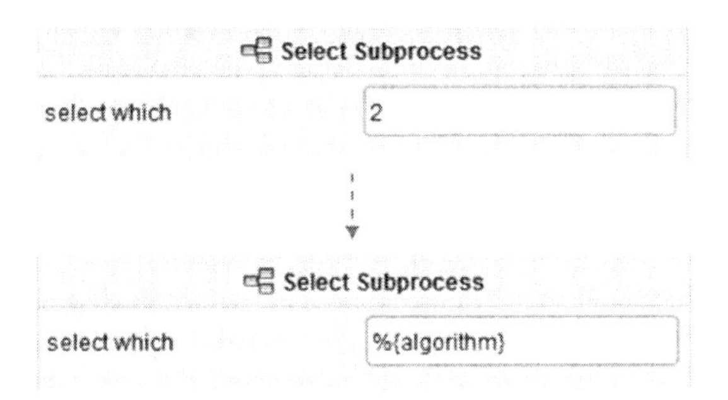

FIGURE 24.5: Parameter for the *Select Sub-process* operator.

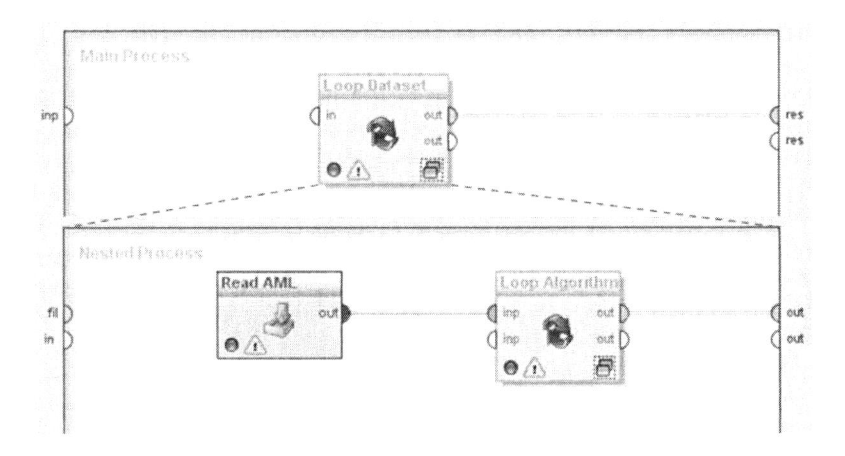

FIGURE 24.6: Loops for datasets and algorithms combined.

help us even draw some exciting visualizations of the results; however, we would need our results to be in a "dataset format" instead of "log format". This is exactly what operator *Log to Data* does.

One way to use this operator is shown in Figure 24.9, where we transformed the resulting log to a dataset, then used the operator *Pivot* to show the data in a more appropriate form, and renamed the attributes to match the chosen algorithms. Since this is not of major importance, we will skip the descriptions of these operators, but will show the processed results, both in a table and graphically, in an even more useful format (See Figure 24.10).

24.3.5 Statistical Analysis of the Results

To this point, we already made a somewhat complex process, which helps us to conduct a larger number of experiments and to present the results in a more useful way. Still, we would want to push the analysis a bit further (since the RapidMiner allows it), to incorporate a very important aspect of research, the statistical analysis of the results.

When examining the results from Figure 24.10, we could say that three algorithms (*Decision Tree*, *Naive Bayes*, and *Random Forests*) are competing to be the best. Among them, we could say that *Decision Tree* is best for the *Australian* dataset, but not that good for the *German* dataset. Now we cannot be too certain of these conclusions, because they represent only average accuracy over all repetitions within cross-validation. This is why we need to also consider the variance of these results, because if the performance overlapped a

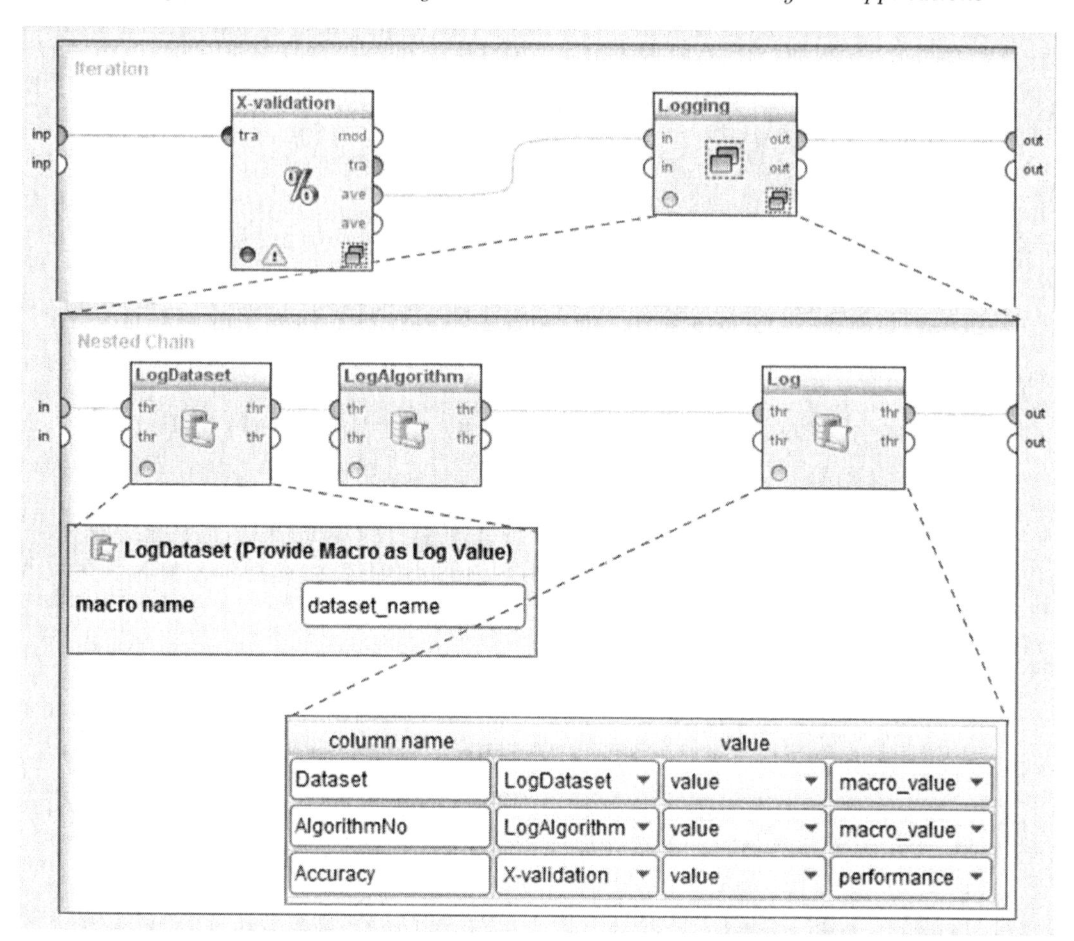

FIGURE 24.7: Operators for logging the results.

lot during these trials, we cannot say that one is better that the other, merely because the average is higher.

This kind of statistical analysis of variance is possible using operators directly in RapidMiner. We will put our statistical analysis (grouped using *Sub-process* operator named "Stat Analysis") right after all evaluations for a dataset are made, i.e., after our named *Loop Algorithm* operator. This is shown in Figure 24.11.

Since we have for the input a collection of *PerformanceVectors* (holding the accuracy measurements), we first need to isolate several performances, which we do with the *Select* operator. Three *Select* operators isolate the performance of the top three algorithms, which we want to analyze statistically. We then feed these accuracies into *T-Test* operator, which will perform statistical t-test, to determine the statistical significance of the differences in accuracy. If the difference between two average accuracies is not significant, we must conclude that their accuracies are indistinguishable;[4] and if the difference is significant, we can say that one algorithm outperforms the other on this dataset. The results of running this process are given in Figure 24.12.

Bolded values in Figure 24.12 are ones that could be considered significant, since the

[4]The term used in the literature is that these algorithms, with no statistically significant difference in performance, are "comparable".

Table View Plot View

Log (10 rows, 3 columns)

Dataset	AlgorithmNo	Accuracy
Australian.aml	1	0.841
Australian.aml	2	0.658
Australian.aml	3	0.772
Australian.aml	4	0.414
Australian.aml	5	0.751
German.aml	1	0.688
German.aml	2	0.608
German.aml	3	0.756
German.aml	4	0.300
German.aml	5	0.700

FIGURE 24.8: Logging results.

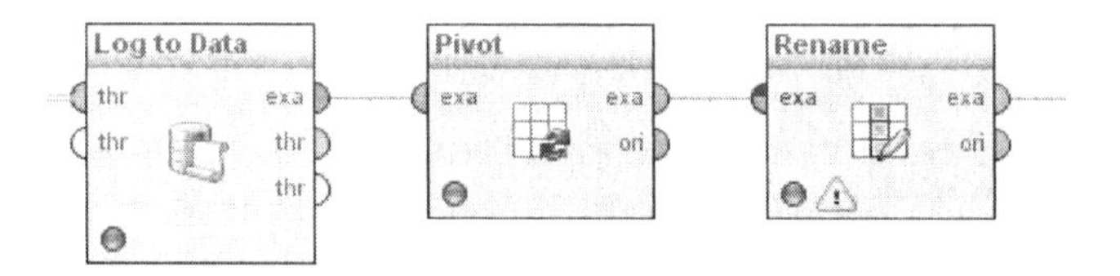

FIGURE 24.9: Post-processing of the results.

p-value is less than 0.05. We see that, for the dataset *Australian* (upper table), there are significant differences between *Decision Tree* VS *Naive Bayes*, and *Decision Tree* VS *Random Forests*, but could not say that *Naive Bayes* is better than *Random Forests*.

Figure 24.11 also features a disabled *ANOVA* operator, also used for statistical analysis. The difference is that *ANOVA* is giving an overall answer, if there are significant differences between all algorithms, while *T-Test* operator does that in a pair-wise manner.

24.3.6 Exception Handling and Parallelization

The given process so far is quite useful for experimental comparisons between algorithms. Still, we want to point to a few caveats, usually arising when the experiments we want to run are numerous.

Experiments, if done on large or numerous datasets, could take some time to run to completion. Occasionally, something can cause a part of the experiment to fail, which, by default stops the entire execution in RapidMiner, rendering a pop-up screen with an error message. This is why it is desirable to have some mechanism that catches these errors, and gives reports quietly, after completing the rest of the process. For this, one can use the *Handle Exception* operator, and wrap whatever operators are prone to failing. Since this is an auxiliary operator, we will not describe it in detail here, but it will be included in the process available with this chapter for exercise.

The problem of large experiments could partially be solved by parallelizing the overall process. If the experiment is conducted on a larger number of datasets, the simplest thing to do is to split datasets into groups, and run the process on several computers, each on

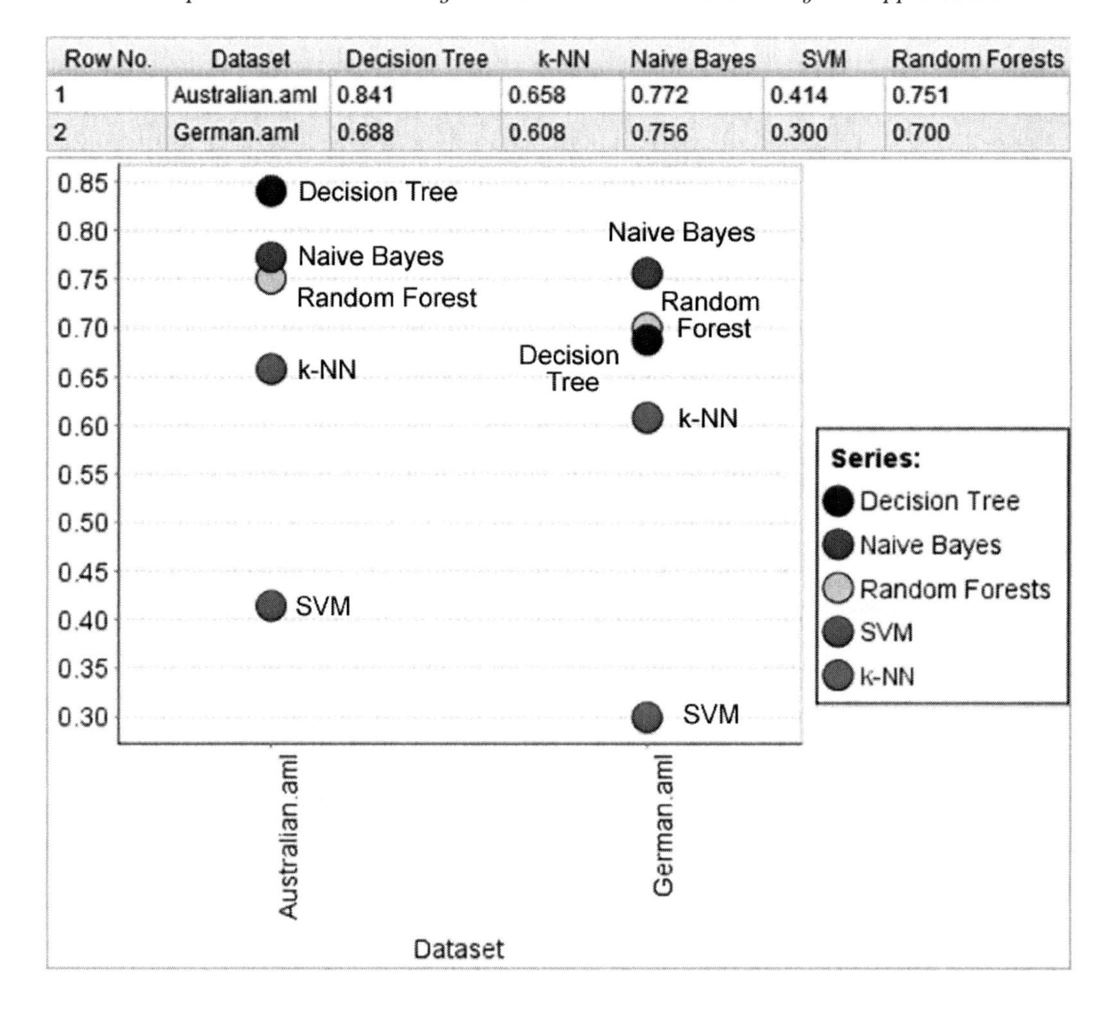

Row No.	Dataset	Decision Tree	k-NN	Naive Bayes	SVM	Random Forests
1	Australian.aml	0.841	0.658	0.772	0.414	0.751
2	German.aml	0.688	0.608	0.756	0.300	0.700

FIGURE 24.10: Results in the final format.

a different subset of datasets. Sometimes, the bottleneck for execution time is one dataset which is quite large to evaluate the algorithm. One way to address this is to use the *Parallel Processing* extension for RapidMiner, which adds an option to parallelize many operators, for example cross-validation, loop, or even learning algorithms. The execution is still done on one computer, but utilizes multiple processor cores that a computer (or server) might have. Finally, if the execution is done on a very large dataset, one might consider using the *Radoop* extension for RapidMiner [5], which features a very powerful distribution mechanism based on *Apache Hadoop*, but is much more complex to use.

24.3.7 Setup for Meta-Learning

We mentioned at the beginning that we are evaluating different algorithms, because we cannot say, a priori which algorithm would be the best option to use on our dataset. However, when analysts do a lot of data analytics, they start to notice and pick up cues that help them focus on some algorithms rather than others. Having a lot of experience in using algorithms, they are able to use this experience, even when they do not understand the inner workings of the algorithms.

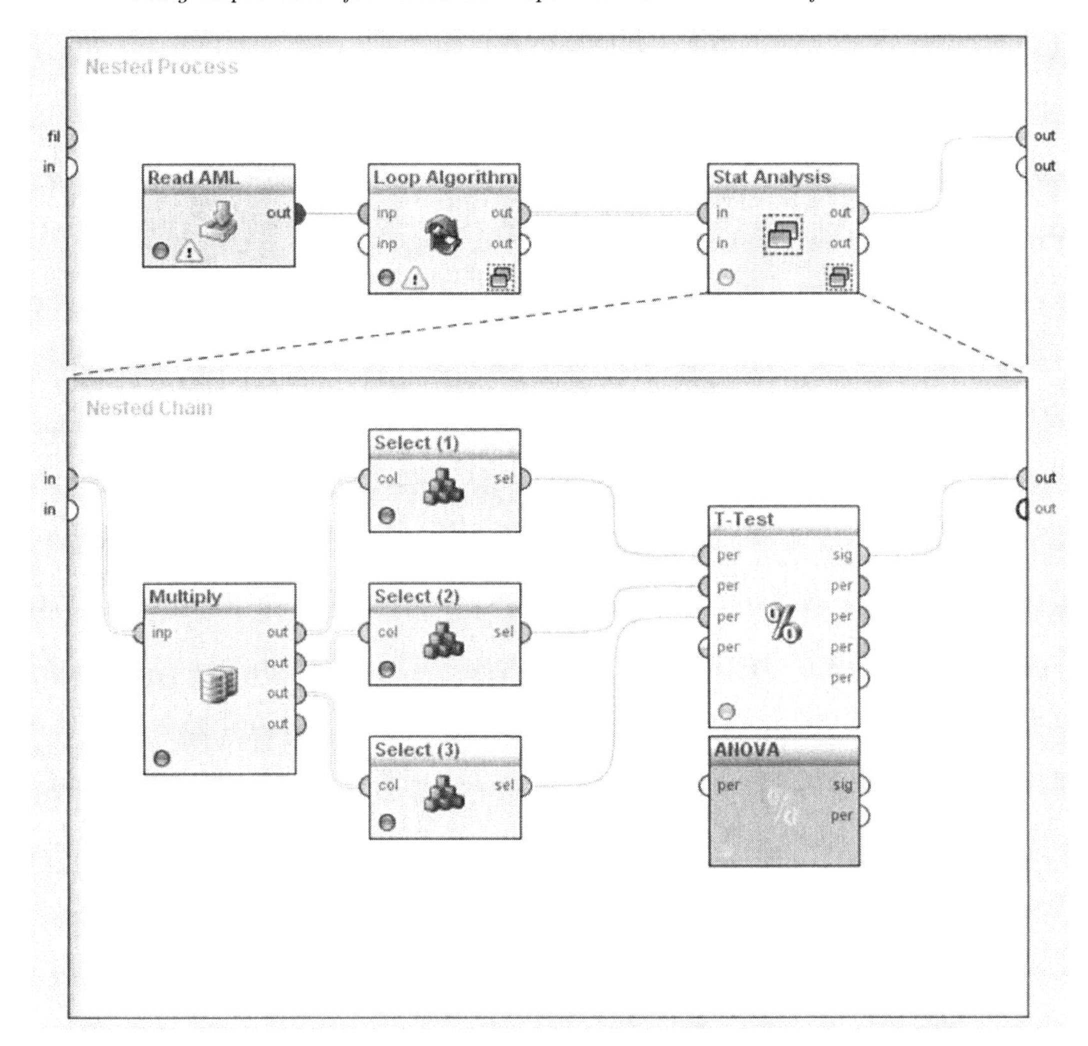

FIGURE 24.11: Statistical analysis of the results.

This is exactly the idea behind meta-learning discipline [6], which tries to learn when a learning algorithm performs well, and when it does not. The idea is to log each experiment, and try to describe the setting, in detail, including which algorithms are used, and most importantly, properties of the dataset which was being solved. These dataset properties describe the kind of attributes it has, the size, dimension, and any statistical measure related to that dataset. If we had such a recorded log of experiments, we would be able to run the same learning algorithms to predict if the process we wish to run will perform poorly or effectively.

Here, we will show how one can create such logs of experiments, which could be utilized for meta-learning. We already were creating logs of how each of the algorithms performed on each dataset. We only need a way to collect dataset properties, which we want to relate to the success or failure of a learning algorithm. Figure 24.13 shows how we can collect such information.

All operators are grouped in a *Sub-process* operator named "Dataset Metadata". Firstly, two operators in the chain are instances of *Extract Macro* operator, which give users a list of properties they want to extract regarding the inputted dataset. This is the easiest way to

T-Test Significance

	0.841 +/- 0.038	0.772 +/- 0.040	0.751 +/- 0.096
0.841 +/- 0.038		**0.001**	**0.014**
0.772 +/- 0.040			0.521
0.751 +/- 0.096			

T-Test Significance

	0.688 +/- 0.024	0.756 +/- 0.043	0.700 +/- 0.000
0.688 +/- 0.024		**0.000**	0.154
0.756 +/- 0.043			**0.001**
0.700 +/- 0.000			

FIGURE 24.12: Results of the statistical t-test for *Australian* (upper) and *German* (lower).

gather information on a dataset, however, it currently has very few properties to calculate. We used it to extract the number of examples ("#Examples"), and the number of attributes ("#Attributes") in the dataset.

Since no other operator can be used to extract other dataset features, we can opt for the *Execute Script* operator, which offers a way to build any custom calculation, but does require programming skills. It is definitely an advanced operator, and not intended for all users, but in this case, offers a solution for our problem. Figure 24.13 shows one way to calculate an additional dataset feature, namely the class imbalance, which is often an important factor in the success (or failure) of some algorithms. The code is given only for illustration on how this operator could be utilized. For less advanced users, we hope that more of these properties will soon be available using the *Extract Macro* operator.

One thing to note is that all these operators place the properties they calculate into a designated macro, which must be extracted using the *Provide Macro as Log Value* operator, as before, prior to logging these properties. After running the process, the final log, with these additional properties, now looks as depicted in Figure 24.14.

Finally, we may even use the success of our learning algorithms for prediction of how another algorithm will perform. For example, we could learn to predict how good Random Forests will be based on how these other algorithms performed. This, of course, makes sense only if these algorithms are quite fast to calculate. This approach is called *landmarkers*, and is the logic behind the already available meta-learning extension for RapidMiner, namely the *PaReN* extension [7].

24.4 Conclusions

Research within data mining is a process that is somewhat different from a simple application of a data mining method on some domain. To answer research questions, one needs to usually run fully customized processes, which often makes researchers turn their

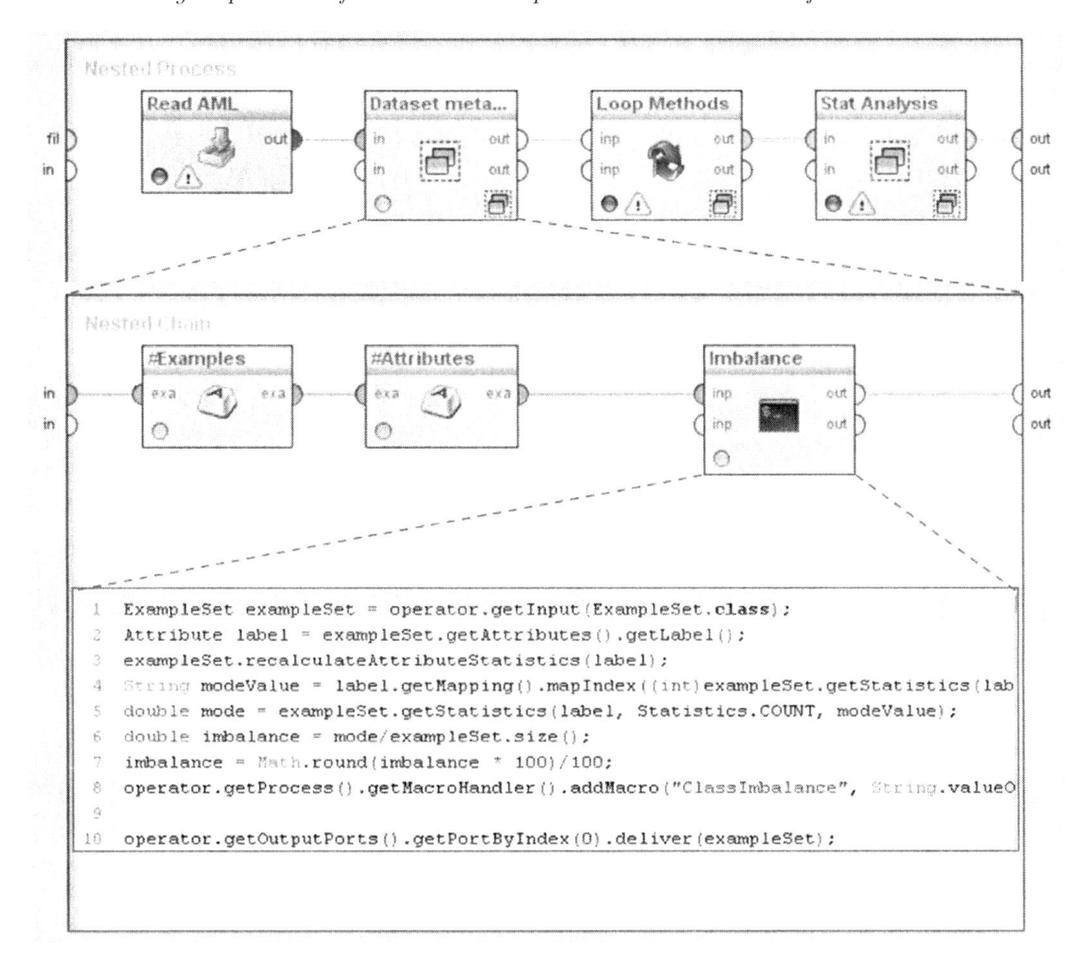

FIGURE 24.13: Extraction of dataset properties for meta-learning.

back on data mining software suites, and turn to programming languages that offer more flexibility. However, RapidMiner proves many of them wrong, offering fully customizable processes that by far surpass simple application scenarios. RapidMiner is perhaps the only tool available in the market that offers this kind of flexibility all through its graphical user interface, with no programming involved at the lower level. This way, RapidMiner could expand the research community, making research easier and more accessible.

In this tutorial, our aim was to demonstrate how RapidMiner could be used for research purposes. Although our example showed only the aspect of evaluation of learning algorithms, it is easy to imagine how similar processes could be used to answer other research questions. For example, one could very similarly test the effect of different parameters that algorithms offer, and give empirical evidence that help to better understand these parameters and suggest the ways to use them more effectively. In addition, one could research some pre-processing steps, e.g., selection of attributes, and also give some empirical support for understanding their effect on the results.

Our example experiment was small-scale, covering only two datasets for credit scoring. This is not so unusual [8], [9], [10], especially in areas where data is hard to collect. The consequence is that we cannot draw general conclusions, but merely offer some experimental evidence, which could label this kind of scale of experiments as "preliminary".

Dataset	Decision T...	k-NN	Naive B...	SVN	Random F...	OK?	Exception	NoExamples	NoAttributes	ClassImbalance
Australian.aml	0.841	0.658	0.772	0.414	0.751	true	OK	690	14	0.56
German.aml	0.688	0.608	0.756	0.300	0.700	true	OK	1000	20	0.7

FIGURE 24.14: Log output with additional dataset properties.

This tutorial comes with all the used RapidMiner processes attached, which could be used for learning features of RapidMiner, give ideas how to use it for research, or even act as templates for some research questions.

Bibliography

[1] Ingo Mierswa, Michael Wurst, Ralf Klinkenberg, Martin Scholz, and Timm Euler. Yale: Rapid prototyping for complex data mining tasks. In *Proceedings of the 12th ACM SIGKDD International Conference on Knowledge Discovery and Data Mining*, pages 935–940. ACM, 2006.

[2] Ranjit Kumar. *Research Methodology: A Step-by-Step Guide for Beginners.* Sage Publications Limited, 2005.

[3] CR Kothari. *Research Methodology: Methods and Techniques.* New Age International, 2009.

[4] Arthur Asuncion and David J Newman. Uci machine learning repository [http://www. ics. uci. edu/~ mlearn/mlrepository. html]. irvine, CA: University of California. *School of Information and Computer Science*, 2007.

[5] Zoltán Prekopcsák, Gábor Makrai, Tamás Henk, and Csaba Gáspár-Papanek. Radoop: Analyzing big data with rapidminer and hadoop. In *Proceedings of the 2nd RapidMiner Community Meeting and Conference (RCOMM 2011)*, 2011.

[6] Pavel Brazdil, Christophe Giraud Carrier, Carlos Soares, and Ricardo Vilalta. *Metalearning: Applications to Data Mining.* Springer, 2008.

[7] Sarah Daniel Abdelmessih, Faisal Shafait, Matthias Reif, and Markus Goldstein. Landmarking for meta-learning using rapidminer. In *RapidMiner Community Meeting and Conference*, 2010.

[8] Cheng-Lung Huang, Mu-Chen Chen, and Chieh-Jen Wang. Credit scoring with a data mining approach based on support vector machines. *Expert Systems with Applications*, 33(4):847–856, 2007.

[9] Iain Brown and Christophe Mues. An experimental comparison of classification algorithms for imbalanced credit scoring data sets. *Expert Systems with Applications: An International Journal*, 39(3):3446–3453, 2012.

[10] Andrew Secker, Matthew N Davies, Alex A Freitas, Jon Timmis, Miguel Mendao, and Darren R Flower. An experimental comparison of classification algorithms for hierarchical prediction of protein function. *Expert Update (Magazine of the British Computer Society's Specialist Group on AI)*, 9(3):17–22, 2007.

Subject Index

Operator Index